A Modern Primer in Particle and Nuclear Physics

A Modern Primer in Particle and Nuclear Physics

Francesco Terranova

University of Milano-Bicocca and INFN

OXFORD
UNIVERSITY PRESS

OXFORD
UNIVERSITY PRESS

Great Clarendon Street, Oxford, OX2 6DP,
United Kingdom

Oxford University Press is a department of the University of Oxford.
It furthers the University's objective of excellence in research, scholarship,
and education by publishing worldwide. Oxford is a registered trade mark of
Oxford University Press in the UK and in certain other countries

Published in the United States of America by Oxford University Press
198 Madison Avenue, New York, NY 10016, United States of America

British Library Cataloguing in Publication Data

Data available

Library of Congress Control Number: 2021936304

ISBN 978–0–19–284524–5 (hbk.)
ISBN 978–0–19–284525–2 (pbk.)

DOI: 10.1093/oso/9780192845245.001.0001

Printed and bound by
CPI Group (UK) Ltd, Croydon, CR0 4YY

A Roberta

Contents

Preface

In most universities, students are exposed to an introductory course on nuclear and particle physics at the undergraduate level during the second or third year. It comes as no surprise that this course represents a challenge for most teachers. Modern particle physics is a discipline grounded on quantum field theory (QFT): a tool that is not available to students at the sophomore or junior level. Even when we explain the Standard Model of particle physics focusing on the "discoveries of new particles," we struggle against the complexity of the experimental techniques devised in the last few decades. As a consequence, we often resort to teaching the fundamentals of particle and nuclear physics building on the historical development of the field: from non-relativistic nuclear systems and the discovery of the first elementary particles up to the Standard Model. Although this history-based approach has stood the test of time, it is a source of confusion for a student joining the field in 2021. Students at the undergraduate level, for instance, are unable to appreciate the straightforward link between nuclear physics and non-perturbative quantum chromodynamics (QCD) because QCD is mentioned at the very end of the course, while nuclear physics and non-relativistic quantum mechanics is presented as the foundation of our understanding of the microscopic world. In a sense, this situation resembles the paradox of many textbooks on classical electromagnetism, where the Maxwell equations are presented without reference to special relativity up to the last chapter of the book.

For particle physics textbooks, there is an additional reason to move away from the historical development of the field. Our students do not discover what a quark is or what the Large Hadron Collider (LHC) is doing by our courses: they hear about them in high school or, more frequently, on the Internet. They attend our courses because they want to *understand* what a quark is or why the LHC has been built. Possibly with the tools they already have: special relativity (SR) and non-relativistic quantum mechanics (QM). In 2016 – the first year I gave this course – I tried to match these expectations and put aside most of the standard textbooks, where either the subject was presented following the historical development or the author sketched the basics of QFT at the beginning of the course. It took a few years to cook up lectures that devise a primer in particle and nuclear physics employing just special relativity and quantum mechanics. Still, most of the students loved this experiment since its inception and helped me to fine-tune the subjects, sharpen arguments and proofs, and prune what was not necessary (or not cute enough to be presented to undergrads). From 2016 to 2020 my folder has grown from a collection of notes to a real textbook, which is the outcome of the teacher's efforts plus the ideas (and patience) of many students and colleagues.

On my side, I was delighted to discover how much particle physics can be introduced using, for instance, relativistic kinematics or symmetries in QM, picking up from excellent textbooks or constructing new demonstrations. I had to negotiate between intuitive explanations and rigour, and sometimes I had to give up saying "No way. I cannot explain this stuff using elementary physics." The narrative of this book has two aims: to provide a solid ground for students willing to join our field at the graduate level and give a modern view of particle physics to students who will not continue to study it any further. The two goals are often in competition and negotiations were harder than expected. The result matches the expectations of a broad readership: undergraduate students in physics in all countries where this course is mandatory, second or third-year physics students where the undergraduate course is optional but chosen by a large audience as an introduction to fundamental physics, Masters students in nuclear or energy and power engineering and, of course, physics enthusiasts at any level.

An overview of the book

We open the course asking "what is an elementary particle" and we set up the scene in Chapter 1, introducing the Standard Model particles and the fundamental interactions – that is, the core of our present knowledge of the microscopic world. We also explain natural units, which are a great tool in particle physics. In general, we urge the readers to go through the examples and the exercises. Some of them are tame, others (e.g. the Planck scale in natural units or – in Chapter 2 – the explanation of why "light is so special in physics") will give them food for thought. The second chapter is devoted to the language of particle physics: the covariant formalism and the way we describe particle decays and production. This is done gently, keeping the emphasis on similarities with atomic physics without resorting to creation and annihilation operators or other advanced tools. The concepts and, sometimes, the maths of Chapter 2 may be a challenge for students who do not have a background in special relativity: this is why we describe relativistic kinematics and the covariant formalism starting from the basics, relying both on intuitive explanations and a careful derivation of all formulas. The chapter is complemented by Appendix A, where special relativity is briefly summarized. Chapter 3 describes how particles interact with matter and how we observe them. We do not review all particle detectors but we pick a few standard detectors to show what is crucial in the design of particle instrumentation from the basic charge formation mechanisms to the Shockley–Ramo theorem. Once the fundamental concepts of detector physics are covered, we introduce modern techniques in tracking and calorimetry and present the most prominent analysis tools for the interpretation of particle physics data: the Monte Carlo methods and the corresponding data analysis techniques. We introduce particle accelerators and colliders in Chapter 4. The phase stability theorem – which is the cornerstone of accelerator science – is demonstrated in a very simple way for linear accelerators and the conditions for stability and the limiting parameters are discussed for all types of modern accelerators, with special emphasis on synchrotrons and colliders. This chapter extends the detection techniques of Chapter 3 explaining the methods employed in modern collider experiments. We introduce symmetries in quantum mechanics in Chapter 5. We address this fundamental topic presenting the symmetries as peculiar features of quantum systems that give us information even if we are unable to solve the corresponding equations of motion. This definition emphasizes the prominent role of symmetries in particle physics, where the Hamiltonian is either not known (physics beyond the Standard Model) or leads to equations of motion that are nearly intractable (strong interactions). We then explore the link between symmetries and conservation laws, introduce the P- and C-parities up to admitting that there is no way to define rigorously what the parity of the photon is in non-relativistic QM. We present two classical experiments to determine the parity of the photon and the pion. In this way, we show that we can establish many fundamental properties of the particles even if we do not know yet the underlying theory. C-parity brings us to the world of antiparticles and we introduce, for the first time, relativistic quantum mechanics through a careful analysis of the Dirac equation. During this discussion, we clarify historical interpretations about the nature of antiparticles and introduce the most general and the only consistent explanation of antimatter, which is grounded on QFT and the CPT theorem.

Electromagnetic interactions are discussed in Chapter 6. The chapter is largely aimed at presenting gauge symmetries and explaining why they play such a key role in the description of fundamental interactions. We do not need QFT to display the strength of the gauge principle: we show that even in non-relativistic quantum mechanics, we can build the semi-classical Hamiltonian employing the $U(1)$ gauge symmetry. The main principle of this book is waived in Chapter 6 when we introduce Feynman diagrams as graphical representations of the perturbative series that describe scatterings in quantum electrodynamics (QED). In this case, nothing is proven rigorously but students can appreciate the power of perturbative techniques in one of its simplest and most spectacular applications, the positronium.

Strong interactions are discussed in Chapter 7 as a generalization of QED. We emphasize the difference between QED and QCD using the corresponding Feynman diagrams. We introduce the concept of asymptotic freedom and present the latest findings on confinement. We explain why quarks are real physical particles and how QCD can be experimentally tested, despite the confinement of quarks inside the hadrons. For the first time, here we attack the fundamental problem of the origin of mass, which we will investigate more deeply in later chapters.

We employ symmetries in QM to build the two- and three-quark models in Chapter 8. We stress that even a rough approximation like neglecting the mass of all heavy quarks provides a powerful symmetry for particle classification. This symmetry successfully predicts the C- and P-parities of mesons and baryons, as well. However, we also underline the limitations of the quark models both on mass predictions and the existence of "exotic" bound states, and how these limitations have been superseded in the last ten years.

Nuclear physics is introduced in Chapter 9 from the discovery of the neutron, the isospin symmetry, and the lattice-QCD results on the nuclear potential. In this way, we establish a direct connection between QCD and nuclear reactions. We then analyze the proton–neutron scattering and the classification of nuclear bound states. We do not review all the time-honored nuclear models but we present in detail the shell model that still provides a deep insight into the nuclear structure. In the chapters devoted to nuclear physics, we study not only the science but also the outstanding applications of this discipline. In Chapter 9, in particular, we discuss nuclear reactions as energy generators for (reactors) and against (nuclear weapons) humankind.

Weak interactions are presented in Chapter 10 starting from the conservation of flavor in QED and QCD and its violation in the decay of the lightest particles. The only particle without electric charge and color (the neutrino) is discussed on the empirical ground starting from the three-body decay of the muon. We glimpse the possibility of parity violation using a thought experiment with photons and present the classical experiment of Chien-Shiung Wu quite in detail. The Goldhaber experiment is another great example of ingenuity: it measures the helicity of the neutrino without seeing neutrinos at all. We show in Chapter 10 that the Dirac equation, combined with the findings of Fermi, Wu, and Goldhaber, gives birth to the modern theory of weak interaction, the V-A theory.

In Chapter 11 we discuss nuclear beta decays as an example of weak interactions in bound systems. We extend the discussion to alpha and gamma decays, which are due to an interplay between strong and electromagnetic interactions, and we introduce the radioactive chains. We point out the countless applications of radioactivity and fusion reactions: from radiocarbon dating to the lifecycle of the stars.

The Standard Model of particle physics is introduced in Chapter 12 starting from the contradictions of the V-A theory. We do not attempt any derivation of the electroweak unification but we explain why a unified theory of weak and electromagnetic interactions is unavoidable. Building on the findings of Chapters 2 and 4, we describe the most important discoveries associated with it: the observation of neutral currents in 1973 and massive spin-1 bosons in 1983. Chapter 13 is a bridge to Masters courses. It is devoted to the least known land of the Standard Model: the Higgs and Yukawa sectors, and their observables (CP violation in neutrinos and mesons, particle mixing, and the Higgs boson). These advanced topics are described in some detail using (classical) relativistic fields and this is the only chapter I would place at the advanced-undergraduate level. Even if it is not an easy reading, all my students loved the chapter because – by theory and experiments – it led them to the frontier of research and shows science in the making. The chapter culminates with the three most important discoveries of the 21st century in our field: the discovery of the Higgs boson, the observation of neutrino oscillations, and the direct detection of gravitational waves. While mentioning these successes, I raise doubts about the beauty and elegance of the Standard Model compared with general relativity and substantiate those doubts pointing an accusing finger to the limitations of both theories. These limitations will be challenges for the next generation of researchers.

Using this book

The prerequisites of this book are a basic knowledge of special relativity and non-relativistic quantum mechanics. As noted above, students lacking this background can fully enjoy the book with little effort: all concepts needed to follow the explanations and the proofs are summarized in Appendix A (Special Relativity), Appendix B (The Principles of Quantum Mechanics) or recalled in Chapter 2. The Further Reading section of Chapter 1 provides suggestions to extend this background (if needed) or deepen knowledge about subjects that are not directly related to particle and nuclear physics.

Each chapter has a set of exercises graded as "easy," "confidence-building" (one asterisk), and "challenging" (two asterisks). I would like to stress, once more, the importance of these tools since, as with any book that introduces a new discipline, a full understanding of the concepts can be achieved only by practising with concrete problems. In some cases, these exercises are used to introduce new material or fill a gap in proof but this is done only if the exercise contributes to give further insight into the main concepts.

In my experience, about 70–80% of the material contained in the book can be presented in a one-semester course. The additional material offers a high degree of flexibility to the teacher, who can tune the course according to the audience's background and teacher's style and goals. Two common examples are either a course where the students have completed the classes on quantum mechanics or a course where they are taking quantum mechanics concurrently. In the first case, the teacher is free to choose any subject, including the most advanced ones, which are labeled with an asterisk. I suggest selecting topics from Chapters 1–4 and covering more in detail in Chapters 5–12. Note that Chapter 13 is an invitation to Master's courses and offers a glimpse of the frontier of research in particle physics. In the second case, I recommend placing emphasis on the main concepts of Chapters 1–11 and resorting to the Appendices every time there is a gap between the lectures and the background of the students. Chapters 12 and 13 can be used as an invitation to more specialized courses at Master's level. In both cases, I encourage the teacher to emphasize Chapters 9 and 11 because they constitute an essential background for undergaduate-level students and give a sense of unity in our discipline (high-energy physics, rare event searches, astroparticle physics, and nuclear physics) that cannot be obtained from any other undergraduate-level course.

I considered this textbook ready for publication when my graduate students headed for these notes every time they got lost in the technicalities of quantum field theory. As with any introductory text, sooner or later, this book will be superseded by your favourite master/graduate textbook. How many times you pick it up from your shelf after the final exam will be the most reliable measurement of its success, and the author is looking forward to your feedback on it.

Acknowledgements

I am indebted to the students that followed my introductory course of particle and nuclear physics in the last five years. The ideas behind this book came from those lectures and I greatly appreciated the comments, suggestions, and enthusiasm I received from them. I am delighted to thank G. Lavizzari, A. Perego, S. Manzini, M. Zanirato, and V. Zito for careful proofreading of the manuscript. I owe special thanks to my colleagues of the Department of Physics "G. Occhialini." Many of them reviewed parts of the manuscript and gave me very useful suggestions. I hope they will be satisfied with the outcome. I especially thank S. Alioli, M. Calvi, L. Giusti, P. Govoni, C. Oleari, E. Previtali, T. Tabarelli de Fatis, A. Zaffaroni together with N. Charitonidis, C. Giunti, M. Pari, and F. Vissani from INFN and CERN. I wish to express my gratitude to S. Adlung, always unfailingly helpful, and to the OUP team – especially G. Lipparini, F. MacMahon, and J. Walker – that followed the production phase of the manuscript and helped me with figures, language, style, and LaTeX issues.

There is no physics book whose acknowledgements do not include thanks to the author's family for their patience, and this book is no exception. I wish to thank my wife Roberta not only for her patience but especially for the enthusiasm and encouragement to do what I really love – including the writing of this textbook. I thank Alice for her curiosity regarding the content of the book and because, together with her sister, she forbade me from writing more than 450 pages. You will also appreciate Valentina, who drew the funniest figures in this book. My final thank goes to a friend, who reminded me that one does not need to be a Grandmaster to write a good chess manual.

Setting the scene

1.1 The dawn of particle physics

We are so used to thinking of matter as made of tiny particles (atoms, electrons, protons, etc.) that we can barely imagine how nature appeared to humans before the inception of the atomic theory. All cultures have been seeking the "fundamental constituents" of matter but, save for a few (and renowned) exceptions, natural philosophers believed that those constituents were continuous media. In ancient Greece, the basic elements were Earth, Air, Water, and Fire. The Vedas added the Ether, while the Chinese replaced Air with Wood. Despite these differences, all things were thought to be the outcome of mixing and interactions of these media. In 1821, A.L. Cauchy provided the first modern definition of continuous medium employing the concept of limit. According to Cauchy, a medium is described by a continuous function – for example, the mass density $\rho(\mathbf{x})$ – that does not abruptly change if you inspect matter with a microscope. Whatever is the magnification power of the microscope, the water looks like water around a space point $\mathbf{x}$ and the density tends to a well-defined value for $\mathbf{x}' \to \mathbf{x}$. The definition is based on the formal concept of continuity of a function. The famous Cauchy's "epsilon-delta definition" (Riley *et al.*, 2006) reads:

> in a continuous medium, for every $\epsilon > 0$, it exists a $\delta > 0$ and a region of space where the change of density is smaller than ϵ. In particular, for every point $\mathbf{x}'$ located in a neighbourhood of radius δ centered in $\mathbf{x}$ $\implies |\rho(\mathbf{x}') - \rho(\mathbf{x})| < \epsilon$.

The empirical falsification of this definition for *any* continuous medium marks the inception of particle physics. During the 19th century, physicists gathered overwhelming evidence that continuity breaks down at $\delta < 10^{-10}$ m. This tiny distance is the scale of the granular structure of matter: its atomic size.

What is the difference between atomic and particle physics? From the conceptual point of view, there is no difference between the two disciplines. The former embraced the latter at the beginning of the 20th century when Thomson and Rutherford demonstrated that $\delta = 10^{-10}$ m is not the smallest conceivable scale (Buchwald *et al.*, 2013). New granular structures were observed at $\delta = 10^{-15}$ m, which we presently call "nuclei." The old atomic structures (atoms) were interpreted as bound states of nuclei and electrons. It soon became clear that neither the bound states nor their more fundamental constituents obey the laws of classical physics. In less than 20 years, quantum mechanics and

A Modern Primer in Particle and Nuclear Physics. Francesco Terranova, Oxford University Press.
© Francesco Terranova (2021). DOI: 10.1093/oso/9780192845245.003.0001

relativity were raised to the status of paradigms for the description of the microscopic world.

Modern particle physics is the discipline that studies the fundamental components of matter (particles) and their interactions. To the best of our knowledge, the interactions among particles, the processes of particle creation, annihilation, and decay can be described only by theories that are compliant with special relativity and quantum mechanics. A class of theories that provide full compliance with special relativity (SR) and quantum mechanics (QM) is quantum field theories (QFTs), where particles are described by quantized fields. The Standard Model (SM) is one of the many possible models of the microscopic world built upon the principles of the QFTs. To date, it is the only one that works.

We start our journey in particle and nuclear physics from the most basic question: what is an elementary particle?

1.2   Elementary particles

In classical physics and special relativity, we do not make a distinction between an elementary particle and a point-like particle. The two definitions are equivalent, and macroscopic bodies are assemblies of point-like particles. This is not in contradiction with the existence of the continuous media. To collapse a medium down to a point-like particle, the mass density function $\rho(\mathbf{x})$ must be a function that differs from zero only in one point and is such that:

$$\int_{\mathbb{R}^3} d^3\mathbf{x}\, \rho(\mathbf{x}) = m. \tag{1.1}$$

The Dirac function does the job:[1]

$$\int_{\mathbb{R}^3} d^3\mathbf{x}\, m\, \delta^3(\mathbf{x} - \mathbf{x_0}) = m \tag{1.2}$$

where the three-dimensional Dirac delta function $\delta^3(\mathbf{x} - \mathbf{x_0})$ is defined as:

$$m\, \delta^3(\mathbf{x} - \mathbf{x_0}) = m\, \delta(x - x_0)\delta(y - y_0)\delta(z - z_0) \tag{1.3}$$

and describes the mass density of a point-like particle.

Classical point-like particles are by definition elementary particles because they are geometrical points associated with an inertial mass that cannot be destroyed or split.

In non-relativistic quantum mechanics, we can still define a point-like particle but the Heisenberg uncertainty principle introduces a trade-off between particle localization and the determination of the trajectory. If the measurement of the particle position localizes the particle down to a point, we are perfectly measuring the position in space. In this case, the uncertainty on the conjugate variable – the momentum, which includes the direction of motion – is infinite by the Heisenberg principle. We, therefore, know where the particle is but we cannot say anything

[1]Well, at the expense of some mathematical subtleties. No ordinary function attains eqn 1.1 if it is zero everywhere but in a point. The Dirac delta function is what we need here and is discussed in Appendix B.

about its trajectory even if we can solve the corresponding equation of motion. We thus need to follow a different path than classical physics.

The motion of a point-like particle in quantum mechanics[2] is described by the non-relativistic **Schrödinger equation** for a single particle of mass m:

$$i\hbar\frac{\partial}{\partial t}\psi(\mathbf{x}, t) = \left[-\frac{\hbar^2}{2m}\nabla^2 + V(\mathbf{x})\right]\psi(\mathbf{x}, t) \tag{1.4}$$

where $V(\mathbf{x})$ is the potential energy (in the following, potential - see Sec. B.3) describing the forces acting on the particle. The localization is performed with finite precision: the initial wavefunction is a wavepacket that always takes up a non-zero volume. The trajectory is replaced by the time-dependent probability of finding the particle in a finite volume in space. Unlike classical physics, this definition is not equivalent to the definition of elementary particles. The motion of an atom in a gravitational field fulfills this definition but an atom is a composite system and its wavefunction results from the wavefunctions of the constituents. A free **elementary particle** in quantum mechanics should be defined as a system whose wavepacket can *only* be expressed by a superposition of plane waves. But who can ensure a priori that another representation does not exist? For instance, the atom could be a bound state of an electron and a proton (hydrogen) and in this case, the wavefunction at sub-atomic scales is the solution of the Schrödinger equation for the hydrogen. A proton is even more malicious although in most experiments it appears as a harmless point-like particle. The proton is a complex bound system of quarks and gluons (see Chap. 7), whose equations of motion are too complicated to be solved analytically.

The issue does not change in relativistic quantum mechanics and QFTs. The elementary constituents of matter in the Standard Model are the fundamental fields that appear in the Lagrangian of the Standard Model. This definition is as void as the definition based on the wavepacket. We are quite convinced that the Standard Model is part of an unknown general theory and we do not know the fundamental fields of its Lagrangian. To be honest, we do not even know if this theory admits a Lagrangian representation at all. Together with S. Weinberg, we can just say that "the task of physics is not to answer a set of fixed questions about Nature, such as deciding which particles are elementary. We do not know in advance what are the right questions to ask, and we often do not find out until we are close to an answer" (Weinberg, 1997).

All said and done, there is no doubt that we can demonstrate experimentally that a particle is *not* elementary. Experimenters try to decompose candidate elementary particles smashing them or observing properties that demonstrate that the system is composite. To date, we know that all particles are composite except for 25: 6 quarks (and their antiparticles), 6 leptons (and their antiparticles), 12 spin-1 bosons, and 1 spin-0 boson: the Higgs boson.

[2] If you are not acquainted with these quantum mechanics concepts, I suggest reading Appendix B, where you find all the material needed in this book.

1.3   Forces, potentials, and fields

Before looking at the particles of the Standard Model, I clear the air on the meaning of "fundamental forces" or, more precisely, **fundamental interactions** among particles.

The paradigm describing interactions among particles has changed (at least) three times from the 17th century to the 20th century and students are exposed to all these paradigms in the course of their academic career. In classical mechanics, the "natural" state of a point-like particle is a uniform motion on a straight line. This motion includes the particle at rest in the laboratory frame as a special case (zero velocity). The origin of a change in the motion of the particle from the natural state is called **force**. The force perturbs the motion under Newton's second law: $\mathbf{F} = m\mathbf{a}$, where m is the inertial mass of the particle, $\mathbf{a}$ the acceleration and $\mathbf{F}$ the force: a three-component vector applied to the particle. There are many types of forces that can be described by classical physics. Some of them are fundamental forces originating from the intrinsic properties of the particles, others are empirical forces that simplify the description of many-body systems at the macroscopic level.

A renowned example is a gravitational force between two particles that originates by the interplay between the inertial mass of the first particle ($m_1^i = m_1^g$) and the gravitational mass of the second particle (m_2^g). The force on the first particle is:

$$\mathbf{F_1} = G\,\frac{m_1^i m_2^g}{r^2}\frac{\mathbf{r}}{r} \qquad (1.5)$$

where $\mathbf{r} \equiv \mathbf{r_2} - \mathbf{r_1}$ is the vector from particle 1 to particle 2 and $r = |\mathbf{r}|$. G is the gravitational constant. This force is attractive, proportional to the product of the masses and inversely proportional to the squared distance between the particles. The direction is along the line that links the two particles: $\mathbf{r}/r$. In this case, $\mathbf{F_1}$ causes the perturbation of the motion of the particle 1 due to the gravitational force generated by the particle 2. General relativity ensures and explains why $m^g = m^i$ for both particles, while this equivalence is accidental in Newton's theory of gravitation (Cheng, 2010). Using this equivalence, $\mathbf{F_1}$ reads:

$$\mathbf{F_1} = G\,\frac{m_1 m_2}{r^2}\frac{\mathbf{r}}{r}\;. \qquad (1.6)$$

The electric Coulomb force generated by a particle 2 on a particle 1 when both are at rest is:

$$\mathbf{F_1} = \frac{1}{4\pi\epsilon_0}\,\frac{q_1 q_2}{r^2}\frac{\mathbf{r}}{r}\;. \qquad (1.7)$$

In eqn 1.7, $\mathbf{r}$ is defined as $\mathbf{r_1} - \mathbf{r_2}$. The Coulomb force has the same functional form as the gravitational law and, therefore, is proportional to the product of the charges and inversely proportional to the squared distance. Unlike gravitation, if the signs of the charges are different, the force is attractive and if the signs of the charges are the same, the force is repulsive. The issue of the gravitational mass is immaterial, here, because the force is mass-independent. The only mass of eqn 1.7 is the inertial mass of particle 1, which appears in the definition of $\mathbf{F_1} = m_1\mathbf{a}$ under Newton's second law. Note that even today, both forces are considered fundamental laws of nature.

On the other hand, the Hooke force is an empirical force that perturbs the motion of a body of mass m through the mechanical contact with a spring (Serway and Jewett, 2018). The force is attractive if the spring is elongated and repulsive if the spring is compressed: $F = -k\Delta x$. k is the stiffness of the spring and Δx is the elongation: if $\Delta x < 0$ the spring is compressed. Hook's force is not a fundamental force of nature and parameterizes the mean effect of the electromagnetic (e.m.) interactions between the body and the spring.

Similarly, the kinetic (or dynamic) friction is defined as the force produced by the friction of two bodies [3] – for example, a floor and a moving rigid body with mass m – and is defined as:

$$\mathbf{F} = -\mu_d|\mathbf{N}| \ . \tag{1.8}$$

In eqn 1.8 and Fig. 1.1, $\mathbf{N}$ is the force perpendicular to the velocity of the body, for example, the weight $|\mathbf{N}| = mg$ of the body. The direction of the frictional force is always opposite to the direction of motion of the body. Again, this is an empirical formula resulting from e.m. interactions and is a typical example of non-conservative forces since the friction reduces the kinetic energy of the body and may bring it to rest.

Newtonian mechanics is suitable to describe a large class of systems, especially at the macroscopic level ($\delta \gg 10^{-10}$ m). Note, however, that the two fundamental forces mentioned above are also the most troublesome for Newton's paradigm: the gravitational law requires an "action-at-a-distance" between bodies that are not in contact. The Coulomb law is in contradiction with Newtonian mechanics when extended to moving charges because the corresponding force – the Lorentz force – is not invariant for Galilean transformations and contradicts Newton's second law (see Chap. 6).

These difficulties can be overcome by the introduction of **fields**: a field is a function that associates a vector to each point in space and may depend on time. If a particle is located in $\mathbf{x}$ at time t, the force perturbing its motion will be uniquely determined by the field at $(\mathbf{x}, t)$ and the intrinsic properties of the particle like the electric charge. Fields were first introduced in electrodynamics by M. Faraday. They are mandatory tools to comply with special relativity, where the speed of any perturbation cannot exceed the speed of light in vacuum. The perturbations are due to the propagation in space and time of the electric ($\mathbf{E}$) and magnetic ($\mathbf{B}$) fields (Griffiths, 2017). In this context, we can define the **potentials** as the minimum number of functions that uniquely define the fields. For the electromagnetic fields, these are the scalar and vector potentials $V(\mathbf{x}, t)$ and $\mathbf{A}(\mathbf{x}, t)$ and:

[3] Frictional forces tend to stop a moving body and deceived natural philosophers for centuries. Aristotle and most ancient thinkers believed that the natural state of a body without forces was the rest state.

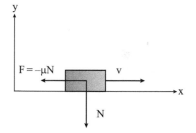

Fig. 1.1 Frictional force $\mathbf{F} = -\mu|\mathbf{N}|$ slowing down a body that moves horizontally on a floor with speed v along the x-axis. $\mathbf{N}$ is the sum of all forces perpendicular to the body velocity. In most cases, $\mathbf{N}$ is just the weight of the body ($|\mathbf{N}| = mg$) and lies along the y-axis.

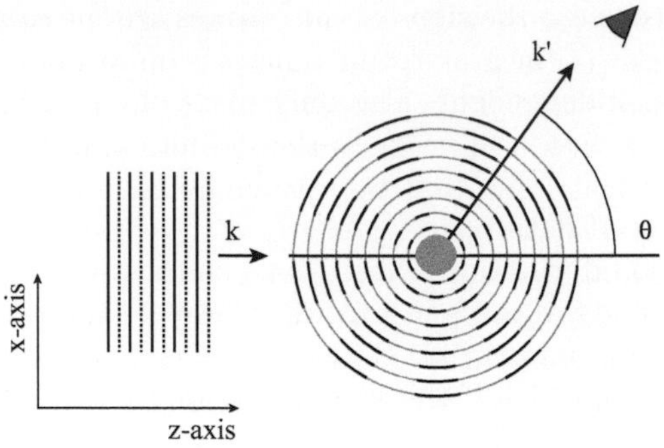

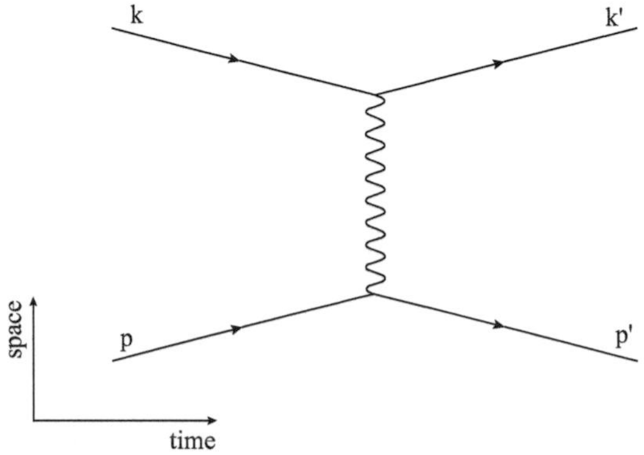

Fig. 1.2 Two-body scattering in QM (top) and QFT (bottom). In the first case, the incoming wavefunction is deformed by the potential of the Schrödinger equation. In the second case, the motion is perturbed by the momentum transferred by the force mediators (see text).

$$\mathbf{E} = -\nabla V - \frac{\partial \mathbf{A}}{\partial t} \qquad (1.9)$$

$$\mathbf{B} = \nabla \times \mathbf{A} \qquad (1.10)$$

To the best of our knowledge,

> all fundamental forces are mediated by fields and are described by a finite number of potentials.

Classical fields and potentials replace the forces in relativistic and non-relativistic QM. The motion of a particle is defined by the Schrödinger equation, which is a function of classical fields and potentials without any reference to forces. The downfall of the concept of forces in physics followed the replacement of classical physics with SR and QM. The new potential-based paradigm describes many particle physics processes employing only the Schröedinger equation or its relativistic extensions. You will love this approach for its simplicity and effectiveness.

The third change of paradigm occurred in the late 1940s when even the concept of wavefunction and potential dropped out of particle physics. The new theory was conceived to account for the creation and destruction of elementary particles: an effect that is ubiquitous in high energy physics. Besides the original motivations, **Quantum Field**

Theories (QFTs) provide a profound understanding of particle interactions at the expense of conceptual (and computational) simplicity. In
QFT, wavefunctions and potentials are cast aside by describing particles as **quantized fields** and interactions – sometimes inappropriately
called forces – as scatterings among quantum fields. In Fig. 1.2 you can
appreciate the difference between the QM description of a scattering
and the corresponding description in the QFT framework. Here, the
forces are mediated by new particles that transfer momentum to other
particles. The reader should be aware that Fig. 1.2-bottom is not only
a pictorial representation of the change of the motion of a particle but
an actual tool to compute the scattering amplitude in perturbation
theory. These kinds of tools are called Feynman diagrams and will be
introduced later. In the QFT paradigm:

> all fundamental interactions are mediated by quantum fields and
> the basic observable of any physical process is the transition am-
> plitude between an initial state of N particles and a final state
> of M particles. In general $N \neq M$ and particles can be created
> and destroyed during the scattering process.

We will show that these interaction fields are spin-1 particles called
mediators. Loosely speaking, in QFT the abstract Newtonian forces
are replaced by "kicks" given to the particles by the mediators.

1.4   The elementary particles of the Standard Model

Even if nearly all matter in the known universe is composed of protons,
neutrons, and electrons, the actual types (in jargon, **flavor**) and the
number of elementary particles are significantly larger. Indeed, protons
and neutrons do not belong to that list.

The elementary particles of the Standard Model can be classified
into four categories: leptons, quarks, spin-1 elementary boson (the force
mediators of QFT), and the Higgs boson.

1.4.1   Leptons

Leptons are a set of six spin-1/2 fermions, whose charge and mass are
very different among flavors. The first elementary particle discovered
belongs to this class and is called the electron. Three leptons are electrically charged. Their charge Q is equal to $-e$, e being the size of the
proton charge: 1.6×10^{-19} C. The charged leptons are:

- the **electron** (e^-), a negatively charged fermion with a mass of
  511 keV/c^2. As discussed in Sec. 1.5, 1 eV/c^2 corresponds to about
  1.78×10^{-36} kg. The electron mass is thus 9.11×10^{-31} kg. The
  electron is the only stable charged lepton. The vast majority of

electrons in the universe are located in atomic bound states or move freely inside the interior of the stars (plasma electrons).

- the **muon** (μ^-), a negatively charged fermion with a mass of 106 MeV/c^2 or, equivalently, 1.88×10^{-28} kg. It has the same interactions as the electron but a larger mass. The muon is unstable and has a lifetime of ~ 2.2 μs in its rest frame.[4]

- the **tau lepton** (τ^-) is a negatively charged lepton with the same properties of the muon but a much bigger mass (1777 MeV/c^2) and shorter lifetime: 2.9×10^{-13} s in its rest frame.

[4]Due to time dilation in special relativity, the lifetime in any other frame is increased by $\gamma = (1 - v^2/c^2)^{-1/2}$. Time dilation is discussed in Sec. A.2.1 and Chap. 2.

The electrons are among the most abundant particles in the universe because they are stable. Muons are visible even with simple detectors since they are often produced by the interaction of cosmic rays (protons or light nuclei) in the earth's atmosphere and reach the surface of the earth due to time dilation. Tau leptons are much more difficult to be detected and are generally studied using accelerators.

The neutral spin-1/2 leptons are all stable particles and are called neutrinos:

- the **electron neutrino** (ν_e) is a light neutral fermion. It has only two types of interactions: the weak interactions and the gravitational interactions. The mass of the electron neutrino is not defined because the neutrino wavefunction is a linear superposition of three mass eigenstates with mass m_1, m_2, and m_3. The exact value of these masses are not known but – as we will see in the forthcoming chapters – are significantly smaller than 1 eV/c^2(see Fig. 1.3).

- the **muon neutrino** (ν_μ) is a light neutral fermion corresponding to a different linear combination of m_1, m_2, m_3. It can be distinguished from ν_e because the scatterings that destroy a ν_μ always produce a μ^- in the final state: $\nu_\mu + X \rightarrow \mu^- + Y$, where X and Y are an arbitrary set of particles.

- the **tau neutrino** (ν_τ) is a light neutral fermion corresponding to another linear combination of m_1, m_2, m_3. The scatterings that destroy a ν_τ always produce a τ^- in the final state.

1.4.2 Quarks

The quarks are spin-1/2 massive fermions with a fractional electric charge. Fractional charges are in clear disagreement with the outcome of Millikan's experiment. They have never been observed in this experiment because they can be found only inside bound states with integer electric charge that are called hadrons. We will provide a profound explanation of this phenomenon in Chap. 7.

- **up quark** (u) is a fermion with a charge equal to $+2/3$ in units of e and a mass of about 2 MeV/c^2 (3.6×10^{-30} kg). In the universe, it is mostly found in protons, which are composed of two quark up and one quark down (uud) and in neutrons (udd). The only stable particle with an up quark is the proton.

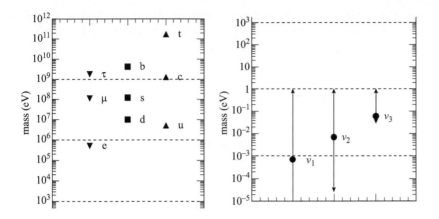

Fig. 1.3 Left: Masses (in eV/c^2) of the elementary fermions of the Standard Model. Downward triangles represent charged leptons. Upward triangles show the "up-type" quarks, i.e. the quarks with electric charge equal to +2/3. The squares show the "down-type" quarks with electric charge equal to −1/3. Right: current values and uncertainties of the mass m_1, m_2, and m_3 of the neutrinos. Redrawn from (de Gouvea, 2009).

- **down quark** (d) is a fermion with a charge equal to −1/3 in units of e and a mass of about 5 MeV/c^2. Again, the only stable particle with a down quark is the proton.

- **strange quark** (s) is a fermion with the same charge as d ($-e/3$) and a mass of about 93 MeV/c^2. No stable bound states contain a strange quark.

- **charm quark** (c) is a fermion with a charge equal to $+2e/3$ and a mass of about 1.3 GeV/c^2. Bound states containing a quark c are always unstable.

- **beauty quark** (b) is a fermion with a charge equal to $-e/3$ and a mass of about 4.2 GeV/c^2 without stable bound states.

- **top quark** (t) is the heaviest known particle (173 GeV/c^2). It is a fermion with charge equal to +2/3 in units of e and decays so quickly ($\tau \simeq 5 \times 10^{-25}$ s) that cannot form any bound state before decaying.

The masses of the elementary fermions (leptons and quarks) are plotted in Fig. 1.3. The current range allowed for the neutrino mass eigenvalues are labeled ν_1, ν_2 and ν_3. The fermions can be divided into $N_f = 3$ **families** made by two leptons of the same type ("electron," "muon," or "tau") and two quarks. The first family is composed by e^-, ν_e, u, d, the second family by μ^-, ν_μ, c, s, and the third one by τ^-, ν_τ, t, b. The Standard Model is mathematically consistent whatever the number of families but it breaks down if a family is not complete. We will explain in Chap. 12 why it is very unlikely that additional families exist. Nonetheless, we do not have explanations for the existence of multiple families ($N_f > 1$) nor any special meaning to ascribe to $N_f = 3$.

1.4.3   Antiparticles

Each elementary fermion has its **antiparticle**: a particle with the same mass, spin, and, if unstable, lifetime but the opposite electric charge. The antiparticles are labeled as $e^+, \mu^+, \tau^+, \bar{\nu}_e, \bar{\nu}_\mu, \bar{\nu}_\tau, \bar{u}, \bar{d}, \bar{s}, \bar{c}, \bar{b}, \bar{t}$. We will discuss their origin in Chap. 5. In the Standard Model, the neutrinos and the antineutrinos are different particles. It is remarkable that after more than 90 years of research we do not have yet an experimental proof that this is actually the case.

1.4.4   Elementary bosons and fundamental interactions

As mentioned in section 1.3, the interactions among fermions are mediated by the exchange of force mediators, which are spin-1 bosons, also called **gauge bosons**. As a consequence, each fundamental force of nature can be classified by its bosons and by the strength of the interaction with the elementary fermions. In the QFT description, the interaction terms are shown in Fig. 1.4 and give rise to:

- The **electromagnetic interactions** mediated by a spin-1 massless particle called the **photon**. This particle is stable and the strength of the interaction is given by the **fine structure constant** α. The photon is also its antiparticle.

- The **strong interactions** mediated by eight spin-1 massless particles called **gluons**. These bosons are stable and the strength of the interaction is given by the fine structure constant of the strong forces: **alpha strong** (α_s). Each gluon coincides with its antiparticle.

- The **weak interactions** are mediated by three spin-1 massive bosons: the W^+ (mass: 80.4 GeV/c^2), its antiparticle W^- and the Z^0 (mass: 91.2 GeV/c^2). The Z^0 is also its antiparticle. The strength of the interaction is different between the $W^\pm$ and Z^0 and also depends on some properties of the scattered fermion. The lifetimes of the weak gauge bosons are extremely short and are of the order of 10^{-25} s.

The strength of an interaction between a boson and a fermion depends on the properties of the fermion. The electromagnetic interactions arise because elementary fermions have an electric charge. Strong interactions originate from strong charges called **colors**, which are owned *only by quarks*. As a consequence,

> the most important difference between leptons and quarks is due to strong interactions. Leptons do not interact strongly, while strong interactions are the dominant interaction mode for quarks. The reason for this remains a mystery in particle physics.

[5] The weak charges are called "weak isospin" and "weak hypercharge" and will be discussed in Chap. 12.

All particles undergo weak interactions if they have non-zero "weak charges." [5] For neutrinos, weak (and gravitational) interactions are the only possible interaction mode because they are free of electric and color charges. Hence, the strength of the interaction depends on the α's, the "charge" of the fermion, and the mass of the gauge boson. In Chap. 12 we will provide a straightforward explanation of why weak interactions are so weak compared with strong and e.m. processes. This is due to the

Question		Refs
Why does each particle have an antiparticle?	Yes	Chap. 4
Why are weak interactions weak?	Yes	Chap. 12
Why are strong interactions strong?	Yes	Chap. 8
Why are the mass of the fermions so different?	No	Chap. 13
Why are neutrinos so light?	No	Chap. 11
Why can't quarks be seen as free particles?	Yes	Chap. 8
Why is the charge quantized in units of $e/3$	No	–
Why have the leptons no color charge?	No	–
Why has the SM three families?	No	Chap. 12

Table 1.1 A (quite arbitrary) selection of questions arising by a first inspection of the elementary particles of the Standard Model. The second column (yes/no) indicates if the Standard Model has an explanation for them. The chapter in this book where the topic is discussed is cited in the third column.

huge mass of the W's and Z^0 mediators. Fig. 1.4 shows the three types of interactions between fermions and gauge bosons and the approximate values of the αs for a center of mass energy of 300 MeV.

It is interesting to inspect Figs. 1.3 and 1.4 from a fresh perspective: this may be difficult for experienced researchers who have been exposed to these topics for years but it works well for a novice who wants to grasp the strength and limitations of the Standard Model. In Table 1.1, I collected – quite anecdotally – the most frequent questions I was asked after this inspection. At the time of writing, the Standard Model can answer just a fraction of them.

1.4.5 The Higgs boson

The Standard Model also predicts a spin-0 neutral elementary particle: the Higgs boson. Its existence has been disputed for decades because it does not descend straightforwardly from the basic principles of the theory. The **Higgs mechanism** is a technique inherited from other fields (material science and hadron physics) and adapted to particle physics to provide mass to the spin-1 bosons of the weak interactions. The interaction of the Higgs field with the elementary fermions provides masses to these fermions as well. The Higgs is, therefore, a **mass generation mechanism**. Without it, all elementary particles would be very similar to the photon. The mechanism gives firm predictions of the mass of the Z^0 and the Ws but lacks predictive power about the mass of the fermions and the Higgs boson itself. The mass generation mechanism for elementary fermions due to the Higgs is called the **Yukawa sector** of the SM and adds a rich set of phenomena that are at the frontier of modern research and will be recapped in the last chapter of this book. The Higgs boson was discovered in 2012 at the LHC collider at CERN

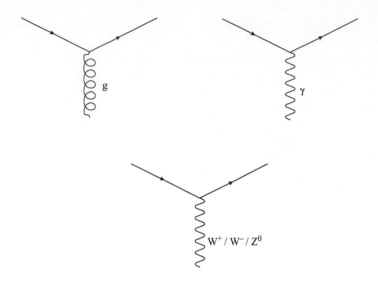

Fig. 1.4 The hierarchy of forces in nature. The figure shows a comparison among the interactions between the fermions (continuous lines) and the mediators (curled lines) that originate the strong (top left), electromagnetic (top right), and weak (bottom) forces. The interactions are mediated by a gluon, a photon, and three massive spin-1 bosons, respectively. The corresponding values of the αs are about 1/137, 1, and 1/30 if measured at a center of mass energy of 300 MeV. The actual strength of weak interactions is further reduced by the mass of the W^+, W^-, and Z^0 bosons.

Interaction	Boson	Fermions	Strength
Strong	gluons	quarks	1
Electromagnetic	photon	quarks and charged leptons	10^{-3}
Weak	Z^0, $W^\pm$	quarks and leptons	10^{-8}

Table 1.2 Fundamental interactions of the SM. The corresponding mediators are shown in the second column ("Boson"). Fermions are the types of particles that undergo the interaction. The (relative) strength is an approximate value of the force experienced by two unit-charged particles at a distance of 10^{-15} m normalized to the force due to strong interactions.

and its mass is 125 GeV/c^2. Its lifetime is predicted by the SM to be 1.56×10^{-22} s, although the direct measurements are still limited by experimental uncertainties.

All known fundamental interactions of the Standard Model, its gauge bosons, and the particles that undergo these interactions are summarized in Table 1.2.

1.5 Natural units

Over the years, particle physicists have developed a notation and a system of units that suit their needs and ease compliance with SR. The most widespread system of units in the field is called the **Natural Unit system** and labeled NU.

1.5.1   Metric systems

Physical laws are constraints between physical quantities as length, time, velocity, electric charge, and the like. Only a fraction of these quantities are independent and are used as basic units in metric systems. For instance, the velocity (v) is a ratio between space x and time t and therefore x, t, and v cannot be used simultaneously to define three basic units.

A **metric system** is fixed by the choice of the basic units and the definition of the prototypes that are associated with these units. The first metric system, formalized during the French Revolution, used the meter as the basic unit for space and the prototype was one ten-millionth of the distance from the equator to the North Pole along the earth's circumference.

By far the most important metric system is the International System of Unit (SI), which is enforced by law in nearly all countries (with the notable exception of the United States). The basic units of SI are seven: length, mass, time, electric current, thermodynamic temperature, amount of substance, and luminous intensity. The advantage of modern metric systems is that the prototypes can be reproduced very precisely and do not rely on physical laws. For instance, the modern definition of the meter is based on the fact that it is much easier to measure the velocity of light in vacuum c than the distance from the Equator to the North Pole. Special relativity definitely helps since c is the same for any inertial system, while the Equator to North Pole distance depends on a plethora of subtle geological effects. The current definition of the meter is, therefore, the length of the path traveled by light in vacuum in 1/299792458 second. In practice, SI employs a velocity as a basic unit to define the length and c is an exact constant (no experimental errors) because is a metric prototype. As expected, time is an independent basic unit in SI. The prototype (1 second – see Fig. 1.5) is defined as the duration of 9192631770 periods of the radiation corresponding to the transition between the two hyperfine levels of the ground state of the ^{133}Ce atom. Once more, the frequency of the photon emitted by the corresponding transition in ^{133}Ce is expressed without errors because is a metric prototype. Any other frequency $(\nu = 1/T)$ or period (T) will have its own statistical and systematic error. Metric systems are optimized to achieve maximum precision and reproducibility. On the other hand, physicists know that the luminous intensity is not as fundamental as time, and temperature is linked to the mean kinetic energy of a macroscopic body. SI is overly complicated because it ignores well-established physical laws and metric units make the equation of physics more cumbersome than needed.

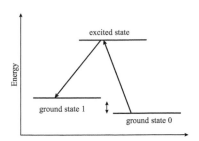

Fig. 1.5 Energy levels of ^{133}Cs employed for the SI definition of time. The hyperfine transition between the lowest states (the double arrow at the bottom of the figure) produces e.m. waves with a frequency of 9192631770 Hz. The "second" is defined by this frequency because of its narrow intrinsic width, which produces a nearly monochromatic line. On the other hand, NU uses a metric prototype only for energy and does not employ a dedicated prototype for time, which results in a loss of precision.

1.5.2   Natural systems of units

Natural units offer an elegant way out. One of the first natural systems was proposed in the 19th century in the framework of Gaussian cgs (the Gaussian unit system: centimeter–gram–second). It employs the Coulomb law to remove a complicated conversion factor between the electric charge prototype and the other mechanical units. As a consequence, the electric charge is no more a basic unit but a derived unit, and its prototype disappears. The Coulomb law in SI reads:

$$F = k\frac{q_1 q_2}{r^2} = \frac{1}{4\pi\epsilon_0}\frac{q_1 q_2}{r^2} \tag{1.11}$$

while Gaussian cgs enforces $k = 1$ so that:

$$F = \frac{q_1 q_2}{r^2} \; . \tag{1.12}$$

Since in the Gaussian cgs, space, time, and mass are measured in cm, s, and g, we have now that one unit charge ("StatCoulomb") is:

$$1 \text{ statC} = 1 \text{ g}^{1/2} \text{ cm}^{3/2} \text{ s}^{-1}. \tag{1.13}$$

The **natural system**[6] that is used in particle physics enforces two constraints: the first is inherited from special relativity ($c = 1$) and the second from quantum mechanics ($\hbar = 1$). $c = 1$ removes the unit of space (meter in SI) and its prototype. As a consequence, space could be expressed in seconds because $x = ct$ for a light ray. Similarly, energy could be expressed in kg since $E = mc^2$. However, the basic unit of time is removed by $\hbar = 1$ because $E = h\nu = \hbar\omega$ for a light ray in quantum mechanics (a photon). The prototype of the electric charge can be removed following the procedure of the Gaussian cgs. Particle physicists put $k = 1/4\pi$ (the "Lorentz–Heaviside" choice of k) and the Coulomb law reads:

$$F = \frac{1}{4\pi}\frac{q_1 q_2}{r^2} \; . \tag{1.14}$$

We are then left with only one unit, which - for practical reasons - is the **electronvolt** (eV). It is the amount of energy gained by the charge of a single electron moving across an electric potential difference of one volt. This unit is commonly used with the metric prefixes milli-, kilo-, mega-, giga-, tera-, peta-, or exa- (meV, keV, MeV, GeV, TeV, PeV, and EeV, respectively). In summary:

> the natural system of unit (NU) fixes $c = 1$, $\hbar = 1$, $k = 1/4\pi$, and all physical quantities are expressed in powers of electronvolts (eV).

We warn the reader that there are other natural systems of units currently in use. You may want to get rid of the prototype of the eV setting $G = 1$ in Newton's law of gravitation. In this case, all units are dimensionless. This system (Planck units), which belongs to the class of "geometrized systems" ($c = G = 1$), is often used in quantum gravity calculations but it is not very common in particle physics and cosmology because G is the most poorly measured fundamental constant.

Natural units simplify enormously the equations of physics but they come with shortfalls we should acknowledge. The most important one is that these units are not well suited for metrology, that is for precision tests of fundamental physics, because they imply the validity of some physics laws. Furthermore, their units are not based on *the best* quantities that can be measured in a lab. What does the "best quantity" in metrology mean? Sometimes, we can measure with outstanding precision observables that are not particularly meaningful for physics like macroscopic currents instead of the electron charge or the hyperfine atomic transitions instead of time differences (see Fig. 1.5). These ultra-precise measurements are not made to enter the Guinness World Records but to extract the best estimate of a fundamental unit indirectly. The SI prototypes are chosen among these special observables. We cannot do the same in NU if we decide to remove all prototypes except energy. To kill two with one stone and get the best out of metric and natural units, we perform our calculations in NU and, then, we return to SI using a set of conversion rules.

1.5.3 Conversion rules

Conversions between NU and SI can be done easily employing dimensional analysis. A formula in NU is proportional to the corresponding formula in SI:

$$(NU) = \hbar^a c^b (SI) \tag{1.15}$$

where a and b are rational numbers that come from the dimensional analysis of NU and SI. For instance, the relativistic energy of a particle is $E = \gamma m c^2$ in SI and $E = \gamma m$ in NU (see Chap. 2). Hence, $a = 0$ and $b = -2$. This can be derived from eqn 1.15 since γ is dimensionless and E is expressed in Joule (Kg m^2/s^2) in SI. The dimension of $\hbar$ ($[\hbar]$) is $[M][L]^2[T]^{-1}$ and $[c]$ is $[L][T]^{-1}$. $[M]$, $[L]$ and $[T]$ are the dimensions of mass, length, and time, respectively. Then,

$$
\begin{aligned}
[M] &= [M][L]^2[T]^{-2}[ML^2T^{-1}]^a[LT^{-1}]^b \\
&= [M]^{1+a}[L]^{2+2a+b}[T]^{-2-a-b}.
\end{aligned}
$$

This equation is satisfied if $a = 0$ (from $1 = 1 + a$) and $b = -2$ (from $2 + 2a + b = 0$). The third constraint ($-2 - a - b = 0$) is automatically satisfied. The reader should not be surprised to find an over-constrained system of three equations and two unknowns. The third equation signifies that $E = \gamma m$ is dimensionally correct in NU!

To convert numerical values from NU to SI, we just need the conversion factor between Joule and eV:

$$1 \text{ eV} = 1.6 \times 10^{-19} \text{ J}. \tag{1.16}$$

A particle with a 511 keV (kilo-eV) mass (the electron) corresponds to an energy of $511 \times 10^3 \cdot 1.6 \times 10^{-19} = 817 \times 10^{-16}$ J and a mass of $817 \times 10^{-16}/(3 \times 10^8)^2 = 9.08 \ 10^{-31}$ kg. Giving up the approximations and using the world best measurements of the electron mass, $m_e = 510.9989461(13)$ keV, corresponding to $9.10938356(11) \times 10^{-31}$ kg. In this expression, (13) and (11) indicate the experimental uncertainty in the last two digits: 61 and 56, respectively. Particle physicists use both the parenthesis notation and the $\pm$ notation ($m_e = 510.9989461 \pm 0.0000013$ keV) to specify the uncertainty. The former is particularly neat for high-precision measurements and is widely employed in metrology.

Another drawback of NU is that the system automatically subsumes dimensional analysis and it is impossible to distinguish among, say, energy, momentum, and mass from the dimension of their units. A useful trick that sidesteps this ambiguity is to write down the c factors that appear in SI: the energy is expressed in eV, the mass in eV/c^2, and the momentum in eV/c. A 1 GeV$/c$ particle is thus a particle with a *momentum* of 1 GeV. Note that the "eV$/c$" unit is just a symbol. In NU, the momentum is measured in eV like the energy and the mass.

Converting NU and SI values is straightforward in most of the cases and you do not need to go through eqn 1.15 if you remember a couple of conversion factors:

$$\hbar c \simeq 197 \text{ MeV} \cdot \text{fm} \tag{1.17}$$

$$\hbar \simeq 6.58 \times 10^{-16} \text{ eV} \cdot \text{s} \tag{1.18}$$

where fm (femtometers or Fermi) corresponds to 10^{-15} m. Equation 1.17 converts lengths from NU (eV^{-1}) to SI (m) and eqn 1.18 converts times from NU (eV^{-1}) to SI (s). You can find two examples below: the former is trivial and does not require eqn 1.15. The latter is more sophisticated.

Example 1.1

The Compton wavelength of a particle is equal to the wavelength of a photon, whose energy is the same as the mass of the particle. For the electron case,

$$\lambda = \frac{h}{m_e c} = \frac{2\pi \hbar}{m_e c} \tag{1.19}$$

In SI, the wavelength is 6.63×10^{-34} J$\cdot$s$/(9.11 \times 10^{-31}$ kg $\cdot 3.00 \times 10^8$ m/s$) = 2.42 \times 10^{-12}$ m. In NU:

$$\lambda = \frac{2\pi}{m_e} = \frac{2\pi}{511 \text{ keV}} = 12.29 \text{ MeV}^{-1} \tag{1.20}$$

You can convert NU into SI using directly eqn 1.17 and sidestep the full analysis of eqn 1.15.

$$\lambda_{SI} = 197 \text{ MeV} \cdot \text{fm} \cdot 12.29 \text{ MeV}^{-1} = 2421 \text{ fm} = 2.42 \times 10^{-12} \text{ m}. \tag{1.21}$$

The next example is a case where we need to employ eqn 1.15.

Example 1.2

The Planck mass is a cornerstone of quantum gravity. It is defined in NU as:

$$M = \frac{1}{\sqrt{G}} \tag{1.22}$$

with G being the gravitational constant. We know G in SI ($6.67430 \pm 0.00015 \times 10^{-11}$ m^{3}kg^{-1} s^{-2}) and we are using a natural system (NU) where $G \neq 1$. What is the value of G in NU and the value of M in SI? In this non-trivial case, we resort to eqn 1.15: $M = \hbar^a c^b G^{-1/2}$. The dimensional analysis provides the values of a and b:

$$[M] = \left[\frac{ML^2}{T}\right]^a \left[\frac{L}{T}\right]^b \left[\frac{L^3}{MT^2}\right]^{-1/2} = [M]^{a+1/2}[L]^{2a+b-3/2}[T]^{-a-b+1} \tag{1.23}$$

which gives

$$a + 1/2 = 1$$
$$2a + b - 3/2 = 0$$
$$1 - a - b = 0$$

This is an over-constrained system of equation solved by $a = 1/2$, $b = 1/2$. The Planck constant in SI is then:

$$M = \sqrt{\frac{\hbar c}{G}} = \sqrt{\frac{1.05 \times 10^{-34}\, \text{J s} \cdot 3.00 \times 10^8\, \text{m/s}}{6.67 \times 10^{-11}\, \text{m}^3\text{kg}^{-1}\text{s}^{-2}}} = 2.17 \times 10^{-8}\, \text{kg} \tag{1.24}$$

The Planck mass is just a few tens of micrograms but it is enormous if expressed in NU. 2.17×10^{-8} kg corresponds to an energy of 2.17×10^{-8} kg $\cdot (3.00 \times 10^8$ m/s$)^2 = 1.95 \times 10^9$ J. Since 1 eV is 1.6×10^{-19} J, the Planck mass corresponds to 1.22×10^{28} eV, which is about 15 orders of magnitude the energy reached by the most powerful accelerator in the world (the LHC). In turn, eqn 1.24 provides G in NU: 6.7×10^{-57} eV^{-2}.

The SI value of M shows how difficult is to compress the energy of a macroscopic body - in this case, the weight of a few thousand human cells - into the volume filled by an "elementary particle."

Further reading

This book requires a basic knowledge of special relativity and quantum mechanics but can be enjoyed even without such background with little effort. All SR concepts needed in the forthcoming chapters are recalled in Appendix A and Chap. 2. If you want to appreciate the elegance of SR, I suggest reading a few introductory chapters among the vast literature on SR. For instance, Chaps 2, 3, 5, and 6 of Rindler, 2006 provide a clear introduction to the subject. Similarly, Chaps 2 and 3 of Cheng, 2010 give a succinct but effective background. A gentler introduction is given by Wittman, 2018. Appendix B provides the basic concepts and tools of quantum mechanics that are used in this book. To delve deeper, there are many outstanding textbooks. A solid background is given in Chaps 1–7 of Griffiths, 2018, for example. If you already have a good knowledge of QM, you may want to dig into the links between the founding principles of QM (symmetries, measurements, perturbation theory, entanglement, etc.) and the corresponding applications in particle and nuclear physics. You will find inspiration and alternative points of view in Weinberg, 2012; Griffiths, 2018; Sakurai and Napolitano, 2017; Manousakis, 2015; and Ballentine, 2014.

Exercises

(1.1) How many types of interactions contribute to the scattering of two quarks?

(1.2) Suppose that quarks have the same color charge. Is the total interaction probability of a quark u and a quark d the same as for two d quarks?

(1.3) Consider the following processes:

$$n \to p + e^- + \bar{\nu}_e$$
$$e^- + \nu_e \to e^- + \nu_e$$
$$e^- + p \to e^- + p$$
$$e^- + e^+ \to e^- + e^+$$

How many of these processes need a full QFT description?

(1.4) If we assume the kinetic theory of gases, what is the unit of temperature in NU?

(1.5) The proton Compton wavelength is defined as

$$\lambda_p = \frac{h}{m_p c}$$

in SI. Compute its value both in SI and NU.

(1.6) Compute the e.m. force between two protons located a 10^{-10} m and compare this value with the corresponding gravitational force.

(1.7) Demonstrate that λ_p/λ_e in SI has the same numerical value of λ_p/λ_e in NU.

(1.8) What is the unit of the wavefunctions in NU? [Hint: inspect the definition of wavefunction in Appendix B.]

(1.9) Is it possible to extend NU in such a way that the unit of temperature is eV, even beyond the kinetic theory of gases?

(1.10) * What is the value of the electron charge in NU?

(1.11) * Demonstrate that electric and magnetic fields have the same dimensions in NU.

(1.12) ** In an imaginary universe, suppose that the maximum velocity of a body is c, as in SR, but the maximum energy of a photon must be finite according to this hypothetical formula:

$$E = \hbar\omega \quad \text{if} \quad \hbar\omega \ll \Lambda$$
$$E \to \Lambda \quad \text{if} \quad \hbar\omega \to \Lambda$$

where Λ is an energy. Can you employ NU to describe this universe? Is there a better natural system that can be used to get only dimensionless units?

Scattering and decay

2

2.1 Covariance

Particle physics is rooted in special relativity (SR) and the principles of the theory are embedded in the units ($c = 1$) and language of particle physicists. This choice comes as no surprise: SR requires a generalization (general relativity, GR) only when gravitational interactions play a role in the dynamics of a system. This situation is extremely rare in particle physics, where other interactions are overwhelming and the distances traveled by particles are so small that space-time can always be considered locally flat. In this framework, SR can be easily employed to handle non-inertial systems and accelerating particles too (see Sec. A.7). Exceptions arise, for instance, when considering the motion of a particle at cosmological distances or in the proximity of a black hole, where a description based on GR is mandatory.

The concept of covariance can be introduced on a general ground to cope with SR, GR and quantum gravity and can be restricted to SR in all cases of interest for this book.

> **Covariance** is a prescription on the form of the physical laws that ensures compliance with the axioms of relativity:
> (A1) The physical laws retain the same functional form in all inertial frames.
> (A2) The speed of light in vacuum is the same for all inertial observers.

Without a covariant formalism,[1] we should check explicitly that all quantities describing a system (constants, vectors, or any other combination) linked by a physical law such as $f(a, b, c, \ldots) = 0$ change in the right way after a Lorentz transformation (LT). The "right way" is:

$$f(a, b, c, \ldots) = 0 \xrightarrow{LT} f'(a', b', c', \ldots) = 0 \ \text{ AND } \ f' = f \qquad (2.1)$$

so that the physical law in the new reference frame reads:

$$f(a', b', c', \ldots) = 0 \qquad (2.2)$$

where a', b', c' are the transformed quantities.

[1] In some SR textbooks, the term covariant is used more subtly. A "covariant law" is a law consistent with SR. An "explicitly-covariant law" is a law written in such a way that the consistency with SR can be assessed at first sight, i.e. without using the Lorentz transformations. For sake of clarity and generality, we prefer to use "covariant" and "explicitly covariant" as synonyms. This choice implies that the physical laws can be classified into three groups: laws compatible with SR, incompatible with SR, and covariant. We can thus recast in a covariant form any law compatible with SR.

A Modern Primer in Particle and Nuclear Physics. Francesco Terranova, Oxford University Press.
© Francesco Terranova (2021). DOI: 10.1093/oso/9780192845245.003.0002

2.1.1 Covariant and contravariant quantities

Modern geometry describes and classifies objects based on how they change under a general class of transformations (rotations, translations, projective transformations, etc.). The quantities that are interesting in physics may change if the reference system is changed by one of these transformations but – to the best of our knowledge – reference frames and coordinates do not exist in nature: they are practical tools used to describe the physical systems. Physical laws should always be independent of the specific choice of the coordinates.

> **General covariance** is the invariance of the form of the physical laws under the action of differentiable (smooth) coordinate transformations.

For concreteness, consider a vector $\mathbf{v}$ in the origin of a coordinate system of a three-dimensional (3-D) Euclidean space. The coordinates depend on the choice of the **versors** or **unit-vectors**[2] describing the points in space (see Fig. 2.1):

$$\mathbf{v} = v_1\mathbf{i} + v_2\mathbf{j} + v_3\mathbf{k}. \tag{2.3}$$

A different choice of the versors corresponds to changing the basis of the Euclidean space and, then, of $\mathbb{R}^3$. The new vector $\mathbf{v}'$ expressed in the new basis is:

$$\mathbf{v}' = v_1'\mathbf{i}' + v_2'\mathbf{j}' + v_3'\mathbf{k}'. \tag{2.4}$$

The real numbers v_1, v_2, v_3 are called **coordinates**. In classical mechanics, coordinates are often labeled as v_x, v_y, v_z. The basis is described by three perpendicular versors such that $\mathbf{i} \cdot \mathbf{j} = \mathbf{j} \cdot \mathbf{k} = \mathbf{k} \cdot \mathbf{i} = 0$. In linear algebra, this is equivalent to changing the "orthonormal basis" $\mathbf{i}$, $\mathbf{j}$, and $\mathbf{k}$ with a new "orthonormal basis" $\mathbf{i}'$, $\mathbf{j}'$, and $\mathbf{k}'$ (Larson and Falvo, 2009).

In a n-dimensional space, the coordinates of $\mathbf{v}$ are an n-tuple of real numbers $v_1 \cdots v_n$ associated to the basis $\mathbf{e}_1 \cdots \mathbf{e}_n$. We can write $\mathbf{v}$ in a neater form as:

$$\mathbf{v} = \sum_{i=1}^{n} v^i\mathbf{e}_i. \tag{2.5}$$

Even better, we can employ the

> **Einstein summation convention**: when an index variable appears twice in a single term, it implies summation of that term over all the values of the index.

The vector is then:

$$\mathbf{v} = v^i\mathbf{e}_i \tag{2.6}$$

which is a shortening of eqn 2.5. Ignore for the moment that the index of the coordinates is a superscript (v^i) and not a subscript (v_i): it will be clearer in a minute. A change of basis in $\mathbb{R}^n$ is a linear transformation that maps each versor $\mathbf{e}'_i$ of the new basis to a linear combination of the

[2]A versor in physics (not to be confused with versors in mathematics) is a unit-norm vector that has the same direction of the vector. A versor is labeled in bold letters with a hat: for instance, $\hat{\mathbf{r}}$. In this book, the versors of the reference-frame axes are indicated without the hat as $\mathbf{i}$ (x-axis), $\mathbf{j}$ (y-axis), and $\mathbf{k}$ (z-axis); see Fig. 2.1. For a general n dimensional space, the axes are n and the corresponding versors are labeled $\mathbf{e}_1, \ldots \mathbf{e}_n$.

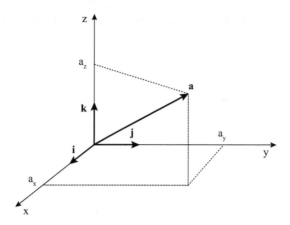

Fig. 2.1 A vector **a** starting from the origin of the reference frame. The coordinates of the vector are the projections of **a** in three arbitrary and perpendicular axes: x, y, and z. These projections are called $a_x \equiv a_1$, $a_y \equiv a_2$, and $a_z \equiv a_3$. The unit vectors (versors) parallel to the axes are $\mathbf{e_1} \equiv \mathbf{i}$, $\mathbf{e_2} \equiv \mathbf{j}$, and $\mathbf{e_3} \equiv \mathbf{k}$, respectively.

old versors $\mathbf{e}_i$:

$$\mathbf{e}'_i = \sum_{j=1}^{n} b^j{}_i \mathbf{e}_j \equiv b^j{}_i \mathbf{e}_j \tag{2.7}$$

We can also describe a change of basis using the matrix formalism. Equation 2.7 reads

$$\mathbf{e}' = B\mathbf{e} = b^j{}_i \mathbf{e}_j \tag{2.8}$$

where B is a $n \times n$ matrix, $b^j{}_i$ are the coefficients of the matrix – that is, the number in row j and column i – and $\mathbf{e}$ is a column of n versors.

If the vector $\mathbf{v}$ must be independent of the choice of the base, we cannot leave the coordinates unchanged, otherwise the position of the vector in space would change too. The coordinates must change to compensate for the change of the versors or, equivalently,

---
the components of the vector that corresponds to the coordinates must *contra-vary* to compensate for the changes in the basis and keep the vector fixed in space.

Since the change of basis makes:

$$\mathbf{v} = v^i \mathbf{e}_i = \mathbf{v}' = v'^i \mathbf{e}'_i = v'^i b^j{}_i \mathbf{e}_j, \tag{2.9}$$

the coordinates must change as:

$$v'^i = a^i{}_j v^j \tag{2.10}$$

to have a vector fixed in space. In eqn 2.10, $a^i{}_j$ are the $n \times n$ elements of the matrix A, which is the *inverse* of B ($A = B^{-1}$). In summary:

- after a change of basis described by the matrix B, the coordinates of a vector transform as $\mathbf{v}' = A\mathbf{v}$, where $A = B^{-1}$. All n-tuples that transform in this way are called **contravariant vectors** or simply **vectors**, and are written with upper indexes (v^i).

- all n-tuples that transform as the versors – which means as one particular choice of the basis for a given space – after a change of basis ($v'_i = Bv_i$) are called **covariant vectors** and are written with lower indexes (v_i).

So covariance and contravariance describe how certain geometric or physical objects change after a change of basis. We require a vector to be base-independent to get a good candidate for a real physical quantity. The constraint of base-independence implies the covariance of the versors and the contravariance of the coordinates. These are non-trivial constraints: a vector is no more a simple n-tuple of real numbers but must transform in a well-defined way if we decide to change the reference frame.

2.1.2   Classification of covariant objects

We can further generalize this approach classifying all base-independent objects by their behavior after a change of basis.

Scalars are quantities that do not change after a change of basis
Contravariant vectors, or simply **vectors**, are n-tuples that transform as $x^{i'} = a^{i'}_{\ i} x^i$.
Contravariant tensors of rank 2 are a $n \times n$ set of numbers that transform as $T^{i'j'} = a^{i'}_{\ i} a^{j'}_{\ j} T^{ij}$.
Covariant vectors are n-tuples that transform as $x_{i'} = b^i_{\ i'} x_i$.
Covariant tensors of rank 2 are a $n \times n$ set of numbers that transform as $T_{i'j'} = b^i_{\ i'} b^j_{\ j'} T_{ij}$.

A rank-2 tensor looks like a $n \times n$ matrix but the n^2 coefficients are carefully chosen to fulfill the definitions above. Further generalizations (e.g. tensors of rank-n, spinors, etc.) are also available but are not needed at this stage. Just out of pedantry, note that the definition of a rank-n tensor is straightforward: a contravariant tensor of rank n is a set of numbers that transforms as:

$$T^{i'_1 i'_2 \ldots i'_n} = a^{i'_1}_{\ i_1} a^{i'_2}_{\ i_2} \cdots a^{i'_n}_{\ i_n} T^{i_1 i_2 \ldots i_n} \tag{2.11}$$

and we can also build mixed covariant–contravariant tensors like:

$$T^{i'_1 i'_2, \ldots i'_n}_{j'_1 j'_2, \ldots j'_m} = a^{i'_1}_{\ i_1} a^{i'_2}_{\ i_2} \cdots a^{i'_n}_{\ i_n} b^{j_1}_{\ j'_1} b^{j_2}_{\ j'_2} \cdots b^{j_m}_{\ j'_m} T^{i_1 i_2 \ldots i_n}_{j_1 j_2 \ldots j_m} \tag{2.12}$$

Physicists are particularly interested in changes of basis that represent smooth (i.e. differentiable) coordinate transformations. The Lorentz transformations in SR and the Galilean transformations in classical physics (see Sec. A.1) are two notable examples. For a generic smooth coordinate transformation between $\mathbf{x}'$ and $\mathbf{x}$, we have:

$$a^{i'}_{\ i} = \frac{\partial x^{i'}}{\partial x^i} \tag{2.13}$$

and

$$b^i_{\ i'} = \frac{\partial x^i}{\partial x^{i'}} \tag{2.14}$$

so a (contravariant) vector transforms as:

$$v^{i'} = \frac{\partial x^{i'}}{\partial x^i} v^i.$$ (2.15)

You may wonder whether there are interesting physical objects that transform as covariant vectors or contravariant tensors. There are many. Quantities that involve *odd derivatives* of vectors are often covariant. For instance, the **gradient** of a generic function $f(x^1, \ldots, x^n)$ of the coordinates is a covariant vector. This statement follows from:

$$\frac{\partial f}{\partial x^{i'}} = \frac{\partial f}{\partial x^i} \frac{\partial x^i}{\partial x^{i'}} = \frac{\partial f}{\partial x^i} b^i{}_{i'}.$$ (2.16)

In classical mechanics, the inertia tensor is a rank-2 contravariant tensor that transforms as a tensor for rotations in space (Goldstein, 1980). Relativistic electrodynamics is also built upon covariant and contravariant rank-2 tensors under Lorentz transformations, as discussed in Chap. 6.

2.2 Particles in space-time

General covariance can be applied to the specific case of SR. Here, the space is four-dimensional (time and the three directions in space) and the smooth changes of basis correspond to the Lorentz transformations. For a reference system O' moving with a velocity $v\mathbf{i}$ in the x direction relative to the system O (Fig. 2.2), the Lorentz transformations in SI read:

$$
\begin{aligned}
t' &= \gamma(t - \frac{\beta}{c}x) \\
x' &= \gamma(x - \beta ct) \\
y' &= y \\
z' &= z
\end{aligned}
$$ (2.17)

where β is equal to v/c and called the $\boldsymbol{\beta}$ **Lorentz factor**.[3] In NU, $\beta = v$.

$$\gamma \equiv \frac{1}{\sqrt{1 - \beta^2}}$$ (2.18)

is called the $\boldsymbol{\gamma}$ **Lorentz factor** and is $\sqrt{1/(1 - v^2)}$ in NU.

SR mixes space with time but only in the direction of motion of the system O' relative to O. The coordinates in the plane perpendicular to this direction are the same in both systems. To embed the covariant formalism in eqn 2.17, we define the coordinates of (contravariant) vectors in $\mathbb{R}^4$ starting from 0 instead of 1:

$$x^\mu = (x^0 \equiv ct, x^1 \equiv x, x^2 \equiv y, x^3 \equiv z)$$ (2.19)

x^μ is called a **four-vector**. In NU,

$$x^\mu = (x^0, x^1, x^2, x^3)$$ (2.20)

[3] Equation 2.17 is just an example of Lorentz transformations and is called a **Lorentz boost along x**. The general Lorentz transformations are described and classified in Sec. A.1.

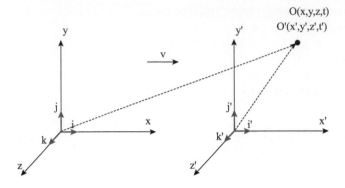

Fig. 2.2 A reference system O' moves in the rightmost direction along the x axis ($\mathbf{v} = v\mathbf{i}$) with respect to a system O. The point on the right shows an event and its coordinates in O and O'.

and

$$\begin{pmatrix} x'^0 \\ x'^1 \\ x'^2 \\ x'^3 \end{pmatrix} = \begin{pmatrix} \gamma & -\beta\gamma & 0 & 0 \\ -\beta\gamma & \gamma & 0 & 0 \\ 0 & 0 & 1 & 0 \\ 0 & 0 & 0 & 1 \end{pmatrix} \begin{pmatrix} x^0 \\ x^1 \\ x^2 \\ x^3 \end{pmatrix}. \tag{2.21}$$

Using the Einstein summation convention, we can write eqn 2.21 as:

$$x'^\mu = \Lambda^\mu{}_\nu x^\nu \tag{2.22}$$

where $\Lambda^\mu{}_\nu$ is the number located in row μ and column ν of the 4×4 matrix of eqn 2.21. As a consequence, the matrix A that represents the change of coordinates in $\mathbb{R}^4$ has a very specific form given by $\Lambda^\mu{}_\nu$. In the language of general covariance, x^μ is a contravariant vector for the change of coordinates represented by the Lorentz transformations $\Lambda^\mu{}_\nu$. We can get the inverse Lorentz transformation from O to O' using eqn 2.17 by changing the sign of the velocity:

$$\begin{pmatrix} x^0 \\ x^1 \\ x^2 \\ x^3 \end{pmatrix} = \begin{pmatrix} \gamma & \beta\gamma & 0 & 0 \\ \beta\gamma & \gamma & 0 & 0 \\ 0 & 0 & 1 & 0 \\ 0 & 0 & 0 & 1 \end{pmatrix} \begin{pmatrix} x'^0 \\ x'^1 \\ x'^2 \\ x'^3 \end{pmatrix}. \tag{2.23}$$

As expected, the matrix associated with eqn 2.23 is Λ^{-1}.

If the direction of motion of the two systems is not along the x-axis but on a generic unit-vector $\mathbf{v}/v$, we can define $\boldsymbol{\beta} = \mathbf{v}/c$ (or $\boldsymbol{\beta} = \mathbf{v}$ in NU). The corresponding Lorentz transformations, called **Lorentz boosts**, are now:

$$x'^0 = \gamma(x^0 - \frac{\mathbf{v} \cdot \mathbf{x}}{c})$$

$$\mathbf{x}' = \mathbf{x}_\perp + \gamma(\mathbf{x}_\parallel - \frac{\mathbf{v}x^0}{c}) \tag{2.24}$$

in SI. Here, $\mathbf{x}'_\perp$ and $\mathbf{x}'_\parallel$ are the spatial components projected in the direction perpendicular and parallel to the motion. We can write eqn 2.17 in NU in a compact form employing the matrix notation of linear algebra:

$$\begin{pmatrix} x'^0 \\ \mathbf{x}' \end{pmatrix} = \begin{pmatrix} \gamma & -\gamma\boldsymbol{\beta}^T \\ -\gamma\boldsymbol{\beta} & \mathbb{I} + (\gamma - 1)\frac{\boldsymbol{\beta}\boldsymbol{\beta}^T}{\beta^2} \end{pmatrix} \begin{pmatrix} x^0 \\ \mathbf{x} \end{pmatrix} \tag{2.25}$$

where $\mathbb{I}$ is the 3×3 identity matrix,

$$\boldsymbol{\beta} = \begin{pmatrix} \beta_1 = v_x \\ \beta_2 = v_y \\ \beta_3 = v_z \end{pmatrix}$$

is a column-vector and

$$\boldsymbol{\beta}^T = (\beta_1, \beta_2, \beta_3) \tag{2.26}$$

is a row-vector. These compact notations may be puzzling for a novice. In the covariant formalism of SR, we should just use covariant quantities like the x^μ vectors. However, it is useful to distinguish the space components of x^μ from the time component because space can be measured with rulers, time with clocks, and real experiments do not provide straightforward space-time measurements. In this sort of mixed notation, the space components are written with *Latin indexes*: $\mathbf{x} = (x^1, x^2, x^3)$. The position of a particle at a time t is labeled $\mathbf{x}$ as in classical physics and corresponds to the SR "event" $x^\mu = (ct, \mathbf{x})$. The **event** $x^\mu = (ct, \mathbf{x})$ is the actual measurement in space-time and it tells us where and when we find the particle: in the position $\mathbf{x}$ at the time t. Events are always labeled with *Greek indexes* ($\mu = 0, 1, 2, 3$) to avoid any source of confusion with space coordinates.

2.2.1 The Minkowski space-time

The covariant formalism makes very clear the rich mathematical structure of SR. It also provides simple means to construct relativistic dynamics in analogy to classical physics, as shown in Sec. A.3.

The first observable we can build using covariance is the **trajectory** of a point-like particle. In classical physics, the trajectory is a function that maps the position in space as a function of time. Since time is universal, the distance between two points is the same for any observer even if the coordinates of the points change among observers. The (squared) distance traveled between t and $t + dt$ is thus the same in any inertial frame and is equal to:

$$ds^2 = dx^2 + dy^2 + dz^2 \tag{2.27}$$

where

$$d\mathbf{x} = dx\mathbf{i} + dy\mathbf{j} + dz\mathbf{k}. \tag{2.28}$$

In SR, the observers disagree on time intervals and on the distance between two points in space.[4] However, they agree on the squared difference between time and space. If an observer in O releases a particle at time t_1 in $\mathbf{x}_1$, records the passage of the particle in a detector located in $\mathbf{x}_2$ at time t_2, and compares her findings with an observer O', SR ensures that:

$$c^2(t_2 - t_1)^2 - (x_2 - x_1)^2 - (y_2 - y_1)^2 - (z_2 - z_1)^2 =$$
$$c^2(t'_2 - t'_1)^2 - (x'_2 - x'_1)^2 - (y'_2 - y'_1)^2 - (z'_2 - z'_1)^2$$

[4]This is a consequence of time dilation and space contraction, as shown in sec. A.2.

(see Exercise 2.3). We can then replace eqn 2.27 with

$$ds^2 = c^2 dt^2 - dx^2 - dy^2 - dz^2 \qquad (2.29)$$

and be certain that every observer measuring ds^2 gets the same number, whatever frame she trusts. In the language of covariance, ds^2 is a Lorentz scalar, that is it is **invariant under Lorentz transformations**. Similarly, the distance between two events is a scalar: $c^2(t_2 - t_1)^2 - (x_2 - x_1)^2 - (y_2 - y_1)^2 - (z_2 - z_1)^2$. If $t_1 = 0$ and $\mathbf{x}$ is the origin of the system, $c^2 t^2 - x^2 - y^2 - z^2 = (x^0)^2 - (x^1)^2 - (x^2)^2 - (x^3)^2$ is a scalar too.

The quantity $(x^0)^2 - (x^1)^2 - (x^2)^2 - (x^3)^2$ can be expressed for any contravariant vector as:

$$g_{\mu\nu} x^\mu x^\nu \qquad (2.30)$$

employing the Einstein summation convention. The 4×4 real matrix $g_{\mu\nu}$ is the **metric tensor** of SR:

$$g_{\mu\nu} = \begin{pmatrix} 1 & 0 & 0 & 0 \\ 0 & -1 & 0 & 0 \\ 0 & 0 & -1 & 0 \\ 0 & 0 & 0 & -1 \end{pmatrix}. \qquad (2.31)$$

The metric tensor of eqn 2.31 distills the essence of the covariant formalism *for the particular case of SR*. $g_{\mu\nu}$ fulfils three important properties:

- $g_{\mu\nu}$ is constant. Unlike, for example, GR (Cheng, 2010), eqn 2.31 holds in any point of the space and at any time, so that $g_{\mu\nu}(x^\mu) = g_{\mu\nu}$.

- $g_{\mu\nu}$ is a diagonal matrix and differs from the identity matrix only by a sign in the space $\mu = 1, 2, 3$ components.

- $g_{\mu\nu}$ is equal to its inverse.

The peculiar form of the metric tensor provides a rule to transform a contravariant quantity into a covariant quantity:

> for a generic contravariant four-vector a^μ,
>
> $$a_\mu = g_{\mu\nu} a^\nu \qquad (2.32)$$
>
> is covariant. As a consequence, the SR rule that transforms a contravariant vector in a covariant vector is:
>
> $$a_0 = a^0$$
> $$a_i = -a^i \qquad (2.33)$$
>
> for $i = 1, 2, 3$.

Of course, such a simple rule holds in SR, only. We can use the same rule to transform a contravariant tensor in a covariant tensor:

$$T_{\mu\nu} = g_{\mu\alpha} g_{\nu\beta} T^{\alpha\beta} \qquad (2.34)$$

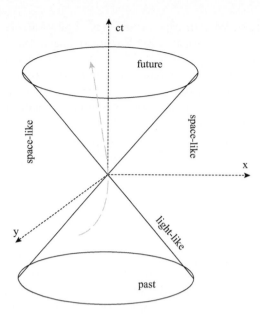

Fig. 2.3 A particle moves in space-time from past to future. The trajectory (dashed line) is contained in the time-like cone because the particle travels slower than light. The possible locations of space-like and light-like events are also shown.

which, in practice, corresponds to changing the sign for all space indexes ($\mu = 1, 2, 3$) and leaving the time ($\mu = 0$) as it is. Inverting eqn 2.33, we get the covariant-to-contravariant transformation rule of SR:

$$a^0 = a_0$$
$$a^i = -a_i \tag{2.35}$$

These rules are very simple because $g_{\mu\nu}$ is very simple: a diagonal matrix, which is equal to its inverse. You can also check that $g_{\mu\nu} = g^{\mu\nu}$. Equation 2.30 can then be rewritten as:

$$g_{\mu\nu} x^\nu x^\mu = x_\mu x^\mu \tag{2.36}$$

The interpretation of eqn 2.36 for a point-like particle released at $t = 0$ and $\mathbf{x} = 0$ is particularly informative. If the passage of the particle in a detector corresponds to an event x^μ, SR provides only two possibilities:

> if $(x^0)^2 - (x^1)^2 - (x^2)^2 - (x^3)^2 > 0$, the event is **time-like**, and the particle reaches the point $\mathbf{x}$ at the time t traveling slower than light in vacuum (see Fig. 2.3).
> If $(x^0)^2 - (x^1)^2 - (x^2)^2 - (x^3)^2 = 0$, the event is **light-like**, and the particle reaches the point $\mathbf{x}$ at the time t traveling exactly at the same speed as the light in vacuum.

If $(x^0)^2 - (x^1)^2 - (x^2)^2 - (x^3)^2 < 0$ the event is **space-like**. In SR, this condition is never fulfilled by a point-like particle because it would imply particles moving faster than light. Similarly, if two events x_a^μ and x_b^μ are such that $(x_a^0 - x_b^0)^2 - (x_a^1 - x_b^1)^2 - (x_a^2 - x_b^2)^2 - (x_a^3 - x_b^3)^2 < 0$, the two events are not causally connected. The first event cannot communicate with the second event and vice versa.[5]

[5]If you are intrigued by the meaning of causality in SR and QM, you will find further insight in Secs A.1 and B.4.

The quantity $(x^0)^2 - (x^1)^2 - (x^2)^2 - (x^3)^2$ plays a role similar to the (squared) norm of a vector in a 3D Euclidean space: $|\mathbf{x}|^2 = x^2 + y^2 + z^2$. The Euclidean norm is invariant for rotations and translations of the reference frame and it is the same for any inertial observer in classical physics. In the language of general covariance, the Euclidean norm is a scalar for the Galilean differentiable coordinate transformations:

$$
\begin{aligned}
t' &= t \\
x' &= x - vt \\
y' &= y \\
z' &= z
\end{aligned}
\tag{2.37}
$$

We can thus define a "norm" for x^μ corresponding to:

$$
x^2 \equiv x_\mu x^\mu = (x^0)^2 - (x^1)^2 - (x^2)^2 - (x^3)^2
\tag{2.38}
$$

[6] In this book, I use the most common notation for the norm of a four-vector in particle physics, even if this symbol is slightly ambiguous. Note that x^2 could be confused with the third component of x^μ, that is x^2. Anyway, it is very easy to grasp the correct meaning from the context and the notation is less cumbersome than $\|x\|^2$ or others.

which is a scalar for the Lorentz transformations.[6] Note that we are using the term "norm" quite loosely since in mathematics the norm should be ≥ 0. This is not the case for space-like events. As we will see in the following, scalars – quantities that are invariant for Lorentz transformations – are a key tool to study the motion of relativistic particles.

The "distance" between a point in the $\mathbb{R}^4$ space that describes an event in SR and the event located at the origin is defined by eqn 2.38. $\mathbb{R}^4$ equipped with the norm of eqn 2.38 is called the **Minkowski space-time** or, in short, the Minkowski space. The trajectories of the particles are curves in the Minkowski space. The curves are maps that uniquely associate the event (ct, x, y, z) to a given curve parameter k: $k \to x^\mu(k)$. In some cases, you can still parameterize the curve as a function of time in a given reference frame ($k = t$), as in classical physics, but this is not practical because the observers do not agree on time. A better choice is to prescribe a reference frame where the motion of the particle is stationary and parameterize the curve by the time measured in that frame. For a single particle propagating in space, this system is the **proper** (or **comoving**, or **rest frame**, RF) of the particle: the system whose origin moves together with the particle. The time measured in the RF is called **proper time** and indicated as τ. The particle is stationary in this frame and this trick simplifies the calculations. We can finally write the definition of the trajectory in SR:

> the **trajectory** of a particle is a function that maps the position of the particle in the Minkowski space-time as a function of its proper time: $\tau \to x^\mu(\tau)$

2.2.2 Notations for special relativity

It is easy to get confused when classical and relativistic quantities are mixed up and we do it pretty often for the reasons mentioned in Sec. 2.2. These habits are annoying for novices but very much appreciated by experienced users, provided that a consistent notation is employed. By happy chance, particle and nuclear physicists use a nearly universal notation, which is summarized here and employed in the rest of the book.

- Classical three-vectors are always written in **boldface**. The position in space of a particle is written as $\mathbf{x}$. The scalar product of classical three-vectors is $\mathbf{a} \cdot \mathbf{b}$ and should be interpreted as $a_x b_x + a_y b_y + a_z b_z$. The norm of $\mathbf{a}$ is written as $|\mathbf{a}|$ and is the standard Euclidean norm: $|\mathbf{a}| = (a_x^2 + a_y^2 + a_z^2)^{1/2}$.

- Four-vectors are written with Greek indexes as a^μ or simply a. The squared "norm" of a^μ in the Minkovsky space is $a^2 \equiv a_\mu a^\mu = g_{\mu\nu} a^\mu a^\nu = (a^0)^2 - (a^1)^2 - (a^2)^2 - (a^3)^2$. For convenience, you may want to write it in a neater form using three-vectors: $a^2 = a_\mu a^\mu = (a^0)^2 - |\mathbf{a}|^2$.

- The "scalar product" between four-vectors is written as $a \cdot b$ or $a_\mu b^\mu$, and:

$$a_\mu b^\mu = g_{\mu\nu} a^\nu b^\mu = a^0 b^0 - a^1 b^1 - a^2 b^2 - a^3 b^3 = a^0 b^0 - \mathbf{a} \cdot \mathbf{b} \quad (2.39)$$

Equation 2.39 testifies to the popularity of mixed notations among experienced physicists, who love easy-to-remember formulas. An example is $(a+b)^2$ that can be written as $a^2 + b^2 + 2a \cdot b$ like a high-school "square of sum" formula. Note, however, that $a \cdot b$ is a scalar product in the Minkowski space and $a \cdot b = a^0 b^0 - \mathbf{a} \cdot \mathbf{b}$.

We also have a few deprecated notations that may be popular in other fields. For instance, $|\mathbf{p}|$ is often written as p to avoid too cumbersome formulas. In this case, we should check that we are not confusing p with the four-vector p. In cosmology and gravitation, the metric tensor is defined as:

$$\eta_{\mu\nu} = \begin{pmatrix} -1 & 0 & 0 & 0 \\ 0 & 1 & 0 & 0 \\ 0 & 0 & 1 & 0 \\ 0 & 0 & 0 & 1 \end{pmatrix}$$

and cosmologists have the opposite sign convention for time-like and space-like events. You will not find these notations in this book but you may find them in the literature.

2.3 Kinematical constrains

We can extend to SR the conservation of energy and momentum introducing the concept of relativistic energy and four-momentum (see Appendix A). In particular,

the **four-momentum** is defined (in NU) as:

$$p^\mu = (m\gamma, m\gamma \mathbf{v}) \equiv (E, \mathbf{p}) \quad (2.40)$$

where m is the mass of the particle in its rest frame. γ and $\mathbf{v}$ are the gamma Lorentz factor and the three-velocity of the particle in the laboratory frame, respectively.

E and $\mathbf{p}$ are called the relativistic energy and three-momentum and will be discussed in detail in Sec. 2.3.2.

Kinematics and, in particular, the constraints coming from four-momentum conservation provide non-trivial information on particle interactions even if the interaction laws are unknown. A system of N_i particles with $p_1^\mu \dots p_{N_i}^\mu$ four-momenta can interact creating a new system of N_f particles with $p_1^\mu \dots p_{N_f}^\mu$ four-momenta (Fig. 2.4) only if the total four-momentum is conserved:

$$\sum_{i=1}^{N_i} (p_i^{ini})^\mu = \sum_{j=1}^{N_f} (p_j^{fin})^\mu \tag{2.41}$$

that is:

$$\sum_{i=1}^{N_i} (p_i^{ini})^\mu - \sum_{j=1}^{N_f} (p_j^{fin})^\mu = 0 \tag{2.42}$$

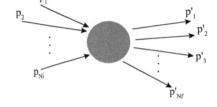

where the first sum extends over all particles in the **initial state** and the second sum over all particles in the **final state**. This conservation law holds in any reference frame. To lighten the formulas, we drop the "ini" and "fin" labels because we can guess whether the four-vectors belong to the initial or final state looking at the sum. The sum ranges from 1 to N_i for "ini" and from 1 to N_f for "fin." In this case, eqn 2.42 is written as:

$$\sum_{i=1}^{N_i} p_i^\mu - \sum_{j=1}^{N_f} p_j^\mu = 0 \tag{2.43}$$

Fig. 2.4 N_i particles in the initial state scatter into N_f particles in the final state. Unlike classical and relativistic mechanics, N_i is in general $\neq N_f$. p_k (p_k') represents the four-momentum of the k-th particle in the initial (final) state. Note that p_k is a contraction for p_k^μ.

Equation 2.43 is tidy and easy to remember but please remember that i and j are the indexes of the particles and *not* indexes of the Minkowski space.

2.3.1 Physical laws and covariance

The conservation of four-momentum is the first relativistic law we encounter in this chapter. Due to covariance, it is straightforward to show that this law has the same functional form in any inertial reference frame. In a system O' linked to O by the Lorentz transformation $\Lambda^\mu{}_\nu$, a four-vector changes to:

$$p'^\mu \equiv \Lambda^\mu{}_\nu p^\nu \tag{2.44}$$

and, in O', eqn 2.42 looks like:

$$\sum_{i=1}^{N_i} \Lambda^\mu{}_\nu p_i^\nu - \sum_{j=1}^{N_f} \Lambda^\mu{}_\nu p_j^\nu = 0. \tag{2.45}$$

Equation 2.45 keeps the same functional form of eqn 2.42 by defining the transformed four-momenta as:

$$p_i'^\mu \equiv \Lambda^\mu{}_\nu p_i^\nu \quad ; \quad p_j'^\mu \equiv \Lambda^\mu{}_\nu p_j^\nu \tag{2.46}$$

so that

$$\sum_{i=1}^{N_i} \Lambda^\mu{}_\nu p_i^\nu - \sum_{j=1}^{N_f} \Lambda^\mu{}_\nu p_j^\nu = \sum_{i=1}^{N_i} p_i'^\mu - \sum_{j=1}^{N_f} p_j'^\mu = 0. \qquad (2.47)$$

We do not need to show that $p_i'^\mu$ or $p_j'^\mu$ is a (contravariant) vector because:

$$p_i'^\mu \equiv \Lambda^\mu{}_\nu p_i^\nu \qquad (2.48)$$

is exactly the definition of a vector in SR: an object that changes as eqn 2.44 for a generic Lorentz transformation.

We can then clarify the most important advantage of the covariant formalism as follows:

---

any physics law compatible with SR can be expressed as

$$T^{\mu_1, \mu_2 \ldots \mu_n}_{\nu_1, \nu_2 \ldots \nu_m} = S^{\mu_1, \mu_2 \ldots \mu_n}_{\nu_1, \nu_2 \ldots \nu_m} \qquad (2.49)$$

where T and S are tensors. In turn, an SR law is a function of Lorentz scalars (a), vectors (b^μ or d_μ) and tensors ($c^{\mu\nu}$ or $e_{\mu\nu}$) in the form:

$$f(a, b^\mu, c^{\mu\nu}, d_\mu, e_{\mu\nu}, \ldots) = 0 \qquad (2.50)$$

where f is a generic constraint among physical quantities.

For all practical purposes, the rigorous definition of eqn 2.49 employing mixed tensors of rank $n + m$ is equivalent to eqn 2.50. This is because a rank-0 tensor is a scalar, a rank-1 tensor is a vector, and the sum, product, and powers of tensors are tensors too. So, if f is a regular function that can be expressed, for example, as a Taylor series, we will always be able to recast eqn 2.50 in the form of eqn 2.49. Still, eqn 2.49 brings an important message to us: at the end of the day, the number and position of the indices on the left must match the ones on the right. Any mismatch signals an inconsistency with the axioms of SR.

Example 2.1

Equation 2.43 fulfills the definition of physical law in SR. It is already written in the form of eqn 2.50 and is the sum of rank-1 tensors (four-vectors). We can recast eqn 2.43 defining

$$T^\mu \equiv \sum_{i=1}^{N_i} p_i^\mu \quad ; \quad S^\mu \equiv \sum_{j=1}^{N_f} p_j^\mu \qquad (2.51)$$

so that:

$$T^\mu = S^\mu. \qquad (2.52)$$

Note that in this case we only have one matched index in a contravariant position.

[7]As discussed in Appendix A, the relativistic extension of Newton's force K^μ is the derivative of the four-momentum by the proper time:

$$K^\mu \equiv \frac{dp^\mu}{d\tau}. \qquad (2.53)$$

The relativistic force is a vector because is the ratio of a vector with a scalar and the scalar is unchanged after a Lorentz transformation.

The reader is acquainted with the fact that Newton's second law of mechanics ($\mathbf{F} = m\mathbf{a}$) is not compatible with SR,[7] while the Lorentz force on a charged particle moving in an electromagnetic field $\mathbf{F} = q(\mathbf{E} + \mathbf{v} \times \mathbf{B})$ describes a genuine relativistic effect. Therefore, the Lorentz formula can be recast as an SR law.

Example 2.2

You can go through this process the hard way using the Lorentz transformations or recast the Lorentz formula in covariant form. The result is:

$$K^\nu = \frac{q}{m} p_\mu F^{\mu\nu}. \qquad (2.54)$$

[8]An index is free if it is not summed with another index in Einstein's summation convention. The name of a **free index** is arbitrary and we can choose any (greek) letter that is not already employed in the formula. The only free index in $(q/m)\, p_\mu F^{\mu\nu}$ is ν. Choosing e.g. σ for that index, eqn 2.54 would look like:

$$K^\sigma = \frac{q}{m} p_\mu F^{\mu\sigma},$$

which is equivalent to eqn 2.54.

Equation 2.54 depends on the relativistic force K^ν (a vector),[8] the electric charge divided by the rest mass of the particle q/m (a scalar), the four-momentum of the particle p_μ in covariant form (a covariant vector), and a rank-2 tensor that depends on the electric and magnetic fields and is defined in eqn 6.7. Due to the Einstein summation convention, $(q/m)\, p_\mu F^{\mu\nu}$ has only one free index ν in contravariant position. This index matches the index of the relativistic force K^ν. Defining $(q/m)\, p_\mu F^{\mu\nu} \equiv S^\nu$ and $K^\nu \equiv T^\nu$, we get eqn 2.49. Even though we deepen our exploration of this topic in Chap. 6, I encourage the reader to look at eqn 6.7 and compute explicitly S^i for $i = 1, 2, 3$. You will stare at the result in awe because $\mathbf{S} \equiv (S^1, S^2, S^3)$ turns out to be $q\gamma(\mathbf{E} + \mathbf{v} \times \mathbf{B})$. Since the relation between the relativistic and ordinary force is just:

$$\mathbf{K} = (K^1, K^2, K^3) = \frac{dt}{d\tau}\frac{d\mathbf{p}}{dt} = \gamma\mathbf{F} \qquad (2.55)$$

we end up with a formula discovered 10 years before the inception of SR: the Lorentz force

$$\mathbf{F} = q(\mathbf{E} + \mathbf{v} \times \mathbf{B}). \qquad (2.56)$$

Equation 2.45 has the same functional form of eqn 2.42 but do not forget that the numerical values of $p_k'^\mu$ are different from p_k^μ. As a consequence, a judicious choice of the reference frame may ease substantially the kinematical calculations. A very special frame to compute eqn 2.41 is the **center-of-mass frame** (CM). CM is the frame comoving with the center of mass of the system. In the CM not only the total momentum is conserved but the total three-momentum is zero in the initial and final states. In most cases, it is the system of choice to perform the computations. Nonetheless, you will need a Lorentz boost to transform the CM quantities into quantities measured in the **laboratory frame** (LAB). For a system made of just one particle, CM is the rest-frame of the particle and, in that case, CM is equivalent to RF.

2.3.2 One particle kinematics

First, let us study the kinematics of a single particle propagating in space and time. The relation between energy, three-momentum, mass, and Lorentz factors (γ and β) can be derived from the norm of p^μ in CM:

$$p^2 = g_{\mu\nu}p^\mu p^\nu = (p^0)^2 - |\mathbf{p}|^2. \tag{2.57}$$

In the proper frame of the particle, that is the CM of the system, $\mathbf{p} = 0$ and, by definition of four-momentum, $p^2 = m^2$. On the other hand, p^2 is a scalar in the covariant formalism, that is it is a Lorentz invariant quantity.

> If p^μ is the four-momentum of the particle, $p^2 = m^2$ in any reference frame.

In the CM frame, $p^0 = m$. An observer located in the comoving frame of a particle not only measures the shortest time interval (time dilation) but also the minimum particle energy. The **energy**[9] E of a particle in any other system is given by eqn 2.57:

$$E \equiv p^0 = (m^2 + |\mathbf{p}|^2)^{1/2}. \tag{2.58}$$

This formula also provides the Lorentz γ factor of a particle as a function of the particle energy since $E = p^0 = \gamma m$:

$$\gamma = \frac{E}{m}. \tag{2.59}$$

Similarly, $\mathbf{p} = \gamma m \mathbf{v}$ and $|\mathbf{p}| = \gamma m (v_x^2 + v_y^2 + v_z^2)^{1/2}$. Since β is the norm of $\mathbf{v}$ in NU and $\beta = |\mathbf{v}| = |\mathbf{p}|/\gamma m$, we get:

$$\beta = \frac{|\mathbf{p}|}{E}. \tag{2.60}$$

These formulas hold only for the case of a 1-particle system. Here, β is both the Lorentz factor of the system and the speed of the particle. For a system of particles, β represents the speed of the center-of-mass of the system and γ is the Lorentz factor that brings the CM to the laboratory frame.

Relativistic kinematics solve a deep question in modern physics. Both Einstein and his contemporaries wondered why nature has chosen light to set the ultimate speed limit. From the formulas above, we can divide the particles into two broad classes:

> **Massive particles** have $m \neq 0$ and $\beta = |\mathbf{p}|/E = |\mathbf{p}|/\sqrt{|\mathbf{p}|^2 + m^2}$. Since $|\mathbf{p}| < \sqrt{|\mathbf{p}|^2 + m^2}$ in all inertial frames, β is always smaller than 1. As a consequence, massive particles travel slower than light for any observer in any reference frame.
> **Massless particles** have $m = 0$, $\beta = |\mathbf{p}|/E = 1$ and $p^\mu = (|\mathbf{p}|, \mathbf{p})$. Massless particles move at the speed of light for any observer in any reference frame.

[9]Note that the energy and momentum we are referring to are the relativistic generalizations of the classical energy and momenta. They should be called "relativistic energy and momenta" but we drop the word "relativistic" for the sake of fluency. The relationships between classical and relativistic quantities for $|\mathbf{v}| \ll c$ are discussed in Appendix A, Sec. A.5.

That statement is pretty amazing. The light is by no means a "special physical quantity," as Einstein's contemporaries believed. Light is made of photons and photons are just one kind of massless particle. Still, detecting and measuring photons – the mediators of electromagnetic interactions – is very easy and photons are the ideal tools to understand SR. The mediators of strong interactions (the gluons) are massless, too. We will discuss strong interactions in Chap. 7 and you will then see how difficult is to study the propagation of the gluons. Replacing the photons with the gluons to build the theory of SR is fine from the conceptual point of view but it would have been a tremendous challenge for the experimentalists. Similarly, if we had a quantum version of gravity, its mediator would be a spin-2 massless particle (the **graviton**). The graviton could also be a standard candle to define the ultimate speed limit. Even if we cannot observe gravitons, we can study the propagation of classical gravitational fields in space. The observation of gravitational waves in 2016 (Abbott *et al.*, 2016*a*) allows us to replace the speed of light waves with the speed of gravitational waves as a standard reference for SR. On the other hand, light detectors employ technologies established for centuries, while gravitational wave detectors are still in their infancy and are described at the end of this book. All in all, we can bet that light will remain the king of SR for a long time.

2.3.3 A note on the Einstein energy–mass relation*

*The sections labeled with an asterisk present advanced topics and may be omitted in a first reading.

In SR, the energy of a particle has a lower limit that corresponds to the energy in the comoving frame. Here, the particle is at rest and $E = m$. You may wonder whether an interaction can transform a massive particle into a massless one. Kinematics forbids such a transformation for a single particle. If a particle has a mass m, $p^2 = (m, 0, 0, 0)^2 = m^2$ in CM. In this case, the comoving frame of the final state particle is not defined because massless objects travel at speed c (1 in NU) in any frame and no observer can see them at rest. We then need to work in the laboratory frame, where $p_f^\mu = (|\mathbf{p}|, \mathbf{p})$. Since p_f^2 is a scalar and must be equal to p_i^2 for momentum conservation, we compute p_i^μ in the CM and p_f^μ in the LAB frame, and the two values must be the same.

$$p_i^2 \text{ [in CM]} = p_f^2 \text{ [in LAB]} \qquad (2.61)$$

implies

$$(m, 0, 0, 0)^2 = m^2 = (|\mathbf{p}|, \mathbf{p})^2 = |\mathbf{p}|^2 - |\mathbf{p}|^2 = 0, \qquad (2.62)$$

which holds only in the trivial case $m = 0$.

On the other hand, SR does not forbid the transmutation of a massive particle into two massless ones. Remarkably, Einstein noted this possibility when there was absolutely no experimental hint of it. As noted in Rindler, 2006, SR could have been consistently formulated to forbid particle creation and destruction using the definition of kinetic energy. For a massive particle, we define[11] the **kinetic energy** of a particle as the difference between the energy and the energy at rest:

$$E_{kin} = E - m = (\gamma - 1)m. \qquad (2.63)$$

[11] The definitions of energy, kinetic energy, momentum, etc. in SR and the non-relativistic limit (classical physics) is discussed in Sec. A.3.

In the non-relativistic limit, $E_{kin} = m[(1 - \beta^2)^{-1/2} - 1] \simeq m + mv^2/2 - m = mv^2/2$. The kinetic energy produces an increase of the inertia of the particle but does not produce the transmutation of the rest mass m into the kinetic energy of new particles. Einstein, inspired by the work of H. Poincaré and J.J. Thomson, imagined the possibility of transforming part or all of the rest mass in energy. These transformations were observed 28 years later by nuclear physics experiments (see Chap. 4). In the 1950s, particle physicists observed a complete transmutation of rest mass into electromagnetic energy in the $\pi^0 \to \gamma\gamma$ decay of the neutral pion. Since photons are massless particles, this decay represents a full conversion of rest mass into energy.

2.4 Decays and two-particle kinematics

Spontaneous transmutations of particles were observed for the first time in 1896. A full understanding of them, however, required the development of quantum field theories, which were conceived to explain the "creation" and "destruction" of the particles. A transmutation is possible only if is consistent with the kinematics of SR and the conservation laws of the Standard Model (see Chaps 10 and 11). We have already seen that an electron cannot decay into a photon due to the conservation of four-momentum. This transition would also be forbidden by the conservation of electric charge. If the particles in a final state are two or more, the decay is possible provided that the mass of the initial state particle is larger than the sum of the rest masses of the final state particles. For instance, a negative pion (rest mass $m_\pi \simeq 139$ MeV) can decay into a muon (rest mass $m_\mu \simeq 106$ MeV) and a muon-antineutrino: $\pi^- \to \mu^- \bar{\nu}_\mu$. Neutrinos are massive particles but their rest mass – whose precise value is still unknown – is extremely small and can be safely neglected. The decay of the pion into a muon and a muon-antineutrino is thus kinematically allowed. In this case, the conservation of four-momentum provides the values of the muon and the antineutrino momenta in any reference system. In general, for the decay of a particle a with four-momentum p_a^μ into two particles with four-momenta p_1^μ and p_2^μ,

$$p_a^\mu = p_1^\mu + p_2^\mu. \tag{2.64}$$

The CM frame is the rest frame of particle a (Fig. 2.5). In CM, $p_a^\mu = (m_a, 0, 0, 0)$, $p_2^\mu = p_a^\mu - p_1^\mu$ and its norm in the Minkowski space is:

$$p_2^2 = (p_a - p_1)^2 = p_a^2 + p_1^2 - 2\, p_a \cdot p_1. \tag{2.65}$$

We have already shown that the norm of a vector for a single particle is its rest mass. So,

$$m_2^2 = m_a^2 + m_1^2 - 2p_a^\mu\, p_{1\mu}. \tag{2.66}$$

In CM, we call the energy and three-momentum of the first particle E_1 and $\mathbf{p}_1$, respectively. Hence,

$$p_a^\mu\, p_{1\mu} = g_{\mu\nu}p_a^\mu p_1^\nu = m_a E_1. \tag{2.67}$$

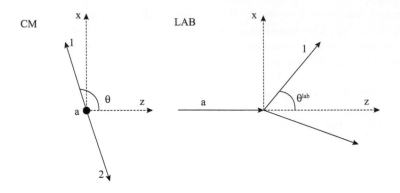

Fig. 2.5 A particle a moves in the rightmost direction in the LAB frame along the z axis and, then, decays into two particles "1" and "2." The decay is seen as a back-to-back decay in the CM of a but the two final-state particles are no more aligned in the LAB frame due to the Lorentz boost.

From eqn 2.66, we get

$$E_1 = \frac{m_a^2 + m_1^2 - m_2^2}{2m_a} \quad ; \quad E_2 = \frac{m_a^2 + m_2^2 - m_1^2}{2m_a} \tag{2.68}$$

where E_2 is the energy of the second particle in CM. The momenta of the final state particles in CM are thus $|\mathbf{p_1}| = (E_1^2 - m_1^2)^{1/2}$ and $|\mathbf{p_2}| = (E_2^2 - m_2^2)^{1/2}$. To compute the momentum of the particles in any reference frame, we need the direction of motion of the particle a. If a moves along the z axis of the LAB system with speed β, the γ factor of the CM is the γ of the proper frame of the particle a: $\gamma = E_a^{lab}/m_a$, being E_a^{lab} the particle energy in the LAB frame. β is thus $|\mathbf{p}_a^{lab}|/E_a^{lab}$. The momenta in the LAB frame are the momenta after the Lorentz transformation from CM to LAB.

$$[\text{CM frame}] \; |\mathbf{p_1}| \sin\theta = p_{1x} \quad ; \quad |\mathbf{p_1}| \cos\theta = p_{1z}$$
$$[\text{LAB frame}] \; p_{1x}^{lab} = p_{1x} \quad ; \quad p_{1z}^{lab} = \gamma(p_{1z} + \beta E_1). \tag{2.69}$$

θ is defined as the angle between the direction of flight of the initial state particle a as seen in LAB (the z-axis of Fig. 2.5) and the direction of the final state particle 1 in CM. If the particle 1 is emitted in CM with an angle θ, the experimentalist measures a contracted angle:

$$\tan\theta^{lab} = \frac{p_{1x}^{lab}}{p_{1z}^{lab}} = \frac{p_{1x}}{\gamma(p_{1z} + \beta E_1)}. \tag{2.70}$$

If γ is very large, the decay products are nearly collinear in LAB even if they were produced back to back in CM. Sometimes this complication is good news because it eases the task of associating the decay particles to the parent particle, especially if the detector is crowded with many tracks.

If particles 1 and 2 are unknown, we may ask what is the minimum value of the masses of the final state particles needed to produce a decay. This value is called the **decay threshold** and can be computed assuming all final state particles to be produced at rest. In CM, $p_1 = (m_1, 0, 0, 0)$ and $p_2 = (m_2, 0, 0, 0)$. Four-momentum conservation gives $m_a \geq m_1 + m_2$, as suggested by common sense. We cannot create particles in CM, whose rest mass is heavier than m_a.

SR does not fully constrain the two-body decay of particle a. The conservation of four-momentum provides four equations but the unknown variables are six: the three-momenta of particle 1 and the three-momenta of particle 2. E_1 and E_2 are not free parameters and can be fixed if the rest mass of particle 1 and 2 are known because $E_1^2 = m_1^2 + |\mathbf{p_1}|^2$. The two variables that are not fixed by kinematics are the decay angles in the CM, which can be parameterized with the two Euler angles θ^{CM} and ϕ^{CM} in space.

2.5   Three-body decays

For a three-body decay $a \to 1+2+3$, the number of constraints is again four but the variables are $12 - 3 = 9$. In this decay, SR cannot fix the final state energies but can provide the maximum and minimum energy of the emitted particles. The minimum energy for particle 1 corresponds to the particle being produced at rest in CM, as shown in Fig. 2.6. In this case $E_1 = m_1$. The maximum energy for particle 1 is achieved when the particle is produced in the direction opposite to the sum of the three-momenta of the second and third particles. This configuration is shown in Fig. 2.6 and corresponds to a two-body decay of the particle 1 and the center-of-mass of particle 2 and 3: since the second and third particles move together and are perfectly collinear, they can be treated as a single object. Applying eqn 2.68:

$$m_1 \leq E_1^{CM} \leq \frac{m_a^2 + m_1^2 - (m_2 + m_3)^2}{2m_a}. \tag{2.71}$$

A well-known three-body decay is the decay of the muon $\mu^- \to e^- \bar{\nu}_e \nu_\mu$ in three stable particles: an electron, an electron antineutrino, and a muon neutrino. Since the rest mass of the muon is much larger than the mass of the electron ($m_e \simeq 511/c^2$ keV) and of the neutrinos,

$$E_e^{max} = \frac{m_\mu^2 + m_e^2 - (m_{\nu_\mu} + m_{\bar{\nu}_e})^2}{2m_\mu} \simeq \frac{m_\mu}{2} \simeq 53 \text{ MeV}. \tag{2.72}$$

The observer then measures a continuous electron energy spectrum that is sharply cut off at 53 MeV. This spectrum is shown in Fig. 2.7 and was measured well before the discovery of neutrinos.

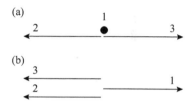

Fig. 2.6 Three-body scattering configurations where the particle 1 has the minimum (a) or maximum (b) energy.

2.6   Decay amplitudes in quantum mechanics

The decay of unstable states is very common in atomic physics. A hydrogen atom in the $n = 2$, $l = 1$ state reaches the ground state $n = 1$, $l = 0$ in about 1.6 nanoseconds (ns) because the electric dipole transitions follow the $\Delta l = \pm 1$, $\Delta s = 0$, $\Delta j = 0, \pm 1$ selection rule (Foot, 2005). Here l and s are the angular and spin quantum numbers and j

is the quantum number of the total angular momentum ($j = 0 \to 0$ is forbidden). If the atom is in the $n = 2$, $l = 0$ state, it takes much longer to reach the ground state because the corresponding direct decay is forbidden and the decay proceeds through the emission of two photons. The lifetime of this $2s$ state is thus very long because the two-photon decay is a second-order transition and is strongly suppressed than the one-photon $2l \to 1s$ decay.

Particles decay following similar rules with one major exception. In atomic physics, the elementary particles reshuffle their positions and release energy in the form of electromagnetic waves (photons). In particle physics, a decay destroys the initial state particle and creates new particles. Since this is impossible in non-relativistic quantum mechanics, people originally believed that the decaying particles were bound states of the final state particles. In Chap. 10, we will see that no known interaction can bound a neutrino in the volume occupied by the wavefunction of a pion. Hence, in a $\pi^- \to \mu^- \bar{\nu}_\mu$ decay the neutrino is created as soon as the pion is annihilated. QFTs are equipped with creation and annihilation operators to cope with particle decay, scattering, and production, and can compute the transition amplitudes. Once the amplitude is known, the process follows the same rules of atomic physics.

2.6.1   The Fermi golden rule

The most powerful tool to evaluate the lifetime of excited states in atomic physics is the Fermi golden rule that reads in SI:

$$\Gamma_{i \to f} = \frac{2\pi}{\hbar} \int_{im} \prod_k \frac{V d^3 \mathbf{p}_k}{(2\pi\hbar)^3} \, |\mathcal{T}|^2 \delta \left(E_i - E_f - \sum_{j=1}^{N_\gamma} E_j \right)$$

$$\times \, \delta^3(\mathbf{p}_i - \mathbf{p}_f - \sum_{j=1}^{N_\gamma} \mathbf{p}_j) \tag{2.73}$$

and can be shortened as:

$$\Gamma_{i \to f} = \frac{2\pi}{\hbar} \, |\mathcal{T}|^2 \, \rho(E). \tag{2.74}$$

$\rho(E)$ is called the **final state density** and "*im*" means **independent momenta** because the momentum of one of the final state particles (for instance, the decayed atom) is uniquely fixed by the three-momentum conservation law: $\mathbf{p}_i - \mathbf{p}_f - \sum_{j=1}^{N_\gamma} \mathbf{p}_j = 0$. In eqn 2.73, the index k runs over all final-state particles: the N_γ photons and the final-state atom, while three-momentum conservation is enforced by the delta function.

This non-relativistic formula is derived in Gasiorowicz, 2003 for a generic volume V. V represents the normalization volume of the wavefunctions and cancels out in the evaluation of eqn. 2.73 thanks to the volumes contained in $\mathcal{T}$, which depends on the wavefunctions too. The derivation holds for any value of V including $V \to +\infty$,

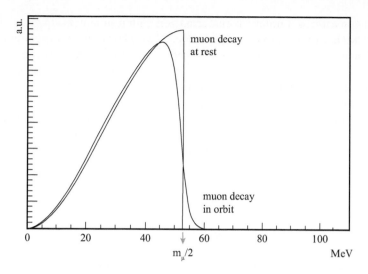

Fig. 2.7 Electron energy spectrum in MeV from a large number of muon decays at rest. The sharp cut-off at 53 MeV ($m_\mu/2$) is smeared because the μ^- can be captured by an atom replacing an electron and decay in orbit.

which is a realistic assumption when dealing with final state particles.

Due to its arbitrariness, we will assume $V = 1$ without loss of generality and express this formula in NU setting $\hbar = 1$. To adapt eqn 2.73 to a a $\rightarrow 1 + 2 + \ldots N_f$ decay in particle physics, we note that the integration over the independent momenta should be replaced by an integration over the independent momenta of the particles in the final state. The momentum of one particle (for example, $k = N_f$) is fixed by the three-momentum conservation: the same law that fixes $\mathbf{p}_f$ in the atom. We can cope with it replacing (in NU):

$$\prod_{k=1}^{N_\gamma} \frac{V d^3 \mathbf{p}_k}{(2\pi\hbar)^3} \delta^3 \left(\mathbf{p}_i - \mathbf{p}_f - \sum_{j=1}^{N_\gamma} \mathbf{p}_j \right) d^3 \mathbf{p}_N$$

$$\longrightarrow \prod_{k=1}^{N_f-1} \frac{d^3 \mathbf{p}_k}{(2\pi)^3} \delta^3 \left(\mathbf{p}_i - \mathbf{p}_f - \sum_{j=1}^{N_f} \mathbf{p}_j \right) d^3 \mathbf{p}_{N_f} \qquad (2.75)$$

where $N_f = N_\gamma + 1$. Equation 2.75 can be written in a more elegant form treating all particles on the same ground and writing the **non-relativistic phase-space** as (Thomson, 2013):

$$(2\pi)^3 \prod_{k=1}^{N_f} \frac{d^3 \mathbf{p}_k}{(2\pi)^3} \delta^3 \left(\mathbf{p}_i - \mathbf{p}_f - \sum_{j=1}^{N_f} \mathbf{p}_j \right). \qquad (2.76)$$

Equation 2.73 describes the transition probability per unit-time (the $\Gamma_{i\rightarrow f}$ transition rate) of an atom in the initial state i to a final state f and the emission of N_γ photons. The non-relativistic matrix element $\mathcal{T}$ can be computed in time-dependent perturbation theory using standard QM techniques (Gasiorowicz, Sakurai; Sakurai, Napolitano, 2017).

Unfortunately, the normalization on the volume of the wavefunction is not compatible with SR since the volume is not Lorentz invariant and changes after a Lorentz transformation. This is a consequence of the contraction of lengths in SR (see Sec. A.2). Different observers measure different volumes, which result in different and inconsistent normalizations of the wavefunction. For a decay of a particle a in N_f particles, the observer moving in the direction of a with speed β measures volumes that are contracted by $1/\gamma$ with respect to CM. Since $\gamma = E/m_a$ and m_a is only a constant, the standard wavefunction normalization:

$$\int_V d^3\mathbf{x} \; \psi^*\psi = 1 \tag{2.77}$$

must be replaced by

$$\int_V d^3\mathbf{x} \; \psi'^*\psi' = 2E. \tag{2.78}$$

Hence, a well-normalized wavefunction ψ' is proportional to ψ multiplied by $\sqrt{2E}$. The factor 2 is a matter of convention but is useful to simplify many formulas. E is the energy of the initial state particle measured in the laboratory frame. $\mathcal{T}$ is $\langle f| \hat{H}' |i\rangle = \langle \psi_1 \cdots \psi_n| \hat{H}' |\psi_a\rangle$ at leading order in perturbation theory.[12] In this formula, i and f are the initial and final states, respectively, and $\hat{H}'$ is the interaction part of the Hamiltonian. As a consequence, the Lorentz-invariant matrix element describing an n-body decay is:

$$\mathcal{M} = (2E_a)^{1/2} \prod_{i=1}^{N_f} (2E_i)^{1/2} \; \mathcal{T}. \tag{2.79}$$

The relativistic version of the Fermi golden rule can be obtained replacing $\mathcal{T}$ from eqn 2.79 in eqn 2.73. The **relativistic golden rule** in NU ($\hbar = 1$) is then:

$$\Gamma_{i\to f} = \frac{(2\pi)^4}{2E_a} \int \prod_{i=1}^{N_f} \frac{d^3\mathbf{p}_i}{(2\pi)^3 2E_i} \; |\mathcal{M}|^2 \delta\left(E_a - \sum_{j=1}^{N_f} E_j \right) \delta^3\left(\mathbf{p}_a - \sum_{j=1}^{N_f} \mathbf{p}_j \right) \tag{2.80}$$

or, equivalently,

$$\Gamma_{i\to f} = \frac{(2\pi)^4}{2E_a} \int \prod_{i=1}^{N_f} \frac{d^3\mathbf{p}_i}{(2\pi)^3 2E_i} \; |\mathcal{M}|^2 \delta^4\left(p_a^\mu - \sum_{j=1}^{N_f} p_j^\mu \right) \tag{2.81}$$

where p_i^μ are the four-vectors of all final particles: $p_i^\mu = (E_i, \mathbf{p}_i)$ and $i = 1, 2, \ldots N_f$. Even if eqn 2.79 is invariant for Lorentz transformations, we do not expect $\Gamma_{i\to f}$ to be invariant because a transition probability per unit time depends on the frame-specific measurement of time. This can be checked explicitly in the covariant formalism. The terms:

$$\frac{d^3\mathbf{p}_i}{(2\pi)^3 2E_i} \tag{2.82}$$

are Lorentz invariant although they are not written as a function of scalars, vectors, or tensors. They can be recast in a covariant form noting that:

$$\frac{d^3\mathbf{p}}{(2\pi)^3 2E} = \delta^4(p^2 - m^2)\,\theta(p^0)\,d^4p \qquad (2.83)$$

where $\theta(p^0)$ is the **Heaviside step function**:[13] $\theta(p^0) = 1$ for $p^0 \geq 0$ and 0 elsewhere. This expression is manifestly Lorentz invariant because a change of variables in d^4p corresponds to d^4p multiplied by the Jacobian of the transformation. For the Lorentz transformations:

$$d^4p' = d^4p\,|\det\Lambda^\mu_{\ \nu}| \qquad (2.85)$$

and the determinant of any Lorentz transformation is $+1$ or -1. The delta is Lorentz-invariant too because $p^2 - m^2$ is a scalar and the delta is non-zero at the same point in every reference frame. $\theta(p^0)$ is Lorentz-invariant because $\Lambda^\mu_{\ \nu}$ cannot change the sign of the energy of a particle. So the product $\delta^4(p^2 - m^2)\,\theta(p^0)\,d^4p$ is Lorentz-invariant, that is it is a Lorentz scalar. Similarly,

$$\delta^4\left(p^\mu_a - \sum_{i=1}^{N_f} p^\mu_i\right) \qquad (2.86)$$

is Lorentz-invariant, as shown in Exercise 2.14. An intuitive explanation is that the delta corresponds to a conservation law: the total four-momentum must be conserved in the final state in any reference frame, whatever happens during the interaction. This statement is universal and, thus, the delta must be a Lorentz scalar. Note that the argument of the delta is not Lorentz-invariant because is the sum of covariant vectors. It means that even if the numerical values of energy and momentum may change among reference frames, the sum of their values is conserved before and after the decay.

It is worth checking explicitly the invariance of eqn 2.82 using the Lorentz transformations (Thomson, 2013) and the invariance of eqn 2.86 by the properties of the delta function (see eqn 2.87 below). This is left as an exercise to the reader (Exercises 2.13 and 2.14).

We can tweak the relativistic golden rule in a very elegant form – where all terms except E_a are explicitly Lorentz invariant – using a property of the delta function:[14]

$$\delta(f(x)) = \frac{\delta(x - x_0)}{\left|\frac{df}{dx}\right|_{x_0}} \qquad (2.87)$$

where x_0 is such that $f(x_0) = 0$. The term $(2E_i)^{-1}$ can thus be written as:

$$\frac{1}{2E_i} = \int \delta(E_i^2 - \mathbf{p}_i^2 - m_i^2)\,dE_i \qquad (2.88)$$

and

[13] The Heaviside step function, or the unit step function, $\theta(x)$ is 0 for negative x and 1 for positive x. It is a powerful tool in QM also because it provides another definition of the Dirac delta function:

$$\delta(x) = \frac{d}{dx}\theta(x). \qquad (2.84)$$

[14] For the convenience of the reader, these properties are summarized in Sec. B.2.2.

$$\int_{\mathbb{R}^3} \prod_{i=1}^{N_f} \frac{d^3\mathbf{p}_i}{(2\pi)^3 2E_i} = \int_{\mathbb{R}^4} \prod_{i=1}^{N_f} \frac{d^3\mathbf{p}_i}{(2\pi)^3} \delta(E_i^2 - \mathbf{p}_i^2 - m_i^2) dE_i$$

$$= \int_{\mathbb{R}^4} \prod_{i=1}^{N_f} \frac{d^4 p_i}{(2\pi)^4} (2\pi) \delta(p_i^2 - m_i^2). \qquad (2.89)$$

The **relativistic golden rule** for the decays is then:

$$\Gamma_{i \to f} = \frac{1}{2E_a} \int \prod_{i=1}^{N_f} \frac{d^4 p_i}{(2\pi)^4} |\mathcal{M}|^2 (2\pi)^4 \delta^4 \left(p_a - \sum_{j=1}^{N_f} p_j \right)$$
$$\times (2\pi) \delta(p_i^2 - m_i^2). \qquad (2.90)$$

The product:

$$\prod_{i=1}^{N_f} \frac{d^3\mathbf{p}_i}{(2\pi)^3 2E_i} = \prod_{i=1}^{N_f} \frac{d^4 p_i}{(2\pi)^4} (2\pi) \delta(p_i^2 - m_i^2) \qquad (2.91)$$

is called the **Lorentz-invariant phase-space** and is the relativistic version of $\rho(E)$ in eqn 2.73.

Equation 2.90 is also called the relativistic Fermi golden rule and was derived by P. Dirac in 1927 both in the relativistic and non-relativistic version. It is named after E. Fermi because he systematically promoted the golden rule in QM and recognized the importance of this result in the derivation of the atomic transition rates. I restrain myself from simplifying the (2π) factors to highlight a bookkeeping rule: in all "golden rules" (eqns 2.90 and 2.135 below), we need to multiply by 2π every Dirac's delta and divide by 2π every dp term.

2.6.2   Phase-space of the decays*

The Lorentz invariant phase-space can be integrated using the delta functions to reduce the number of free variables. A classical example is the following.

Example 2.3

Evaluate the two-body decay phase-space in the RF of the decaying particle. The two-body phase-space is:

$$d\mathrm{PS}_2 \equiv \frac{d^4 p_1}{(2\pi)^4} (2\pi) \delta(p_1^2 - m_1^2) \theta(E_1) \frac{d^4 p_2}{(2\pi)^4} (2\pi) \delta(p_2^2 - m_2^2) \theta(E_2)$$
$$\times (2\pi)^4 \delta^4(p_a^\mu - p_1^\mu - p_2^\mu).$$

This formula embeds both the conditions of positive energy by the Heaviside step functions $\theta(E_1)$ and $\theta(E_2)$, and the four-momentum conservation. We can get rid of the second integration over p_2 using the δ^4 term:

$$dPS_2 = \frac{d^4 p_1}{(2\pi)^4} \, (2\pi) \, \delta(p_1^2 - m_1^2) \, \theta(E_1) \, (2\pi) \, \delta[(p_a - p_1)^2 - m_2^2]$$
$$\times \theta(m_a - E_1)$$

and we go on removing the free variables of particle 1 integrating $d^4 p_1$ in spherical coordinates:

$$dPS_2 = \frac{|\mathbf{p}_1|^2 d|\mathbf{p}_1| dE_1 d^2\Omega_1}{(2\pi)^4} \, (2\pi) \, \delta(E_1^2 - |\mathbf{p}_1|^2 - m_1^2) \, \theta(E_1)$$
$$\times \, (2\pi) \, \delta[(m_a - E_1)^2 - |\mathbf{p}_1|^2 - m_2^2] \, \theta(m_a - E_1)$$
$$= \frac{\sqrt{(m_a^2 - m_1^2 - m_2^2)^2 - 4m_1^2 m_2^2}}{8m_a^2} \frac{d^2\Omega_1}{(2\pi)^2}.$$

As a consequence,

the phase-space of a two-body decay in RF only depends on the scattering angles and the particle masses:

$$dPS_2 = \frac{\sqrt{(m_a^2 - m_1^2 - m_2^2)^2 - 4m_1^2 m_2^2}}{8m_a^2} \frac{d^2\Omega_1}{(2\pi)^2} \qquad (2.92)$$

In eqn 2.92, E_1 is fixed to $(m_a^2 - m_1^2 - m_2^2)/2m_a$ and $|\mathbf{p}_1| = \sqrt{E_1^2 - m_1^2}$, as expected by the results of Sec. 2.4.

It is instructive to estimate the phase-space of massless particles, which is finite too.

Example 2.4

A finite phase-space follows directly from eqn 2.92 setting $m_1 = m_2 = 0$. For instance, the phase-space of the neutral pion decay $\pi^0 \to \gamma\gamma$ is:

$$\int dPS_2 = \frac{1}{8\pi} \qquad (2.93)$$

and the phase-space does not depend on m_a, either.

The general formulas for an n-body decay are rather cumbersome but can be computed recursively from the two-body results (Byckling and Kajantie, 1973).

2.6.3   Lifetimes and branching ratios

In eqn 2.90, the only term that changes after a Lorentz transformation is E_a and the decay rate transforms as:

$$\Gamma'_{i\to f} = \Gamma_{i\to f}/\gamma. \qquad (2.94)$$

An observer in the laboratory frame sees, for example, a muon decaying slower than in CM. Since a muon always decays in $e^-\nu_\mu\bar{\nu}_e$, its lifetime is simply Γ^{-1}. This is a consequence of quantum mechanics: for a system of N particles in the same unstable state, the number of particles that reach the ground state in dt is $dN = -N\Gamma dt$. The minus sign indicates that the number of unstable particles decreases with time and the number of particles in the ground state increases. The unstable particles that survive after a time t is then:

$$N(t) = N(0)e^{-\Gamma t}. \qquad (2.95)$$

The lifetime τ of a particle is the time needed to reach $N(\tau) = N(0)e^{-1}$ of the initial number so

$$\tau = \frac{1}{\Gamma}. \qquad (2.96)$$

This definition, however, is ill-posed. Since Γ is not Lorentz invariant, each observer will measure a different lifetime. The lifetime in LAB is always larger than the lifetime in the CM:

$$\tau_{lab} = \gamma\,\tau_{CM}. \qquad (2.97)$$

This is a consequence of time dilation in SR and the principles of quantum mechanics. The lifetime dilation of muons has been observed for the first time in 1940 by B. Rossi and D.B. Hall studying the rate of decay of the muons in the atmosphere. Since $\tau_{CM} \simeq 2.2~\mu$s and no particles travel faster than light, a flux of muons produced in the atmosphere would travel no more than a few kilometers if Einstein were wrong on time dilation:

$$N(t) = N(0)\exp(-t/\tau) = N(0)\exp(-L/c\tau)$$
$$\simeq N(0)\exp(-L/660~\text{m}) \qquad (2.98)$$

where L is the distance traveled by the muon in LAB. Time dilation reduces the muon losses. A muon moving with speed β decays according to this law:

$$N(t) = N(0)\exp(-t/\tau_{lab}) = N(0)\exp\left(-\frac{L}{\beta c}\frac{1}{\gamma\tau}\right)$$
$$= N(0)\exp\left(-\frac{L}{\gamma\beta c\tau}\right) \qquad (2.99)$$

where τ_{lab} is the lifetime in the laboratory frame and τ is the lifetime in the proper frame of the particle. Muons are produced in the atmosphere at an altitude of about 15 km with a mean energy of $\sim$ 6 GeV. Their mean Lorentz factor $\gamma = E/m$ is thus $\simeq 57$. At 6 GeV, $\beta = 0.9998 \simeq 1$. The lifetime in LAB is, therefore, 57 times greater than the lifetime measured in CM. Even if muons experience energy losses crossing the earth's atmosphere (see Chap. 3), they can easily reach sea level and they are routinely observed in detectors located on the earth's surface.

Most particles decay in multiple modes and the total decay rate is the sum of the partial decay modes. For a particle that decays in N modes:

$$\Gamma = \sum_{i=1}^{N} \Gamma_i. \tag{2.100}$$

The **lifetime** of a particle is defined as the inverse of the total decay width *measured in the rest frame* of the particle:

$$\tau = \frac{1}{\Gamma} = \frac{1}{\sum_{i=1}^{N} \Gamma_i}. \tag{2.101}$$

The **branching ratio** (or branching fraction) of the mode i is defined as:

$$\mathrm{BR}(i) = \frac{\Gamma_i}{\Gamma} \tag{2.102}$$

and ranges from 0 to 1.

For the muon, the BR of the $\mu^- \rightarrow e^- \nu_\mu \bar{\nu}_e$ mode is practically 100%. The charged pion has several decay modes: the $\pi^- \rightarrow \mu^- \bar{\nu}_e$ mode is dominant (BR $\simeq$ 99.99988). The others are strongly suppressed. The $\pi^- \rightarrow \mu^- \bar{\nu}_\mu \gamma$ mode is suppressed because the three-body decay with the emission of a photon is a higher order process in perturbation theory, as for the above-mentioned $2s \rightarrow 1s$ transitions of hydrogen. The corresponding BR is 2.0×10^{-4}. The $\pi^- \rightarrow e^- \bar{\nu}_e$ is suppressed by a conservation law of the Standard Model discussed in Sec. 10.7.

2.6.4 Unstable states

Even if non-relativistic quantum mechanics is unable to describe the creation and annihilation of particles (Bromberg, 1976), it is equipped with a formalism to describe unstable states. The probability of finding an unstable particle in space decreases with time and the wavefunction fades away as:

$$|\psi(\mathbf{x}, t)|^2 = |\psi(\mathbf{x}, 0)|^2 e^{-t/\tau}. \tag{2.103}$$

The damping term is an exponential function compatible with the decay law of eqn 2.95. The stationary states of the Schrödinger equation are no more:

$$\psi(\mathbf{x}, t) = \psi(\mathbf{x}) e^{-iEt} \tag{2.104}$$

but are replaced by states vanishing for $t \rightarrow +\infty$:

$$\psi(\mathbf{x}, t) = \psi(\mathbf{x}) e^{-iEt} e^{-t/2\tau}. \tag{2.105}$$

A vanishing state is not a stationary state and its disappearance at large times brings subtle QM effects. Due to the Heisenberg uncertainty principle of Sec. B.4, the energy of the stationary state can be measured with arbitrary precision provided that $\Delta E \, \Delta t \geq \hbar/2$ or (in NU) $\Delta E \geq 1/(2\Delta t)$. If the particle does not endure the passage of time, the observation time cannot be longer than a few lifetimes. Δt is thus of the order of τ in the rest frame of the particle. As a consequence, the intrinsic uncertainty of the particle energy cannot be smaller than:

$$\Delta E \geq (2\tau)^{-1}. \tag{2.106}$$

Greater uncertainty can be observed if the experimental precision of the energy detector is larger than the right-hand side of eqn 2.106. The scattering theory in quantum mechanics gives the probability to measure the unstable state of energy $\tilde{E}$ with energy between E and $E + dE$. The probability $P(E)$ follows a **Breit–Wigner distribution**:[15]

$$P(E)dE = \frac{dE}{(E - \tilde{E})^2 + \Gamma^2/4}. \tag{2.107}$$

Equation 2.107 can be derived from eqn 2.105 making a Fourier transform of $\psi(\mathbf{x}, t)$. This is due to the fact that time and energy are conjugate variables in QM and the Fourier transform is the QM tool to swap the domain of the wavefunction from the first to the second conjugate variable (Cohen-Tannoudji *et al.*, 1991). The Fourier transform of $\psi(\mathbf{x}, t)$ is:

$$\int_0^\infty \psi(\mathbf{x}, t)e^{iEt}dt = \int_0^\infty \psi(\mathbf{x})e^{-i\tilde{E}t}e^{-\Gamma t/2}e^{iEt}dt. \tag{2.108}$$

The integral can be computed analytically and gives:

$$\int_0^\infty e^{i(E - \tilde{E} + i\frac{\Gamma}{2})t}dt = \frac{i}{(E - \tilde{E}) + i\Gamma/2}. \tag{2.109}$$

The probability density at E is, therefore:

$$P(E) = \left| \frac{i}{(E - \tilde{E}) + i\Gamma/2} \right|^2 = \frac{1}{(E - \tilde{E})^2 + \Gamma^2/4}. \tag{2.110}$$

This is also the reason why the total decay rate Γ of a particle is called the total **decay width**. It is important to grasp the difference between measurement uncertainties and the intrinsic width of an unstable state, that is, the total decay width. For a very heavy particle with plenty of decay modes, the lifetime is extremely short. Observing the decay in flight of a particle with $\tau = 10^{-25}$s is hopeless even if the gamma factor of the particle is large. For $\gamma \sim \mathcal{O}(1)$, the distance traveled before a decay would be 3×10^{-17} m: 100 times less than the size of an atomic nucleus. On the other hand, if the particle is produced by a head-on scattering of two particles as in Fig. 2.8, $P(E)$ can be measured very precisely recording millions of these scatterings. Fig. 2.9 shows the distribution of $P(E)$ for the Z^0 particle, whose mass is 91.2 GeV. The particle is produced by a head-on scattering of an electron and a positron and decays

[15] Atomic physicists call this distribution a "Lorentzian distribution." It represents the intrinsic energy spread of an atomic transition with a finite lifetime. As expected by eqn 2.106, the longer the lifetime, the narrower the spectral line of the transition.

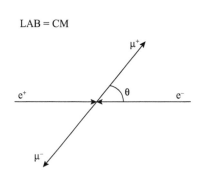

LAB = CM

Fig. 2.8 A head-on scattering $e^+e^- \to \mu^+\mu^-$. This is a two-body scattering where the CM is also the LAB frame and the final state muons are scattered by an angle θ.

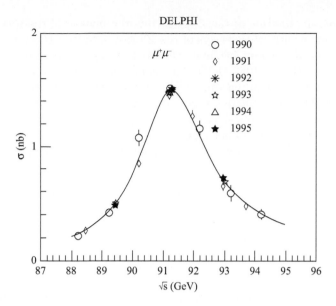

Fig. 2.9 Measurement of the cross-section of $e^+e^- \rightarrow Z^0 \rightarrow \mu^+\mu^-$ as a function of the CM energy. This measurement was performed by the DELPHI experiment over several years (1990-1995) setting the CM in the proximity of M_Z and scanning different positions of the Breit–Wigner curve. Reprinted by permission from Springer Nature: Springer EPJC, Abreu et al. 2000, Copyright 2000.

(for instance) in a muon and an antimuon. Unlike classical mechanics, we do not observe a sharp production peak at 91.2 GeV but a broad distribution. This distribution follows eqn 2.107 and the width of the distribution is the total decay width of the Z^0: about 2.5 GeV. In SI, a lifetime of $(2.5 \text{ GeV})^{-1}$ corresponds to:

$$\frac{6.58 \times 10^{-16} \text{ eV s}}{2.5 \times 10^9 \text{ eV}} = 2.63 \times 10^{-25} \text{ s.} \qquad (2.111)$$

There is no intrinsic limitation on the precision you can achieve on $\tilde{E}$ and Γ if your measurement relies on very large statistics and the experimental systematics are well under control. For the Z^0, the rest mass M_Z was measured in the 1990s at CERN (Fig. 2.9) fitting the Breit–Wigner distribution with millions of $e^+e^- \rightarrow Z^0 \rightarrow X$ events. Here, X is a generic final state observable by the detector. Such large statistics, combined with an exquisite precision in the determination of the CM energy, provided a precise fit of the Breit–Wigner and, therefore, the lifetime of the Z^0. The precisions achieved were $\tilde{E} = M_Z = 91.1876 \pm 0.0021$ GeV and $\Gamma = \tau^{-1} = 2.4952 \pm 0.0023$ GeV. In general,

> the lifetimes of long-living particles are measured by observing the rate of their decays, while the lifetimes of short-living particles are measured fitting the width of the corresponding Breit–Wigner distribution.

A final consideration is in order. The experimenter is free to choose any specific decay mode to measure the parameters of the Breit–Wigner. For instance, if we observe only $Z^0 \rightarrow \mu^+\mu^-$ final states instead of all $Z^0 \rightarrow X$ decay modes, we get exactly the same value of Γ and M_Z as measured from $Z^0 \rightarrow X$. This is because the fit of the Breit–Wigner measures the lifetime of the Z^0, which is the inverse of the total decay width. Observing other decays of the Z^0 (e.g. $e^+e^- \to Z^0 \to \tau^+\tau^-$), we find curves of different heights – different cross-sections – but with the same width ($\Gamma = 2.4952$ GeV). This is obvious because the total width is the linear sum of all partial widths, not just the one we observe in the detector. The only difference from detecting all $Z^0 \to X$ modes is that we observe fewer events since the vast majority of the Z^0 decays remain undetected.[16] The ratio between the events recorded by a detector that selects only $\mu^+\mu^-$ final states and an ideal detector observing all final states X is:

$$\frac{\Gamma_{Z^0 \to \mu^+\mu^-}}{\Gamma} = \mathrm{BR}(Z^0 \to \mu^+\mu^-) = 3.4 \ \%. \qquad (2.112)$$

2.7   Scattering

In classical physics, scattering among particles may occur either due to long-range forces (e.g. the Rutherford scattering) or by impulsive forces in the collision of rigid bodies. Rigid bodies do not exist in SR (see Appendix A) and, hence, **impulsive forces** are never employed in particle physics. They can be replaced by finite range interactions, whose effects are negligible if the minimum distance between the particle and the scattering center is larger than the range of the force. A renowned example is the scattering of a proton on a neutron due to strong interactions. Unlike QM, a scattering in classical physics and SR is a deterministic process. If the initial conditions are perfectly known, the trajectory can be predicted at any time with infinite precision. In most practical cases, we are interested in the trajectory of the final state particles once the interactions among them can be neglected and the particles fly apart from the scattering center. In classical physics, this approximation is driven by practical reasons, that is the challenge of tracking a microscopic object at any time during the scattering process. In QM, this is an intrinsic limitation introduced by the Heisenberg uncertainty principle. As a consequence, scattering theory (Weinberg, 2012) aims to solve the equations of motion for $t \to +\infty$ and determine the scattering angles of the particles in space. The most prominent observable of scattering theory is the cross-section.

> The **differential cross-section** is the ratio between the number of particles that are scattered in a solid angle between Ω and $\Omega + d\Omega$ per unit time divided by the flux:
>
> $$\frac{d\sigma}{d\Omega}(E, \Omega) = \frac{1}{F}\frac{dN_s}{d\Omega}. \qquad (2.113)$$
>
> The **flux** is the rate of incoming particles per unit surface.

A differential cross-section is, therefore, a measurement of the likelihood that a particle is scattered at an angle Ω. The larger the cross-section

---

[16] This is also the reason why the definition of a "partial lifetime" $\tau_i \equiv \Gamma_i^{-1}$ is confusing and highly deprecated, although sometimes used in the past. It makes no sense to define τ_i as a partial lifetime if $\sum_i \tau_i \neq \tau$, unless you really want to trick your students!

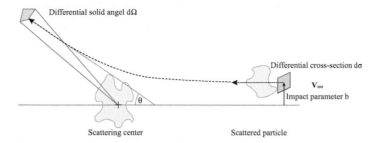

Fig. 2.10 Classical scattering of a body due to the central potential $U(r)$ generated by a scattering center. The body is scattered by an angle θ and an azimuthal angle ϕ. The corresponding solid angle is Ω. b is the impact parameter and v_∞ is the velocity far from the scattering center. Redrawn from (Wikimedia, 2014a) under CC-BY-SA 3.0 licence.

the larger the number of particles that are observed by a detector located at an angle Ω, far from the scattering center. The flux is expressed as particles per unit time (rate) per unit surface and the SI flux unit is $s^{-1}m^{-2}$ (eV^3 in NU). For now, we consider the flux on a surface perpendicular to the velocity of the incoming particles and the scattering as due to another particle or a microscopic continuous medium at rest in LAB. The **total cross-section** is the differential cross-section integrated on the solid angle:

$$\sigma = \int d\Omega \, \frac{d\sigma}{d\Omega}(E, \Omega). \tag{2.114}$$

Figure 2.10 shows the most general scattering case in classical physics. Here, θ is the scattering angle in the plane containing the particle trajectory and Ω the solid angle in space. These angles are uniquely determined once the **impact parameter** b is known. b is defined as the distance between the path of the projectile far from the scattering center and the line crossing the scattering center and parallel to the initial path of the projectile, as shown in Fig. 2.10. For a general time-independent central potential, θ is given by Goldstein, 1980:

$$\theta = \pi - 2b \int_{r_{min}}^{\infty} \frac{dr}{r^2 \sqrt{1 - (b/r)^2 - 2U/mv_\infty^2}} \tag{2.115}$$

where v_∞ is the velocity of the projectile at $t \to \pm\infty$ and r_{min} is its closest distance from the center.

The simplest example of scattering in classical mechanics is the collision of a point-like particle of mass m with an incompressible sphere (**hard sphere**) of mass M. In this case (Fig. 2.11), the center of $U(r)$ is the center of the sphere and the potential is zero outside the sphere. If the potential is infinite inside, the force is impulsive and the collision is between an incompressible rigid body and a point-like particle. Here, $\theta = \pi - 2\alpha$ and $b = R \sin \alpha = R \sin(\pi/2 - \theta/2) = R \cos(\theta/2)$. All particles that are inside the area $d\sigma = b \, db \, d\phi$ will be scattered in a cone around θ. The solid angle of the cone is $d\Omega = \sin \theta \, d\theta \, d\phi$. Hence,

$$\frac{d\sigma}{d\Omega} = \frac{b \, db \, d\phi}{\sin \theta \, d\theta \, d\phi} = \frac{R^2 \cos(\theta/2) \sin(\theta/2)}{2 \sin(\theta)} = \frac{R^2}{4} \,. \tag{2.116}$$

The total cross-section is thus $\sigma = \pi R^2$: the surface of the sphere projected in the plane perpendicular to the incoming particle direction. It is

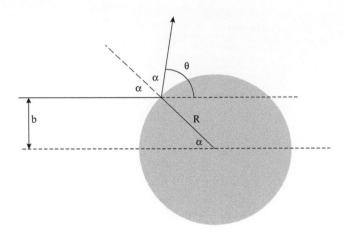

Fig. 2.11 Scattering of a point-like particle on a hard sphere of radius R and infinite mass. The impact parameter and the scattering angle are b and θ, respectively.

also called the **cross-sectional area** of the scattering center. $\sigma = \pi R^2$ fulfills our naïve expectations: for impulsive forces, the interaction takes place only if the projectile reaches the sphere. All particles, whose impact parameter is located outside the disk of radius R, will not be perturbed by the hard sphere.

Cross-sections represent 2D surfaces also from the dimensional analysis of eqn 2.113. σ and $d\sigma/d\Omega$ are measured in eV^{-2} in NU and in m^2 in SI. The best practical unit is the **barn** (b) corresponding to 10^{-28} m^2. The conversions from NU to SI can be done using this formula:

$$1 \text{ GeV}^{-1} = 3.9 \times 10^{-4} \text{ b} = 0.39 \text{ mb}. \tag{2.117}$$

The name "barn" was coined during the Second World War by the physicists of the Manhattan Project, and corresponds to the interaction cross-section between two uranium nuclei. If we consider a uranium nucleus as a hard sphere, $\sigma = \pi R^2 \simeq 10^{-28}$ m^2 (100 fm^2). By modern standards, a barn is a huge cross-section, really as big as a barn! This funny analogy contributed to the popularity of the barn as the unit of choice for cross-sections in nuclear and particle physics.

The definition of cross-section in eqns 2.113 and 2.114 holds both in SR and quantum mechanics. Unlike classical physics, relativistic cross-sections are not Lorentz-invariant because an observer moving in the direction parallel to the cross-section surface experiences a length contraction. The definition of a Lorentz-invariant cross-section is given in Sec. 2.8.2.

Finally, do not forget that scattering in QM is a stochastic (non-deterministic) process. Even for the hard sphere, there is a finite probability of finding a particle with $b = 0$ beyond the sphere.[17]

[17]This effect is the quantum equivalent of the Poisson spot in classical optics (Sharma, 2006). The intensity of an electromagnetic wave moving toward a black disk along the disk axis ($b = 0$) has a maximum in a point located beyond the disk. The Poisson (or Arago) spot is a bright point that appears at the center of the disk's shadow caused by Fresnel diffraction.

2.7.1 Fixed-target collisions

Nearly all scatterings of interest in particle physics can be classified into two groups: fixed-target and head-on (or colliding beam) collisions. The definition of cross-section given in eqn 2.113 holds for a single scattering

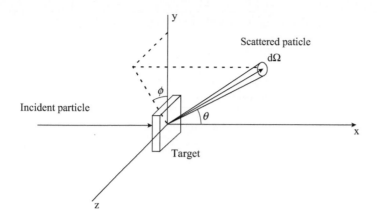

Fig. 2.12 Scattering of a particle on a fixed target. The scattered particle is contained in a cone corresponding to the solid angle Ω. Redrawn from Leo, 1994.

center. In most applications, however, we steer the particles toward macroscopic bodies called **targets**. The targets are made up of a large number of atoms that act as scattering centers (Fig. 2.12). If the target is at rest in LAB, the scattering is called a **fixed-target collision**. The number of particles per unit time that are scattered by a thin target is

$$\frac{dN_s}{d\Omega} = F A \, \tilde{N} dx \, \frac{d\sigma}{d\Omega} \tag{2.118}$$

where F is the incoming particle flux (particles per unit time per unit surface), $\tilde{N}$ is the density of the scattering centers, and dx is the thickness of the target. A is the cross-sectional area of the beam if this is smaller than the surface of the target. Otherwise, A is the target surface projected in the plane perpendicular to the particle direction. In practice, FA represents the number of incoming particles per unit time that undergo a collision with the atoms of the target. $\tilde{N}$ depends on the type of interaction. For instance, the interactions and energy losses of heavy charged particles in matter are dominated by the electromagnetic interactions with the atomic electrons (see Sec. 3.1). In this case, $\tilde{N}$ is the electron density $\tilde{N} = \rho Z N_A / A$ where ρ is the density of the target, N_A the Avogadro number, Z and A are the atomic and mass numbers of the target atoms.

The generalization to thick target is straightforward. The number of particles that have not interacted yet between the depth x and $x + dx$ is

$$N(x + dx) = N(x) - \sigma \tilde{N} N(x) dx \tag{2.119}$$

where σ is the total cross-section and $N(x) = FA$ at a given depth x. Since $N(x + dx) \simeq N(x) + (dN/dx)\, dx$, we get $dN(x) = -\sigma N(x) \tilde{N} dx$. Solving this differential equation, we can compute the number of particles at the depth x that has *not* interacted yet.

> The survival rate of a particle beam in a thick target is
>
> $$N(x) = N(0)e^{-x\sigma\tilde{N}} = N(0)e^{-\mu x}. \qquad (2.120)$$
>
> $\mu \equiv \sigma\tilde{N}$ is called the **attenuation coefficient** and $\lambda = \mu^{-1}$ is the **attenuation length**.

In eqn 2.120, the intensity of a particle beam is reduced to $e^{-1} \simeq 1/3$ of its initial value at a depth $x = \lambda$. The attenuation is stronger if the interaction cross-section is large or the medium has a high density of scattering centers.

2.7.2   Colliding beams

In classical mechanics, the most effective way to transfer momentum in a scattering is through a **head-on collision** (also called a **collider collision**). Here, the two initial state particles travel in opposite directions and collide in a point at rest in LAB, as shown in Fig. 2.13. The laboratory system is then the CM system, too. The most powerful accelerators in the world are operated in collider mode to produce the heaviest particles, as we will see in Chap. 4. The flux is not well suited to describe cross-sections at colliders because the rate of particles per unit surface does not convey information on how well the particle beams are aligned. The beam alignment is immaterial in fixed-target experiments but it is the name of the game when designing a collider.

In a collider or, in general, in any head-on collision, the **instantaneous luminosity** is the ratio between the scattered particles per unit time and the cross-section at a given time t:

$$\frac{dN_s}{d\Omega} = \mathcal{L}(t)\frac{d\sigma}{d\Omega}. \qquad (2.121)$$

The luminosity decreases with time because the particles are injected into a vacuum tube (**beampipe**) located inside the collider and stay there for tens of hours or more. During this time, the beams deteriorate due to the head-on collisions and the interactions with the residual atoms inside the beampipe. The total number of interactions after a run of a collider is provided by the **integrated luminosity**:

$$\mathcal{L} = \int dt \, \mathcal{L}(t) \qquad (2.122)$$

and the number of scattered events at any angle is $N_s = \mathcal{L}\sigma$. The instantaneous luminosity has the same unit as the flux, particles per unit time per unit surface. It is measured in $\mathrm{m}^{-2}\mathrm{s}^{-1}$ in SI and eV^3 in NU. The integrated luminosity is measured in m^{-2} or eV^2. Accelerator physicists are used to expressing $\mathcal{L}$ in barn^{-1} because this unit provides a guideline for the design of the collider. An integrated luminosity of 1 inverse-millibarn ($1 \ \mathrm{mb}^{-1}$) allows the experimentalist to observe events that have typical cross-sections of 1 mb. If we want to observe very rare events, for example the Higgs boson production ($\sigma \simeq 1$ nb), we need a high integrated luminosity, which in turn requires well-designed accelerators and/or a very long period of data-taking. As expected, the luminosity depends on the performance of the collider and its main parameters: the cross-sectional area of the beams, the number of particles in the beams, etc. We will discuss these quantities in Chap. 4.

2.8 Scattering in quantum mechanics

Most of the consideration done for the decay holds for the quantum mechanical description of the scattering. A scattering between two particles $a + b \rightarrow 1 + 2 + \cdots n$ is a two-body generalization of the single-particle decay $a \rightarrow 1 + 2 + \cdots n$ and the relativistic golden rule can be adjusted to handle the two-body initial states.

2.8.1 Flux

In a fixed-target experiment, the flux was defined as the number of incoming particles per unit time and unit surface. This definition holds only if the scattering centers are at rest in the laboratory frame. A particle a interacts with the scattering center b only if its unperturbed trajectory goes through the plane perpendicular to the direction of motion of b. For instance, if the particle a moves at a velocity $\mathbf{v} = (0, 0, v)$ and b moves at a velocity $\mathbf{v} = (0, 0, 2v)$, the two particles do not interact because a will never be in the proximity of b. It is worth stressing that we are considering the *unperturbed* motion of a: the real trajectory may not reach the plane perpendicular to b due to the interactions with b. For instance, the projectile that approaches the hard sphere at $b < R$ never reaches the vertical plane of the sphere. This is already encoded in the cross-section but if you want to count how many times the interactions take place, you have to count how many unperturbed projectiles reach such a plane.

If a single particle a moves along the z direction (Fig. 2.13) with a velocity $v_a \mathbf{k}$ toward b and b moves toward a with a velocity $v_b \mathbf{k}$, the flux will be proportional to $|v_a - v_b|$. The rate of interactions will be greater than a process where the scattering center is at rest ($v_b = 0$). Clearly, this is true only if the two particles move in opposite directions ($v_b < 0$). For a scattering of a with several particles b located in a region with an area A and density $\tilde{N}$, the number of scattering centers crossed by a in dt will be:

$$dN = \tilde{N}|v_a - v_b|A dt \tag{2.123}$$

and the **flux** defined as the number of incoming particles per unit time and unit surface is

$$F = \frac{dN}{A dt} = \tilde{N}|v_a - v_b|. \tag{2.124}$$

For a single scattering center in a volume V, $\tilde{N} = 1/V$ and the flux is:

$$F = \frac{1}{V}|v_a - v_b|. \tag{2.125}$$

Fig. 2.13 Two particles in a head-on collision. In this case, the flux is larger than a process with a scattering center at rest in LAB.

This quantity is not covariant and depends on an arbitrary volume V that must disappear in physical quantities like the cross-sections.

2.8.2   Cross-sections and Lorentz invariance

The number of scatterings per unit time of the particles a on b is given by the product of the flux and the cross-section. In QM, the number of transitions per unit time (the **transition rate**) is given by the non-relativistic Fermi golden rule, so:

$$\Gamma_{i \to f} = \frac{|v_a - v_b|}{V} \, \sigma \tag{2.126}$$

which links the cross-section to eqn 2.73. In the non-relativistic case, V is canceled by the volume that appears in $\mathcal{T}$ due to the normalization of the wavefunction: $\mathcal{T} = \langle f| \hat{H}' |i\rangle = \langle \psi_1 \cdots \psi_n| \hat{H}' |\psi_a\rangle$ for the decay and $\mathcal{T} = \langle f| \hat{H}' |i\rangle = \langle \psi_1 \cdots \psi_n| \hat{H}' |\psi_a \psi_b\rangle$ for the scattering. The volume V associated with ψ_b cancels the volume coming from the definition of the flux. The relativistic version of the Fermi rule is obtained, as for eqn 2.79, from

$$\mathcal{M} = (2E_a)^{1/2}(2E_b)^{1/2} \prod_{i=1}^{n} (2E_i)^{1/2} \mathcal{T} \tag{2.127}$$

where the additional $2E_b$ term comes from the additional initial state particle b.

> The **relativistic golden rule for the cross-sections** in a $a + b \to 1 + 2 + \cdots n$ scattering is:
>
> $$\sigma_{i \to f} = \frac{(2\pi)^4}{2E_a \, 2E_b |v_a - v_b|} \int \prod_{i=1}^{n} \frac{d^3 \mathbf{p}_i}{(2\pi)^3 2E_i}$$
>
> $$\times |\mathcal{M}|^2 \delta \left(E_a + E_b - \sum_{j=1}^{n} E_j \right) \delta^3 \left(\mathbf{p}_a + \mathbf{p}_b - \sum_{j=1}^{n} \mathbf{p}_j \right). \tag{2.128}$$

As for the decay, the integral is Lorentz-invariant but the $E_a E_b |v_a - v_b|$ term is not, and, therefore, the cross-section is not a Lorentz scalar. We can define a Lorentz-invariant cross-section if we embed in its definition the frames where the cross-section has to be computed. We have already done this for the proper time τ. Time is not Lorentz-invariant but the proper time is defined as the time measured in the comoving frame: all observers agree on its numerical value because the definition forces the measurement to be performed in a specific reference frame. As a consequence, the proper time can be expressed in covariant form as a Lorentz scalar (see Exercise 2.4). The same procedure can be employed for the **Möeller flux factor**:

$$F = 4E_a E_b |v_a - v_b| \tag{2.129}$$

requiring F to be computed in any reference frame moving in the direction of motion of particles a and b (the z-axis in Fig. 2.13). F can thus be rewritten as:

$$F = 4\left[(p_{a\mu}p_b^\mu)^2 - m_a^2 m_b^2\right]^{1/2} = 4\left[(p_a \cdot p_b)^2 - m_a^2 m_b^2\right]^{1/2}. \quad (2.130)$$

Equation 2.130 defines the **Lorentz invariant flux factor**, which is equal to the Möeller flux factor in the above-mentioned frames.

A very useful Lorentz-invariant that can be employed in relativistic calculations is the **first Mandelstam variable "s."** For a two-body scattering:

$$s \equiv (p_a + p_b)^2 = (p_{tot})_\mu (p_{tot})^\mu \quad (2.131)$$

where $p_{tot}^\mu = p_a^\mu + p_b^\mu$ is the total initial four-momentum. $\sqrt{s}$ is often called the **invariant mass** because it can be interpreted as the rest mass of a particle that could be created if the whole energy of a and b were transformed in rest energy. Sometimes this happens. For instance, the Z^0 can be produced by a two-body scattering of an electron and a positron after a head-on collision. The two particles merge and create a particle of mass M_Z: $e^+ e^- \to Z^0$. In this case $s = M_Z^2$. The Lorentz-invariant flux factor can be written as a function of s because:

$$s - (m_a + m_b)^2 = p_a^2 + p_b^2 + 2p_a \cdot p_b - m_a^2 - m_b^2 - 2m_a m_b$$
$$= m_a^2 + m_b^2 + 2p_a \cdot p_b - m_a^2 - m_b^2 - 2m_a m_b = 2\left[p_a \cdot p_b - m_a m_b\right].$$

Similarly, $s - (m_a - m_b)^2 = 2\left[p_a \cdot p_b + m_a m_b\right]$. Then,

$$[s - (m_a + m_b)^2][s - (m_a - m_b)^2]$$
$$= 4\left[p_a \cdot p_b - m_a m_b\right]\left[p_a \cdot p_b + m_a m_b\right] = 4\left[(p_a \cdot p_b)^2 - m_a^2 m_b^2\right].$$

As a consequence:

$$F = 4\left[(p_a \cdot p_b)^2 - m_a^2 m_b^2\right]^{1/2}$$
$$= 2\left[(s - (m_a + m_b)^2)(s - (m_a - m_b)^2)\right]^{1/2} \quad (2.132)$$

and

$$F \to 2s \quad \text{for } \sqrt{s} \gg m_a, \, m_b. \quad (2.133)$$

This is an amazing result:

> if the energy of the particles is much larger than their rest mass, the Lorentz-invariant flux factor is just twice the squared invariant mass.

The **Lorentz-invariant cross-section** is thus:

$$\sigma_{i \to f} = \frac{(2\pi)^4}{4\left[(p_a \cdot p_b)^2 - m_a^2 m_b^2\right]^{1/2}} \int \prod_{i=1}^n \frac{d^3\mathbf{p}_i}{(2\pi)^3 2E_i}$$
$$\times |\mathcal{M}|^2 \, \delta\left(E_a + E_b - \sum_{j=1}^n E_j\right) \delta^3\left(\mathbf{p}_a + \mathbf{p}_b - \sum_{j=1}^n \mathbf{p}_j\right). \quad (2.134)$$

The neatest rearrangement of eqn 2.134 recasts the Lorentz-invariant phase-space using an additional delta (see eqn 2.91) and merges the delta functions that represent the energy and momentum conservation in a four-dimensional delta. The final result is the:

Lorentz-invariant cross-section written in covariant form:

$$\sigma_{i \to f} = \frac{1}{4\left[(p_a \cdot p_b)^2 - m_a^2 m_b^2\right]^{1/2}} \int \prod_{i=1}^{n} \left[\frac{d^4 p_i}{(2\pi)^4} \, (2\pi) \, \delta(p_i^2 - m_i^2) \right]$$

$$\times |\mathcal{M}|^2 (2\pi)^4 \delta^4 \left(p_a + p_b - \sum_{j=1}^{n} p_j \right). \qquad (2.135)$$

Equation 2.135 is extremely useful because it decouples the kinematic constrains (phase-space and conservation of four-momentum) from the transition amplitudes $\mathcal{M}$. The computation of $\mathcal{M}$ is the main aim of QFT and in most cases $|\mathcal{M}|^2$ is a smooth and slowly varying function of s. Strong variations of $|\mathcal{M}|^2$ signal the production of new particles or the appearance of new interactions. Kinematics, however, suppress the cross-sections at high energy. If $|\mathcal{M}|^2 \simeq$ constant,

$$\sigma \sim \frac{1}{s} \qquad (2.136)$$

for $\sqrt{s} \gg m_a, \, m_b$. This suppression also occurs in non-relativistic physics and has a simple interpretation: at high velocities, the interaction time between the two particle is shorter and the cross-section is reduced by $|v_a - v_b|^{-1}$ which can be arbitrarily small because there is no upper limit on the velocities. In SR, $|v_a - v_b| \to 2c$ for ultra-relativistic particles but the cross-section is now suppressed by $E_a E_b |v_a - v_b| \sim s$.

Exercises

(2.1) Consider a (fake) physical law:

$$\sum_{i=1}^{N_i} (p_i^{ini})^\mu - \sum_{j=1}^{N_f} 2(p_j^{fin})^\mu = 0 \qquad (2.137)$$

that describe an N-body scattering. Is eqn 2.137 consistent with the axioms of SR? Provide an example that falsifies this law.

(2.2) The energy of a particle is given by $E = \gamma m$ in NU. Is this equation a physical law in SR? Recast it in a covariant form.

(2.3) Demonstrate that the "distance" between two events defined as:

$$c^2 (t_2 - t_1)^2 - (x_2 - x_1)^2 - (y_2 - y_1)^2 - (z_2 - z_1)^2$$

is invariant for the Lorentz transformations, i.e. is a **Lorentz scalar**. Perform the proof in two ways: applying the Lorentz transformations and using the covariant formalism.

(2.4) * Write the proper time $d\tau$ in covariant form and show that $d\tau$ is a scalar for Lorentz transformations.

(2.5) The most updated values of the particle properties as measured by the experiments are provided every two years by the **Particle Data Group** (PDG). You can find this information in https://pdg.lbl.gov. Using the latest PDG data, compute the maximum energy of a positron in a $K^+ \to \pi^0 e^+ \nu_e$ neglecting the neutrino mass.

(2.6) Compute the kinematic threshold for π^0 production in a $p+p \to p+p+\pi^0$ scattering occurring in a fixed-target experiment. [Hint: use the invariant mass s]

(2.7) Demonstrate the **threshold formula** for a fixed-target experiment, i.e. the minimum energy of the projectile a needed to produce N particles of mass $m_1, m_2, \ldots, m_N$:

$$E^a_{min} = \frac{\left(\sum_{i=1}^{N} m_i\right)^2 - m_a^2 - m_b^2}{2m_b} \qquad (2.138)$$

(2.8) Compute the equivalent formula of eqn 2.138 for a head-on collision $a + b \to X$ producing N particles of mass $m_1, m_2, \ldots, m_N$.

(2.9) Show that a $\mu^- \to e^-$ decay cannot occur.

(2.10) An inertial observer records a pion decay $\pi^- \to \mu^- \bar{\nu}_\mu$ at x_1^μ, i.e. at a time t_1 and in a position $\mathbf{x}_1$. Another observer records the same event in x_2^μ. Is it possible to find a third observer that records the event at t_3 with $t_1 < t_3 < t_2$? Can we find an observer that records the decay as $\pi^- \to e^- \bar{\nu}_e$?

(2.11) * Demonstrate that the one-body Lorentz-invariant phase-space

$$\int \frac{d^3\mathbf{p}}{(2\pi)^3}\bigg|_{E=\sqrt{|\mathbf{p}|^2+m^2}} \qquad (2.139)$$

grows as the square of the particle energy. [Hint: perform the integration in spherical coordinates]

(2.12) ** Using eqns 2.90 and 2.92 show that

$$d\Gamma = \frac{1}{32\pi^2}|\mathcal{M}|^2 \frac{|\mathbf{p_1}|}{M^2} d\Omega \qquad (2.140)$$

for a two-body decay $a \to 1+2$ in the RF of a particle with mass M.

(2.13) * Check explicitly the invariance of eqn 2.82 using the Lorentz transformations.

(2.14) ** Using eqn 2.87, show that

$$\delta^4\left(p_a^\mu - \sum_{i=1}^{n} p_i^\mu\right) \qquad (2.141)$$

is a Lorentz scalar.

(2.15) ** Derive eqn 2.25 from 2.24.

3 Measurements in particle physics

All information on the dynamic of elementary particles are encoded in the Lorentz-invariant matrix element of eqn 2.79. To decode this information, we need to measure the trajectory and momenta of the particles in the final state. Measurements are performed by macroscopic systems called **particle detectors** that identify the particle and generate a signal proportional to the particle energy, momentum, or position. Particle detectors are based on physical processes that exhibit large cross-sections. The larger the cross-section, the higher the efficiency of the detector and, quite often, the precision of the measurement. The vast majority of the detectors exploit the electromagnetic interactions of particles with matter since the cross-section of these processes generally exceeds 10^{-21} m^2 (10 Mb). Electromagnetic interactions are strongly enhanced if the particle is charged and propagates in a continuous medium. For neutral particles, we employ scattering processes that produce charged particles in the final state, which in turn generate an observable signal in the detector. To design high-resolution and high-efficiency detectors, we need a full understanding of electromagnetic interactions of charged particles with the electrons and the nuclei of the medium. In 1914, N. Bohr showed that the energy transfer from the charged particle to the detector is due to the interactions with the electrons of the medium. Nuclei may change the trajectory of the incoming particles by **multiple Coulomb scatterings** (MS) but the energy transfer from the charged particle to the nuclei is always negligible. Therefore, the atomic nuclei play a role in tracking systems but are of limited interest to estimate the energy loss of a charged particle and discussion of them will be neglected until Sec. 3.10.

3.1 Heavy charged particles

The energy loss of a charged particle, whose mass is much heavier than the electron, can be estimated using classical electrodynamics and corrected for quantum effects. The simplest derivation of the Bohr formula is credited to E. Fermi and is based on three crude approximations: the trajectory of the particle is not perturbed by the medium, the electrons of the medium are at rest during the passage of the particle, and every scattering between the particle and the electrons are non-relativistic.

A Modern Primer in Particle and Nuclear Physics. Francesco Terranova, Oxford University Press.
© Francesco Terranova (2021). DOI: 10.1093/oso/9780192845245.003.0003

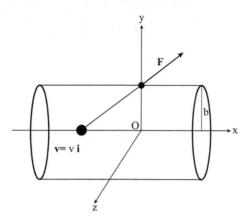

Fig. 3.1 A heavy particle of mass M (large dot) and speed v moves along the x axis in the proximity of an electron (small dot). The impact parameter is b and the force **F** experienced by the electron is repulsive if the charge z of the particle is negative (as in the figure) or attractive if it is positive.

For a particle approaching an electron with an impact parameter b (see Fig. 3.1), the force experienced by the electron is given by the Coulomb law. In SI, it reads:

$$\mathbf{F} = \frac{ze^2}{4\pi\epsilon_0}\frac{\widehat{\mathbf{r}}}{r^2} \tag{3.1}$$

where ze is the charge of the particle (e is the absolute value of the charge of the electron), $\widehat{\mathbf{r}}$ is the unit vector along the distance between the two particles, and r is the size of the distance. Since the system has a cylindrical symmetry, we locate the origin of the LAB frame at the point O in the figure so that the coordinates of the electron are $(x, r, \phi) = (0, b, 0)$. The force experienced by the electron when the particle is at $\mathbf{x} = -x\mathbf{i}$ is $\mathbf{F}_1 = F_{1x}\mathbf{i} + F_{1y}\mathbf{j}$. The force at $\mathbf{x} = x\mathbf{i}$ is $\mathbf{F}_2 = F_{2x}\mathbf{i} + F_{2y}\mathbf{j}$ but $F_{1x} = -F_{2x}$. As a consequence, the net force on the electron is along the vertical axis: $\mathbf{F} = F\mathbf{j}$. The scattering is non-relativistic and the change of the kinetic energy of the electron can be derived from the impulse theorem. The impulse $\mathbf{I}$ is the time integral of a force. The impulse applied to an object produces a change of momentum along the direction of $\mathbf{I}$. In this case,

$$\mathbf{I} \equiv \int \mathbf{F}dt = e\int_0^\infty \mathbf{E}_\perp dt. \tag{3.2}$$

Due to the approximations made by Fermi, the trajectory of the heavy particle is not perturbed by the interaction with the electron and the particle velocity $\mathbf{v} = v\mathbf{i}$ is constant:

$$\mathbf{I} = e\int \mathbf{E}_\perp \frac{dt}{dx}dx = \frac{e}{v}\int_{-\infty}^{+\infty} \mathbf{E}_\perp dx. \tag{3.3}$$

We are assuming that the electron is at rest during the passage of the particle, so the electric field is purely electrostatic and the magnetic field produced by the particle does not change the trajectory of the electron. Equation 3.3 can thus be solved by Gauss law. The flux of the electric field over the infinite cylinder of radius b, whose axis is along the particle trajectory, is:

$$\Phi \equiv \int_S \mathbf{E}_\perp \cdot \hat{\mathbf{n}} \, dA = 2\pi b \int_{-\infty}^{+\infty} |\mathbf{E}_\perp| dx. \tag{3.4}$$

In eqn (3.4), $\hat{\mathbf{n}}$ is the unit-vector perpendicular to the surface element dA. From Gauss law:

$$\Phi = \frac{ze}{\epsilon_0} \tag{3.5}$$

where ze is the charge inside the cylinder, that is, the charge of the heavy particle. Hence, replacing eqns (3.4) and (3.5) in (3.3):

$$\mathbf{I} = I\mathbf{j} = \frac{ze^2}{2\pi\epsilon_0 vb}\mathbf{j}. \tag{3.6}$$

The electron is at rest and, therefore, the initial momentum of the electron is zero. The final momentum is $\mathbf{p}_f$ and the non-relativistic kinetic energy gained by the electron is $|\mathbf{p}_f|^2/2m_e = I^2/2m_e$. The kinetic energy E lost by the particle interacting with a single electron is:

$$\Delta E(b) = \frac{I^2}{2m_e} = \frac{z^2e^4}{8\pi^2\epsilon_0^2 b^2 v^2 m_e}. \tag{3.7}$$

The medium is composed of a large number of electrons, which are uniformly distributed with density N_e. The total kinetic energy transferred from the charged particle to the electrons enclosed in a volume $d^3x = dx \, bdb \, d\phi$ is then:

$$-d^3E = \Delta E(b)N_e d^3x. \tag{3.8}$$

The minus sign signals that the particle is reducing its energy due to the scattering with the electrons in d^3x. Integrating on the ϕ coordinate, we get:

$$-d^2E = \int_0^{2\pi} \Delta E(b)N_e dx \, bdb \, d\phi = 2\pi b \, \Delta E(b)N_e dx \, db = \tag{3.9}$$

$$= \frac{z^2e^4N_e}{4\pi\epsilon_0^2 v^2 m_e}\frac{db}{b}dx \tag{3.10}$$

$$\tag{3.11}$$

which gives:

$$-\frac{dE}{dx} = \frac{z^2e^4N_e}{4\pi\epsilon_0^2 v^2 m_e} \log\left(\frac{b_{max}}{b_{min}}\right). \tag{3.12}$$

Equation 3.12 without quantum and relativistic corrections is divergent since b goes from 0 to $+\infty$. SR dictates $b_{min} > 0$ because the maximum kinetic energy that can be transferred to an electron is

$$\Delta E(b_{min}) = \frac{z^2e^4}{8\pi^2\epsilon_0^2 b_{min}^2 v^2 m_e} = 2m_e\gamma^2 v^2 \tag{3.13}$$

where $\gamma = [1 - (v/c)^2]^{1/2}$ is the Lorentz gamma of the heavy particle. The right-hand side of eqn 3.13 follows from relativistic kinematics – see Exercise 3.16 – and provides a good approximation for b_{min}. Besides, the electron is bound to the nucleus. To get a higher level of accuracy compared with the Fermi approximations, we can define the semi-classical mean orbital frequency $\bar{\nu}$: this is the mean frequency of the electrons in the Bohr atomic model averaged on all bound states. If the interaction time of the charged particle is longer than $\bar{\nu}^{-1}$, the mean Coulomb force averages to zero and the energy transfer is negligible.[1] In the proper frame of the charged particle, the typical interaction time is $\simeq b/v$, which reads $b/\gamma v$ in the laboratory frame. Hence,

$$b_{max} \simeq \frac{\gamma v}{\bar{\nu}}. \tag{3.14}$$

Replacing eqns 3.13 and 3.14 in eqn 3.12, we get the **classical Bohr formula**:

$$-\frac{dE}{dx} \simeq \frac{z^2 e^4}{4\pi\epsilon_0^2 m_e v^2} N_e \log\left(\frac{4\pi\epsilon_0 \gamma^2 m_e v^3}{z e^2 \bar{\nu}}\right). \tag{3.15}$$

Bohr's formula provides the correct relation between the energy loss and the velocity for non-relativistic particles. It fails in predicting the energy loss of ultra-relativistic particles or particles with energy comparable with the binding energy of the electrons ($v \to 0$).

The first complete quantum-mechanical calculation of the energy loss of a heavy charged particle in a continuous medium was carried out by H. Bethe in 1932.

The **Bethe formula**, also called the **Bethe–Bloch formula**, is (Peter, 2006):

$$-\frac{dE}{dx} = \frac{z^2 e^4}{4\pi\epsilon_0^2 m_e c^2 \beta^2} N_e \left[\log\left(\frac{2 m_e \gamma^2 c^2 \beta^2}{I}\right) - \beta^2 - \frac{\delta(\gamma)}{2} - \frac{C(\beta\gamma, I)}{Z}\right]. \tag{3.16}$$

Note that the formula is expressed in SI, and I is the mean excitation potential of the atom, which corresponds to $h\bar{\nu}$ in the Bohr model. The energy loss of a heavy particle is large if the particle is slow, that is non-relativistic. This follows from the impulse theorem: the slower the particle, the longer the interaction time, the larger the energy transfer. For $v \to 0$ the classical Bohr formula diverges and breaks when the velocity is smaller than the orbital velocity of the electron. The approximation of electrons at rest during the interaction time does not hold in this regime and the derivation of eqn 3.15 is invalid. For $v \to 0$, eqn 3.16 is dominated by an empirical function of $\beta\gamma$ that depends on N coefficients $\alpha_i(I)$ ($i = 1, \ldots N$):

$$C(\beta\gamma, I) = \sum_{i=1}^{N} \alpha_i(I)(\beta\gamma)^{-2i}. \tag{3.17}$$

This function is called the *shell correction*. Since the behavior of a charged particle at low speed is very important in many applications of particle physics and, in particular, in medical physics, the shell correction and other dominant corrections are measured and tabulated for a large number of molecules and compounds. If the velocity is larger

[1] The demonstration relies on classical electrodynamics but is quite lengthy. See, for example, Chap. 13 in Jackson, 1998, where you can find the original Fermi derivation presented in modern language.

than the orbital velocity, the Bohr formula properly describes the energy loss per unit length $-dE/dx$. The energy loss decreases as v^{-2} until the particle reaches the minimum of the Bethe formula at $v \simeq 0.96\,c$ (see Fig. 3.2). Particles at the minimum of the dE/dx curve are called **minimum ionizing particles** (mip). The Bohr formula diverges as $\log\gamma$ for $v \to c$, that is $\gamma \to \infty$. The logarithmic growth of the energy loss for particles faster than a mip is called "relativistic rise" because it is a genuine relativistic effect. The minimum impact parameter b_{min} decreases as γ because the maximum kinetic energy increases. From eqn 3.14, b_{max} increases as γ, as well. Hence $\log b_{max}/b_{min} \sim \log\gamma^2 = 2\log\gamma$. In the rest frame of the charged particle this effect can be interpreted as an increase of $\mathbf{E}_\perp$. This is not surprising because a Lorentz transformation contracts space and momentum in the direction parallel to the relative motion of the frames, while transverse quantities do not change. Electric and magnetic fields are changed the other way round: the dilation occurs in the transverse plane and the longitudinal field is Lorentz-invariant.[2] For the frames of Fig. 2.2, the Lorentz transformations of the $\mathbf{E}$ and $\mathbf{B}$ fields are due to the motion of the heavy charged particle in LAB and are:

$$\mathbf{E}'_\parallel = \mathbf{E}_\parallel$$
$$\mathbf{B}'_\parallel = \mathbf{B}_\parallel$$
$$\mathbf{E}'_\perp = \gamma(\mathbf{E}_\perp + \mathbf{v} \times \mathbf{B})$$
$$\mathbf{B}'_\perp = \gamma(\mathbf{B}_\perp - \frac{1}{c^2}\mathbf{v} \times \mathbf{E}) \qquad (3.18)$$

where $\mathbf{E}$ and $\mathbf{B}$ are the fields in the RF of the heavy particle. SR dilates the fields in the direction perpendicular to the motion, increasing $\mathbf{E}_\perp$ by γ and, in turn, it inflates the energy loss of the particle. Such an increase shows up as a small **relativistic rise** in the Bethe formula.

Inside a dielectric, the relativistic rise is compensated by polarization effects. In the laboratory frame, if γ increases, b_{max} in eqn 3.14 increases, too, and the contributions of distant collisions intensify. On the other hand, the electric field of the particle polarizes the medium and the polarization partially shields the fields of distant electrons. This effect is accounted for by the empirical function $\delta(\beta\gamma) \simeq \delta(\gamma)$ that represents the leading term in eqn 3.16 for $\beta \to 1$. The asymptotic behavior of $\delta(\gamma)$ is $\log\gamma$ and cancels the $\log\gamma$ term in the Bethe formula. As a consequence, $-dE/dx$ is constant for $\gamma \to \infty$.

The Bethe formula can be expressed in practical units replacing the electron density N_e with $\rho N_A Z/A$, where N_A (atoms/mole) is the Avogadro number, and Z and A (g/mole) are the atomic and mass number of the material:

$$-\frac{dE}{dx} = \left(0.153\,\frac{\mathrm{MeV\,cm^2}}{\mathrm{g}}\right)\rho\frac{Z}{A}\frac{z^2}{\beta^2}\left[\log\left(\frac{2m_e\gamma^2c^2\beta^2 W_{max}}{I^2}\right) - 2\beta^2\right.$$
$$\left. -\delta(\gamma) - 2\frac{C(\beta\gamma,I)}{Z}\right]. \qquad (3.19)$$

[2]We further discuss field dilation in relativistic electrodynamics in Chap. 6. A thorough derivation is available in Rindler, 2006.

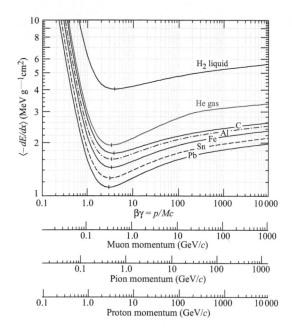

Fig. 3.2 Mean energy loss expressed in mass thickness $dE/d\tilde{x}$ (MeV·cm^2/g) for several media from liquid hydrogen (liquid H_2) to lead (Pb) as a function of the $\beta\gamma$ of the particle. Below the plot, $\beta\gamma = p/M$ is converted in momentum for several types of particles of mass M where the Bethe formula is accurate (heavy particles as muons, pions, and protons). Reproduced with permission from Zyla *et al.*, 2020.

W_{max} is the maximum kinetic energy transfer and $W_{max} \simeq 2m_e\gamma^2\beta^2c^2$ if the mass of the particle is $\gg m_e$. For a heavy particle whose mass is $\gg m_e$ eqn 3.19 is then equivalent to eqn 3.16. In eqn 3.19, ρ is expressed in g/cm^3 and dE/dx in MeV/cm. The higher the density of the medium, the bigger the energy loss since $-dE/dx \sim N_e$. In addition, $Z/A \simeq 0.5$ for the larger part of materials and compounds. Hence, if we define the **mass thickness** $\tilde{x}$ as the product $\rho \cdot x$

> the energy loss per mass thickness $dE/(d\tilde{x})$ expressed in MeV·cm^2/g is about the same in any material and a mip has an energy loss of $\simeq 2$ MeV·cm^2/g.

This rule of thumb is a gem for the experimentalist. Since the relativistic rise is a small correction to the energy loss of a mip and can often be neglected, any relativistic heavy particle of charge ± 1 deposits $\simeq 2$ MeV cm^2/g independently of its energy.

Note that the energy loss of a particle is the sum of many scatterings of the particle with the electrons and each scattering occurs with a certain quantum mechanical probability. The energy loss is thus a stochastic process and the Bethe formula provides the *mean* energy loss of a particle.

The total length traveled by a particle in a medium is called **range** and is given by:

$$R(\text{cm}) = \int_{T_0}^{0} \left(\frac{dE}{dx}\right)^{-1} dE \qquad (3.20)$$

where T_0 is the initial kinetic energy of the particle. The energy loss as a function of the depth is shown in Fig. 3.3. As expected from eqn 3.19, most of the energy is deposited at the end of the path, once the particle becomes non-relativistic. The maximum of the energy deposit is called the **Bragg peak** and

> the penetration depth of a heavy particle grows with the initial kinetic energy but most of the energy is deposited in the final part of the range. If the particle is non-relativistic, $R \sim T_0^2$.

This property is the cornerstone of cancer therapy with heavy charged particles. Accelerated heavy particles (protons or light nuclei) can destroy tissues like malignant neoplasms (cancer) located at a given depth without damaging the cells between the skin and the neoplasm. The energy deposited in the tissues before the target cells is negligible and the target can be destroyed with high precision if the particles are monochromatic, that is the uncertainty on T_0 is small and T_0 is tuned to reach the actual position of the neoplasm.

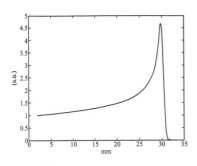

Fig. 3.3 Relative energy deposit (in arbitrary units) as a function of the depth (in mm) for a 60 MeV proton in water. A relative energy deposit of 1 corresponds to about the energy loss of a minimum ionizing particle. The energy deposit peak at the end of the particle range is the Bragg's peak.

3.2   Energy loss of electrons

Unlike heavy charged particles, electrons that enter a medium have a mass identical to the target particles, and the energy that is transferred ranges between 0 and half the kinetic energy of the incoming electron: $(E - mc^2)/2$. The maximum kinetic energy that can be transferred is different from $2m_e\gamma^2c^2\beta^2$ and the Bethe formula must be corrected accordingly. Besides, the formula must account for the Pauli exclusion principle since the incoming particle and the scattering center are two identical fermions. The corresponding Bethe formula can be written analytically with empirical corrections similar to the shell and density effects (Leo, 1994).

The electrons, however, show an additional mechanism of energy loss in matter. **Radiation loss** or **bremsstrahlung**, which means braking radiation in German, is the radiation produced by the acceleration or deceleration of a charged particle. Radiation losses follow from the Maxwell equations and take place in any charged particle. The irradiated power (Griffiths, 2017), however, is $\sim \gamma^6$ if the acceleration is parallel to the velocity and $\sim \gamma^4$ if it is perpendicular like, e.g., in circular accelerators. Since $\gamma = E/m$, this source of losses is negligible for heavy particles up to TeV energies but is the leading energy-loss mechanism for electrons. In quantum mechanics (QM), bremsstrahlung corresponds to the spontaneous emission of photons by an electron in a medium. Four-momentum conservation forbids the $e^- \to e^-\gamma$ transition in vacuum but the process can occur in the proximity of another particle and the transition probability linearly increases with the density of the medium. Even if the quantum mechanic treatment of radiation losses is complicated, its empirical description is rather simple. Radiation losses become dominant above the **critical energy**, which is the energy when the energy loss due to radiation overtakes the **ionization loss**, the loss due to the interactions with the electrons of the medium described by the Bethe formula. Unlike heavy charged particles, the medium plays a crucial role because it changes the trajectory of the incoming electron, causing the breaking radiation. The critical energy is often measured and tabulated but some empirical formulas are available, too. For solid materials,

$$E_c \simeq \frac{610 \text{ MeV}}{Z + 1.24}. \tag{3.21}$$

The critical energy in a gaseous material like the air is about 84 MeV and goes down to 7 MeV in solid lead. Well above E_c, the energy of an electron decreases following an exponential law:

$$E = E_0 e^{-x/X_0}. \tag{3.22}$$

In this formula we can express the depth and X_0 either in mass thickness or in standard SI length units. The parameter X_0 is crucial in the physics of particle detectors and it is called the **radiation length**. For most of the media, X_0 in mass thickness units can be approximated to:

$$X_0 = \frac{716.4 \ A}{Z(Z+1) \log\left(287/Z^{1/2}\right)} \frac{\text{g}}{\text{cm}^2} \tag{3.23}$$

where A is the mass number in grams per mole.[3] The radiation length in air is 37 g/cm^2, which corresponds to about 30 m ($\rho_{air} \simeq 1.2 \times 10^{-3}$ g/cm^3), and drops to 0.56 cm in lead. Such a large variation results both from the low density of the air and the high atomic number of lead ($Z = 82$).

[3]Unfortunately, the symbol X_0 is used to indicate the radiation length in unit of mass thickness and in length. You can distinguish the two definitions by dimensional analysis since the units of X_0 are g/cm^2 and cm, respectively.

3.3 The discovery of antimatter

The energy loss of charged particles with matter was established at the beginning of the 20th century for all particles known at that time: electrons, protons, and light nuclei. The energy released to the atomic electrons can either bring the atoms to an excited state or move the electron to the continuum ionizing the atom (see Sec. 3.5). If the air temperature and humidity are properly tuned, highly ionized air acts as a condensation center for the formation of water droplets and, eventually, the clouds. Even if the dynamic of cloud formation is extremely complex and still at the focus of modern research in chemistry and environmental science, the basic formation principle can be exploited to visualize the trajectory of charged particles. An **expansion cloud chamber** consists of a vessel containing a supersaturated vapor of water. If the gas mixture is at the point of condensation, a trail of small droplets forms in the volume where the density of ions is high. The droplets are visible along the trajectory of the particle for several seconds while they fall through the vapor. The detector is called an expansion chamber because we use a diaphragm to perform the adiabatic expansion that cools the air and starts the condensation of the vapor. The detector is sensitive to particles only after the expansion of

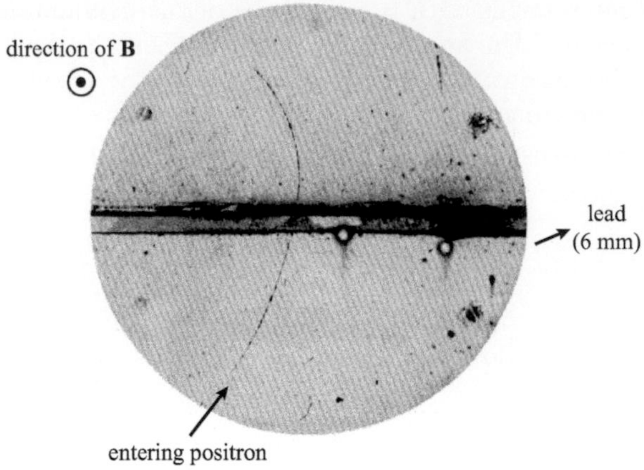

Fig. 3.4 Anderson's celebrated picture of a positron entering from the lateral side of the cloud chamber and stopping ~ 5 cm after the lead plate. The direction of the magnetic field is perpendicular to the picture and is shown on the top of it. Reprinted figure with permission from Anderson, 1933. Copyright 1933 American Physical Society.

the diaphragm, which is set in coincidence with a camera that takes pictures of the tracks. Cloud chambers have been used since 1911 to observe tracks produced by cosmic rays and radioactive decays. The most celebrated application is the discovery of the first anti-particle in 1932 by C.D. Anderson (Anderson, 1933) confirmed nearly at the same time by P. Blackett and G. Occhialini (Blackett, 1933).

The experimental setup of Anderson was a $17 \times 17 \times 3$ cm^3 cloud chamber equipped with a constant magnetic field of 1.5 T (see Fig. 3.4). The field bends the trajectory of the charged particle and the size of the bending is proportional to the momentum of the particle. The direction of the bending depends on the direction of the velocity and the sign of the charge. The relation between charge, momentum, and curvature radius is well known in classical electrodynamics (Griffiths, 2017) and is derived below using the covariant formalism for a relativistic particle. The formula can be computed starting from eqn 2.54 in Example 2.2. For a particle propagating inside the cloud chamber in the direction perpendicular to the magnetic field $\mathbf{B} = B\mathbf{k}$, the relativistic force experienced by the particle is:

$$\mathbf{K} \equiv (K^1, K^2, K^3) = \frac{d\mathbf{p}}{d\tau} = \frac{dt}{d\tau}\frac{d\mathbf{p}}{dt} = \gamma\frac{d\mathbf{p}}{dt} = q\gamma(\mathbf{E} + \mathbf{v} \times \mathbf{B}) \quad (3.24)$$

so that the ordinary force is the good old Lorentz force:

$$\mathbf{F} = \frac{d\mathbf{p}}{dt} = q(\mathbf{E} + \mathbf{v} \times \mathbf{B}) \quad (3.25)$$

and $\mathbf{p}$ is the relativistic momentum:

$$\mathbf{p} = \gamma m \mathbf{v}. \quad (3.26)$$

The first component ($\mu = 0$) of eqn 2.54 is the relativistic version of the work-energy theorem in classical physics and, in NU, reads:

$$\frac{dE}{d\tau} = q\mathbf{v} \cdot \mathbf{E}. \quad (3.27)$$

In Anderson's experiment, $\mathbf{E} = 0$, $q(\mathbf{E} + \mathbf{v} \times \mathbf{B}) = q|\mathbf{v}|B$ and

$$\frac{dE}{d\tau} = \dot{\gamma}m = 0. \tag{3.28}$$

γ is, therefore, a constant during the motion of the particle. This is equivalent to the classical statement that magnetic fields do no work and that the variations of the particle kinetic energy are only due to the work of the electric field. Since γ is constant, eqns 3.25 and 3.26 give:

$$\frac{d\mathbf{v}}{dt} = \mathbf{v} \times \frac{e\mathbf{B}}{\gamma m} \equiv \mathbf{v} \times \boldsymbol{\omega}_{\mathbf{B}} \tag{3.29}$$

where $\boldsymbol{\omega}_B$ is called the **precession** or **gyration frequency**. We can rewrite eqn 3.29 in SI as:

$$\frac{d\mathbf{v}}{dt} = \mathbf{v} \times \frac{e\mathbf{B}}{\gamma mc} \equiv \mathbf{v} \times \boldsymbol{\omega}_{\mathbf{B}} \tag{3.30}$$

and define $\boldsymbol{\omega}_{\mathbf{B}}$ as $(eB/\gamma mc)\,\mathbf{k}$. The motion of the particle is a uniform circular motion in the $x-y$ plane. The radius of the circular trajectory is R and $\omega_B \equiv |\boldsymbol{\omega}_{\mathbf{B}}| = |\mathbf{v}|/R$. In a uniform circular motion, the acceleration is centripetal ($\mathbf{a} = \mathbf{a}_\perp$) and is given by $|\mathbf{v}|^2/R$. Therefore, $\mathbf{F} = \gamma m|\mathbf{v}|^2/R$ and:

$$q|\mathbf{v}|B = \gamma m \frac{|\mathbf{v}|^2}{R}. \tag{3.31}$$

Taking the modules:

$$|\mathbf{p}| = \gamma m|\mathbf{v}| = qBR. \tag{3.32}$$

This SI formula can be expressed in practical units converting the momentum of the particle from kg m/s (i.e. J·s/m) to GeV/c and considering q in units of the electron charge ($q = ze$). A GeV is 10^9 e Joule where e is the size of the electric charge in SI units (1.6×10^{-19}). So

$$|\mathbf{p}|[\text{GeV/c}] \cdot 10^9 \cdot \frac{e}{c} = zeBR \tag{3.33}$$

$$|\mathbf{p}|[\text{GeV/c}] = \frac{c}{10^9}\, zBR \simeq 0.3\, z\, B[\text{T}]\, R[\text{m}]. \tag{3.34}$$

In conclusion,

---

for a particle of charge z in units of e, running perpendicularly to $\mathbf{B}$:
$$|\mathbf{p}|(\text{GeV/c}) = z\, 0.3\, B(\text{T})\, R(\text{m}). \tag{3.35}$$

Note that eqn 3.35 provides the momentum relying on neither the knowledge nor the measurement of the particle mass but only on the **curvature radius** R. If the particle momentum is not perpendicular to the magnetic field, eqn 3.35 gives only the size of the momentum projection in the plane perpendicular to $\mathbf{B}$:

$$|\mathbf{p}_\perp|(\text{GeV/c}) = z\, 0.3\, B(\text{T})\, R(\text{m}). \tag{3.36}$$

The cloud chamber used by Anderson can determine $|\mathbf{p}_\perp|$ by measuring the radius of curvature of the trajectory.

Anderson mounted a 6 mm lead plate in the middle of the chamber. He randomly opened and closed the diaphragm taking hundreds of pictures and analyzed the photographs. In those pictures, Anderson found several particles of charge $z = +1$, whose energy loss was incompatible with a proton. One of those tracks, nearly perpendicular to $\mathbf{B}$ (Fig. 3.4), entered from the lateral side and crossed the chamber from below. The value of $|\mathbf{p}_\perp| \simeq |\mathbf{p}|$ before crossing the plate was 63 MeV/c. After crossing the plate, $|\mathbf{p}|$ went down to 23 MeV/c. If the particle were a proton, the final kinetic energy (in NU) would have been $E_{kin} = (|\mathbf{p}|^2 + m_p^2)^{1/2} - m \simeq 282$ keV. This is the kinetic energy of a very slow ($\beta = 0.02$) non-relativistic proton, whose range in air is ~ 5 mm. This range was incompatible with the length of the track after the plate, which exceeded 50 mm. The particle observed by Anderson resembled a "positive electron." A 63 MeV/c electron is relativistic and the energy loss in lead is dominated by radiation losses since $E_c \simeq 7$ MeV and $E = (|\mathbf{p}|^2 + m_e^2)^{1/2} \equiv E_0 \simeq 63$ MeV. After the plate, a 23 MeV momentum electron has an energy of $E = (|\mathbf{p}|^2 + m_e^2)^{1/2} \equiv E_f \simeq 23$ MeV. The observed energy loss was compatible with the radiation loss in lead:

$$E = E_0 e^{-6 \text{ mm}/X_0} \simeq 22 \text{ MeV} \tag{3.37}$$

with $X_0 = X_0^{lead} = 5.6$ mm. In addition, a 23 MeV electron has a range in air of about 50 mm, in agreement with the observed length of the track. This "positive electron" was later named **positron** and we know now that is the antiparticle of the electron.

A full understanding of the origin of the positron took more than 20 years and the origin of antiparticles has been clarified in the late 1950s, as discussed in Chap. 5. In general, the **antiparticle** of a particle of mass m and charge q is a particle with mass m and charge $-q$. The antiparticle of the electron is called an anti-electron or positron and its charge is $q = +e = 1.6 \times 10^{19}$ C. Note that the charge of the positron is incidentally the same as the charge of the proton but the positron and the proton are two very different objects.

3.4   Interaction of photons with matter

The interaction of light with matter strongly depends on the wavelength of the associated photon. If the wavelength is macroscopic (radio waves, sub-mm, infrared), we can observe a direct interaction of the electric field of the wave with the medium, which produces a wealth of optical effects. At smaller wavelengths, quantum optical effects are dominant and the interaction of light is better described by the scattering or absorption of the photons. The absorption of photons may lead to a chemical change of the medium (photography, the eye), modulate the electric current in a material (photodiodes, superconductors, etc.), or can dislodge an electron from an atom. The latter is called **photoelectric effect** or photoemission and is the most efficient process to transfer energy from a neutral to a charged particle.

3.4.1 Photoelectric effect

A transmutation of a photon into an electron is forbidden both by SR kinematics and by conservation of charge, but a photon can be absorbed by an atom bringing a bound electron to a state belonging to the continuum and, hence, producing an ion–electron pair. The kinetic energy of the outgoing electron is $E_{kin} = h\nu - E_b$, where E_b is the binding energy of the electron in the atom. Photoemission takes place in any material, including noble gases and insulators, even if it was discovered for the first time in metals (Serway and Jewett, 2018). The binding energy in metals is called the **work function**, which is the minimum energy required to remove an electron from the metal surface.

The photoelectric effect is a genuine quantum effect. Unlike charged particles, the energy of the photon is not degraded along its path in the medium but the photons can be absorbed. Hence, the intensity of the photon beam is degraded following an exponential law:

$$I = I_0 e^{-\mu x}. \tag{3.38}$$

In this case, I_0 is the initial number of photons per unit time and x is the thickness of the medium. The μ coefficient is called the **absorption coefficient** and is equal to the total absorption cross-section (see Sec. 3.4.3) multiplied by the density of the scattering centers. For the photoelectric effect, the scattering centers are the atoms of the medium and the cross-section depends on the wavefunction of the bound atomic electrons. Its derivation is quite complicated. The shape of the photoelectric cross-section in lead is shown in Fig. 3.5. The cross-section decreases as the energy of the incoming photon increases but it is not a smooth function: we observe a sharp increase of the cross-section when the energy of the photon corresponds to the binding energy of the electron in the atom and the photon undergoes resonant absorption. The resonance at the largest energy corresponds to electrons that were located at the $n = 1$, $l = 0$ atomic shell. This resonance is named the **K-edge**. Other resonances are visible at lower energies. The resonances corresponding to $n = 2$ are called **L-edges**. For photon energies well above the K-edge, the photoelectric cross-section decreases as $1/E$. The cross-section is a smooth function away from resonances and the asymptotic behavior is:

$$\sigma_{pe}(E) \simeq 0.665 \, \frac{\sqrt{32} \, \alpha^4 Z^5}{(E/m_e c^2)^{7/2}} \, \text{barn} \quad \sim \frac{Z^5}{E^{7/2}} \quad \text{if } E \ll m_e \tag{3.39}$$

and

$$\sigma_{pe}(E) \simeq 0.665 \, \frac{3\alpha^4 Z^5}{(2E/m_e c^2)} \, \text{barn} \quad \sim \frac{Z^5}{E} \quad \text{if } E \gg m_e. \tag{3.40}$$

In these formulas, α is the fine structure constant and Z the atomic

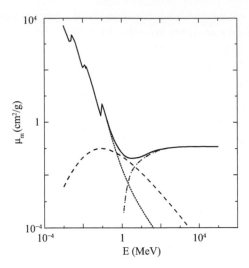

Fig. 3.5 The mass attenuation coefficient $\mu_m \equiv \mu/\rho$ (see eqns 3.38 and 3.53) in cm^2/g versus the photon energy in MeV. At low energy, the total cross-section and, hence, μ_m (thick line) are dominated by the photoelectric effect (dotted line). The Compton and pair production μ_m are depicted with a dashed and dot-dashed line, respectively. The pair production cross-section is the leading contributions at high energy.

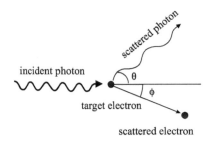

Fig. 3.6 Kinematics of the Compton scattering: a scattering of a photon off an electron at rest in LAB.

[4]In classical optical diffraction, the incoming and outgoing wave frequencies are the same (Brooker, 2003).

number of the atom that absorbs the photon. The photoelectric effect is, then, strongly enhanced if the medium has a high atomic number Z.

3.4.2 Compton scattering

In classical electrodynamics, 0.665 barn corresponds to the cross-section of the elastic scattering of a photon to a free electron. This number is called the **Thomson cross-section**:

$$\sigma_{Th} = \frac{8}{3}\pi r_e^2 \tag{3.41}$$

where r_e is the classical electron radius. In SI, $r_e = e^2/4\pi\epsilon_0 m_e c^2$ and $\sigma_{Th} \simeq 6.65 \times 10^{-25}$ cm^2 (0.665 barn). The Thomson scattering is the classical limit of the **Compton scattering**, which is the scattering of photons off free atomic electrons (see Fig. 3.6).

In most materials, the Compton scattering is a competing process of the photoelectric effect if the energy of the photon is above the K-edge. Like photoemission, the Compton scattering is a genuine quantum effect because the frequency of the scattered light is smaller than the incoming light[4] and there is a net energy transfer between the photon and the electron. The energy of the outgoing photon is a function of the initial photon energy and the scattering angle θ can be derived by four-momentum conservation in SR (Exercise 3.13). In NU,

$$E' = \frac{E}{1 + \frac{E}{m_e}(1 - \cos\theta)} \tag{3.42}$$

and the maximum energy transfer occurs when $\theta = \pi$. The $\theta = \pi$ scattering corresponds to a photon bouncing back after a head-on collision with a free electron. The corresponding cross-section was derived in 1928 by O. Klein and Y. Nishina and is considered one of the first achievements of quantum electrodynamics. The total Klein–Nishina cross-sections over an atom of atomic number Z is:

$$\sigma_c^{atom} = \frac{3}{4} Z \sigma_{Th} \left[\frac{1+\epsilon}{\epsilon^2} \left(\frac{2(1+\epsilon)}{(1+2\epsilon)} - \frac{\log(1+2\epsilon)}{\epsilon} \right) \right. \tag{3.43}$$

$$\left. + \frac{\log(1+2\epsilon)}{2\epsilon} - \frac{1+3\epsilon}{(1+2\epsilon)^2} \right] \tag{3.44}$$

and the asymptotic behavior is

$$\sigma_c^{atom} \sim Z \frac{\log E}{E} \quad \text{if } E \gg m_e \tag{3.45}$$

and

$$\sigma_c^{atom} \sim Z \sigma_{Th} \quad \text{if } E \ll m_e \tag{3.46}$$

where $\epsilon = E/m_e$. Note that, unlike the photoelectric cross-section, the total Compton cross-section per atom is simply the cross-section on a single electron summed over all the atomic electrons: $\sigma_c = Z \sigma_c^{electron}$. The Compton scattering is a zero threshold process. It means that $\gamma + e^- \rightarrow \gamma + e^-$ can occur even if the energy of the photon is very small because $\sigma_c^{electron} \rightarrow \sigma_{Th}$ if $E \rightarrow 0$. This is not the case for the photoelectric effect where the energy must be larger than the binding energy of the electrons in the outermost atomic shell.

3.4.3 Pair production

If the energy of the photon is significantly larger than the electron mass, a new interaction channel opens up. The **pair production** of an electron and a positron is a $\gamma \rightarrow e^+ e^-$ conversion. Pair production is kinematically forbidden in vacuum by four-momentum conservation (see Fig. 3.7). If it were possible, the CM frame of the $e^+ e^-$ pair would see the photon at rest, which contradicts SR. Such a contradiction can be appreciated also in the LAB frame, where the incoming photon four-momentum is $p^\mu = (p, 0, 0, p)$ and p is a positive number. After the conversion, the four-momenta of the electron and positrons are

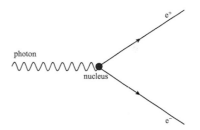

Fig. 3.7 A conversion of a photon in a $e^+ e^-$ pair is kinematically forbidden in vacuum. The plot shows the conversion in the presence of a nucleus at rest in LAB.

$$p^- = \left(\sqrt{(p_x^-)^2 + (p_y^-)^2 + (p_z^-)^2 + m^2}, p_x^-, p_y^-, p_z^- \right)$$

$$p^+ = \left(\sqrt{(p_x^+)^2 + (p_y^+)^2 + (p_z^+)^2 + m^2}, p_x^+, p_y^+, p_z^+ \right).$$

The conservation of the $i = 1, 2, 3$ components gives $p_x^- = -p_x^+$, $p_y^- = -p_y^+$ and $p = p_z^- + p_z^+$. The conservation of energy (the $\mu = 0$ component) gives:

$$p = \sqrt{(p_x^-)^2 + (p_y^-)^2 + (p_z^-)^2 + m^2} + \sqrt{(p_x^+)^2 + (p_y^+)^2 + (p_z^+)^2 + m^2}$$

$$= p_z^- + p_z^+. \tag{3.47}$$

Equation 3.47 never holds because:

$$\sqrt{(p_x^-)^2 + (p_y^-)^2 + (p_z^-)^2 + m^2} > p_z^-$$

and

$$\sqrt{(p_x^+)^2 + (p_y^+)^2 + (p_z^+)^2 + m^2} > p_z^+ \ .$$

The process is thus forbidden by kinematics. In the equations above, we labeled the electron mass m and we exploited the equality of the rest mass of particles and antiparticles: $m = m_{e^-} = m_{e^+}$. Pair production can take place in matter because the recoil momentum is transferred to either one of the atomic electrons or to the nucleus. The threshold of the process depends on the recoiling particle. As discussed in Chap. 2, the threshold for a process corresponds to the energy that produces all final state particles at rest in CM. For a $\gamma + X \to e^+ e^- X$ conversion, the four momenta of the final state particles are $p^+ = (m, 0, 0, 0)$ (positron), $p^- = (m, 0, 0, 0)$ (electron), $p'_X = (M, 0, 0, 0)$ (recoiling particle). The squared invariant mass of the final state is thus $s = (2m + M)^2$. In the laboratory frame, the four-momenta of the initial state particles are $p_\gamma = (p, 0, 0, p)$ (photon) and $p_X = (M, 0, 0, 0)$. The squared invariant mass of the initial state is then $s = (p + M)^2 - p^2 = 2pM + M^2$. Since the invariant mass is a Lorentz scalar, it has the same numerical value in all inertial frames. As a consequence:

$$s_{ini}^{CM} = s_{ini}^{LAB} = (2m + M)^2 = s_{fin}^{CM} = s_{fin}^{LAB} = 2pM + M^2 \qquad (3.48)$$

and, therefore,

$$p = \frac{2m(m + M)}{M}. \qquad (3.49)$$

If the recoiling particle is a nucleus, $M \gg m$ and, then, $p \simeq 2m$. In the laboratory frame, the recoil momentum of the nucleus N is negligible with respect to its mass and the observer sees the nucleus at rest even after the conversion. The $\gamma + N \to e^+ e^- N$ process is similar to the scattering of a classical projectile on a hard sphere (see Sec. 2.7). In this case, a photon can convert to an $e^+ e^-$ pair if the energy is larger than the sum of the rest energy of the electron and the positron. In CM, it corresponds to complete conversion of energy into rest mass and provides another spectacular test of the Einstein energy–mass relation. If the recoiling particle is an atomic electron, $M = m$, and $p = 4m$. In this case, the initial state electron recoils after the scattering in the laboratory frame and a significant fraction of the energy of the photon is transferred in the form of e^- kinetic energy. Pair production in the Coulomb field of the electrons is, however, strongly suppressed compared with pair production in nuclei, where the electric charge is concentrated in a $(10^{-15}$ m$)^3$ volume.

The pair production threshold can be considered $2m$ for any practical purpose.

The $\gamma \rightarrow e^+e^-$ cross-section increases as Z^2 and the asymptotic behavior is:

$$\sigma_{pair}(E) \simeq 4\alpha^2 r_e^2 Z^2 \left(\frac{7}{9} \log 2\epsilon - \frac{109}{54} \right) =$$
$$= \frac{3}{2\pi} \sigma_{Th} Z^2 \left(\frac{7}{9} \log 2\epsilon - \frac{109}{54} \right) \text{ cm}^2$$
$$\text{if } E \simeq 2m_e \qquad (3.50)$$

and

$$\sigma_{pair}(E) \simeq 4\alpha^2 r_e^2 Z^2 \left(\frac{7}{9} \log \frac{183}{Z^{1/3}} - \frac{1}{54} \right) \text{ cm}^2$$
$$\text{if } E \gg 1/(\alpha Z^{1/3}). \qquad (3.51)$$

In practice, if the photon has an energy of several MeV, the pair production cross-section is energy independent:

$$\sigma_{pair}(E) \simeq 4\alpha^2 r_e^2 Z^2 \left(\frac{7}{9} \log \frac{183}{Z^{1/3}} \right) \simeq \frac{7}{9} \frac{A}{N_A} \frac{1}{X_0} \qquad (3.52)$$

(N_A is the Avogadro number in atoms/mole, X_0 the radiation length in g/cm^2, and A is the mass number in g/mole), and pair production completely dominates over the Compton and photoelectric cross-sections. In summary:

> well below the threshold for pair production, the photoelectric effect is the leading photon interaction process. Compton scattering is the most probable process in the 100 keV-1 MeV energy range, especially for low-Z materials. Above a few MeV, the interactions of photons are mostly $\gamma \rightarrow e^+e^-$ conversions.

The total cross-section per atom σ_t is the sum of the photoelectric, Compton, and pair production cross-section. The **total absorption coefficient** μ of a photon is thus σ multiplied by the density of atoms in the material. The mass attenuation coefficient mentioned in Fig. 3.5 is $\mu_m = \mu/\rho$. Therefore,

$$\mu = \tilde{N}\sigma_t = \rho \frac{N_A}{A}\sigma_t = \rho\mu_m \qquad (3.53)$$

where N_A is the Avogadro number, A the atomic mass in mole per gram and ρ is the density in g/cm^3. The average distance between two photon interactions is then:

$$\lambda_t = (\tilde{N}\sigma_t)^{-1}. \qquad (3.54)$$

If we consider only the pair production cross-section:

$$\lambda_p = (\tilde{N}\sigma_p)^{-1} = \frac{9}{7}\frac{X_0}{\rho}. \qquad (3.55)$$

if X_0 is expressed in mass thickness (g/cm^2). λ_p is called the **pair production length**. It represents the average depth where a beam of photons reduces its intensity to $e^{-1} \simeq 1/3$ of the initial value due to $\gamma \to e^+ e^-$ conversions. Note that the physical interpretation of λ_p is somehow different from X_0 in units of length. The former is the distance needed for a $1/e$ reduction of the *intensity* of a photon beam. The latter is the distance needed for a $1/e$ reduction of the *energy* of an electron due to radiation losses. However, both processes are quantum effects and undergo statistical fluctuations. λ_p and X_0 represent average lengths and should not be interpreted as classical trajectories: pair production and radiation losses (bremsstrahlung) are thus genuine stochastic processes.

3.5   The simplest particle detector

The Bethe formula provides the average energy per unit length deposited in matter by the passage of a charged particle. An atomic electron is ejected from the atom (i.e. brought to the continuum) if the transferred energy is larger than the binding energy. In this case, the energy transfer results in the production of an ion–electron pair. If the energy is smaller than the binding energy, the atom may reach an excited state and returns to the ground state emitting one or more photons. **Ionization detectors** are particle detectors that produce an electrical signal proportional to the number of ion–electron pairs. These devices are not fully efficient because atomic excitations are not detected and the corresponding energy is lost. On the other hand, ionization detectors are the simplest particle detectors and are still widely used in our field. The materials employed in ionization detectors are noble gases with small quantities of electronegative compounds. The average energy needed to produce an ion–electron pair, for instance, in argon is $w = 26$ eV. w is much larger than the first ionization potential of argon (15.8 eV) because a large fraction of the particle energy is wasted in excitation processes or used to bring to the continuum electrons from inner shells, which have larger binding energy than 15.8 eV. Noble gases are the material of choice because the outer shells are completely filled and the electron affinity is very small (-1.0 eV for Ar). If the detector volume is located inside a static electric field, an electron produced in a noble gas can drift up to macroscopic distances without being absorbed by electronegative atoms. The simplest particle detector can be built from a flat capacitor replacing the dielectric of the capacitor with argon. The capacitor provides a static electric field. Electrons drift toward the anode of the capacitor and ions toward the cathode. If the electric field is ~ 1 kV/cm, the drift velocity of an electron is of the order of 1 cm/μs. If the field is sufficiently large, the **drift velocity** saturates and does not depend on the intensity of the field. As a consequence, the electrons and the ions drift at constant speed and are not accelerated inside the device. We can get large and stable drift velocities if we add to argon small quantities of electronegative compounds as isobutane or methane. The drift velocity of ions is about three orders of magnitude slower than electrons: ~ 1 cm/ms.

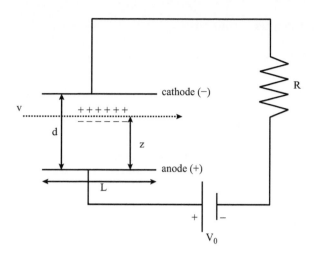

Fig. 3.8 A particle crosses an ionization chamber with a velocity **v** at a distance z from the anode. The chamber is composed by a plane capacitor, a load resistance R, and a voltage generator V_0. d is the distance between the cathode and the anode.

An isolated charged capacitor cannot be operated as a particle detector for a long time because the charge drifting to the plates neutralizes the capacitor and cancels out the electric field. If the capacitor is in series with a voltage generator and a resistance, the voltage drop at the resistance is proportional to the charge produced by the passage of the particle in argon. This detector is one of the first particle detectors ever build. It was developed by P. Villard in 1908 and is called a **ionization chamber**. If a charged particle crosses the capacitor at a distance z from the anode (Fig. 3.8), the number of ion–electron pairs produced in argon is:

$$n_e = n_{ion} \equiv n = \frac{dE}{d\tilde{x}} \frac{L\rho}{w} \qquad (3.56)$$

where L is the length of the capacitor, ρ is the density of the gas, and $dE/d\tilde{x}$ is given by the Bethe formula in mass thickness.[5]

For a mip in argon, $dE/d\tilde{x} \simeq 1.5$ MeV/(g/cm^2) and $dE/d\tilde{x} \cdot \rho \simeq 2.5$ keV/cm. In a 10 cm capacitor, $n \simeq 10^3$. This is a tiny amount of charge compared with the macroscopic charge stored in a capacitor. As a consequence, the passage of a single particle does not change the voltage at the capacitor even if the detector is completely isolated. The electric field sensed by the electron–ion pairs during the drift is thus constant: $\mathbf{E}(t) \simeq |\mathbf{E}|\mathbf{k}$. In a time t, the electrons travel a distance v_-t and the ions v_+t. Inside the plates of a flat capacitor, the electric field is $\mathbf{E} = E\mathbf{k}$ and $E = V(t)/d$, where $V(t)$ is the voltage difference at the capacitor and d is the distance between the two plates. The drift of the particle over a distance d of a 1 cm takes about 1 μs for the electrons and $\sim$ 1 ms for the ions. From the electrical point of view, an ionization chamber is an RC circuit connected to a voltage generator. C is given by the detector capacitance and, if needed, can be tuned by adding in parallel other parasitic capacitors. If the time constant of the circuit $\tau = RC$ is larger than a few ms, no appreciable current can flow in the circuit during the drift of the charges and the energy needed to move the particles in argon must come from the energy stored in the capacitor: $CV_0^2/2$. The energy

[5]The energy loss $dE/d\tilde{x}$ is generally expressed in MeV·cm^2/g and, therefore, the density is given in g/cm^3, the length in cm and w in MeV.

of an electron after a time t is

$$q \int \mathbf{E} \cdot d\mathbf{x} = (-e) \cdot -|\mathbf{E}|v_- t = e|\mathbf{E}|v_- t \qquad (3.57)$$

and the energy of an ion is:

$$q \int \mathbf{E} \cdot d\mathbf{x} = e \cdot |\mathbf{E}|v_+ t = e|\mathbf{E}|v_+ t. \qquad (3.58)$$

The conservation of energy in the (isolated) capacitor gives:

$$\frac{1}{2}CV_0^2 = ne|\mathbf{E}|v_- t + ne|\mathbf{E}|v_+ t + \frac{1}{2}CV_f^2 \qquad (3.59)$$

where V_f is the voltage at time t on the capacitor. The voltage drop measured at the resistance is $V_R(t) = V_0 - V_f$ and for $n \simeq 10^3$, $V_R(t) \ll V_0$. Therefore,

$$\frac{1}{2}C(V_0^2 - V_f^2) = \frac{1}{2}C(V_0 - V_f)(V_0 + V_f) = ne|\mathbf{E}|(v_- + v_+)t. \quad (3.60)$$

Here, $|\mathbf{E}| \simeq V_0/d$ and $(V_0 - V_f)(V_0 + V_f) \simeq 2V_0 V_R(t)$ so that:

$$V_R(t) = \frac{ne}{dC}(v^+ + v^-)t. \qquad (3.61)$$

At the time $t_- = z/v_-$ all electrons reach the anode (i.e. the plate at higher potential) and do not contribute anymore to the signal (see Fig. 3.9). Note, however, that these electrons do not flow through the circuit, either, as long as $t_- \ll RC$. For $t > t_-$

$$V_R(t) = \frac{ne}{dC}(v^+ t + z) . \qquad (3.62)$$

For $t > t_+ = (d - z)/v^+$ all ions reach the cathode and

$$V_R(t) = \frac{ne}{dC}[(d - z) + z] = \frac{ne}{C} \qquad (3.63)$$

Equation 3.63 is a remarkable result: once all charges reach the plates, the voltage drop at the resistance is the total charge produced by the passage of the particle divided by the capacitance of the detector. The initial conditions $V_f = V_0$ and $V_R(t) = 0$ are restored for $t \simeq RC$ because the voltage generator neutralizes the charges collected in the plates through a current flowing in the circuit.

If the rate of the incoming particles is high, we may want to avoid the pile-up of the electrical signals produced by each particle. In this case, the rate of charged particles in the capacitor must be smaller than $(RC)^{-1}$. An ionization chamber sensitive to all charges produced by the particle (electrons and ions) cannot be operated at rates much larger than $(1 \text{ ms})^{-1} = 1$ kHz. If RC is chosen between t_- and t_+, the signal induced by the ions is lost but the chamber can be operated at $\sim (1 \text{ } \mu s)^{-1} = 1$ MHz. This operation mode is called electron-sensitive mode.

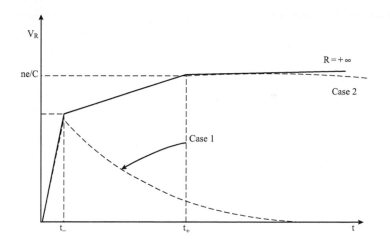

Fig. 3.9 Voltage drop at the load resistance R as a function of time for an ionization chamber. The first segment corresponds to the drift inside the chamber of both ions and electrons. The second segment starts at $t = t_-$ and ends at $t = t_+$ (see text). For $t > t_+$ the voltage is constant except for the recovery time of the circuit τ ("Case 2" in the figure). Shorter recovery times ("Case 1" in the figure, corresponding to $t_- \ll RC \ll t_+$) can be used to increase the rate capability of the detector by recording only the signal of the electrons or part of the ions. For an infinite recovery time ($R \to +\infty$), $V_R = V_R^{max} = ne/C$.

If the resistance of the circuit were zero, the voltage generator would neutralize immediately any voltage drop at the capacitor so that $V_R(t) \to 0$ for $R \to 0$. If the resistance is too large, the signals of the particles pile up because V_0 is never restored. $R \to \infty$ is equivalent to an open circuit and the capacitor is isolated from the voltage generator forever. This is why R is called the **quenching resistance**: it quenches the accumulated charge in the detector at a rate that depends on RC (the **discharge time**).

Ionization chambers are simple and effective. Their main drawback is the smallness of n, which results in a charge of $ne \simeq 10^{-16}$ C (0.1 fC). Nowadays, these detectors are used mainly for dosimetry: if the particle flux is large, the signals pile up and the average current on R provides an estimate of the incoming flux. In 1912, **cosmic rays** (see Chap. 11) were discovered using an early version of the ionization chamber, where the steady current was measured by an electrometer. V. Hess showed that the flux of charged particles at 5 km height increases fourfold over the flux at sea level. Measuring during a solar eclipse, he demonstrated that the flux was not originating from the sun. We know now that this flux is mainly due to protons produced in our galaxy that reach the upper atmosphere from any direction and generate secondary particles: most of these particles are pions and muons (heavy charged particles), electrons, high-energy photons, and the memorable positrons observed for the first time by Anderson, Blackett, and Occhialini.

3.6 Detector technologies

Detector science is a major subfield of particle and nuclear physics and there are a plethora of devices that suit the needs of pure and applied science. Most of them are specialized detectors to measure the passage of the particles (trackers), the energy (calorimeters, range meters, etc.), the mass-to-charge ratio (mass spectrometers), the charge sign, and momentum (trackers in magnetic fields, range meters), or perform particle identification. A systematic review is outside of the scope of this book but it is useful to know that the detector technologies can often be classified looking at the w parameter introduced in Sec. 3.5. w is the average energy needed to create one **elemental event** in a device that produces a visible signal. For the ionization chamber, the elemental event is the creation of an ion–electron pair, which requires an average energy deposit of $w \simeq 26$ eV in argon. If it were possible to design detectors with a much smaller w, both the detection efficiency and the energy resolution would improve substantially because these quantities depend on the statistical fluctuations in the number of elemental events. A rigorous demonstration is given in Sec. 3.7.1. We can guess that the smaller the w, the higher the complexity of the detector so that a careful choice of the technology is paramount in the design phase of the experiments. Looking at w, we can group the technologies into four broad classes:

- **Gaseous detectors**: like the ionization chamber, the elemental event of the gaseous detectors is the formation of an ion-electron pair in a noble gas. These devices operate with electric fields that drift the ion and the electrons toward a cathode and an anode, respectively. $w = \mathcal{O}(30)$ eV in this class of devices.

- **Scintillators**: Scintillators are insulators that produce visible or ultraviolet (UV) light after the passage of a charged particle. This effect is called luminescence and – in most cases – originates from additional energy levels between the valence and conduction band (i.e. in the **forbidden band** of the insulator). The energy deposited by the particle moves valence electrons into the conduction band. These electrons decay in one of the levels of the forbidden band. Then, the electrons go back to the valence band producing photons that are not reabsorbed in the scintillator because their energy is smaller than the band-gap of the insulator. The scintillators can be based on organic and inorganic compounds within a broad range of w, which is generally larger than gaseous detectors. The best scintillators need $w \simeq 25$ eV per visible photon but generally $w_{scint} = \mathcal{O}(100)$ eV. The scintillators are solid (or liquid) state devices and, therefore, surpass gaseous detectors in efficiency thanks to the higher density because $dE/dx \sim \rho$. Besides, the visible photons can be observed by **photomultipliers** (PMT) or other photosensors that can detect even a single photon with an efficiency of 30–50%. The first scintillator ever built produced light visible by the naked eye or by microscopes and was used in the famous Geiger–Marsden experiments that were fundamental to the Rutherford atomic model.

- **Semiconductor detectors** are based on semiconductors, where the energy losses produce **electron-hole pairs** as elemental events. In this case, an electron in the valence band moves to the conduction band leaving an excess of positive charge (a "hole") that drifts like a real positive particle. Electrons and holes drift to the anode and the cathode like in gaseous detectors and induce a signal in the load resistance, which is proportional to the number of electron–hole pairs. The w for the production of an electron–hole pair is 3.6 eV in silicon and 2.9 eV in germanium. Silicon detectors are by far the most common devices because they can be operated at room temperature, while Ge detectors must be cooled and are employed especially for high-precision photon measurements. The strengths of semiconductor detectors are the low w, the possibility of miniaturization, which is essential in detectors with a high spatial resolution (see Sec. 4.6.4), and the small operation voltage (tens of volts compared with $\mathcal{O}(1)$ kV in gaseous detectors and PMTs). To date, only good old photographic films (**nuclear emulsions**) can beat the micrometric precision of silicon devices. Unlike nuclear emulsions, silicon detectors generate as output an electrical signal that can be read and analyzed in real time and can withstand rates up to a few GHz/cm^2.

- **Thermal detectors** use as an elemental event the production of phonons in solid-state crystals, that is the heating that results from the energy loss of the incoming particle. Their w is incredibly small (tens of μeV) and thermal detectors are potentially the most precise devices for energy measurements. The phonons produce an increase of temperature in the device that is sensed by high-precision thermometers. Since they must be operated at extremely low temperatures (about 10 mK), the cooling infrastructure is very complex and these detectors are used only for special applications that require an outstanding energy resolution (low-energy rare event searches, the study of the cosmic microwave background, metrology, etc.)

The reader is warned that this classification is oversimplified. Some detectors (e.g. the Cherenkov detectors described in Sec. 13.6.1, transition radiation detectors, devices based on superconductors, etc.) do not fit any of these classes.

3.7 Gaseous detectors

To overcome the limitation of ionization chambers, n – the number of ion–electron pairs – must be amplified inside the gas. Signal multiplication can be achieved in argon if the electric field is so large that the kinetic energy gained by an electron in a mean-free path is larger than the first ionization potential. Above this value, an electron can ionize an atom and produce secondary electron–ion pairs. The number of e^- and ions, then, grows exponentially after just a few mean-free paths. In this **avalanche multiplication** regime:

$$n = n_0 e^{\alpha x} \qquad (3.64)$$

after a length x. α is a constant that depends on the gas and is called the **first Townsend coefficient**. The simplest way to reach such a

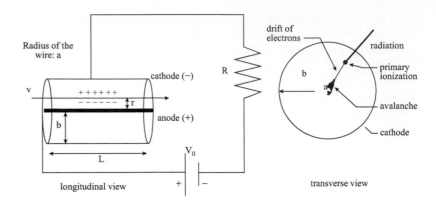

Fig. 3.10 Longitudinal (left) and transverse (right) view of a proportional counter. The first ion–electron pairs are produced by the passage of the particle with velocity v ("radiation" in the figure on the right) inside the counter. The electrons move toward the wire of radius a and the avalanche occurs in the proximity of the wire.

high field is to replace the flat capacitor with a cylindrical capacitor (see Fig. 3.10). The anode of the capacitor is a metallic wire with a diameter of $\sim 10~\mu$m. The electric field is thus:

$$\mathbf{E} = \frac{V_0}{\log(b/a)} \frac{1}{r} \hat{\mathbf{r}} \tag{3.65}$$

where a is the inner radius (the radius of the wire) and b the outer radius of the cylinder. $\hat{\mathbf{r}}$ is the radial unit-vector in cylindrical coordinates and V_0 is the voltage at the capacitor. If a is very small, the maximum field is very large. Since $\mathbf{E} = -\nabla\phi$

$$\phi(r) = -\frac{V_0}{\log(b/a)} \log\left(\frac{r}{a}\right). \tag{3.66}$$

If the dielectric of the cylinder is a noble gas or a gas mixture, the capacitor can be operated as a **proportional counter**. For the sake of simplicity, let us assume that a charged particle crosses the cylinder in the direction parallel to the wire at a distance r_0, as shown in Fig. 3.10. The electrons drift radially toward the anode and the ions move toward the outer radius b. In the proximity of the wire, the avalanche multiplication takes place and:

$$n = n_0 e^{\alpha r'} = n_o G. \tag{3.67}$$

$r' + a$ is the radius where the avalanche starts and in most proportional counters $a \simeq a + r' \ll b$ and $r' = \mathcal{O}(1)~\mu$m. G ranges between 10^3 and 10^6, and is called the **gain** of the counter. The computation of the signal amplitude V_R at the resistance R of Fig. 3.10 can be done as in the ionization chambers. The energy transferred from an isolated capacitor to a particle of charge q after a displacement dr is:

$$dE = q\,d\phi = q\left(\frac{d\phi}{dr}\right) dr \tag{3.68}$$

and the energy of the capacitor is reduced by $d(CV_0^2/2) = CV_0 dV$.

Unlike the ionization chamber, in a proportional chamber nearly all n charges are produced in the proximity of the wire. The path traveled by the electron is very small $(r' - a)$ compared with the path of the ion $(b - r' - a \simeq b)$ and the travel time is even smaller since $v_- \ll v_+$. As a consequence, the voltage drop $V_-(t)$ produced in R by the drift of the electrons is completely negligible with respect to the voltage drop $V_+(t)$ due to the ions. We show it starting from the drop dV due to a small displacement of the electrons:

$$dV = \frac{(-e)}{CV_0} \frac{d\phi(r)}{dr} dr; \tag{3.69}$$

the voltage drop at R when all the electrons reach the anode at $t_- = r' - a/v_-$ is:

$$V_-(t_-) = -\frac{e}{CV_0} \int_{a+r'}^{a} \frac{d\phi}{dr} dr. \tag{3.70}$$

The capacitance of the cylinder is $C = 2\pi\epsilon l / \log(b/a)$ where l is the length of the cylinder and $\epsilon = \epsilon_0 \epsilon_r$ is the absolute permittivity of the gas. From eqn 3.66:

$$V_-(t_-) = -\frac{e}{2\pi\epsilon l} \log\left(\frac{a+r'}{a}\right). \tag{3.71}$$

For an ion:

$$V_+(t_+) = -\frac{e}{2\pi\epsilon l} \log\left(\frac{b}{a+r'}\right). \tag{3.72}$$

Since $n = n_- = n_+ = Gn_0$, the total contribution is

$$V^{fin} = V_-^{fin} + V_+^{fin} =$$

$$-\frac{ne}{2\pi\epsilon l} \left[\log\left(\frac{a+r'}{a}\right) + \log\left(\frac{b}{a+r'}\right) \right] = -\frac{ne}{C}. \tag{3.73}$$

The final voltage drop $(t > t_+)$ is the same as the ionization chamber but the amount of charge is increased by a factor G. This is the reason why G is called the *gain* of the detector: the proportional counter works as an intrinsic signal amplifier with a gain G. As expected, V^{fin} is completely dominated by the ions since:

$$\frac{V_-^{fin}}{V_+^{fin}} = \frac{\log\left(\frac{(a+r')}{a}\right)}{\log\left(\frac{b}{a+r'}\right)} \simeq 0 \tag{3.74}$$

for $a < r' \ll b$. The voltage drop at the resistance R is depicted in Fig. 3.11. During the drift of the n_0 primary electrons produced by the particle, the voltage drop is the same as for the ionization chamber (not visible in the figure). When the avalanche multiplication takes place, the voltage drop grows as

$$V_-(t) + V_+(t) \simeq V_+(t) = -\frac{q}{4\pi\epsilon l} \log\left(1 + \frac{\mu C V_0}{a^2 \pi \epsilon} t\right) \tag{3.75}$$

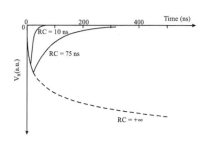

Fig. 3.11 Signal in arbitrary units (a.u.) at the load resistance R in a proportional counter as a function of time for different values of $\tau = RC$. Redrawn from Leo 1994.

where $\mu = v_+/|\mathbf{E}|$ is the **mobility** of the gas (Leo, 1994; Knoll, 2010). All ions reach the cathode at the time $T = (b^2 - a^2)\pi\epsilon/\mu C V_0$ and $V_+(T) = -ne/C$. In real life, $V_-(t)$ never reaches the value of eqn 3.73 because for $t \simeq RC$ the voltage generator restores the initial conditions. Since the signal grows logarithmically, choosing $RC < T$ is a wise compromise between a good signal efficiency and a fast recovery of the detector to sustain high particle rates.

3.7.1   Energy resolution

The energy resolution is one of the most important parameters of particle detectors. If a particle deposits an energy E_d, the detector produces a signal proportional to E_d. In ionization detectors, the main sources of uncertainty in E_d are the fluctuations in the number of ion-electron pairs produced by the particle while crossing the detector. Noise in the electronic circuits that records V_R also contributes to the uncertainty on E_d. If the distribution of the measured E_d is Gaussian, the (relative) **energy resolution** is defined as

$$R = \frac{\sigma}{E_d}. \tag{3.76}$$

This definition is mostly used in high-energy physics. Nuclear physicists prefer a definition based on the full width at half maximum (FWHM) of the distribution:

$$R = \frac{\text{FWHM}}{E_d} = \frac{2\sqrt{2\log 2}\,\sigma}{E_d} \simeq \frac{2.355\sigma}{E_d}. \tag{3.77}$$

Unfortunately, both quantities are called "energy resolution" and misunderstandings may arise. In an ionization chamber, the production of n is a Poisson process. Since $E_d = wn$ and the error on the mean energy per ionization in a gas (w) is negligible, $\sigma = w\sqrt{n}$ and

$$R = \frac{\sigma}{E_d} = \frac{1}{\sqrt{n}}. \tag{3.78}$$

You may wonder whether the energy resolution of a proportional counter is better than the resolution of an ionization chamber. It is not. The gain amplifies n_0 by a constant factor but a statistical fluctuation on n_0 is amplified, too. Hence, the one-sigma error of $E_d = wn = wGn_0$ is $\sigma \simeq wG\sqrt{n_0}$ if we neglect the (small) fluctuations in the gain, and

$$R = \frac{\sigma}{E_d} = \frac{1}{\sqrt{n_0}}. \tag{3.79}$$

On the other hand, the avalanche multiplication increases the size of the signal by three to six orders of magnitude. Such a large signal is not swamped by the electronic noise and a proportional counter can detect even a single particle crossing the counter using standard electronic components. This feature explains the popularity of proportional counters in particle and nuclear experiments.

3.7.2 Geiger–Müller counters

No proportional counters can exceed a gain of 10^6. For such a G, the amount of charge produced in the gas perturbs the electric field of the capacitor and E_d is no more proportional to n. If the gain is larger than 10^9, the detector is operated as a **Geiger–Müller counter** or Geiger counter/tube.

> The signal produced by a charged particle in a Geiger–Müller counter is the same regardless of the initial value n_0. The gain of the detector is $10^9 - 10^{10}$.

This device is of great practical value because it counts the number of particles irrespective of fluctuations in E_d. It comes at the price of losing information about the energy deposited into the detector. Geiger tubes are cheap, light, and require only standard voltmeters to be read. Even if they were discovered in 1928, they are still widely in use as hand-held survey-type meter or for personnel protection. Even today, they are considered the best portable devices for photon detection in the keV (**X-ray**) and MeV (**γ-ray**) energy range.

How can we get a constant signal for any value of n_0? If the electric field is increased well above the proportional counter regime, the capacitor discharges. The discharge in noble gases is due to UV photons that reach either the gas or the cathode and produce additional avalanches. The electrons drift swiftly toward the wire while the ions are nearly still. As a consequence, the concentration of ions in the gas increases with time when the electrons are drifting to the anode and new avalanches are produced. Sooner or later, the number of ions in the gas are so many that their space charge screens the electric field in the capacitor. If the screening is so strong that the field is brought below the threshold to initiate an avalanche, the avalanche production stops. Such a screening eventually terminates the discharge. Since the number of ions needed to stop the discharge process is always the same for a given detector geometry and voltage, and corresponds to ~ 20 avalanches in most counters, $n(n_0) \simeq 20\, G$ and the signal does not depend anymore on n_0.

3.8 Detection of photons

Photons are neutral particles and their detection relies on the energy transfer to the electrons of the medium. The most straightforward transfer process below 1 MeV is the **photoelectric effect** discussed in Sec. 3.4.1 because all the energy is transferred to the electron, which ranges out inside the detector. If the medium has a high value of Z, the photoelectric cross-section is large compared with Compton scattering, and the photon is promptly absorbed. Gas detectors are poor photon detectors at the MeV scale because the density of the medium is very small: $\rho \simeq 10^{-3}$ g/cm^3. Semiconductor detectors and, in particular,

germanium detectors are better suited because they are high-density solid-state detectors. Germanium detectors can achieve an FWHM resolution of about 0.1 % for a 1 MeV γ-ray but the thickness of these detectors cannot exceed a few mm. If the detection efficiency is more important than the energy resolution, the experimentalists resort to high-Z **scintillator detectors**. The most common scintillator for γ ray detection is a crystal of sodium iodide (NaI) doped with thallium. For NaI, w is similar to argon: we need on average 25 eV to produce a luminescence photon. Still, NaI is a solid-state crystal with $\rho = 3.67$ g/cm^3 and $Z = 53$ for iodine, so that the photoelectric cross-section is strongly enhanced compared with an ionization chamber.

If a photoelectric effect takes place in the detector, the photon is absorbed and the electron deposits all its energy in the detector because the electron range is shorter than the detector thickness. The voltage drop V_R at the detector resistance signals the absorption of the photon. The peak of the voltage pulse

$$V^{max} = \max_{0 < t \ll RC} [V_R(t)] \tag{3.80}$$

recorded by the detector is proportional to the kinetic energy of the electron and, hence, to the full energy of the photon. If a large flux of monochromatic photons of energy E impinges on the detector, the distribution of the reconstructed energy $E_R \sim V^{max}$ is (nearly) Gaussian and the resolution is $\mathcal{O}(0.1\%)$ for germanium detectors and of $\mathcal{O}(10\%)$ for NaI. The peak of this distribution, which corresponds to the full photon absorption, is called the **photopeak**.

3.8.1   Compton scattering and pair production

The reconstruction of the photon energy is subtler for Compton scatterings. From eqn 3.42, the kinetic energy of the scattered electron in NU is

$$E_{kin} = E - E' = E \frac{(1 - \cos\theta)E/m}{1 + (1 - \cos\theta)E/m} \tag{3.81}$$

and the photon transfers only a fraction of its energy to the electron. If the final state photon escapes from the detector, the energy deposited in the detector (E_R) is a function of θ, which is generally not measurable. If a large flux of monochromatic photons of energy E undergoes Compton scattering, the detector records a continuum distribution ranging from $E_R = 0$ ($\theta = 0$) to

$$E_R = \frac{2E^2}{m + 2E} \quad \text{for} \quad \theta = \pi. \tag{3.82}$$

The maximum energy is called the **Compton edge** and corresponds to a head-on collision of a photon with a free electron in the detector. If the final state photon interacts again in the detector, it might be absorbed by the photoelectric effect. In this case, all the energy is transferred to two electrons: the final state of the first interaction (Compton scattering) and

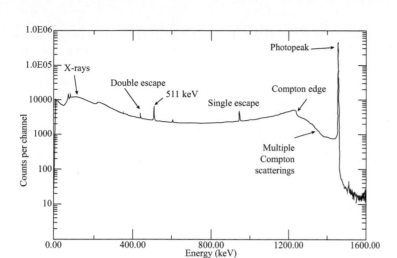

Fig. 3.12 Reconstructed energy spectrum of a germanium detector for 1460 keV photons originating from the decay of ^{40}K. The photopeak at 1460 keV corresponds to the full energy transfer from the photon to the electrons of the detector. The single and double escape peaks are visible at $1460-511 = 949$ keV and $1460-1022 = 438$ keV. A 511 keV peak due to photons from e^+ annihilations is also visible between the double and single escape peaks. Redrawn from Knoll, 2010.

the final state of the second interaction (photoelectric effect). Since E_R is the sum of the two kinetic energies, the detector records the full photon energy and, in this special case, E_R contributes to the photo-peak.

A high-energy photon can also produce an electron–positron pair. The electron ranges out in the detector and its kinetic energy is recorded in E_R. The positron ranges out, as well. Once at rest, it forms a e^+e^- bound state with an electron of the medium, which is called **positronium**. We will see in Chap. 6 that the positronium is an unstable state that mostly decays in two photons. Since the positronium is at rest in the laboratory frame and its rest mass is $2m_e$, the two photons are emitted back to back and the energy of each photon is $m_e = 511$ keV. If both photons are absorbed in the detector by the photoelectric effect, all the initial energy of the photon is recorded in E_R, and – once more – E_R contributes to the photo-peak.

For a photon of energy E, the position of the peak resulting from a pair production with full absorption of the two 511 keV photons is identical to the photo-peak at E. This may sound puzzling because, during the slow down of the electron originally created by the photon, the rest energy m_e of the electron is not recorded by the detector. However, inside the detector, there is also a positron and when the positron stops, it annihilates with an electron of the medium. The rest mass of this electron is converted into a photon and compensates the missing rest mass of the original electron. This consideration holds only if both photons produced by the positronium undergo a photoelectric effect inside the detector. On the contrary, if one of the two 511 keV photon escapes from the detector, we miss 511 keV and the detector records another peak at $E - m_e$. This is called the **first escape** peak (see Fig. 3.13). If both photons escape, they produce a **second escape** peak at $E - 2m_e$. In conclusion,

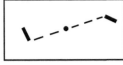

Fig. 3.13 Bottom: two photons (dashed lines) are produced by the annihilation of the positronium inside a detector and are absorbed by photoelectric effect producing a stopping electron (thick continuous tracks). The event contributes to the photopeak at the full energy E. Top: one of the two 511 keV photons escapes from the detector. The event contributes to the first escape peak at $E - m_e$. Image credit: V. Terranova.

a photon contributes to the photo-peak ($E_R = E$) only if is totally absorbed in the detector. Full photon absorption can take place in a single interaction (photoelectric effect) or in multiple interactions. If the photon interacts by single or multiple Compton scatterings and the final state photons escape, $E_R < E$. If the photon undergoes pair production and one photon escapes, then, $E_R = E - m_e$. If two photons escape, $E_R = E - 2m_e$.

All these possibilities are shown in Fig. 3.12, which depicts E_R for a medium-size detector where photon escapes can occur.

3.8.2 High-energy photons and calorimeters

What happens if the energy of the photon is very large ($E \gg 10$ MeV)? The photon converts into an electron–positron pair but both the electron and the positron are well above the critical energy. These charged particles can emit secondary photons by radiation, which in turn create other e^+e^- pairs as in Fig. 3.14. The number of particles increases exponentially until the energy becomes lower than the critical energy. Then, all charged particles range out (see Fig. 3.3) and the low-energy photons are absorbed by the photoelectric effect or multiple Compton scatterings. This exponential process is called the development of an **electromagnetic shower**. The dynamics of the shower can be described using a simple approximation due to B. Rossi:

above the critical energy, a photon converts into an electron–positron pair every X_0, and the photon energy is equally shared between the two particles. An electron (or positron) emits a photon with half its energy every X_0.

Within the Rossi approximation, the number of particles produced by a photon of energy E after $t = nX_0$ is $N = 2^t$. No particles are produced if the energy goes below the critical energy E_c. At each step, the energy of the particles is $E_t = E_{ini}/2^t$. Hence the maximum depth of the shower t_{max} corresponds to

$$E_{t_{max}} = \frac{E}{2^{t_{max}}} = E_c \tag{3.83}$$

giving

$$t_{max} = \frac{\log E/E_c}{\log 2} . \tag{3.84}$$

Note that an electromagnetic shower can also be initiated by an electron or a positron with the emission of an energetic photon of energy $E/2$ after 1 X_0. In practice, we can use the Rossi approximation to describe high-energy photons, electrons, and positrons in the same way. MS is the process that drives the development of the shower in the transverse plane, that is the plane perpendicular to the direction of the first particle of the shower. In this case, the scale parameter is the **Moliere radius** $\mathbf{R_M}$, which is a property of the material and depends on X_0 and the

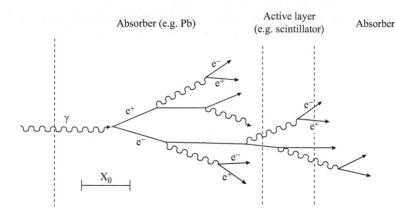

Absorber (e.g. Pb) Active layer (e.g. scintillator) Absorber

Fig. 3.14 Development of an electromagnetic shower inside a sampling calorimeter. The active layer (e.g. a small tile of scintillator material) can detect only the Bethe energy loss deposited by the electrons and the positrons. The absorber is chosen among materials with small radiation lengths to stop the particle in a small volume and build compact detectors even in high-energy accelerators.

atomic number Z of the material. The Moliere radii are tabulated for many materials.

> A cylinder with a radius of $2R_M$ and the axis located in the direction of the first incoming particle contains about 95% of the energy of the shower.

For instance, R_M in iron is just 1.7 cm and the containment size at 95% is a circle with a 6.8 cm diameter. On first approximation, the transverse profile of the shower does not depend on the energy, unlike the longitudinal profile, which scales as $\log E$. Fluctuations in the size of the shower can appear close to the critical energy, where the absorption length of the photons may inflate the size of the low-energy shower tail. This is why there is always a large contingency in the size of the e.m. calorimeters to avoid any energy leakage.

Equation 3.84 is the formula that grounds the physics of **electromagnetic calorimeters** (Fabjan and Gianotti, 2003; Wigmans, 2017). Calorimeters in particle physics are particle detectors that measure the energy of photons, electrons, and positrons by converting their energy into showers. The measurement destroys the particle by absorption in the medium. The name is somehow misleading: they are called calorimeters because eventually the energy is absorbed and converted into heat. However, particle calorimeters do not measure the increase of temperature of the medium but the total E_R deposited by the charged particles. Pair production and bremsstrahlung create the particle shower but the energy transfer to the medium occur through the ionization losses described by the Bethe formula and the calorimeters are often made of high-density and high-Z scintillator detectors (**homogeneous calorimeters**). **Sampling calorimeters** represent a cheaper alternative: in these detectors sheets of lead or iron (called "absorbers") are interleaved with conventional detectors (low-Z scintillators, gas detectors, etc.; see Fig. 3.14). The absorbers cause the showering of the particle and the active detectors measure the dE/dx of all charged particles produced inside the shower. Since this is only a fraction of the total energy, sampling calorimeters must be accurately calibrated and their energy resolution mostly depends on fluctuations in the visible energy inside the active layers. This effect deteriorates the energy resolution of sampling calorimeters compared with homogeneous devices and the resolution loss is called the **sampling term**. The energy resolution of the electromagnetic calorimeters can be described by:

$$\frac{\sigma}{E} = \sqrt{\left(\frac{a}{\sqrt{E[\text{GeV}]}}\right)^2 + \left(\frac{b}{E[\text{GeV}]}\right)^2 + c^2} \qquad (3.85)$$

or, using a compact notation

$$\frac{\sigma}{E} = \frac{a}{\sqrt{E[\text{GeV}]}} \oplus \frac{b}{E[\text{GeV}]} \oplus c \qquad (3.86)$$

where $\oplus$ indicates that the contributions must be summed in quadrature and, then, squared. The term a is called the **stochastic term** and, in sampling calorimeters, is dominated by the sampling term. It is not surprising that its contribution depends on $E^{-1/2}$ because it results from Poisson fluctuations of the elemental events, whose standard deviation is proportional to $\sqrt{E}$ (see Sec. 3.7.1). The **constant term** c mostly depends on uncertainties in calibrations, non-uniformities, or dead material inside the detector. It is the leading term for $E \to +\infty$ and can remarkably grow if the shower is not fully contained in the detector. The b term is usually negligible and depends on the noise of the readout electronics or pile-up of multiple particles. The same formula holds for homogeneous calorimeters but the value of a is much smaller. For instance, the homogeneous calorimeter of the CMS experiment at the Large Hadron Collider (LHC) has $a = 3\%$, $b = 250$ MeV and $c = 0.5\%$, a record in experimental particle physics! One of the best sampling calorimeter, the liquid argon–lead calorimeter of ATLAS at the LHC, has $a \simeq 10\%$, much larger than CMS, but achieves an outstanding constant term: $c = 0.3\%$.

3.9   Physical measurements[*]

Measurement in QM is the coupling of a quantum system with a macroscopic body, which results in the collapse of the wavefunction, as discussed in Sec. B.4. The origin of the collapse is not fully understood yet, but it is properly described by the generalized statistical interpretation of QM. If A is an observable associated with a hermitian operator $\hat{A}$ with eigenvalue a_k and eigenvector $|k\rangle$, the probability of recording a_k as an outcome of the measurement of A is $|\langle k|\psi\rangle|^2$ if ψ is the wavefunction of the system. If the system is a particle and, for instance, the observable is discrete, ψ may be an entangled state of the eigenvectors. If, for instance, the eigenvalues are 0 and 1 ("qubit"), $|\psi\rangle = \alpha\,|0\rangle + \beta\,|1\rangle$. We can also label the quantum state of the macroscopic body (the detector) as $|d\rangle$. If the system is in the state 0, the detector evolves in the state $|d_0\rangle$; if it is in the state 1, the detector evolves in the state $|d_1\rangle$.

The particle+detector system is thus an entangled state:

$$\alpha \left|0\right\rangle \left|d_0\right\rangle + \beta \left|1\right\rangle \left|d_1\right\rangle \tag{3.87}$$

When we look at the detector, we actually disregard the particle. The effect of disregarding part of an entangled quantum system can be described using the density matrix formalism because eqn 3.87 is a straightforward generalization of a two-qubit system $\psi = (\left|00\right\rangle + \left|11\right\rangle)/\sqrt{2}$ (Ballentine, 2014). We briefly describe this powerful tool below.

Consider a quantum system, whose probability of being in the $\left|\psi_i\right\rangle$ state is p_i. Then, the density matrix of the system is defined as:

$$\rho = \sum_i p_i \left|\psi_i\right\rangle \left\langle\psi_i\right|. \tag{3.88}$$

In the two-qubit system, disregarding one qubit is equivalent to considering a system whose density matrix is:

$$\rho = \frac{1}{2} \left|0\right\rangle \left\langle0\right| + \frac{1}{2} \left|1\right\rangle \left\langle1\right|. \tag{3.89}$$

For the particle+detector system, looking only at the detector produces a density matrix like:

$$\rho_d = \alpha^2 \left|d_0\right\rangle \left\langle d_0\right| + \beta^2 \left|d_1\right\rangle \left\langle d_1\right|. \tag{3.90}$$

Equation 3.90 represents the probability distribution function over all possible detector configurations. The same considerations hold for a continuous variable: a detector that measures the x position of a particle evolves in the state $\left|d_x\right\rangle$.

If you perform twice the measurement almost simultaneously you will get the same result, as if QM were a deterministic theory. If you got 0 the first time, you will get 0 the second time. If you got 1 at the first trial, you get 1 at the second trial, too. This is a consequence of the **collapse of the wavefunction** once the first measurement is performed. These (ideal) measurements are called **projective measurements** because they are equivalent to projecting the state of the particle to one eigenstate of the hermitian operator $\hat{A}$ by measuring the physical quantity A associated with the operator. Projective measurements are the least disruptive measurements we can perform and are only limited by the Heisenberg uncertainty principle: precise measurement of x compromises only the knowledge of the conjugate variable p_x so that $\Delta x \Delta p_x \geq \hbar/2$.

Similarly, if we measure twice a system with a detector, the first measurement produces the collapse $\psi \rightarrow \left|0\right\rangle \left|d_0\right\rangle$ (or $\psi \rightarrow \left|1\right\rangle \left|d_1\right\rangle$ if the outcome of the measurement is 1) and the second measurement changes neither the system wavefunction nor the reading of the detector.

In particle physics, projective measurements are rare but possible. The Stern–Gerlach experiment is a well-known example. However, the vast majority of the measurements are disruptive and there is no way to perform the measurement twice if the particle does not exist anymore after the first measurement. We gave extreme examples in the sections above: the measurement of the full energy of a photon always destroys the photon. On the other hand, the measurement of the momentum of a heavy charged particle is a reasonable approximation of a projective measurement and is performed by high-precision particle trackers.

3.10   Tracking particles

A detector capable of visualizing trajectories without perturbing the motion of the particle would achieve a perfect ("projective") measurement of the momentum perpendicular to the magnetic field. The cloud chamber employed by Anderson to detect positrons is an approximation of such an ideal device. Modern detectors designed to reconstruct the trajectory of a particle and record the data in the form of electric signals are called **trackers**. They employ the same physical principles as the energy measurement detectors discussed in Sec. 3.6 but the volume of a tracker is divided into small cells and the energy deposited in the cells is used to signal the presence of the particle. The trajectory is reconstructed lining the activated cells as bright pixels on a screen. Early trackers were based on images in photographic films and the analysis was extremely cumbersome since it was mostly done by visual inspection. Modern trackers are fully digital and based on gas chambers, scintillators, or pixelized semiconductor detectors.

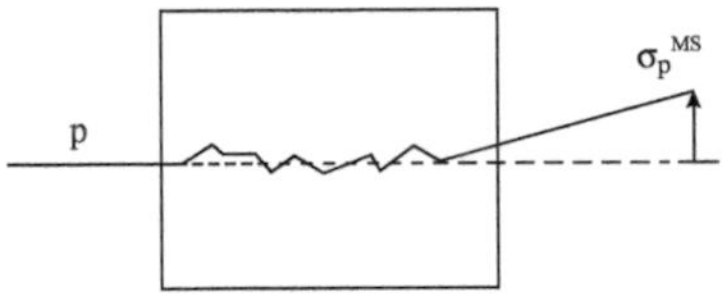

Fig. 3.15 Uncertainty σ_p^{MS} introduced by MS after the passage of a particle in a slab of material.

3.10.1   Precise reconstruction of trajectories

Loosely speaking, precision in tracking depends on reducing as much as possible the perturbations in the particle motion and employing devices divided into many small pixels, that is with very high **granularity**. The main source of nuisance is the material the tracker is made of. The particle interacts not only releasing a small fraction of its energy to allow for detection but can scatter on the tracker nuclei changing the direction. This effect is the aforementioned MS. A large random set of scatterings – as the ones shown in Fig. 3.15 – can be handled by the central limit theorem (James, 2006): on average the motion of the particle driven by all these scatterings is unperturbed. However, the outgoing angle of the particle that crosses the slab of material is Gaussian-distributed in a given plane and the sigma of the Gaussian is (Grupen and Shwartz, 2008):

$$\sqrt{\langle \theta_{2D}^2 \rangle} = z \frac{13.6 \text{ MeV}}{\beta c p} \sqrt{\frac{x}{X_0}} \left(1 + 0.038 \log \frac{x}{X_0} \right) \tag{3.91}$$

where p (in MeV/c) is the momentum, βc the velocity, and z the charge of the particle. x/X_0 is the thickness of the slab in units of radiation lengths (X_0). We compute the sigma in space $\sqrt{\langle \theta_{3D}^2 \rangle}$ in Exercise 3.11.

The detector granularity is the minimum size of the points that the tracker can identify to reconstruct the trajectory. It is analogous to the size of the pixels in a television or a laptop screen. The granularity impacts on the determination of $\mathbf{x}(t)$ and, in turn, on the measurement of the particle momentum. This is extracted by eqn 3.35: $p = 0.3BR$ for a particle of unit charge. If the tracker is located in a volume with a constant magnetic field B, the momentum perpendicular to B is given by the curvature radius of the particle produced by the Lorentz force. The precision depends on the number of points where the tracker samples the trajectory (N) and the spatial precision of each point (σ). The corresponding uncertainty on the momentum is (Gluckstern, 1963):

$$\frac{\sigma}{p} \simeq \frac{\sigma[\text{m}]}{0.3B[\text{T}] \cdot (L[\text{m}])^2} \sqrt{\frac{720}{N+4}} \cdot p[\text{GeV}/c] \qquad (3.92)$$

and linearly grows with p. The total uncertainty on p is the sum in quadrature of the errors due to the MS:

$$\frac{\sigma_p^{MS}}{p} \simeq \frac{0.0136 \ [\text{GeV}/c]}{0.3 \ B[\text{T}]} \sqrt{\frac{1}{xX_0[\text{m}^2]}} \qquad (3.93)$$

and the uncertainty in the reconstruction of the trajectory (eqn 3.92). Note that the right-hand side of eqn 3.93 does not depend on p. Therefore,

---

the momentum precision of the tracker increases as $1/B$ if B and $\mathbf{p}$ are perpendicular. However, at large momenta, the error due to the tracking increases as p (eqn 3.92) while the error due to multiple scattering remains constant (eqn 3.93).

The name of the game of modern high-precision tracking is devising detectors with a very small material budget (small X_0 or small size) and good precision in the reconstruction of the trajectory inside a volume that senses a large magnetic field.

3.10.2 Tracking detectors

Gaseous detectors (Sauli, 2014) naturally provide a small X_0 to mitigate the effect of multiple scattering because the volume is filled with a gas. These devices are still used for tracking purposes in moderate-rate experiments. The simplest large-volume position-sensitive gas chamber was developed by G. Charpak in 1968 and is called the **multi-wire proportional chamber** (MWPC). It consists of two parallel planes and an array of parallel wires running in the mid-plane between the plates, as in Fig. 3.16. The plates are generally at ground and the (anodic) wires are positively charged, like in a proportional counter. The volume is filled with an argon–isobutane mixture. The field lines of a MWPC are drawn in Fig. 3.17. When a charged particle crosses the chamber, the primary electrons n_0 produced by the ionization losses drift toward the wires and the primary ions move toward the cathode. We have shown in Sec. 3.7 that the motion of the primary electrons and ions give a negligible electrical signal but when the primary electrons reach the neighborhood of

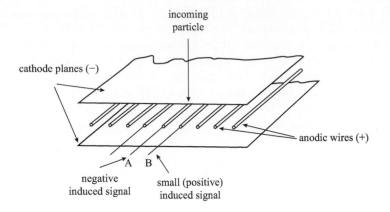

Fig. 3.16 Layout of a multiwire proportional chamber. The cathode planes are usually grounded and the anodic wires are positively charged. The wire closest to the incoming particle is A. The next to closest is B (see text).

a wire and trigger an avalanche, we record a large signal. Equation 3.74 shows that the signal is dominated by the motion of the secondary ions toward the plates. The signal produced at the resistance between the wire closest to the particle trajectory and the ground (cathode) is *negative* as in a conventional proportional counter (eqn 3.73). But what happens to the wires next to it? The signal in the neighboring wires is small and *positive* so that the identification of the wire closest to the trajectory is unique.

The sign of the signal induced by the motion of the ions follows from the **Shockley–Ramo theorem** that we state without demonstration (Barr *et al.*, 2019; Spieler, 2005):

> the instantaneous current i induced on a given electrode A due to the motion of a charge is given by $i(t) = -q\mathbf{v}(t) \cdot \mathbf{E_w}(t)$, where q is the charge of the moving particle and $\mathbf{v}$ is the instantaneous velocity. $\mathbf{E_w}$ is the "weighted electric field". It is computed by the Laplace equation $\nabla^2 V(\mathbf{x}) = 0$ removing the charge q, grounding all electrodes except A, and bringing A to unit potential (1 V).

In the case of the MWPC, for $t \ll RC$ the ions are in the proximity of the wire A closest to the trajectory and they move (see Fig. 3.17) slightly toward the neighboring wires before drifting along the field lines perpendicular to the cathode. We first apply the Shockley–Ramo theorem to compute the current flowing in A. In this wire, the velocity and the weighted field have the same direction and $-\mathbf{v}(t) \cdot \mathbf{E_w}(t) < 0$. The current flowing in A is thus negative. If B is next to A, we can compute the current in B grounding all electrodes but B. The ions produced near A and closest to B experience the largest $\mathbf{E_w}$ but they go away from B to reach the cathode.[6] Still, at the very beginning of the drift, they get slightly closer to B as shown in Fig. 3.17. Hence, for small t, $-\mathbf{v}(t) \cdot \mathbf{E_w}(t) > 0$ and the current flowing in B is positive. Later on (vertical arrow in Fig. 3.17), the ions move toward the cathode and the weighted field in B decreases. Then, the size of the current in B fades away. In conclusion, the only wire with a negative signal is A, which

[6]This is because the size of the weighted field decreases with the distance between the ion and the wire B. So, the ions that give the largest contribution to the current in B are the ions closest to B.

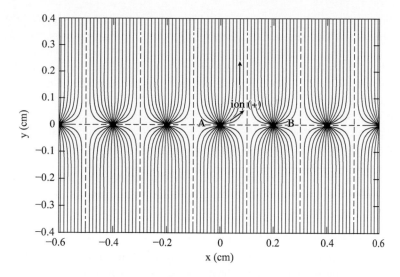

Fig. 3.17 Electric field lines of a MWPC (transverse view in cm). The motion of a ion created in A and moving toward the cathode is shown as a black arrow.

provides the x coordinate of the trajectory. All other wires have either a small positive signal that goes swiftly to zero (neighboring wires) or a null current since the beginning (wires far from the crossing particle). The MWPC is therefore a one-dimensional (1D) tracker because measures the x position with a precision of about the distance between the wires (see Exercise 3.12 for a better estimate).

Two stacked MWPC with orthogonal wires provide both the x and the y position and make the device a 2D tracker. The z coordinate is measured either stacking additional MWPC or measuring the drift time of the primary electrons up to the anode as in the **drift** and **time projection chambers** (TPC).

TPC are the paradigm of 3D trackers and are still used if the rate of the events is moderate. In this case, the tracks do not overlap and the electrons that migrate to the anode can be clearly identified. A TPC is a hollow cylinder with a radius of tens of cm or even meters.[7] It is filled with an argon-based mixture and has a static magnetic and electric field both parallel to the axis of the cylinder. The cathode is positioned on one end of the cylinder (see Fig. 3.18). A particle crossing the TPC is bent by the magnetic field and produces ions moving toward the cathode and electrons toward the anode. The anode is made of wires similar to an MWPC, where the avalanche multiplications occur: the newly produced electrons reach the wire, while the corresponding ions are collected in small pads located a few cm after the wires: these pads act as the cathode of an MWPC but they are segmented and can be read out separately. The detector fulfills all the requirements cited in Sec. 3.10.1: multiple scattering is reduced by the low density of the medium (Ar-based gas) and the momentum is measured by the bending radius. Furthermore, the thermal diffusion of the electrons along their path toward the anode is reduced by the fact that the magnetic field is parallel to the electric field, and the electrons spiral around the magnetic field lines. The position in the plane perpendicular to the axis of the

[7]The TPC of the ALEPH experiment at the LEP collider was build in the late 1980s with a diameter of 3.6 m and a length of 4.4 m. The momentum resolution achieved was $\sigma_p/p \simeq 1.2 \times 10^{-3} p[\text{GeV}/c]$ (Atwood *et al.*, 1991).

[8]Fast detectors that signal the passage of a particle and activate the rest of the apparatus are called a **trigger**. For instance, in the Anderson experiment, the cloud chamber was activated randomly raising the piston, while in the more sophisticated experiment by Blackett and Occhialini, they used Geiger-Müller counters as triggers to select only real physical events. We will discuss triggers in Sec. 4.6.1 when speaking about detectors for collider physics.

cylinder is given by the pads with a few mm precision and the dE/dx is measured by the amount of signal in the wires. In many cases, we know the time when the particle crosses the TPC: in collisions with particles artificially produced by accelerators or when the passage of the particle is measured by dedicated fast detectors.[8] In this case, the z coordinate is just $v_e(t_f - t_0)$, where v_e is the drift velocity of the electrons, t_0 is the time when the particle crosses the TPC, and t_f is the time when the primary electrons produced by the particle end their journey toward the anode and start the avalanche multiplication in the wires.

> A time projection chamber is a modern example of 3D tracker where the trajectory is reconstructed in cylindrical coordinates $\mathbf{x}(r, \theta, z)$, the momentum by the bending radius $R = p/0.3B$ and the energy loss per unit length dE/dx by the anode wires.

TPCs give their best in particle colliders (see Chap. 4), that is in head-on collisions where the particles originate close to the high voltage cathode. They are also used in fixed target and neutrino experiments. TPCs are considered "slow detectors" due to the time taken by the electrons ($\simeq 1$ cm/μs) to reach the anode and, at a high particle rate, may be plagued by the accumulation of ions in the gas volume that distort the electric field. So, very high-rate experiments do not use TPCs. 3D trackers better suited for these applications will be reviewed in Chap. 4.

3.11   Simulating detectors

All effects discussed so far are genuine quantum processes and are probabilistic in nature. For instance, a heavy particle crossing iron deposits a mean energy given by the Bethe formula of eqn 3.19. If the medium is thick, the central limit theorem ensures that the distribution of the dE/dx losses is Gaussian but for thin media, the distribution of dE/dx follows other distributions. In any case, the dE/dx is never uniquely determined. The same consideration holds for the trajectory and, in turn, for the determination of the momentum. Modern experiments are so precise as to account for the entire probability distributions. For more than 50 years, the response of particle detectors is handled numerically by figuring out the actual probability distribution function of each observable. By far the most powerful technique used in particle physics to accomplish this task has been invented during the Second World War by S. Ulam and N. Metropolis and formally developed by J. von Neumann and N. Metropolis from 1946 to 1953 (Metropolis *et al.*, 1953). This set of algorithms and techniques are referred to as **Monte Carlo (MC) simulations** because they are based on random sampling, as in a gambling game. This is why the technique was named after a European city that hosts a very popular casino.

All MC techniques are equivalent to performing the integration of a known function without solving analytically the integral. Suppose, for

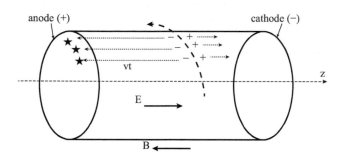

Fig. 3.18 Layout of a TPC. The cathode plane is at high negative voltage and the anode wires are positively charged (or at ground). If a charged particle (dashed line) crosses the TPC, the ionization electrons drift toward the anode (dotted lines) at a speed v. The z-coordinate is given by v multiplied by the time it takes to reach the anode. The anode is made of wires and pads (not shown) and produces an avalanche (stars) like a standard MWPC, which gives the position of the track in the $x - y$ plane and the energy deposited per unit length (dE/dx).

instance, that we want to compute:

$$\int_0^1 x^2 dx = \int_{\mathcal{D}} x^2 dx = \frac{1}{3} \tag{3.94}$$

but we do not know that the primitive of x^2 is $x^3/3$. A trivial way to perform the integration is to compute or estimate the maximum of x^2 (the integral kernel) in the domain $\mathcal{D}$ – $M \equiv \max_{\mathcal{D}}\{x^2\} = 1$ – and generate with a computer a random number uniformly distributed in $\mathcal{D}$. In this case, between 0 and 1. At our first trial, we may get $(x_1, y_1) = (0, 3, 0.01)$. If $y_1 < x_1^2$ we are inside the area spanned by the integral: so our first hit is a "good hit" to compute the full area. If we get instead $(0.3, 0.5)$, we are off the area of the integral because $0.5 > 0.3^2 = 0.09$. That is a bad move and does not contribute to the size of the area. We can repeat the procedure 100 times and the ratio between good hits and total (good + bad) hits provides an estimate of eqn 3.94. The larger the number of hits the better the precision of the estimate and it is easy to show that:

$$\lim_{N \to \infty} \frac{N_{good}}{N_{good} + N_{bad}} = \frac{N_{good}}{N} = \frac{1}{3} = \int_0^1 x^2 dx. \tag{3.95}$$

This method[9] is called the "hit-and-miss" MC and you can guess that it is a safe but extremely ineffective way to solve eqn 3.94. MC techniques are algorithms that improve the convergence rate of the hit-and-miss MC so that it can tackle problems that cannot be handled with other analytical or numerical methods. These methods not only compute the sought-for integral but also provide a high statistics sample of events (good hits) that follow a given probability density function (p.d.f.) $P(x)$. In our dummy example, $P(x) = x^2$.

[9] The technique has been pioneered by G.L. Leclerc, Comte de Buffon in 1777 replacing the computer with random throws of mechanical objects and, in mathematics, it is known as the Buffon needle method to compute π.

> MC tools generate the (fake) data expected by the experimentalist from the detector to compare with the actual data.

Any physical process has its own probability density function. The p.d.f. of a scattering that produces a particle is given by the cross-section. The energy loss in a material with thickness δx is given by the distribution of dE/dx: a Gaussian centered around the dE/dx of the Bethe formula if δx is large and a **Landau distribution** (Leo, 1994) if δx is small. In general, the probability density functions can be computed either analytically or numerically for most physical processes. As a consequence, a computer simulation of a physical event in a detector can be performed very precisely in the following way:

- generate a particle with a given position and momentum according to the p.d.f. of the physical process
- if the particle reaches the detector at $\mathbf{x}$, generate all physical processes that occur in a small region between $\mathbf{x}$ and $\mathbf{x}+\Delta\mathbf{x}$ according to their p.d.f. In practice, we roll a die according to these p.d.f. and check the outcome. If for that particular outcome, the physical process changes the energy or the trajectory of the particle, we correct the particle kinematics accounting for the change
- compute the next step of the particle trajectory from $\mathbf{x} + \Delta\mathbf{x}$ to $\mathbf{x} + 2\Delta\mathbf{x}$ and repeat the process until the particle either leave the detector or is absorbed
- if during a step a new particle is generated (e.g. a decay, a photon emission, a photon pair production, etc.), all new particles must be followed in parallel with the main particle until they reach a minimum energy cutoff set by the experimentalist. The cutoff is a compromise between precision and computing time and is chosen empirically
- record all steps and all changes per step, and check if the detector is able to produce a signal for such energy deposit (in jargon, this signal is called a "detector hit")
- add possible detector inefficiencies or instrumental effects (e.g. read out electronics) from calibration data or other experimental inputs collected before the run of the detector. It creates a "detector digitized hit" or, in short, a "detector digi" that corresponds to the actual signal from the detector
- repeat the entire process for a new particle

The final result is a full-fledged copy of the raw data (detector digi) that the detector collects during a run. Discrepancies with the actual data can be due to unaccounted systematic errors or they can provide a hint of a new discovery! The popularity of MC techniques is due to the exponential growth of computing power over the last decades and the possibility to improve the precision nearly at will, adding new information on the p.d.f. of the physical process or the detector as soon as they become available. There are many specialized codes for physics process generation and for detector simulation. At the time of writing, the most common detector simulation tool is a set of C++ libraries called GEANT4 (Allison *et al.*, 2016). I believe that most readers will get acquainted with it during their graduate courses.

Further reading

The bibliografy of particle detector science is very large. For a general introduction at Master's level I recommend Grupen and Shwartz, 2008 and Leo, 1994 for high-energy physics, and the classical textbook from Knoll, 2010 for low-energy physics. Comprehensive reviews that cover nearly all aspects of particle detection, including particle identification, trigger and collider detectors (see Chap. 4) are Kolanoski, 2020 and the second volume of the Particle Physics Reference Library: Myers and Schopper, 2020.

Two classical book on gaseous and semiconductor detectors are Sauli, 2014 and Spieler, 2005, respectively. Electronics plays a central role in this field. You can find a gentle introduction in Suits, 2020. The classical books – also aimed at electronic engineers – are Horowitz and Hill, 2015 and Sedra *et al.*, 2020. Detector simulation and analysis technique are well-covered by Lista, 2017 and Frühwirth *et al.* 2008.

Exercises

(3.1) A minimum ionizing particle crosses 1 cm of iron (iron density: $\rho = 7.874$ g/cm^3). What is the amount of energy that is lost by this particle? What is the corresponding energy if the velocity of the muon is $\beta = 0.99$? Finally, what is the energy lost by a proton with the same velocity as the muon?

(3.2) A muon crosses a TPC with an energy of 50 GeV. Compute the maximum energy that can transfer from a single collision to an electron. Is such electron visible in the TPC?

(3.3) Can you make a rough estimate of how many electrons a tomato has? Is a tomato able to stop a 2 MeV proton?

(3.4) Compute the number of mips produced by a 100 GeV photon in iron in the Rossi approximation and the depth of the corresponding electromagnetic shower.

(3.5) What is the ratio between the photopeak area and the Compton area of a 1 MeV photon? Can you give an approximate numerical estimate if the material is lead?

(3.6) Compute the maximum rate that can be sustained by an ionization chamber as a function of the load resistance and of the detector capacitance.

(3.7) * The **recovery time** of a detector is the time needed to restore the initial detector condition after the passage of a particle. If a detector has a recovery time of 100 ns and the rate of particles is 10 kHz, how many particles are missed after a 1-hour run?

(3.8) * The **efficiency** of a detector is the number of particles that is able to detect divided by the total number of particles crossing it. We can measure the efficiency putting the detector between two other identical detectors, one above and one below it. In this way, a vertical mip or a high-energy muon crossing the upper and lower detectors must cross the detector in the middle and release the same energy in all of them. If the middle detector misses 2 muons over 100, what is its efficiency? What is the error on the efficiency?

(3.9) Suppose we need to measure the energy of a 100 GeV electron. Is it better to use a sampling or a homogeneous calorimeter? What is the expected resolution in the ATLAS and CMS calorimeters?

(3.10) What is the ratio between the area of the Compton distribution and the photo-peak in a detector of infinite volume for a 5 MeV photon?

(3.11) * Why the standard deviation of the multiple Coulomb scattering in a plane (eqn 3.91) is $\sqrt{2}$ smaller than the one in space?

(3.12) * If the distance between two wires of a MWPC is 1 cm in the x direction, what is the precision σ of the tracker along x?

(3.13) * Derive eqn 3.42 using the four-momentum conservation in a $\gamma\, e^- \to \gamma\, e^-$ scattering.

(3.14) * A track crosses a TPC parallel to the high-voltage cathode. If the momentum of the track is 1 GeV, the radius of the TPC is 1 m and $B = 1.2$ T, what is the precision on the particle momentum if the readout pads have a size of 1 cm^2?

(3.15) * A 2D tracker is made of two perpendicular MWPC and two particles cross the planes at the same time. Are the crossing points of each particle uniquely determined? Are multiple solutions possible?

(3.16) ** Derive eqn 3.13 using the four momentum conservation in SR. Perform the calculation under the hypothesis that the mass of the particle M is $\gg m_e$ and generalize the result for any M.

Accelerators and colliders

4

4.1 Particle accelerators

Accelerators are the workhorse of modern particle physics. High-energy particles can be used to smash quantum systems down to their elementary constituents, measure cross-sections, and produce new particles by electromagnetic, strong, or weak interactions. They prepare the system in well-defined initial conditions and serve detectors capable of recording the final states and reconstruct the event kinematics. Accelerator physics is a fascinating discipline that employs relativistic electrodynamics, electric engineering, multi-body dynamics, and information technology.

Still, this science is based on a few fundamental principles that have been established from 1931 (E. Lawrence, J.D. Cockcroft, E.T. Walton) to 1952 (E. McMillan, V. Veksler, E. Courant, H. Snyder) and spread open in a very large number of applications. This chapter aims to introduce beginners to the two most important principles of accelerator science: phase stability and focusing. We will illustrate the principles in the most important classes of accelerators: static, linear, and circular accelerators, with special emphasis on colliders and their detectors.

4.2 Static accelerators

A charged particle in vacuum experiencing a static electric field is accelerated by a force proportional to the particle charge and the field: $\mathbf{F} = q\mathbf{E}$. As a consequence, a simple plane capacitor may act as an accelerator. The definition of electronvolt (eV) is given in this framework as the energy gained by a particle with unit charge (i.e. with the same charge of the electron or the proton) crossing a region, whose potential difference is 1 V. A set of conductors can produce an accelerating field with two important limitations:

- the trajectory of the particle must be open, that is electrostatic accelerators cannot keep particles in stable closed orbits; and

- the electrostatic accelerator draws energy from the environment because transfers part of its electrostatic energy to the particles and because a collection of point charges (the accelerator components) cannot be maintained in stable equilibrium solely by the electrostatic forces.

A Modern Primer in Particle and Nuclear Physics. Francesco Terranova, Oxford University Press.
© Francesco Terranova (2021). DOI: 10.1093/oso/9780192845245.003.0004

The first statement directly follows from Maxwell's equations and, in particular, from Faraday's law:

$$\nabla \times \mathbf{E} = -\frac{\partial \mathbf{B}}{\partial t} = 0 \tag{4.1}$$

for static electric and magnetic fields. The Stokes theorem applied to eqn 4.1 zeroes the total amount of work done by the electric field on the charged particles:

$$q \oint_{\substack{\text{closed}\\\text{curve}}} \mathbf{E} \cdot d\mathbf{l} = 0 \tag{4.2}$$

where q is the particle charge. As a consequence, the particle does not gain any net acceleration since $V(\mathbf{x}^{ini}) = V(\mathbf{x}^{fin})$. The second statement follows from the conservation of the electrostatic energy and Earnshaw's theorem.[1] Despite these limitations, **electrostatic accelerators** can achieve energies of hundreds of keV or more. They are still used for industrial applications and several high-precision nuclear physics experiments (Wilson, 2001).

[1] This theorem states that a set of point charges cannot be maintained in stable stationary equilibrium only by the electrostatic interaction of the charges. S. Earnshaw derived the theorem in 1842 and a modern proof is given, for instance, in Griffiths, 2017.

4.2.1 The Greinacher voltage rectifier

The core of an electrostatic accelerator is a **Greinacher multiplier**. This is a circuit that transforms an AC voltage into DC, doubling its peak value. The multiplier is then a rectifier circuit for voltage doubling. To understand how it works, let us start with the basic one-stage rectifier shown in Fig. 4.1. The circuit consists of a standard AC source coupled with a transformer that increases the peak voltage. On the right of the circuit, the transformer is connected to a diode and a load – such as a resistance or a vacuum pipe where the particles are accelerated by $V_{out} - V_B$ – in parallel with a capacitor. In most cases, B is at ground voltage as in Fig. 4.1 and $V_A(t)$ (V at point A) oscillates between V_{max} and $-V_{max}$. When the voltage difference between A and B increases, the diode lets the current flow from the transformer to the capacitor and the voltage difference of the capacitor increases because $V_A(t) = V_C(t)$. Once $V_A(t)$ reaches its maximum and starts to decrease, the passage of the current from the capacitor to the transformer is stopped by the diode ($V_A(t) \neq V_C(t)$) and the capacitor discharges only through the load resistance R, as shown in Fig. 4.2. The discharge time is driven no

Fig. 4.1 One-stage rectifier driven by a transformer that provides a voltage $V_A(t) = V_{max}\sin\omega t$. The circuit includes a diode (D), a capacitor (C), and a resistive load (Load). Note that, in this case, the negative pole of the transformer is at ground so that $V_B(t) = 0$.

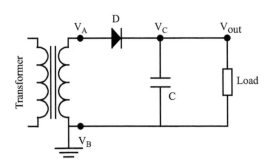

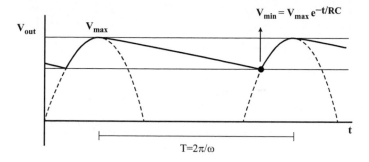

Fig. 4.2 V_{out} as a function of time in a rectifier. R is the load resistance and C the capacitor in parallel with the load. T is the period of the AC voltage of the transformer.

more by the frequency of the AC but by the time constant of the RC circuit. If R is large, the discharge is very slow and the capacitor is still at high voltage when $V_A(t) < V_{max}$. Note, however, that RC is finite and a small voltage drop V_{min} shows up until the transformer goes back to $V_A(t) > V_{min}$. Then, the peak value $V_A(t) = V_C(t) = V_{max}$ is swiftly restored. Therefore,

> in a single stage rectifier, also called a **leveling circuit**, the voltage at the capacitor is nearly constant at $V_C(t) \simeq V_{max}$

as shown in Fig. 4.2.

4.2.2   Voltage doubler

In 1919, H. Greinacher noted that by replacing the load resistance with another diode–capacitor pair as in Fig. 4.3, we can double the V_{max} of the second-stage capacitor. The voltage V_{out} is similar to a DC source with $V_{out} \simeq 2V_{max}$ as in Fig. 4.4. If $V_A(t)$ in Fig. 4.3 is negative, the capacitor C_2 cannot discharge through the diode $D2$, which acts as an open conductor. The current can only flow through C_1 charging this auxiliary capacitor up to V_{max}. On the other hand, when $V_A(t) > 0$, $D2$ acts as a conductor, and the current flows to charge C_1 up to a value corresponding to $V_A(t) + V_{C_2}$. C_2 can be charged to a voltage that is even larger than V_{max} and reaches the maximum value $V_{C_2} = 2V_{max}$. After two AC periods, C_2 is at a constant voltage and the voltage at C_2 is twice the maximum voltage provided by the transformer.

4.2.3   The Cockcroft–Walton accelerators

Thanks to Greinacher's technique, a circuit with n diode–capacitor pairs can reach $V_{out} = nV_{max}$ at the last capacitor C_n. The intuitive explanation for the voltage multiplication is that, during the recharge phase, the capacitors are connected in series and, therefore, the voltage is multiplied with respect to a single capacitor because:

$$\frac{1}{C_{eq}} = \frac{1}{C_1} + \frac{1}{C_2} + \cdots \frac{1}{C_n} > \frac{1}{C_1} \qquad (4.3)$$

Fig. 4.3 A one stage Cockcroft–Walton accelerator or, equivalently, a Greinacher voltage-doubler. The circuit consists of a rectifier (Fig. 4.1) where the resistive load is replaced by another diode–capacitor pair. We can accelerate charged particles connecting a load to V_{out} that provides an energy gain qV_{out} to the particles. Note that, in this case, the negative pole of the transformer is floating.

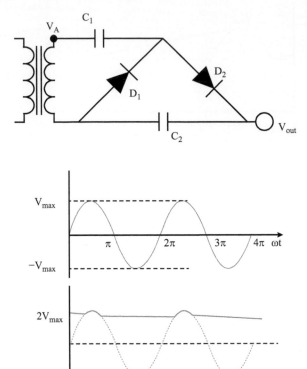

Fig. 4.4 Top: Input voltage at the transformer of Fig. 4.3 providing a $V(t) = V_{max} \sin \omega t$ input wave. Bottom (continuous line): V_{out} as a function of ωt. The diode–capacitor pair generates a rectified voltage of amplitude V_{max}. The size of the small exponential drop at $2V_{max}$ also depends on the value of the load connected to V_{out}.

and $C_{eq} < C_1$. So, for a fixed charge $Q = C_{eq}V$, an equivalent capacitance C_{eq} smaller than C_1 produces an increase of V. In the discharge phase, the diodes change the topology of the network. The capacitors are now in parallel and can only discharge through the resistance of the load (if any) with a large $C_{eq}R = (C_1 + C_2 + \ldots + C_n)R$.

The output terminal of the network is at a nearly fixed voltage $V_{out} \simeq nV_{max}$. The peak-to-peak voltage is twice this value and a particle can be accelerated by this device up to $V_{pp} \equiv 2nV_{max}$. The main drawback of this method is that the performance is poor if the load current is high. The impedance of the capacitors produce a sag in the DC output and the shape of the voltage profile is deformed by the parasitic resistance of the components, especially for many stages (high n). Even worse, if large currents flow through the load, V_{out} shows ripples that increase as n^3. As a consequence, the Greinacher multiplier is not particularly effective in high-power electronics but is an excellent tool for low-power devices. Nowadays, it is used in liquid crystal display (LCD) back-light, laser systems, oscilloscopes, low-current high-voltage (HV) suppliers, etc.

In 1932, J.D. Cockcroft and E.T. Walton realized that a few- ($n = 3-5$) stage device could be used to accelerate ions at MeV energies. This approach works because all accelerators run with very few particles (low

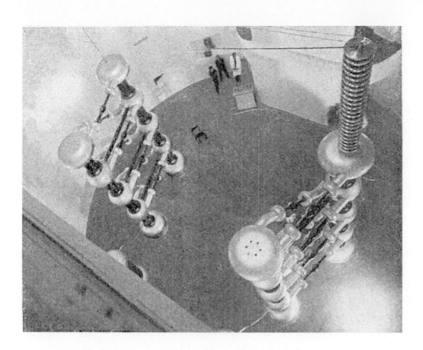

Fig. 4.5 A rare photograph of two 4-stage, 3 MV Cockcroft–Walton accelerators at the Kaiser Wilhelm Institut für Physik (Brown *et al.*, 1937). After the discovery of nuclear fission in 1938 (Hahn, 1958), this lab became the leading center of the German nuclear weapons program from 1939 to 1942.

currents) compared with standard industrial applications and are ideally suited for the Greinacher technique. These electrostatic accelerators are called the **Cockcroft–Walton accelerators** (see Fig.4.5). Before the Second World War, Cockcroft and Walton carried out a wealth of nuclear physics experiments using these devices and they were awarded the Nobel Prize for Physics for a direct demonstration of manmade atom transmutations: the long-sought-after dream of alchemists! The first transmutation reaction observed by Cockcroft and Walton was:

$$p + {}^{7}\text{Li} \rightarrow {}^{4}\text{He} + {}^{4}\text{He} \tag{4.4}$$

which proves the possibility of creating helium from lithium. Another nuclear reaction they observed during this experimental campaign:

$$d + d \rightarrow n + {}^{3}\text{He} \tag{4.5}$$

demonstrates that a small part of the *rest* mass of the deuterium can be transformed into kinetic energy, which is visible in the $n + {}^{3}\text{He}$ final state. Equation 4.5 confirmed the energy–mass equivalence principle put forward by Einstein and discussed in Sec. 2.3.3.

Other types of electrostatic accelerators can achieve higher energies than the Cockcroft–Walton machines at the expense of lower currents. The **Van de Graaff** and **tandem** accelerators reach several tens of MeV and are described, for example, in (Wille, 2001).

Fig. 4.6 Schematics of a linear accelerator. The even hollow conductors ("drift tubes") at time t have opposite polarity with respect to the odd conductors and the electric field oscillates at the frequency of the RF source. The conductors are installed in a vacuum pipe to prevent scattering of the particles in the air.

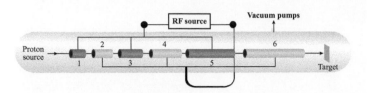

4.3 Linear accelerators and the phase stability principle

In the 1920s, G. Ising and R. Widerøe proposed time-varying fields to overcome the limits of the Cockcroft–Walton devices. The most elegant example still in use is the **linear accelerator** (LINAC). Here, a set of hollow metallic cylinders inside the main vacuum tube is connected to a high-power radiofrequency (RF) generator in the configuration of Fig. 4.6. Each conductor is connected to the positive (odd conductors) and negative (even conductors) poles of the AC voltage source.

At a time t_0, a charged particle in the proximity of a cylinder experiences an oscillating electric field $\mathbf{E} = E_0 \sin \omega t \ \mathbf{k}$ along the direction of motion (z-axis). If a positive particle is properly synchronized (point a in Fig. 4.7), it will be accelerated when it is near the conductor and the electric field is parallel to the z axis. Likewise, a negative particle will be accelerated by anti-parallel fields. When the particle enters the hollow area of the conductor, the conductor acts as a Faraday cage. It thus shields the particle by the electric field which, in the meanwhile, changes sign at the angular velocity ω. If the positive particle exits the conductor when the field is again parallel to $\mathbf{k}$, that is after a half-period, the acceleration continues up to the second conductor. Hence, if the synchronization of the particles with the field can be preserved over many conductors, the energy of the particles increases as nV_{max}. n is the number of the conductors and V_{max} is the peak voltage of the RF field connected to the conductors. In the non-relativistic case, the length of the conductors must increase as:

$$L_n \simeq v_n T \tag{4.6}$$

because the acceleration increases the velocity of the particle. In eqn 4.6, L_n is the length of the n-th conductor, $T = 2\pi/\omega$ is the RF period and v_n is the particle speed at the conductor n. The non-relativistic kinetic energy is:

$$E_n^{kin} = qnV_{max} = \frac{1}{2}mv_n^2 \tag{4.7}$$

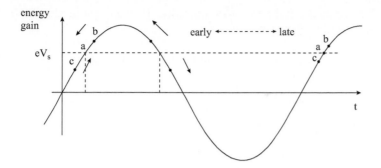

Fig. 4.7 Accelerating potential V or, equivalently, energy gain qV of a particle that reaches one of the hollow conductors at the time t. The synchronous particle is labeled a. Slower and faster particles are labeled b and c, respectively. These particles get closer to a at the next conductor (rightmost part of the plot) since they all lie in the $dV/dt > 0$ region.

so that L_n must increase as $\sim \sqrt{n}$. If the particle is ultra-relativistic, $v_n \simeq c$ and L_n must be constant. For practical reasons, we often inject particles in LINACS when the particles are already ultra-relativistic.

4.3.1 Veksler–McMillan stability

In the 1930s, people were skeptical about achieving such synchronization.[2] The first accelerators built on the synchronization principle performed well beyond expectations but a full understanding of their dynamics was achieved only in the late 1940s, thanks to the work of E. McMillan and V. Veksler who independently discovered the **phase stability principle**. We illustrate the principle at linear accelerators because of their simplicity but we will extend it to circular accelerators in the next section.

Fig. 4.7 shows the voltage in the conductors as a function of time. The first period corresponds to the first conductor and a particle a perfectly synchronized with the field experiences the same field at the entrance of any conductor. This orbit corresponds to the trajectory of the **synchronous particle**. If we inject a continuous flux of particles in the accelerator, the particles in the proximity of a are accelerated only if the derivative of $V(t)$ is positive. This is because a particle slightly out of phase with respect to a (e.g. slower than a) arrives later than a because of the smaller velocity. This particle (b in Fig. 4.7) experiences a stronger field than a if $dV/dt > 0$. Hence, in the next period, it arrives slightly in advance because the velocity is a little greater than earlier. Similarly, a particle that arrives slightly before a because it has a higher velocity than a (particle c in Fig. 4.7) experiences a weaker field if $dV/dt > 0$. In the next period, it arrives slightly later because the velocity is a bit smaller than earlier. Along their path through the conductors, these particles oscillate around a and are automatically synchronized, even if their initial conditions are slightly out of phase. In practice, the vast majority of the particles located in the rising edge of $V(t)$ are accelerated up to the last conductor and cluster around the synchronous particle. Unfortunately, no particle is accelerated if it lies in the falling edge $dV/dt < 0$, where a particle that is slower than a experiences a smaller field than a and runs even slower. These particles are brought out of phase and decelerated down to $v \simeq 0$. They get stuck among the conductors and eventually hit the walls of the accelerator

[2]For the sake of clarity, we are not following the historical development of the phase stability concept. The issue of synchronization was perceived as outstanding because the first non-electrostatic machines were circular (cyclotrons, betatrons, etc). In these devices, the particles must be stored inside the accelerator for a long time and reach the extraction area with high precision. For a historical review see e.g. Chap.1 of Wille, 2001.

or the residual gas in the vacuum tube. In conclusion, thanks to the phase stability principle:

> all particles reaching the conductor in the proximity of the synchronous particle when the derivative of V(t) is positive are automatically synchronized. All other particles are lost.

4.3.2   High-energy LINACS

This self-synchronization mechanism ensures that we can accelerate particles even if they are not perfectly in phase with the electric field. About half of them are lost and the other half clusters in **bunches** around the synchronous particles. The maximum number of bunches we can fill with particles corresponds to the number of periods of the RF field. So, a $L = 1$ km long accelerator working at $\nu = 1$ GHz can have at most:

$$L/(vT) \simeq L\nu/c \simeq 3300 \qquad (4.8)$$

bunches. These potential places for bunches are called **buckets**. For practical reasons, the number of bunches is less than the buckets. Unlike the electrostatic accelerators, there are no intrinsic limitations to the number of conductors that can be connected to the RF field.

> The maximum energy attainable in a LINAC is proportional to the number of the conductors, that is the length of the accelerator, and the maximum achievable electric field.

Current research and development (R&D) projects aim at field gradients of $\mathcal{O}(100$ MV/m$)$ employed in very long LINACS (tens of km) to accelerate electrons up to several (Stapnes, 2019; Michizono, 2019).

The most powerful linear accelerator ever built was the Superconducting Linear Collider (SLC). It was constructed and operated at SLAC (United States) in the 1990s to perform head-on collisions of electrons and positrons at $\sqrt{s} \simeq 90$ GeV. Circular accelerators bend the trajectory of the particles with a centripetal force (see Sec. 4.4) and, therefore, induce radiation losses. As discussed in Chap. 3, radiation losses are nearly immaterial for heavy particles but these losses represent a serious obstacle for the acceleration of electrons at high energy. The acceleration of electrons and positrons is, therefore, the application of choice for LINACS in high-energy physics. At lower energies, linear accelerators are also used for medical applications to boost protons or ions (Karzmark, 1997).

4.4   Synchrotrons

The first LINAC was patented by L. Silard in 1928. One year after, he patented a circular accelerator, too, which was developed in the United States by E. Lawrence and M.S. Livingston, the **cyclotron** (see Fig. 4.8). In this machine, the particles follow a circular path inside a large iron magnet and are accelerated by an AC electric field. A cyclotron stable orbit is a spiral running from the inner to the outer rim of the machine because **B** is constant. At the same time, R. Widerøe proposed a machine with varying magnetic fields that was built in Germany by M. Steenbeck in 1935 and called the **betatron** (Wille, 2001). As already mentioned, Veksler and McMillan elaborated the phase stability principle to understand why these machines were working so well when producing particles in bunches. Their study brought to the proposal of the synchrotron that was built for the first time by M. Oliphant in 1952.

> A **synchrotron** is a circular machine where particles are accelerated in closed orbits using time-varying electric and magnetic fields.

The magnetic field steers the particles around the **synchronous orbit**. This trajectory is the optimal path and is always in phase with the accelerating electric fields. During the acceleration, the momentum of the particle increases and, therefore, the magnetic field $\mathbf{B}(t)$ must increase according to:

$$|\mathbf{p}| = 0.3B_z(t)R \tag{4.9}$$

for a particle with unit charge (see eqn 3.35) in order to keep the particle in the same orbit. In eqn 4.9, R is the radius of the machine and B_z is the component of the field perpendicular to the plane of the orbit. The product $B_z R$ is called **rigidity** because it represents the inertia of the particle against steering. At first sight, one could imagine the best field configuration to be $B_x(t) = B_y(t) = 0$. We will show later that this configuration is dynamically unstable and a small correction for the magnetic field in the $x - y$ plane is mandatory. The accelerator is called **synchrotron** because the magnetic field must be synchronized with the electric field and the rate of momentum transfer. An increase of $|\mathbf{p}|$ is compensated by an increase of $B_z(t)$ to keep the particle in the circle of radius R. A synchrotron requires a "double synchronization": the particle must be in phase with the AC electric field and the increase of $B_z(t)$ must be synchronized with the increase of particle's momentum.

Like LINACS, perfect synchronization is technically impossible and the stabilization of the particles around the synchronous orbit is ensured by the phase stability principle.

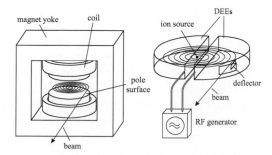

Fig. 4.8 Left: schematics of a cyclotron. The magnetic field (not shown) is vertical and constant. The stable orbit between the magnet poles is shown as a black spiral. Right: the particles are accelerated by an AC electric field in the gap located between the two D-shaped halves of the magnet (the "dees"). From Wille, 2001.

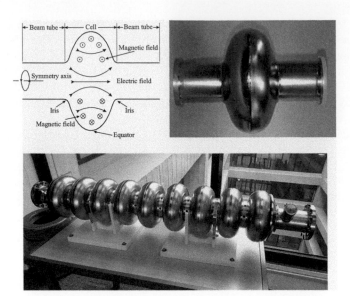

Fig. 4.9 Upper left: schematics of an RF cell. The electric field **E** lies along the cell axis, while the magnetic field **B** produced by the RF cavity curls around **E**. Note that **B** is negligible compared with the fields produced by the magnetic dipoles of the machine. Upper right: a single-cell RF cavity. Bottom: a multi-cell niobium cavity operating at 1.3 GHz. Redrawn from Adolphen *et al.*, 2013.

[3]Superconductors are materials whose electric resistance drop to zero below a given temperature due to collective quantum effects (Annett, 2004). There are hundreds of elements (Pb, Al, Nb, etc.) and alloys that show this behavior. Unfortunately, superconductivity shows up only at low temperature and room temperature superconductors have not been discovered yet.

4.4.1 The components of a synchrotron

A modern synchrotron is a system made up of magnetic dipoles, quadrupoles, and sextupoles, together with several RF cavities. These components are located around a circular vacuum tube called the **beampipe**. A magnetic dipole is an electromagnet, whose field lines are perpendicular to the plane of the synchronous orbit. In general, the accelerators use conventional copper coils to produce the magnetic field. Normalconducting magnets are plagued by the large consumption of energy due to heat dissipation in the coils (**Joule heating**) and produce magnetic fields in the 1–2 T range. Superconducting magnets use coils made of superconductors[3] so that the Joule heating becomes negligible. This smart solution comes at the price of complexity and cost since no superconductors can be operated at room temperature and the magnets must be cooled down to a few Kelvin. The most spectacular use of cooled magnets is employed at the LHC, a proton accelerator located at CERN and capable of reaching an energy of 7 TeV. The LHC employs niobium–titanium (Nb-Ti) alloys in the coils of the magnets. The coils are cooled to 1.9 K and produce a magnetic field of 8.3 T. The protons remain inside the stable orbit (see Sec. 4.5) for tens of hours. Hence, once or twice a day, new protons are injected into the machine at 400 GeV and accelerated from 400 GeV to 7 TeV. In the acceleration stage, the field of the superconducting magnets increases from 0.54 T to 8.3 T.

As in LINACS, the AC electric fields accelerate the particles. The fields are produced by hollow conductors connected to the AC source and called **RF cavities** (see Fig. 4.9). At the LHC, the cavities work at a frequency of 400.8 MHz and a peak voltage of 16 MV. The shape of the conductor is such that the electric field is in the direction of the orbit but oscillates from $-E_{max}$ to E_{max} with a frequency ν. The cavities

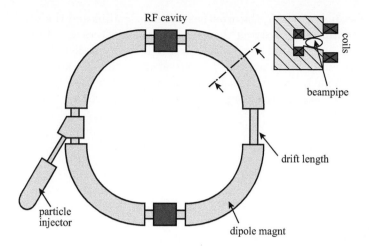

RF cavity

coils

beampipe

drift length

particle injector

dipole magnt

Fig. 4.10 Layout of the simplest synchrotron. In this machine, there are only two RF cavities to accelerate the particles and four bending dipole magnets. Quadrupole and sextupole magnets (not shown) can be used for particle focusing. The particles travel inside the beampipe along a closed orbit. Particles are inserted into the beampipe by an injector and removed after the acceleration by an extractor (not shown) based on electrical kickers. Redrawn under CC-BY-SA from Wikimedia, 2007.

are interleaved with the magnets following a circular path of radius R (Fig. 4.10)

4.4.2 Phase stability in synchrotrons

Phase stability can be achieved also in circular orbits provided that the gradients of the magnetic fields are properly tuned. The synchronous particle crosses the ring of the accelerator in a time $T = 2\pi R/v \simeq 2\pi R/c$ for ultra-relativistic particles. The RF frequency f_{RF} must, then, be a multiple of T^{-1}:

$$f_{RF} = \frac{c}{2\pi R} \cdot n \qquad (4.10)$$

where n is an integer number. Since producing RF fields is technically simpler in the GHz range (radiowaves or microwaves from 100 MHz to 10 GHz), n is typically much larger than 1 ($n \simeq 10^4$ at the LHC).

We can evaluate the stability of a synchrotron firstly in a 2D model that describes the motion in the plane of the synchronous orbit. We call this plane the "$x - y$ plane" and express the particle position in cylindrical coordinates. We label r the distance of a particle from the center of the ring, as shown in Fig. 4.11. Note that $r = R$ for the synchronous particle. θ represents the polar angle of the particle and z its distance from the $x - y$ plane. The position of the particle at time t is thus given by $\mathbf{x}(t) = (r(t), \theta(t), z(t))$.

A particle with momentum $\mathbf{p} = (p_r, p_\theta, 0)$ is synchronous if:

$$qvB_z(R) = \frac{\gamma m v^2}{R}. \qquad (4.11)$$

This equation describes a uniform circular motion in special relativity for a particle of charge q and velocity v (see Sec. 3.3). The ideal field is, then:

$$B_z(R) = \frac{\gamma m v}{qR}. \qquad (4.12)$$

The orbits of Fig. 4.12 are stable in the $x - y$ plane if any perturbation that brings the particle at $r > R$ is compensated by the Lorentz force.

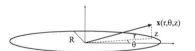

$\mathbf{x}(r,\theta,z)$

R

z

θ

Fig. 4.11 Position of a particle in a synchrotron in cylindrical coordinates. The nominal orbit has a radius R and lies in the $x - y$ plane.

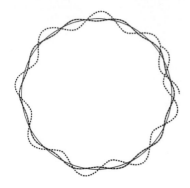

Fig. 4.12 Orbits inside the beampipe of a synchrotron. The continuous line is the synchronous orbit, i.e. the ideal orbit of the machine. Particles that are not perfectly synchronized oscillate around the ideal orbit and produce a closed periodic motion (dot-dashed line). The transverse (horizontal and vertical) oscillations are the betatron oscillations (see text). A horizontal betatron oscillation is shown in the figure with a dashed line.

For instance, if $r < R$, the orbit has a smaller radius and the particle has a momentum slightly lower than the synchronous particle. The standard orbit will be restored only if the Lorentz force is smaller than the nominal force in order to increase the curvature radius. If the momentum is slightly higher, the orbit has a larger radius ($r > R$) and the Lorentz force must be slightly higher than the nominal force to bring r closer to R. In turn,

$$qvB_z(r) < \gamma\frac{mv^2}{R} \quad \text{for } r < R \tag{4.13}$$

$$qvB_z(r) > \gamma\frac{mv^2}{R} \quad \text{for } r > R.$$

If these conditions are fulfilled, the machine reaches a stable acceleration in the horizontal plane. To study the dynamics of the particles in the beampipe, we consider small perturbations from the nominal orbit (Rossbach and Schmuser, 1992):

$$r = R + x = R\left(1 + \frac{x}{R}\right) \tag{4.14}$$

$$\frac{\gamma mv^2}{r} = \frac{\gamma mv^2}{R\left(1 + \frac{x}{R}\right)} \simeq \frac{\gamma mv^2}{R}\left(1 - \frac{x}{R}\right). \tag{4.15}$$

Similarly, a small perturbation of B_z can be written as:

$$B_z(r) \simeq B_0 + \left.\frac{\partial B_z}{\partial r}\right|_{r=R} x \tag{4.16}$$

and we introduce the **field index n** in:

$$evB_z(r) \simeq evB_0\left(1 - n\frac{x}{R}\right) \tag{4.17}$$

defining n as:

$$n \equiv -\frac{R}{B_0}\left.\frac{\partial B_z}{\partial r}\right|_{r=R}. \tag{4.18}$$

Inserting these formulas in eqn 4.13, we see that:

> the motion of a particle close to the synchronous particle is stable in the horizontal plane if $0 \leq n < 1$.

Note, in particular, that if we neglect any possible drift in the vertical axis z, the stability condition can be achieved even with a constant magnetic field ($n = 0$). The $n = 0$ condition is called **geometrical focusing**: particles with a slightly different momentum will run on stable closed orbits, all crossing the same point (see Fig. 4.13). Unfortunately, $n = 0$ is unstable in the vertical plane where a small difference from $p_z = 0$ provokes the particle into spiraling upward or downward until it smashes the walls of the beampipe.

A restoring force $F_z \sim -z$ is needed in the vertical plane, too, and can be obtained by an appropriate gradient of the magnetic field along z. If $F_z = -Cz$, then $B_x = -C'z$ and

$$\frac{\partial B_x}{\partial z} = C'' \neq 0 \tag{4.19}$$

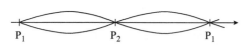

particles around the
synchronous orbit

synchronous orbit

P_1 P_2

P_1 P_2 P_1

Fig. 4.13 Top: particle orbits in a homogeneous magnetic field. The synchronous particle is shown with a thicker line. Two other particles diverging by an angle α are shown with thinner lines. These particles intercept the orbit at the points P_1 and P_2. Bottom: the same orbits expressed on a straight line. This representation clarifies that P_1 and P_2 act as focusing points. The maximum deviation from the nominal orbit is called the **betatron oscillation** and, in this case, corresponds to αR. Redrawn from Rossbach and Schmuser, 1992.

where C and $C' = -C''$ are constants. In a real accelerator, we need simple field configurations to reduce the perturbations introduced by mechanical and electrical imperfections. We therefore seek a magnetic field that does not depend on θ so that the correcting magnets can be uniformly distributed along the synchrotron. Besides, when the accelerator is empty, there are no currents and charges due to the particles, and $\nabla \times \mathbf{B} = 0$. This condition holds even during the running of the machine because the current produced by the particles generates a negligible field compared with the magnetostatic fields of the dipoles. Hence,

$$\mathbf{B}(r,\theta,z) = \mathbf{B}(r,z) \text{ and } \nabla \times \mathbf{B} \simeq 0. \tag{4.20}$$

Recasting the curl in cylindrical coordinates, we have:

$$\nabla \times \mathbf{B} = \hat{\mathbf{r}} \left(\frac{1}{r} \frac{\partial B_z}{\partial \theta} - \frac{\partial B_\theta}{\partial z} \right) + \hat{\boldsymbol{\theta}} \left(\frac{\partial B_r}{\partial z} - \frac{\partial B_z}{\partial r} \right) \\ + \hat{\mathbf{z}} \frac{1}{r} \left(\frac{\partial B_\theta r}{\partial r} - \frac{\partial B_r}{\partial \theta} \right) \tag{4.21}$$

where $\hat{\mathbf{r}}$, $\hat{\mathbf{z}}$ and $\hat{\boldsymbol{\theta}}$ are the unit-vectors in cylindrical coordinates. In a synchrotron, $\mathbf{B}$ does not depend on θ and we require:

$$B_\theta = 0, \quad \frac{\partial B_r}{\partial \theta} = 0 \text{ and } \frac{\partial B_z}{\partial \theta} = 0 \tag{4.22}$$

so that

$$\nabla \times \mathbf{B} = \hat{\boldsymbol{\theta}} \left(\frac{\partial B_r}{\partial z} - \frac{\partial B_z}{\partial r} \right). \tag{4.23}$$

Setting eqn 4.23 equal to zero, we end up with:

$$\frac{\partial B_r}{\partial z} = \frac{\partial B_z}{\partial r} \neq 0. \tag{4.24}$$

Equation 4.24 establishes the stability condition also in the vertical plane so that the Veksler–McMillan principle for synchrotrons reads:

> the motion of a particle close to the synchronous particle is stable if $0 < n < 1$ and
>
> $$\frac{\partial B_r}{\partial z} = \frac{\partial B_z}{\partial r}. \tag{4.25}$$

4.4.3 Weak and strong focusing*

Several dipolar magnets can fulfill eqn 4.25. The most common are **tapered pole magnets** (or skewed magnets) like the one shown in Fig. 4.15. Accelerators designed to fulfill the conditions above are called **weak focusing synchrotrons** because the betatron oscillations are corrected by the custom fringe fields of these magnets. This technique has been used for many years in cyclotrons and early synchrotrons but has a serious drawback. Weak focusing is based on the fact that nearby circles – the orbits of charged particles moving in a uniform magnetic field – intersect once per revolution. Weak focusing is thus a small correction to the geometrical focusing, where the magnetic field is completely uniform. Far from the intersection, the particles diverge by several cm and require a large diameter beampipe to be contained inside the machine. In turn, weak focusing requires the magnetization of large volumes and a big vacuum pipe, which substantially impact on the cost and complexity of the machine.

We can build much cheaper and more powerful accelerators if we manipulate the particles employing **strong focusing**. This method had been developed from 1949 to 1952 by N. Christofilos, E. Courant, and H. Snyder, and is the reference design principle of all modern synchrotrons. It is based on a well-known law of geometrical optics, depicted in Fig. 4.14: a set of convex and concave lenses produces a net focusing effect both in the horizontal and vertical plane. Similarly, Courant and Snyder demonstrated that the net effect of alternating the field gradient is that both the vertical and horizontal focusing of the particles is enhanced. The orbits then shrink toward the synchronous particle and the dimension of the beam in the direction perpendicular to $\hat{\boldsymbol{\theta}}$ is reduced. In this way, the space that must be magnetized inside the dipoles is much smaller because we do not need to collect particles that are far from the ideal orbit. Since the cost and size of the magnets depend on the size of the magnetized volume, strong focusing reduces the cost and complexity of any kind of circular accelerators by orders of magnitude.

Strong focusing is produced by magnetic multipoles. The quadrupolar magnets (quadrupoles) act as linear optical lenses and the motion of a particle through the quadrupoles can be described by linear equations. These magnets are enough to achieve strong focusing and are the main focusing elements of the accelerator. Higher-order multipoles, as sextupoles or octupoles, are introduced in the machine to correct for subtler effects, like the dependence of the quadrupole focusing on the particle

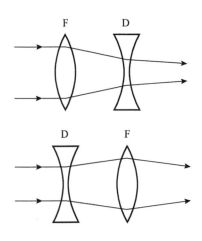

Fig. 4.14 Two light rays cross a convex (F) and concave (D) lens. The rays are focused both in the FD configuration (top) and in the DF configuration (bottom). This principle is employed in strong focusing replacing the rays with particle bunches and the lenses with quadrupoles.

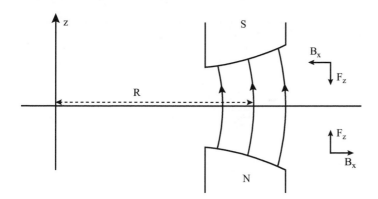

Fig. 4.15 The magnetic field lines of a tapered pole magnet. R is the radius of the synchrotron. N and S are the magnetic poles of the electromagnet. The magnetizing coils are not shown. Redrawn from Rossbach and Schmuser, 1992.

momentum. This effect is akin to chromatic aberration in geometrical optics.

Multipoles are important to increase the power and luminosity of the synchrotrons but introduce non-linearities in the particle motion. They must be carefully simulated, optimized in the commissioning phase of the machine, and are potential sources of non-linear instabilities.

A **quadrupole** like the one in Fig. 4.16 is an electromagnet where the coils produce a magnetic field in a quadrupolar configuration. The field lines of a quadrupole are shown in Fig. 4.18. The dashed lines represent the four poles of the magnet.

The basic unit of a strong focusing accelerator is a combination of alternating-gradient quadrupoles and is called a FODO cell.

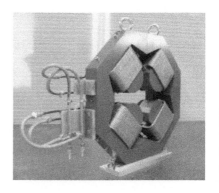

Fig. 4.16 A quadrupole produced for the EMMA experiment at CERN. Reproduced under CC-BY-SA from Wolsky, 2009.

> A **FODO cell** is the simplest structure that realizes the strong focusing mechanism. It is composed of a quadrupole that focuses particles in the horizontal direction and defocuses in the vertical direction (F) followed by a drift length (O) and a quadrupole that focuses in the vertical direction and defocuses in the horizontal direction (D) followed by another drift length (O).

Fig. 4.17 A particle crossing three FODO cells and oscillating around the synchronous orbit (horizontal axis). The quadrupoles are shown as focusing (F) and defocusing (D) lenses in the vertical plane.

At the end of the last drift length,[4] the particles enter the next FODO cell with a net focusing in the transverse plane. The accelerator is interspersed by FODO cells to form a **lattice**. The design of the lattice is aimed at keeping the beam close to the synchronous orbit minimizing the dispersion in the transverse plane and, hence, the dimension of the magnets. A fraction of a FODO lattice is shown in Fig. 4.17.

[4]The drift length is just a part of the accelerator without any field, where the particles propagate in rectilinear motion.

4.4.4 Emittance

The main figure of merit of a lattice design is the **emittance** of the beam, which is a measure of the spread of the particles inside the machine or, equivalently, a measure of the size of the beam in the position-momentum phase space. The emittance of a beam along the x-axis at a given point of the ring depends on the position of the particles in x and the conjugate variable p_x. To ease the notation, we can use a

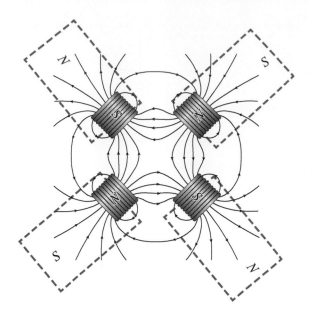

Fig. 4.18 The magnetic field lines of a quadrupolar magnet (quadrupole). The dashed areas represent the North (N) and South (S) poles of the magnet, while the coils producing this field configuration are shown on top of the field lines. The force experienced by a particle entering the quadrupole in the direction perpendicular to the figure is proportional to $\mathbf{v} \times \mathbf{B}$. In the field configuration of this figure, a positive particle is focused in the horizontal direction and defocused in the vertical direction. A particle crossing the quadrupole along its center experiences no focusing.

reference frame whose z-axis is along the axis of the lattice, the y-axis represents the vertical direction and the x-axis is the horizontal direction as in Fig. 4.19. The synchronous particle is thus at $\mathbf{x} = (0, 0, z)$. If we call $x' = dx/dz = p_x/p_z$ the angle of the particle in the $x - z$ plane:

> the (2D) emittance along x is called ϵ_x and is the area occupied by the beam particles in the $x - x'$ plane divided by π.

The SI unit of ϵ_x is m·rad. In real machines, it is difficult to compute the area for the *entire* beam. For instance, if the beam is bi-Gaussian in the $x - x'$ plane, the particle distribution is:

$$f(x, p_x) = \frac{1}{2\pi\sigma_x\sigma_{p_x}} \exp\left(-\frac{x^2}{2\sigma_x^2}\right) \exp\left(-\frac{p_x^2}{2\sigma_{p_x}^2}\right) \qquad (4.26)$$

and we should integrate on $\mathbb{R}^2$ to collect each particle, including small tails far from the synchronous orbit. More often, the emittance is defined with respect to a fixed fraction of the beam (68%, 90%, or 95%) and, in the following, we will mainly consider 68% ("one sigma") emittances. Fig. 4.19 shows the ellipse containing 68% of the beam for three points in the lattice.

The emittance plays a prominent role in accelerator physics because, according to Liouville's theorem (Taylor, 2005), its surface is preserved in a machine without losses. Therefore, no magnetic lens can change its value. If the particles have reached the maximum energy, ϵ_x is an integral of motion, that is its value does not change along the lattice. In general,

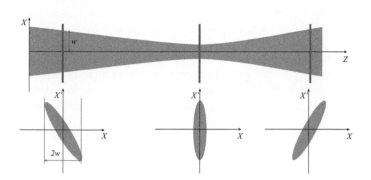

Fig. 4.19 Top: beam profile in three points of the lattice (vertical bars). The beam propagates along z. The gray area is the space occupied by the beam in the $x - z$ plane during its motion and w is the waist of the beam at a given point. Note that the minimum geometrical size is achieved at the second point, which corresponds to the largest spread of x'. Bottom: Distribution of the particles in the $x - x'$ plane. Since the normalized emittance is a conserved quantity (see text), the area of the ellipse is constant at any point.

> if the beam is accelerated along z, the emittance decreases but the **normalized emittance**
>
> $$\epsilon_x^* \equiv \gamma\beta\epsilon_x \qquad (4.27)$$
>
> is an integral of motion, that is it is a conserved quantity.

Here, γ and β are the Lorentz factors at the energy where ϵ_x is computed.[5] Similarly, we can define ϵ_y and ϵ_y^* as the emittance and normalized emittance in the $y - y'$ plane, respectively.

Why is the emittance so important for accelerator scientists? Fig. 4.19 shows that the beam can be shrunk at will using magnetic multipoles but the smaller the waist of the beam along an axis (w in the figure) the larger the spread of x'. As a light ray, if we shrink a bunch of particles in a point z, the bunch starts diverging at $z + dz$ and the divergence rate is proportional to $1/w$. So, if we want to minimize the size of the beam, the lattice will not help: what you gain at a point of the machine will be lost at another point. The normalized emittance is constant in the absence of dissipative or cooling forces and, therefore, is a better quantity to describe a beam than the geometrical size, which varies along the ring.

Even more, the correlation between the planes of motion is very weak in most accelerators. In these cases, we can define the 4D and 6D emittance as the product $\epsilon_x\epsilon_y$ and $\epsilon_x\epsilon_y\epsilon_z$, respectively.[6]

You may guess that there should be a link between the small size of the emittance and the performance of the accelerator because a small emittance eases the collisions between bunches. You are on the right track: we discuss this link in Sec. 4.5.

Finally, what is the maximum energy a synchrotron can achieve? Is there an intrinsic limit?

[5]The decrease of the emittance after the acceleration phase is quite a trivial effect: since $x' = p_x/p_z$, $x' \to 0$ for $p_z \to \infty$. The corresponding grows of the luminosity is called **natural rise** and partially compensates eqn 2.136. On the other hand, the demonstration that the emittance at fixed momentum is an integral of motion is more complicated and is based on the Liouville theorem of Hamiltonian mechanics (Wiedemann, 2015).

[6]A full demonstration of phase stability in accelerator lattices and the development of the theory of emittance is beyond the scope of this book. The interested reader may find a gentle but rigorous introduction in Wilson, 2001, and a comprehensive treatment in Wiedemann, 2015.

> Thanks to the phase stability principle, a particle can be indefinitely accelerated by the same set of RF cavities. In this case, the upper limit to the energy is imposed by the maximum bending field of the dipoles and the radius of the accelerator because a larger $|\mathbf{p}|$ requires either a longer R or a stronger B_z (see eqn 4.9).

Over the years, the energies of the synchrotrons have been raised using higher magnetic fields (superconducting magnets), increasing the size of the machine, and – since 1961 – operating the circular accelerators in collider mode.

4.5 Colliders

The easiest way to exploit a beam of accelerated particles is to set up a fixed-target experiment, that is, steer the beam against a solid-state target. On the other hand, the maximum momentum transfer among particles occurs when the collision is head on: a configuration that can be implemented both in linear and circular accelerators. The first successful experiment of this kind was performed in 1961 in Frascati (Italy). A team led by B. Touschek designed a small circular device accelerating electrons clockwise and positrons anticlockwise. Since the only difference between particles and antiparticles is the electric charge, any circular machine can be used in this configuration, although the design of the colliding point, that is the point where the trajectories of the electron bunches intersect the trajectory of the positron bunches, requires special care. As discussed in Sec. 2.7.2, the main figure of merit of a collider is the luminosity. A high-luminosity machine produces a wealth of interactions at fixed center-of-mass energy or, equivalently, at fixed $\sqrt{s}$. In a **collider**, the detectors are located near the interaction point where strong quadrupoles focus the beam to its minimum waist. The new particles produced after the interactions are scattered far from the synchronous orbit and cross the walls of the beampipe, which is surrounded by particle detectors that we will describe in Sec. 4.6.

The luminosity of a collider depends on the quality of the particle beams and can be expressed as:

$$\mathcal{L} = \frac{N_1 N_2}{A} f \tag{4.28}$$

where A is the average transverse area of the bunches, N_1 and N_2 are the number of particles in the colliding bunches, and f is the collision frequency. All these parameters can be improved:

- increasing the number of particles per bunch. This is often limited by the electrostatic repulsion of same-sign particles squeezed inside the bunch (**space-charge effects**);
- increasing the number of bunches, which is limited by the maximum amount of particles that can be injected and the available buckets;

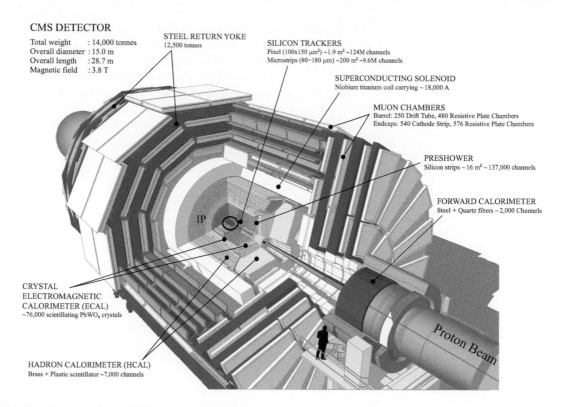

Fig. 4.20 Artist's view of the CMS detector at LHC (Sakuma and McCauley, 2014). The interaction point is labeled IP and located at the center of the detector. The beams of protons reach the IP from opposite directions through the beampipe ("proton beam") and produce head-on proton–proton scatterings at $\sqrt{s} = 14$ TeV. The momentum of the final state charged particles is measured by the bending of the tracks near the IP. The bending is produced by a solenoidal magnet parallel to the beampipe. Reproduced under CC-BY-SA 4.0 from CMS, 2014.

- improving the focusing and the overall transverse size of the bunch to reduce A. The techniques that reduce the momentum of the particles in the plane perpendicular to the ring are called **cooling**. We will discuss one cooling method, namely stochastic cooling, in Sec. 12.5 since it played a central role in the development of the Standard Model.

Example 4.1

It is not surprising that A is somehow linked to the beam emittance. Inside a lattice, the particles oscillate around the ideal orbit according to **Hill's equation**:

$$\frac{d^2x}{ds^2} + K(s)\,x = 0 \qquad (4.29)$$

where s parameterizes the position of the synchronous particle along the ring. $s \in (0, 2\pi R)$ and $s = 0$ corresponds to $\mathbf{x} = (R, 0, 0)$ in Fig. 4.11. $K(s)$ is a function that describes the steering due to the magnetic multipoles (Wilson, 2001). The most general solution of the Hill's equation is a periodic motion in the transverse plane that produces the betatron oscillations of Fig. 4.12. The solution depends on three parameters: the **amplitude function** $\beta(s)$, the **phase advance** $\phi(s)$ and the transverse emittance $\epsilon = \epsilon_x$ in the $x - x'$ plane:

$$x = \sqrt{\epsilon \beta(s)} \cos \phi(s) \tag{4.30}$$

$$x' = -\sqrt{\frac{\epsilon}{\beta(s)}} \sin \phi(s). \tag{4.31}$$

From eqns 4.30 and 4.31 we can easily derive the axes of the beam ellipsis in the $x - x'$ plane: $\sqrt{\epsilon \beta}$ and $\sqrt{\epsilon/\beta}$. The waist $w(s)$ along the ring of the accelerator (see Fig. 4.19) is also called the **beam envelope** and is given by:

$$w(s) = \sqrt{\epsilon \beta(s)}. \tag{4.32}$$

The beam dimension in the $x - y$ plane is:

$$\left(2\sqrt{\epsilon_x \beta_x(s)}\right) \cdot (2\sqrt{\epsilon_y \beta_y(s)}) \simeq 4\epsilon \beta(s). \tag{4.33}$$

In eqn 4.33 we assumed a bunch symmetric in the $x - y$ plane, that is, $\epsilon_x \beta_x(s) \simeq \epsilon_y \beta_y(s)$. This assumption is not always a good approximation but the general formula is a simple extension of eqn 4.30 (see Exercise 4.12). The cross-section of a bunch is the area of the bunch projected in the $x - y$ plane and, if the beam is symmetric in this plane, $\sigma_x = \sigma_y \equiv \sigma$. The cross-section is then:

$$\pi(2\sigma)^2 = 4\pi\sigma^2 = 4\epsilon\beta. \tag{4.34}$$

As a consequence,

the amplitude function $\beta(s)$ is a function of the emittance and can be expressed as:

$$\beta(s) = \frac{\pi\sigma^2(s)}{\epsilon}. \tag{4.35}$$

$\beta(s)$ can be geometrically interpreted as the distance between the point of the minimum waist (the focus of the lattice or a FODO cell) and the closest point where the waist is twice this value.

A higher luminosity results from a tighter squeezing of the bunches at the interaction point (IP). Since $A \simeq 4\pi\sigma^2$, we have from eqns 4.28 and 4.35:

$$\mathcal{L} \simeq \frac{N_1 N_2}{4\pi\sigma^2} f = \frac{N_1 N_2}{4\epsilon\beta*} f \tag{4.36}$$

where β^* is the amplitude function at the IP. As expected:

> the luminosity of a collider is large if the machine emittance is small and the focusing at the interaction point is strong, that is β^* is small.

Equation 4.36 provides a formula that links the expected luminosity at a collider with the machine parameters of the synchrotron.

Despite technical complexity, a collider offers unprecedented opportunities to produce heavy particles and perform precision physics. e^+e^- colliders scatter elementary particles of the same energy and, therefore, the rest frame (RF) of the system is the same as the laboratory frame (LAB). Radiation losses are sizable for the electrons (see Exercise 4.7) and the average energy that can be achieved is generally smaller than, for example, protons. On the other hand, protons are composite particles made of quarks. The RF of the colliding quarks is unknown a priori and is different from the center-of-mass frame of the colliding protons (LAB). As a consequence, proton-antiproton colliders are considered "discovery machines" because they can attain the highest energies. Nonetheless, e^+e^- colliders have a better precision because the interaction energy is uniquely defined by the energy of the incoming $e^\pm$ and there are no other particles in the initial state. We can appreciate the power of colliders by simple kinematics considerations:

Example 4.2

Let us compute the projectile energy that is needed to create a Z^0 particle ($M_Z \equiv M \simeq 91$ GeV/c^2). Producing a heavy particle in a fixed target experiment requires a proton momentum $|\mathbf{p}| = p_z \equiv p$ that can be computed from the invariant mass in LAB. In a $a + b \to c$ fixed target scattering, $p_a^\mu = (E_a, 0, 0, p_a)$ and $p_b^\mu = (m_b, 0, 0, 0)$. Hence,

$$s = p_{tot}^2 = (E_a + m_b)^2 - p_a^2 = m_a^2 + m_b^2 + 2E_a m_b. \tag{4.37}$$

The final state can be computed in the RF so that $p^\mu = (M, 0, 0, 0)$, where M is the mass of the heavy particle. Since s is Lorentz invariant, its numerical value is the same both in RF and in LAB:

$$s_i = m_a^2 + m_b^2 + 2E_a m_b = s_f = M^2. \tag{4.38}$$

In order to produce M, the particle a must have:

$$E_a = \frac{M^2 - m_a^2 - m_b^2}{2m_b} \tag{4.39}$$

and if the new particle is much heavier than the projectiles:

$$E_a \simeq \frac{M^2}{2m_b}. \tag{4.40}$$

Even if we smash a proton against another proton ($m_b \simeq 1$ GeV), we need a projectile of:

$$E_a \simeq \frac{(91 \text{ GeV})^2}{2 \text{ GeV}} \simeq 4 \text{ TeV}. \tag{4.41}$$

So, only a fixed target experiment based, for example, on a beam of the LHC has some chance to discover the Z^0. Fortunately, this particle was discovered decades earlier than the LHC using a proton–antiproton collider (see Sec. 12.5). Precision studies of the Z^0 were performed in the early 90s with a e^+e^- circular collider (the Large Electron–Positron Collider (LEP) at CERN) and with a e^+e^- linear collider (the aforementioned SLC at SLAC) running with electrons and positrons at an energy of $M_Z/2 \simeq 45$ GeV. This is the threshold energy to produce the Z^0 in a head-on collision.

Most of the colliders smash particles against their antiparticles but other options are viable, too: proton–proton colliders (e.g. the LHC), electron–proton colliders (HERA at DESY in Hamburg), and ion–ion colliders (the Relativistic Heavy Ion Collider (RICH) at Brookhaven National Laboratory (BNL) and, again, the LHC).

4.6   Detecting particles at colliders

A head-on collision of two bunches at the interaction point inside the beampipe creates a wealth of final state particles scattered far from the axis of the beampipe. The detectors that trace the trajectories of the final state particles without perturbing significantly the motion of the particles are located just after the beampipe at a small radial distance from the IP. They are used for tracking and momentum measurement, as discussed in Sec. 3.10.2. Detectors that destroy the particles through electromagnetic showers (the e.m. calorimeters of Sec. 3.8.2) are located just after the trackers. They are followed by hadron calorimeters that measure the energy of the hadrons (see Sec. 4.6.3). The only particles that cross the calorimeters are penetrating particles like muons that are recorded by muon chambers installed at the largest radial distance from the IP. Neutrinos are also produced in large quantity but the cross-section is so small that they remain undetected at colliders. Hence, a typical **collider detector** is a set of sub-detectors made of hollow cylinders installed one inside the other like a Russian doll and hosted in the yoke of the electromagnet used to measure the momentum. Particles produced in the forward or backward directions are recorded by two **endcaps** that close up the central cylinder (**barrel**). One of the largest collider detectors in operation is the CMS detector at the LHC, which is depicted in Fig. 4.20.

4.6.1   Triggering a particle detector

Any particle detector is sensitive for a limited amount of time. Historical detectors like Wilson's cloud chambers have a limited live-time due to the physical process used for the detection: the condensation of water droplets, which need several seconds to evaporate and restore the initial conditions. Modern detectors are generally limited by the amount of data we can store between two interactions. If the recording time is too long, the second interaction will be lost or mis-reconstructed. Instead of operating the detectors in continuous mode at a limited rate, we can employ a **trigger**: a set of conditions that initiates the storage of the data. Triggers are extremely useful in colliders because the vast majority of collisions are of no relevance for the experiment. For instance, a e^+e^- collider produces millions of small-angle elastic scatterings:

$$e^+e^- \to e^+e^- \tag{4.42}$$

without creating new particles. Even if this process played an important role in the development of quantum electrodynamics (see Sec. 6.6.3), it is now very well known. We can get rid of it by ignoring events with two back-to-back e.m. showers in the forward region of the detector, provided that the sum of the energy of the two showers is $\simeq \sqrt{s}$. An even smarter way is to record only a tiny fraction of them (**prescaling**) to measure $\mathcal{L}(t)$ in a direct manner as:

$$\mathcal{L}(t) = \frac{N_{e^+e^- \to e^+e^-}}{\sigma_{e^+e^- \to e^+e^-}} \tag{4.43}$$

because $\sigma_{e^+e^- \to e^+e^-}$ is known with a precision better than 0.1%. In eqn 4.43, N is the number of scatterings observed in the forward region. This method is much more precise than relying on eqn 4.36. The same consideration holds for the luminosity measurement at the LHC with $pp \to pp$ elastic scatterings.

The bread and butter of collider physics are interactions with particles that show a large transverse momentum $\mathbf{p}_T$. A large $\mathbf{p}_T$ signals that a head-on collision with a significant momentum transfer has occurred and new particles or new high-energy processes can take place. This is the reason why triggers are generally based on the observation of patterns that can be interpreted as high-$\mathbf{p}_T$ particles.[7] In this case, the information stored in each sub-detector is recorded for further analysis. From time to time we record a tiny fraction of low $\mathbf{p}_T$ events called **minimum bias** events. They are used to evaluate the efficiency of the trigger conditions and single out any source of leakage of interesting events. To get a sense of the challenges that trigger designers must face, note that in the LHC we have about 40 million interactions per second and we store events at no more than 1000 Hz. The interactions originate from bunches with $N \simeq 10^{11}$ protons that intersect every 25 ns. A triggered collision, however, produces about 100 tracks to be recorded and analyzed. This puzzle is further jumbled by other interactions among particles in the same bunch, which create a pile-up of primary vertices.

[7] Trigger conditions employ the rapidity and pseudo-rapidity of the particles, too. These observables are discussed in Exercises 4.10 and 4.11.

Even inside the primary interaction, we may find several secondary vertices that signal the decay of short-living particles. The identification of these vertices and the correction of **pile-up** effects among tracks is essential to reconstruct the kinematics of the event.

Even if the LHC is the most extreme example of a high-energy collider, the progress in digital electronics is so big that we are starting to conceive **triggerless** experiments that perform a preliminary analysis of the events just after the digitization. These novel recording techniques are performed by fast microprocessors and employ machine-learning methods to discard minimum bias events in real time. The most prominent example is the upgrade of the LHCb experiment, which is aimed at working in triggerless mode at the LHC to identify short-living particles containing b and c quarks.

At the time of writing, the frontiers of trigger and particle identification are **triggerless** operation modes, **real-time reconstruction** of tracks, and **4D tracking**. We perform 4D tracking when we associate with a track not only a set of points in space but also the time when the signal of each point has been released. Time tagging helps us to remove the piling up of the tracks and separate the primary vertices belonging to the same bunch but requires detectors with a precision of $\mathcal{O}(10)$ ps.

4.6.2 Detecting jets

A prominent feature of high-energy interactions is that the particles are not produced uniformly in space but are clustered in **jets**. We will provide a profound explanation of the nature of jets in Chap 7, in the framework of quantum chromodynamics. We want to understand here how it is possible to identify the jets and gain information on their kinematics.

The vast majority of particles composing a jet are light hadrons (pions or kaons) that look like heavy charged particles in the tracker: they are bent by the magnetic field and the energy loss is driven by the Bethe formula. There are, however, two notable exceptions: photons produced inside the jets by the decay of neutral pions ($\pi^0 \to \gamma\gamma$) and high-energy neutrons. The first step to reconstruct the jets is to design detectors that can stop the light hadrons and measure their energy. These detectors are called **hadron calorimeters** and are the strong-interaction counterpart of the electromagnetic calorimeters of Sec. 3.8.2.

4.6.3 Hadron calorimeters

A high-energy charged hadron loses energy inside a detector either by electromagnetic interactions with the electrons or by strong interactions with the nuclei of the medium. If the transferred energy is sufficiently high, the nucleus is broken and the hadron creates new particles. A high-energy charged hadron impinging on an absorber interleaved by detectors sensitive to mips generates a shower that resembles the electromagnetic one of Fig. 3.14 (see Fig. 4.21). Unfortunately, **hadron**

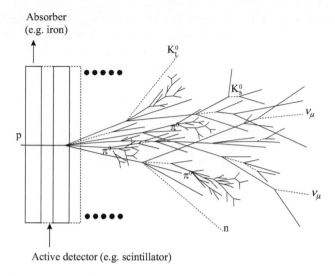

Absorber
(e.g. iron)

K_L^0

K_S^0

ν_μ

p

$\pi^{\pm}$

π^0

$\pi^{\pm}$

ν_μ

n

Active detector (e.g. scintillator)

Fig. 4.21 Hadronic shower developing inside a hadron calorimeter. The absorber and active tiles of the calorimeter are shown in the figure only at the beginning of the shower. Neutral particles do not release energy in the active detectors and are indicated by dashed lines. These particles are neutrons, neutrinos, and neutral hadrons (π^0, K_L^0, etc.) Some hadrons like the π^0 produce e.m. showers that develop inside the hadronic shower (see text). Redrawn from Grupen and Shwartz, 2008.

showers are much more difficult to be simulated than e.m. showers and are prone to larger statistical fluctuations. Nonetheless, it is possible to define a strong counterpart of the radiation length that describes the shower development:

> the **nuclear interaction length** λ_I is the mean distance traveled by a hadron between two strong interactions. The **nuclear collision length** λ_c is the mean distance traveled by a hadron between two strong reactions.

We generally use λ_I to design hadron calorimeters because this quantity also includes the elastic scattering and gives the size of the detector that stops the shower. However, the energy deposition is mostly driven by λ_c. λ_I is always much longer than X_0 and a rough estimate of this quantity is given by (Zyla *et al.*, 2020):

$$\lambda_I \simeq 35 \text{ g/cm}^2 \cdot A^{1/3}. \tag{4.44}$$

The transverse development of a hadronic shower is much more complex than for $e^+/e^-/\gamma$, too. As noted in Sec. 3.8.2, the leading contribution to the lateral spread of an e.m. shower is given by multiple Coulomb scattering and parameterized by the Moliere radius. The strong interactions occurring in the hadronic shower produce hadrons with a large transverse momentum, which, in turn, sets the scale of the transverse size. As a rule of thumb, to guess the transverse size of the shower, we can replace the Moliere radius with the interaction length but a realistic design must account for a complete simulation of the physical processes.

The most common particles produced by inelastic interactions are pions because they are the lightest quark bound states. Next to the leading pion population, we find kaons,[8] neutrons, protons, and other light hadrons. The number of particles produced inside the shower approximately scales as $\log E$ and larger energies increase especially the

[8]Kaons are the lightest states containing a *s* quark. These particles are discussed in Chap. 8 together with all other light mesons and baryons.

longitudinal size of the shower. For instance, a 50 GeV charged pion is stopped in 1.2 m of iron, that is after 7 λ_I. The average transverse momentum of these strong scatterings at high energy ($E \gg 1$ GeV) is about 0.35 GeV/c (Grupen and Shwartz, 2008).

Besides the complexity of strong interactions, e.m. processes introduce an additional source of statistical fluctuations in the development of the hadronic shower. During the formation of the shower, we produce several neutral pions with a cross-section comparable to their charged counterparts ($\pi^\pm$). The π^0's create high-energy photons because they decay into two photons ($\pi^0 \to \gamma\gamma$). The e.m. showers of these photons overlap with the hadronic energy deposited in the detector, giving a contribution that scales with X_0. Since the calorimeter is optimized for hadron detection, the response to e.m. showers may be different and the statistical fluctuations in the number of π^0 produced in the shower further worsen the energy resolution.

Strong interactions inside the absorber also break the nuclei of the material. The corresponding binding energy is released in the form of the kinetic energy of the nuclear fragments and in nuclear excitations, which produce few MeV photons. This energy is usually lost inside the absorber. The nuclear fragments are often neutrons that may escape detection, as well. Light hadrons can decay, producing neutrinos, which are virtually invisible in any calorimeter. Similarly, light nuclear fragments as neutrons may escape the detector and increase the amount of undetected energy. Even charged pions can leak due to statistical fluctuations in the number of reactions (**punch-through** pions) or by early decays that produce muons and neutrinos.

> The **invisible energy fraction f_{inv}** is the fraction of the shower energy that is not recorded because either it leaks outside the calorimeter or cannot reach the active area of the detector. In most cases, it amounts to 30–40% of the total energy. The fluctuation on f_{inv} and on the number of π^0 dominate the energy resolution of the device, which is much poorer than the e.m. calorimeters.

Without special countermeasures, a standard iron-scintillator sampling calorimeter achieves an hadronic energy resolution of about:

$$\frac{\sigma}{E} = \frac{90\%}{\sqrt{E(\text{GeV})}} \oplus 10\%. \tag{4.45}$$

Note that iron is an excellent absorber of hadrons because $\lambda_I \simeq 17$ cm is relatively short and the material is cheap and easy to machine.

Special calorimeters tuned to compensate for the e.m. component of the shower and the loss of soft photons can achieve at most a sampling term of 35%. Two common methods to improve the resolution of hadron calorimeters consist of either enriching the absorber with uranium or segmenting the calorimeter in the longitudinal direction. The first technique is named **compensation** because uranium produces additional neutrons and photons and compensates for the loss of the de-excitation

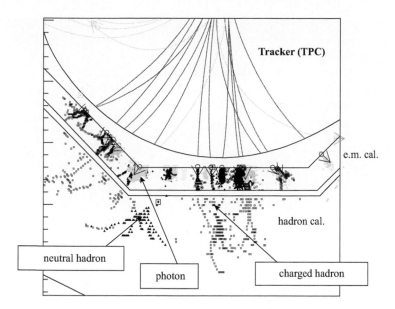

Fig. 4.22 Transverse view of a simulated 100 GeV jet reconstructed using the particle flow technique. The charged particles are tracked by a TPC and reach the e.m. calorimeter (e.m. cal.) and the hadron calorimeter (hadron cal.). Redrawn from Thomson, 2007.

photons in the absorber. The longitudinal segmentation is used to identify regions of the shower rich in e.m. energy due to the decays of the π^0. In this way, the statistical fluctuations on the number of π^0's in the shower can be corrected on an event-by-event basis (Wigmans, 2017).

Despite these limitations, hadron calorimeters are essential to measuring the energy of the jets and we can achieve further improvements if the detector is segmented in the transverse direction, that is it provides a measurement of the impact point of the hadron into the calorimeter. Thanks to the knowledge of the impact point, we can associate the energy deposit of the charged hadrons in the calorimeter to the corresponding trajectory in the tracker. The bending of the trajectory inside the magnetic field provides a very precise measurement of the momentum of the tracked particles (see Sec. 3.10.1) and adds information on the particle content and kinematics of the jet. In the most aggressive design developed for the next generation colliders, the transverse segmentation is so high that we can implement a full **particle flow calorimetry** (Marshall *et al.*, 2013): as shown in Fig. 4.22, the four-momentum of the jet can be estimated measuring the momentum of all charged particles belonging to the jet by the tracker and using the calorimetric data only for photons and neutrons. In this way, we can achieve percent level energy resolutions for jets despite the poor resolution of the hadron calorimeters.

4.6.4 Particle identification

In multipurpose detectors like the ones employed at colliders and made of several sub-detectors, we identify the particles by the signal released in each sub-detector. Table 4.1 summarizes the particles that leave energy in the tracker, e.m. calorimeters, hadron calorimeters, and muon chambers. For instance, an electron is identified by the presence of a

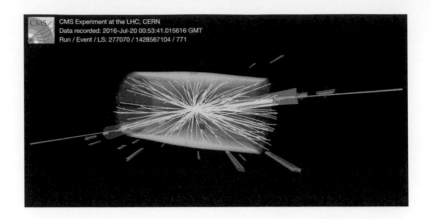

Fig. 4.23 A two-jet event recorded by the CMS experiment at the LHC in 2016 (CMS, 2017). The invariant mass of the two jets (enclosed within cones in the figure) is 8 TeV. The outer boxes show the energy deposits in the calorimeters or muon chambers. The softer particles mostly come from the other quarks of the interacting protons (see text). Copyright: CERN, Reproduced under CC-BY-SA from CMS, 2017.

Particle	tracker	e.m. cal	hadr. cal.	μ ch.
$e^{\pm}$	sign and **p**	E	-	-
γ	-	E	-	-
$p, \pi^{\pm}$	sign and **p**	-	E	-
n	-	-	E	-
$\mu^{\pm}$	sign and **p**	s.s.	s.s.	s.s.
ν	-	-	-	-

Table 4.1 Particle identification in a multipurpose detector. For each type of particle, we indicate whether a signal is visible in the subdetector (tracker, e.m. calorimeter, hadron calorimeter, muon chamber), and what kind of information can be retrieved. E stands for the particle energy; **p** and "sign" are the momentum and electric charge; "s.s." means "small signal", that is a signal released by a mip. ν is a neutrino with a given flavor. Short-living particles like $\tau^{\pm}$, π^0, K_s^0, etc., can be identified by the reconstruction of the decay products and secondary vertices.

charged track impinging in the e.m. calorimeter and the absence of signals in the hadron calorimeters and muon chambers. A photon has the same energy-release pattern as the electron but the impact point in the e.m. calorimeter is not linked to any visible track.

The inner core of a collider detector plays a pivotal role in particle identification. In Chap. 3 we discussed only one type of 3D tracker, the TPC. This device, however, has two serious drawbacks in colliders. First, the spatial resolution is constant in the whole volume. An ideal tracker should give its best near the IP to identify the primary interaction vertices and the secondary vertices due to the decays of short-living particles (tau leptons, hadrons with c and b quarks, etc.). In this case, the space precision requested may be at the level of tens of μm! The second drawback is time recovery. A TPC is based on a slow electron drift, which is swamped by the rate and multiplicity of charged tracks in modern colliders. An elegant solution is the use of high-precision trackers based on silicon detectors (Hartmann, 2017) in the proximity of the IP. These detectors are fast and radiation-tolerant and can reconstruct the origin of the tracks, the primary and secondary vertices with micrometric precision. At larger radii, the constraints on space resolution can be relaxed and, if the rate is tolerable, a TPC can be still employed to find the tracks, measure the momenta and record the pattern of dE/dx along each track.

The dE/dx pattern, corresponding to the amount of energy per unit length released in each point of the track, is a powerful tool to identify non-relativistic particles. It can be used to enhance particle identification because the ionization losses depend on v^{-2} through eqn 3.19. The energy losses per unit length must be compatible with the mass of the candidate particle and the measured momentum since:

$$ v = \frac{|\mathbf{p}|}{\sqrt{m^2 + |\mathbf{p}|^2}}. \tag{4.46} $$

We can then use eqns 3.19 and 4.46 to constrain the value of m and tag the type of particle. This technique is the modern avatar of the method employed by Anderson for the discovery of the positron.

Finally, note that in the most extreme environments, the rates are too high for any TPC and the tracking task is handed over to additional layers of silicon detectors.

An increase in the energy of the collider impacts both the number and the average energy of the final state particles, making pile-up corrections more and more challenging. Fig. 4.23 shows a two-jet events with an invariant mass of 8 TeV and 25 reconstructed vertices. The display only shows the tracks in the core of the detector, the cones enclosing the jets, and the energy deposited in the calorimeters and the muon chambers. Most of the tracks belong to proton quarks that interact loosely compared with the quarks originating the high-$\mathbf{p}_T$ jets. These **underlying events** blur the high-energy part of the collision and must be properly subtracted. Neutrinos are never visible due to their tiny cross-section. Their transverse momentum is generally computed applying the four-momentum conservation constraint of eqn 2.42 after summing the momenta of all visible particles.

4.7 Novel acceleration techniques*

The science and technology of accelerators have been based for nearly a century on two milestones: the phase stability principle and the strong focusing technique. Many other breakthroughs occurred in the meanwhile but the most powerful accelerators are still based on these founding principles. To make a substantial leap toward higher energies or higher intensities, particle physicists are developing several novel approaches. Some of them will replace conventional accelerators in practical applications. Others are much more ambitious and may replace high-energy synchrotrons in a few decades. At the time of writing, the new approaches under investigation can be classified into three groups.

Firstly, a substantial R&D is ongoing to build superconducting magnets capable to reach very strong magnetic fields (16 T, i.e. twice the fields of the LHC) at a relatively high temperature replacing the Nb-Ti coils with new superconducting materials. This is the most conservative approach to reach the 100 TeV domain increasing both the fields and the radius of the LHC (Abada *et al.*, 2019).

A second approach exploits the storage and acceleration of muons to sidestep the limitation of the energy losses in circular electron-positron collider (Palmer, 2014). In this case, radiation losses are suppressed by a factor $(m_\mu/m_e)^4$ compared to a standard electron LINAC. This approach requires a fast acceleration of the particles due to the finite lifetime of the muons and novel cooling techniques.

The third approach is probably the most unconventional: we can achieve electric fields that are hundreds of times bigger than current linear colliders replacing the RF cavities with a region of ionized plasma (**plasma acceleration**). The plasma acceleration area can be created using high-power lasers (Tajima and Dawson, 1979) or protons (Adli *et al.*, 2018). In this case, we must give up the Veksler–McMillan stability because the acceleration of the injected particles takes place in a transient and highly non-linear regime. The laser-plasma acceleration is an excellent candidate to replace low-energy accelerators if the quality of the beam reaches the standards of conventional machines. Both plasma (Tajima and Dawson, 1979) and vacuum (Esarey *et al.*, 1995; Terranova, 2014) acceleration techniques reap the enormous growth of laser power achieved after the invention of the Chirped Pulse Amplification[9] (CPA) technique (Strickland and Mourou, 1985). CPA-driven lasers combined with the wake-field acceleration method of T. Tajima, might boost the laser-based accelerators to the energies of their RF-based competitors (Mourou *et al.*, 2006).

[9]G. Mourou and D. Strickland have been awarded the 2018 Nobel Prize in Physics for the discovery of CPA. D. Strickland thus became the third woman ever to be awarded this Prize, after M. Skłodowska Curie in 1903 and M. Goeppert Mayer in 1963.

Further reading

Two excellent introductions to accelerator physics are Wille, 2001 and Wilson, 2001. The classical reference book at graduate level is Wiedemann, 2015. Collider detectors are introduced in Thomson, 2013 and Barr *et al.*, 2019. For a general overview of particle accelerators, their applications, and the corresponding detection technique, I suggest the second and third volumes of the Particle Physics Reference Library: Myers and Schopper, 2020 and Fabjan and Schopper, 2020

Exercises

(4.1) Compute the production threshold in LAB for the $p + p \rightarrow p + p + \pi^0$ process (see Exercise 2.6) and check if this reaction can be observed in a Cockcroft–Walton accelerator.

(4.2) Can we accelerate a proton with a kinetic energy of 1 GeV in a linear accelerator with constant L_n? Can we do it with electrons?

(4.3) Estimate the maximum number of bunches we can store in a 1 km LINAC operating at a frequency of 500 MHz.

(4.4) How much is the energy stored in the bunches of the LHC at the end of the acceleration phase?

(4.5) Using the acceleration formula A.16 in Appendix A, demonstrate that the rate of momentum growth in a linear accelerator is

$$\frac{d\mathbf{p}}{dt} = m\gamma^3 \dot{\mathbf{u}}. \tag{4.47}$$

(4.6) ** The **Larmor–Liénard** formula is the relativistic equation that gives the power P irradiated by a charged particle as a function of its three-acceleration $\dot{\mathbf{u}}$. In SI, it reads (Griffiths, 2017):

$$P = \frac{e^2}{6\pi\epsilon_0 c^3}\gamma^6 \left[\dot{\mathbf{u}}^2 - \frac{(\mathbf{u} \times \dot{\mathbf{u}})^2}{c^2}\right]. \tag{4.48}$$

Demonstrate that the radiated power in a LINAC does not depend on the particle energy.

(4.7) * If the acceleration of a particle is centripetal along a circle of radius r, eqn (4.48) reads

$$\frac{e^2 c}{6\pi\epsilon_0}\frac{\gamma^4}{r^2} \tag{4.49}$$

and is called the **synchrotron-radiation** loss formula. Demonstrate that the average energy loss per revolution of an electron inside a synchrotron is:

$$P = \Delta E[\text{GeV}] = 8.85 \times 10^{-5}\,\frac{E[\text{GeV}]^4}{R[\text{m}]}. \tag{4.50}$$

(4.8) * The dipoles of the LEP e^+e^- collider reach a maximum value of $B = 0.135$ T. Compute the maximum energy that LEP can achieve. We remind readers that LEP was located in the same tunnel as the LHC, whose length (circumference) is 27 km, and that the dipoles occupied about 2/3 of the LEP tunnel.

(4.9) * The beampipe of the LHC has a vacuum of 10^{-7} Pa. The residual particles are mostly due to gas desorption induced by radiation emitted by the beam. How many particles does a residual atom cross during a beam revolution inside the ring?

(4.10) * The **rapidity** is defined as:

$$y = \frac{1}{2}\log\left[\frac{E + p_z}{E - p_z}\right] \tag{4.51}$$

where $\mathbf{k}$ (z-axis) is the beam axis. Using the Lorentz transformations, demonstrate that rapidity differences do not change for boosts along the z direction

(4.11) * Consider a jet at an angle θ with respect to the beam axis, whose mass is much smaller than the energy E and $p_z \simeq E\cos\theta$. Show that in this case:

$$y \simeq \eta \equiv -\log\left(\tan\frac{\theta}{2}\right). \tag{4.52}$$

The quantity η is called **pseudo-rapidity**.

(4.12) * Generalize eqn 4.35 to the case $\epsilon_x\beta_x(s) \neq \epsilon_y\beta_y(s)$. Assuming the bunches to be Gaussian in both directions of the transverse plane ($x - y$ plane), show that the luminosity of a collider can be expressed as:

$$\mathcal{L} = \frac{1}{4\pi e^2 f_{rev} n_b}\frac{I_1 I_2}{\sigma_x^* \sigma_y^*} \tag{4.53}$$

where $I_{1,2}$ are the beam currents, n_b the number of bunches, f_{rev} is the revolution frequency in the ring. σ_x^* and σ_y^* are the sigma of the Gaussian in the two transverse directions computed at the IP.

(4.13) ** Why are the kinematic properties of the jets expressed in terms of η in hadron colliders and θ in e^+e^- colliders?

(4.14) ** Demonstrate eqn 4.49 from eqn 4.48.

5 Symmetries and anti-matter

5.1 Symmetries and non-integrable systems

The majority of physical systems do not admit analytical solutions for a host of reasons. Classical and quantum multi-body systems are unsolvable for more than two point-like particles and approximate solutions can only be obtained with perturbative or variational methods. Even the two-body scattering of particles may pose major challenges either because the interaction Hamiltonian is unknown or because the corresponding relativistic equation of motion cannot be handled with perturbative techniques. The first occurrence was very common up to the 1970s, notably before the formulation of the Standard Model, and is still the usual working condition for any search of physics beyond the Standard Model. To date, perturbative methods are the most reliable techniques to carry out calculations within the Standard Model (see Chap. 6) but very few strongly interacting systems can be described in perturbation theory. The Hamiltonian of the strong interactions has been known since 1973 and is embedded in the Standard Model through quantum chromodynamics (QCD) but the whole field of nuclear physics is beyond the reach of perturbative QCD.

The inspection of symmetries in non-integrable or non-perturbative systems is the most powerful tool to extract physical information without integrating the equations of motion. The reader is probably acquainted with this technique, which is implemented in atomic physics and quantum chemistry to determine the number and degeneracy of the energy eigenstates without solving the N-body Schrödinger equation. Symmetries provide only partial information on the physical system but can be used to guess the underlying Hamiltonian or test the predictions of a theory even if the equations of motion are intractable. It is therefore a tool that particle physicists must master.

5.2 Symmetries in quantum mechanics

Continuous symmetries in classical physics were rigorously introduced at the beginning of 20th century and connected to conservation laws or the integrals of motion. In 1918, this connection was established for

A Modern Primer in Particle and Nuclear Physics. Francesco Terranova, Oxford University Press.
© Francesco Terranova (2021). DOI: 10.1093/oso/9780192845245.003.0005

any theory that admits a Lagrangian formulation – including quantum mechanics (QM) – by Emmy Noether. In QM,[1] a **symmetry** represents a set of transformations of the state vector of a system that does not change the physical observables. The natural candidates to describe those transformations are **unitary operators** defined as the operators U in the Hilbert space of the system where $U^\dagger U = \mathbb{I}$ or, equivalently $U^{-1} = U^\dagger$. Here, $U^\dagger$ denotes the hermitian conjugate of the operator and $\mathbb{I}$ is the identity operator in the Hilbert space describing the physical system. Given two states $|\alpha\rangle$ and $|\beta\rangle$, their transformed states under U are $|\tilde{\alpha}\rangle \equiv |U\alpha\rangle$ and $\left|\tilde{\beta}\right\rangle \equiv |U\beta\rangle$. The matrix element of the form $\langle\alpha|\beta\rangle$ associated to physical observables stays unchanged because:

$$\left\langle\tilde{\alpha}\middle|\tilde{\beta}\right\rangle = \langle U\alpha|U\beta\rangle = \langle UU^\dagger\alpha|\beta\rangle = \langle\alpha|\beta\rangle. \tag{5.1}$$

A system is invariant for the symmetry described by U if its Hamiltonian is invariant, too. This is equivalent to $U^\dagger H U = H$. The Hamiltonian then commutes with the unitary operator U and its **commutator** $[H, U]$ is zero:

$$[H, U] \equiv HU - UH = 0. \tag{5.2}$$

The most general definition of symmetry in quantum mechanics is given by the **Wigner theorem**:

> all symmetries in QM are represented by unitary (or anti-unitary) operators that commute with the Hamiltonian.

In the following, we will consider only unitary operators.[2] You may wonder what is the link between a symmetry in a physical system and the existence of a conservation law or an integral of motion. This link exists but is subtler than expected. If U is a symmetry of the system, there is no doubt that $\langle U \rangle$ does not change in time. This is due to the fact that:

$$\frac{d}{dt}\langle U \rangle = \int d^3\mathbf{x} \left[\frac{\partial\psi^\dagger}{\partial t}U\psi + \psi^\dagger U\frac{\partial\psi}{\partial t}\right]. \tag{5.3}$$

ψ fulfills the Schrödinger equation, which is written in NU as $i\partial\psi/\partial t = H\psi$. Hence, $\partial\psi/\partial t = -iH\psi$ and, computing $(\partial\psi/\partial t)^\dagger$, we have

$$-i\frac{\partial\psi^\dagger}{\partial t} = \psi^\dagger H^\dagger = \psi^\dagger H \implies \frac{\partial\psi^\dagger}{\partial t} = i\psi^\dagger H \tag{5.4}$$

because H is **hermitian** ($H = H^\dagger$). Therefore,

$$\frac{d}{dt}\langle U \rangle = i\int d^3\mathbf{x}\left(\psi^\dagger HU\psi - \psi^\dagger UH\psi\right) = i\int d^3\mathbf{x}\ \psi^\dagger[H, U]\psi = 0. \tag{5.5}$$

The expectation value of U is then an integral of motion because it does not change in time. Unfortunately, U does not represent a physical observable because observables in QM are described by hermitian operators. As a consequence, $\langle U \rangle$ is not necessarily a conserved physical quantity and eqn 5.5 should not be read as a conservation law.

[1] For readers not acquainted with QM, observables, unitary, hermitian operators, and the notation used in this chapter are discussed in Appendix B.

[2] Unlike a standard linear operator, anti-unitary operators are anti-linear, that is $U(a\alpha + b\beta) = a^*U(\alpha) + b^*U(\beta)$, where a^* and b^* are the complex conjugate of $a, b \in \mathbb{C}$. For anti-unitary operators, $\left\langle\tilde{\alpha}\middle|\tilde{\beta}\right\rangle = \langle\alpha|\beta\rangle^* = \langle\beta|\alpha\rangle$. The most important symmetry in particle physics described by anti-unitary operators is the **time-reversal** or **T-parity** discussed in Sec. 5.8.

5.3   Classification of symmetries

A firm link between symmetries and conservation laws can be established classifying the symmetries in QM into four groups:

- **Continuous external symmetries:** these are transformations represented by unitary (or anti-unitary) operators that change the space or time coordinates of the wavefunction and depend on N real variables. The operators are continuous and differentiable (i.e. smooth) in these variables.

- **Continuous internal symmetries** are represented by unitary (or anti-unitary) operators that do not change the space or time coordinates of the wavefunction and depend on N real variables. Again, the operators are continuous and differentiable in these parameters.

- **Discrete external symmetries:** they are transformations represented by a *finite* number of unitary (or anti-unitary) operators that change the space or time coordinates. The operators, therefore, are *not* continuous in the variables that describe the finite transformations.

- **Discrete internal symmetries** are defined as in the previous case, except that space and time coordinates do not change.

The most prominent example of a continuous external symmetry in physics is the **translation in space** that brings a point $\mathbf{x}$ in $\mathbf{x}+\mathbf{a}$. Here, $\mathbf{a}$ is a three-vector described by the real coordinates $a_x, a_y, a_z \in \mathbb{R}$. A translation in space changes the wavefunction as:

$$\psi(\mathbf{x}) \to \psi(\mathbf{x} + \mathbf{a}). \tag{5.6}$$

The unitary operator that describes the translation is continuous in the three parameters a_x, a_y, a_z in the sense that:

$$\lim_{a_i \to 0} U(a_i)\psi(\mathbf{x}) = U(0)\psi(\mathbf{x}) = \mathbb{I}\psi(\mathbf{x}) = \psi(\mathbf{x}) \tag{5.7}$$

for $a_1 = a_x, a_2 = a_y$ and $a_3 = a_z$. Here, $U(a_i)$ is a unitary operator in the Hilbert space and $\psi(\mathbf{x})$ is an element of the Hilbert space. A translation in space is then:

$$U(\mathbf{a})\psi(\mathbf{x}) = \psi(\mathbf{x} + \mathbf{a}). \tag{5.8}$$

A fundamental result of the theory of Lie groups[3] ensures that:

> if $U(a_j)$ $(j = 1 \dots N)$ is a continuous and smooth operator in N real parameters a_j, then
>
> $$U(a_1 \dots a_N) - e^{i \sum_j a_j T_j} = e^{i a_j T_j} \tag{5.9}$$
>
> where we employ the Einstein summation rule in the last equality. T_j are hermitian operators called the **generators** of the group of the transformations described by U.

[3]In mathematics, the operators describing continuous symmetries are a special case of **Lie groups**: groups where each element can be labeled by continuous and differentiable (smooth) parameters. The Lie groups are differential manifolds and can be studied in the framework of differential geometry. In eqn 5.9, the sign in front of i is a matter of taste because it can be included in the real parameters a_j and, sometimes, physicists employ a minus sign. In this book, we use the standard modern notation shared by physicists and mathematicians, where the sign is plus.

It is straightforward to demonstrate that T_j are hermitian if we recall that the exponential of an operator is just a symbol for its Taylor expansion:

$$U(a_1 \ldots a_N) = e^{ia_j T_j} \equiv \sum_{n=0}^{+\infty} \frac{1}{n!}(ia)^n T_j^n. \qquad (5.10)$$

In this case, if U is unitary $U^\dagger U = e^{-ia_j T_j^\dagger} e^{ia_j T_j} = \mathbb{I}$. If we consider an infinitesimal transformation, $U(a_j) \simeq \mathbb{I} + ia_j T_j$ and $U^\dagger U = \mathbb{I}$ for every a_j. Then, $T_j^\dagger = T_j$. Similarly, if $d\langle U \rangle / dt = 0 \implies d\langle T_j \rangle / dt = 0$. The expectation values of T_j are then integrals of motion and conserved quantities in QM. That's good news because – at least for continuous symmetries – there is a link between conserved physical quantities (the expectation value of the (hermitian) generators) and the existence of a symmetry. We deepen this link looking at concrete examples.

5.4 Translations and rotations in quantum mechanics

Translations in QM can be described in the language of symmetries to unveil the connection between the conservation of momentum and the invariance of a physical system for arbitrary translations. For instance, a simple quantum system is given by a wavefunction that depends on the position of a particle in space. The choice of the position of the origin of the reference frame is arbitrary and should not affect any observable. As a consequence, a rigid translation of all points along a vector $\mathbf{a}$ represents a symmetry for any quantum system. Such translation moves the position of $\mathbf{x}$ to $\mathbf{x} + \mathbf{a}$ and we expect $\psi'(\mathbf{x}) = \psi(\mathbf{x} + \mathbf{a})$, where the primed quantities are the transformed ones. Using eqn 5.10 for $a_j \ll 0$

$$\psi'(\mathbf{x}) = \psi(\mathbf{x} + \mathbf{a}) \simeq \psi(\mathbf{x}) + \mathbf{a} \cdot \nabla \psi(\mathbf{x}) = (\mathbb{I} + i\mathbf{a} \cdot (-i\nabla)) \psi(\mathbf{x}). \quad (5.11)$$

It is amazing that the generators of space translations turn out to be the momentum operators in QM (see Tab. B.1):

$$T_1 = -i\frac{\partial}{\partial x} \; ; \;\; T_2 = -i\frac{\partial}{\partial y} \; ; \;\; T_3 = -i\frac{\partial}{\partial z}. \qquad (5.12)$$

In short, $\widehat{\mathbf{T}} = \widehat{\mathbf{P}} = -i\nabla$ or, in SI, $-i\hbar\nabla$, where $\widehat{\mathbf{P}} = (\hat{P}_x, \hat{P}_y, \hat{P}_z)$ are quantum mechanical momentum operators. In summary,

> if the translations are a symmetry of the system, each component of the total momentum of the system is an integral of motion. In the language of quantum mechanics, the expectation values of the operator $\widehat{\mathbf{P}}$ do not change in time and the total momentum is conserved.

Similar considerations hold for an arbitrary change of the orientation of the axes in space, that is a rigid **rotation** of the reference frame.

Rotations are described by real orthogonal matrices: 3×3 matrices of real numbers such that $\mathbf{R}^T \mathbf{R} = \mathbb{I}$. Like the translations, the rotations are continuous external symmetries since they change the space coordinates to $\mathbf{x} \to \mathbf{R}\mathbf{x}$. As usual, we can write this transformation as a product of matrices or, equivalently, as $x^i = R^i{}_j x^j$ in the covariant formalism. The derivation of the generators of the rotations is the same as before but the algebra is more complicated.

A general rotation in space can be expressed as the product of three real matrices:

$$\mathbf{R} = \begin{pmatrix} 1 & 0 & 0 \\ 0 & \cos\theta_x & \sin\theta_x \\ 0 & -\sin\theta_x & \cos\theta_x \end{pmatrix} \begin{pmatrix} \cos\theta_y & 0 & \sin\theta_y \\ 0 & 1 & 0 \\ -\sin\theta_y & 0 & \cos\theta_y \end{pmatrix}$$
$$\times \begin{pmatrix} \cos\theta_z & \sin\theta_z & 0 \\ -\sin\theta_z & \cos\theta_z & 0 \\ 0 & 0 & 1 \end{pmatrix}. \tag{5.13}$$

In the language of the continuous smooth groups, the three real parameters are $\theta_x, \theta_y,$ and θ_z. In fact, even if $\mathbf{R}$ is not the unitary operator we are seeking for, it is worth noting that it can be expressed in exponential form by a Taylor expansion around $\theta_x, \theta_y, \theta_z = 0$. For instance, if θ_x is small and $\theta_y, \theta_z = 0$:

$$\mathbf{R} = \begin{pmatrix} 1 & 0 & 0 \\ 0 & \cos\theta_x & \sin\theta_x \\ 0 & -\sin\theta_x & \cos\theta_x \end{pmatrix} = \begin{pmatrix} 1 & 0 & 0 \\ 0 & 1 & \theta_x \\ 0 & -\theta_x & 1 \end{pmatrix} + \mathcal{O}(\theta_x^2)$$
$$= \mathbb{I} + i\theta_x J^x \tag{5.14}$$

where J^x is the matrix:

$$J^x = -i \begin{pmatrix} 0 & 0 & 0 \\ 0 & 0 & 1 \\ 0 & -1 & 0 \end{pmatrix}. \tag{5.15}$$

Similarly, we can make small rotations along the y and z axes that give:

$$J^y = -i \begin{pmatrix} 0 & 0 & 1 \\ 0 & 0 & 0 \\ -1 & 0 & 0 \end{pmatrix}; \quad J^z = -i \begin{pmatrix} 0 & 1 & 0 \\ -1 & 0 & 0 \\ 0 & 0 & 0 \end{pmatrix}. \tag{5.16}$$

This notation brings to a tidy expression for $\mathbf{R}$:

$$\mathbf{R} = e^{i\boldsymbol{\theta} \cdot \mathbf{J}} \tag{5.17}$$

so that general rotations in space can be recast using the covariant formalism as

$$x'^i = (\exp[i\boldsymbol{\theta} \cdot \mathbf{J}])^i{}_j \, x^j. \tag{5.18}$$

Note, in particular, that for small rotation angles,

$$x'^i = (\exp[i\boldsymbol{\theta} \cdot \mathbf{J}])^i{}_j \, x^j \simeq x^i + (\boldsymbol{\theta} \times \mathbf{x}). \tag{5.19}$$

Even if eqn 5.17 resembles eqn 5.10, we are considering here quite a different mathematical object. $\mathbf{R}$ are real 3×3 matrices and are *not* operators in the Hilbert space. If we rotate the reference frame of a quantum system, what happens to its wavefunctions? The Wigner theorem ensures the existence of an unitary operator that commutes with the Hamiltonian and can be expressed in the exponential form of eqn 5.10.

The choice of the orientation of the axis of the reference system is arbitrary and should not affect any observable. As a consequence, a rigid rotation of all points through $\mathbf{R}$ represents a symmetry for any quantum system. Such rotation moves the position of $\mathbf{x}$ to $\mathbf{x}' = \mathbf{R}\mathbf{x} = R^i{}_j x^j$ and we expect $\psi'(\mathbf{x}) = \psi(\mathbf{R}\mathbf{x})$, where the primed quantities are the transformed quantities. Expanding for small $\boldsymbol{\theta}$, we get for each coordinate x^i ($x^1 = x, x^2 = y, x^3 = z$):

$$\psi'(x^j) = \psi(x^j + (\boldsymbol{\theta} \times \mathbf{x})^j) \simeq \psi(x^j) + ((\boldsymbol{\theta} \times \mathbf{x}) \cdot \boldsymbol{\nabla}) \psi(x^j)$$
$$= [\mathbb{I} + i\boldsymbol{\theta} \cdot (\mathbf{x} \times (-i\boldsymbol{\nabla}))] \psi(x^j). \tag{5.20}$$

The last equality in eqn 5.20 comes from the triple product rule of vector algebra $(\mathbf{A} \times \mathbf{B}) \cdot \mathbf{C} = \mathbf{A} \cdot (\mathbf{B} \times \mathbf{C})$ if we treat $\boldsymbol{\nabla}$ as a standard three-vector. The angular momentum operator in QM is $\widehat{\mathbf{L}} = \mathbf{x} \times (-i\boldsymbol{\nabla})$ in NU (see Tab. B.1) and the unitary operator that describes the rotations can be recast in the exponential form[4] as:

$$U(\boldsymbol{\theta}) = e^{i\boldsymbol{\theta} \cdot \mathbf{L}}. \tag{5.21}$$

[4]To lighten the notation, we drop now the hat from the symbols of the operators ($\widehat{\mathbf{L}} \equiv \mathbf{L}$, etc.) as we have done for H and U.

The rotational invariance of any quantum system requires $[\mathbf{L}, H] = 0$, which, in turn, implies the conservation of angular momentum. In general,

> the freedom to choose the position of the origin and the orientation of the axes of the reference frame in a quantum system brings to the conservation of the total momentum and the total angular momentum, respectively.

5.4.1 The first Noether theorem*

The conservation of the total momentum and angular momentum are properties of all known physical systems and are not restricted to non-relativistic QM. Since eqn 5.10 is a general property of continuous and smooth groups, every continuous symmetry must bring an integral of motion. The reader may wonder what is the most general relationship between symmetries and conservation laws in physics. The relation is provided by the **first Noether theorem**, which applies to all theories that can be described by local actions. We remind that an **action** (or **action integral**) in analytical mechanics is defined as:

$$S = \int_{t_1}^{t_2} L \, dt \tag{5.22}$$

where L is the Lagrangian and the integration is performed between the initial and final time of an arbitrary trajectory (Goldstein 1980; Taylor 2005). In classical physics, the actual trajectory is the trajectory that minimizes the action. Local action is an action described by a local Lagrangian, a Lagrangian that does not allow for any action at a distance.

The first Noether theorem states that:

> if the (local) action of the theory is invariant for a continuous and smooth symmetry, the theory has a conserved current that corresponds to the symmetry.

Agreed, it sounds quite abstract. But what is a conserved current? In physics, a **conserved current** is a vector j^μ that satisfies the continuity equation:

$$\partial_\mu j^\mu = 0. \tag{5.23}$$

Equation 5.23 says that if a physical quantity (e.g. the electric charge in a conductor) flows from one position to another, the amount that leaves the starting point must be equal to the amount that reaches the final point. We generally call this statement a **local conservation law**.

Example 5.1

In classical electromagnetism, we can see eqn 5.23 at work if we define j^μ as $j^\mu \equiv (c\rho, j_x, j_y, j_z)$ and compute $\partial_\mu j^\mu$ (see Exercise 5.2).[5] The result is familiar to us and reads:

$$\partial_\mu j^\mu = \frac{\partial c\rho}{dt} + \nabla \cdot \mathbf{j} \tag{5.24}$$

where ρ is the charge density in a volume V, $\mathbf{j}$ is the current density and $\nabla \cdot \mathbf{j}$ is the divergence of $\mathbf{j}$. Equation 5.23 is just the **continuity equation** of electromagnetism:

$$\nabla \cdot \mathbf{j} = -\frac{\partial c\rho}{\partial t}. \tag{5.25}$$

The intuitive meaning of this equation is that the charge we lose in a volume $(-\partial\rho/\partial t)$ must flow outside the volume because you cannot create and destroy the electric charge. This statement is a conservation law called **local conservation of charge**. It is a stronger statement than "charge cannot be created from void" because, for instance, we could create an electron in the earth and a positron in the moon at the same time. This event is allowed if the charge must be globally conserved but violates eqn 5.25. And violates our real-life experience, too, because we have never seen an electron popping up alone in a volume V located in the earth. This is why the local conservation of charge is embedded in Maxwell's theory of electromagnetism.

[5] We will discuss all the implications of this definition in the framework of quantum electrodynamics in Chap. 6 and eqn 6.12.

In the examples discussed in Sec. 5.4, the Hamiltonian is invariant for the transformations describing translations and rotations. The Noether theorem, hence, returns six integrals of motions corresponding to the translations along the three axes (**a**) and the rotations around the three angles θ_x, θ_y, and θ_z ($\boldsymbol{\theta}$). It comes as no surprise that:

> the conservation of momentum and angular momentum is a conservation law that holds not only in non-relativistic QM but also in classical physics, special relativity (SR), relativistic quantum mechanics, and in the Standard Model.

Discrete symmetries are not continuous and smooth transformations and we cannot ensure a conservation law associated with them. Most finite transformations lack conserved currents, indeed. A notable exception is given by the inversions.

5.5   Inversions

Inversions are (external or internal) discrete transformations of a physical system. They correspond to a finite group with just two elements. An inversion is a symmetry for a given quantum system if it is described by a unitary operator that fulfills:

$$U^2 = \mathbb{I} \tag{5.26}$$

and if the operator commutes with the Hamiltonian: $[H, U] = 0$. By definition, if we apply twice an inversion to a physical system, we revert to the original system. Inversions can be defined in classical and relativistic theories. A well-known example is the **classical parity**, an external discrete symmetry that changes the coordinate of the system from **x** to $-\mathbf{x}$. This is equivalent to moving from a left-handed to a right-handed reference frame. Similarly, **classical C-parity** changes the sign of the electric charges of a system and corresponds to an internal discrete symmetry of classical electrodynamics. **Classical time inversion** changes the sign of time from t to $-t$ and thus traces the evolution of the system from future to past. For a scattering process, it corresponds to changing the sign of the initial- and final-state momenta so that the final state particles collide and produce the initial particles. Classical mechanics and electrodynamics have this discrete external transformation among their symmetries too.

5.5.1   Parity

The classical parity can be extended straightforwardly to any quantum system starting from a subset of wavefunctions: the eigenstates of the particle position. In a one-dimensional system, the parity operator is defined as:

$$\mathcal{P} \, |x\rangle \equiv |-x\rangle \,. \tag{5.27}$$

$\mathcal{P}$ is clearly an inversion but is also hermitian. This feature follows from the properties of the position eigenfunctions:[6] $\langle x|\mathcal{P}|x'\rangle = \langle x|-x'\rangle = \delta(x+x')$. The complex conjugate is $\langle x|\mathcal{P}^\dagger|x'\rangle = \langle x'|\mathcal{P}|x\rangle^* = \langle x'|-x\rangle^* = \langle -x|x'\rangle = \delta(x + x')$. Since $\langle x|\mathcal{P}|x'\rangle = \langle x|\mathcal{P}^\dagger|x'\rangle$ for every x and x' $\implies \mathcal{P}^\dagger = \mathcal{P}$.

A hermitian inversion is unitary because $\mathcal{PP} = \mathbb{I}$ and if we multiply $\mathcal{PP} = \mathbb{I}$ from the right by $\mathcal{P}^{-1}$, we get $\mathcal{P}^\dagger = \mathcal{P} = \mathcal{P}^{-1}$. Since the position eigenfunctions are a base for the Hilbert space, every wavefunction can be expressed as a sum of $|x\rangle$ functions. As a consequence,

> the **parity** in QM is a unitary, hermitian operator such that:
>
> $$\mathcal{P}\psi(\mathbf{x}) = \psi(-\mathbf{x}) \qquad (5.28)$$
>
> for every wavefunction. Furthermore, if the wavefunction of a system is an eigenstate of the parity operator, the eigenvalues are either $+1$ or -1.

The last statement follows from the definition of inversion. If ψ is an eigenfunction of the parity operator,

$$\psi = \mathcal{PP}\psi = \mathcal{P}a\psi = a\mathcal{P}\psi = a^2\psi. \qquad (5.29)$$

$\mathcal{P}$ is hermitian and, hence, the eigenvalues must be real. This means that a can only be ± 1.

If the wavefunction of a quantum system is an eigenstate of $\mathcal{P}$, we define the **parity of the system** as the value of a.

> The parity of a system is defined *only* if the wavefunction describing the system is an eigenstate of $\mathcal{P}$ and the corresponding eigenvalue can be either $+1$ (even parity) or -1 (odd parity).

In non-relativistic QM we can define the **intrinsic parity** of a particle as the eigenvalue of the wavefunction describing the particle in its rest frame (RF). For instance, if your measurement cannot resolve the composite nature of the hydrogen atom[7] $^1_1\mathrm{H}$, the intrinsic parity of the $^1_1\mathrm{H}$ "particle" will be $(-1)^l$ as shown in Sec. 5.6.1 and in any QM textbook. You may guess that the knowledge of the intrinsic parities of the particles provides useful information on their nature and internal dynamics.

The intrinsic parities are Lorentz-invariant quantities because they are computed by any observer in the RF of the particle. Quantum field theories (QFT) provide the means – not discussed in this book – to define the intrinsic parities of massless particles, too, where the particle RF does not exist.

If the parity is a symmetry of the system ($[H, \mathcal{P}] = 0$), the expectation value of the parity operator is an integral of motion and the parity of the system is conserved. This is a general property of all inversions: even if they do not fulfill the conditions of the Noether theorem, they provide conserved quantities. In this case, the observable is the parity operator itself since $\mathcal{P}$ is hermitian and, unlike continuous symmetries, there is no need to find a corresponding hermitian "generator." The conserved quantity is, of course, the parity of the system.

Until 1957, physicists believed that parity was a symmetry for all physical systems. If this statement were true, we could empirically infer the intrinsic parity of a particle from scattering and decay processes, applying the parity conservation law to the initial and final state. Nonetheless, you should be aware that parity is violated in weak interactions and these empirical techniques can be employed only using parity-conserving processes, the strong and electromagnetic interactions.

5.5.2 Additive and multiplicative quantities*

The algebraic properties of the parity are shared by all inversions as long as we redefine the operator to be hermitian. In particular,

> the conserved quantities associated with the inversions are **multiplicative quantum numbers** and the conserved quantities associated with continuous symmetries are **additive quantum numbers**.

What is an "additive quantum number?" If we have a system made of several *independent* parts, the wavefunction of the system can be expressed as the product of the wavefunction of each part.[8] For example, the two-body final state of Sec. 2.4 is described by:

$$\psi = \psi_1 \psi_2 \tag{5.30}$$

For a quantum system where the wavefunction can be factorized in independent parts,

$$\mathcal{I}\psi = \mathcal{I}_1 \psi_1 \mathcal{I}_2 \psi_2 = a_1 a_2 \psi \tag{5.31}$$

where $\mathcal{I}$ is a generic inversion, $\mathcal{I}_1$ and $\mathcal{I}_2$ are the inversions applied to ψ_1 and ψ_2, and a_1, a_2 are the eigenvalues of $\mathcal{I}$ for ψ_1 and ψ_2, respectively In this case, the eigenvalue of the product is a multiplicative quantum number because it is the product of the eigenvalues of each part.

For a continuous symmetry with one generator G:

$$U\psi = e^{iaG}\psi_1\psi_2 = e^{iaG_1}\psi_1 e^{iaG_2}\psi_2 \simeq$$
$$(1 + iaG_1 + \ldots)\psi_1 \cdot (1 + iaG_2 + \ldots)\psi_2 =$$
$$(1 + iag_1 + \ldots)\psi_1 \cdot (1 + iag_2 + \ldots)\psi_2 = e^{iag_1}\psi_1 e^{iag_2}\psi_2$$
$$= e^{ia(g_1+g_2)}\psi_1\psi_2 \tag{5.32}$$

where G_1 and G_2 are the generator G applied to ψ_1 and ψ_2, respectively. g_1 and g_2 are the corresponding eigenvalues. In short,

$$G\psi_1\psi_2 = (g_1 + g_2)\psi_1\psi_2 \tag{5.33}$$

and the eigenvalues of the parts of the system represented by ψ_1 and ψ_2 add up to give the conserved quantity of the whole system $g_1 + g_2$ Therefore, the eigenvalue of G is an additive quantum number.

[8]This occurrence is common in particle and nuclear physics after a scattering but, in general, is a very strong requirement: the **entangled states** in QM do not admit such factorization, as shown in Sec. B.4.

If we have a system with two non-entangled particles, the momentum of the system is the *sum* of the momentum of each particle but the parity of the system is the *product* of the intrinsic parity of the particles. If the particles are entangled, that is eqn 5.30 does not hold, we need to account also for correlation terms like the angular momentum that may affect the system parity, too.

The intuitive reason for the behavior of the conserved quantities is simple: since the unitary operators of the continuous symmetries are written as an exponential (eqn 5.9), the eigenvalues of the generators are the logarithm of the eigenvalues of the unitary operators and $\log ab = \log a + \log b = g_1 + g_2$. In a sense, additive numbers from hermitian operators are the logarithms of the multiplicative numbers of the corresponding unitary operators.

These considerations provide an elegant demonstration of eqn 2.41: the action of SR is invariant for rigid translations so that the total momentum is conserved thanks to the first Noether theorem. Since the initial- and final-state particles are non-interacting, the wavefunction factorizes and the total initial (final) momentum is the linear sum of the initial (final) particle momenta. This is the reason why eqn 2.41:

$$\sum_{i=1}^{N_i} (p_i^{ini})^\mu = \sum_{j=1}^{N_f} (p_j^{fin})^\mu$$

is not an axiom of SR but a consequence of this theory.

5.6 The intrinsic parity of the particles

The empirical determination of the intrinsic parity of particles played a prominent role in the development of the Standard Model and is still of great practical value for the identification of the particles in scattering experiments and the characterization of bound states. A straightforward example is the experimental determination of the intrinsic parity of the photons. This property is well defined in QFT but cannot be computed in non-relativistic QM due to the lack of the photon RF and must be assessed experimentally.

5.6.1 The parity of the photon

Hydrogen atoms in excited states decay emitting one or more photons. For a single photon decay,

$$^1_1\mathrm{H}^* \to \, ^1_1\mathrm{H} + \gamma \tag{5.34}$$

and the decay amplitude is dominated by electric dipole transitions. These transitions are the most probable because the photon is emitted with zero angular momentum. The wavefunction of the hydrogen atom is known for any excited state because the corresponding Schrödinger equation can be solved analytically (Griffiths, 2018). It is expressed as a function of the electron position in spherical coordinates (see Fig. 5.1):

$$\psi(r,\theta,\phi) = \frac{U_{n,l}(r)}{r} Y_{l,m}(\theta,\phi). \tag{5.35}$$

$Y_{l,m}(\theta,\phi)$ are the Laplace's spherical harmonics and the radial position of the electrons is decoupled (i.e. factorizes) from the functions that describe the angular position.

In the rest frame (RF) of the atom, the proton is at the origin of the reference frame since $m_p \gg m_e$. The parity operator changes the space coordinates of the electron so that:

$$\theta \to \pi - \theta$$
$$\phi \to \pi + \phi \tag{5.36}$$

The spherical harmonics are (Riley *et al.*, 2006; Griffiths, 2018):

$$Y_{l,m}(\theta,\phi) = \left[\frac{(2l+1)(l-m)!}{4\pi(l+m)!}\right] P_l^m(\cos\theta)e^{im\phi} \tag{5.37}$$

where $P_l^m(\cos\theta)$ are the Legendre polynomials. Hence,

$$\mathcal{P}Y_{l,m}(\theta,\phi) \sim (\mathcal{P}P_l^m(\cos\theta))(\mathcal{P}e^{im\phi}) \tag{5.38}$$

but

$$\mathcal{P}e^{im\phi} = e^{im(\pi+\phi)} = (-1)^m e^{im\phi}$$

and

$$\mathcal{P}P_l^m(\cos\theta) = (-1)^{l+m} P_l^m(\cos\theta)$$

so that:

$$\mathcal{P}\psi = (-1)^l \psi. \tag{5.39}$$

Equation 5.39 shows that the parity of the hydrogen atom is even if l is even or zero and odd if l is odd. This is a very general result that will be often employed in this book:

> if the angular part of the wavefunction of a system is described by spherical harmonics, the parity of the space part of the wavefunction is $(-1)^l$.

In eqn 5.34, the initial-state parity is $(-1)^l$. The total angular momentum of the two final-state particles ${}_1^1\text{H}$ and γ is zero and the final-state wavefunction is the product of the wavefunction of the hydrogen and the photon. From eqn 5.39:

$$\mathcal{P}\psi_{{}_1^1\text{H}} = (-1)^{l_{fin}} \psi_{{}_1^1\text{H}} \tag{5.40}$$

and

$$\mathcal{P}\psi_{{}_1^1\text{H}}\psi_\gamma = (-1)^{l_{fin}} \psi_{{}_1^1\text{H}} \cdot \mathcal{P}_\gamma \psi_\gamma \tag{5.41}$$

where l_{fin} is the angular momentum quantum number of the final state atom and $\mathcal{P}_\gamma$ is the intrinsic parity of the photon. In the initial state:

$$\mathcal{P}\psi_{{}_1^1\text{H}^*} = (-1)^{l_{in}} \psi_{{}_1^1\text{H}^*}. \tag{5.42}$$

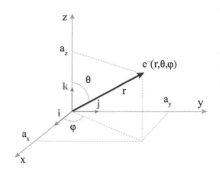

Fig. 5.1 Position of the electron in spherical coordinates r, θ, ϕ in the RF of the proton. In the hydrogen atom where $m_p \gg m_e$, this frame is equivalent to the CM frame and, therefore, to LAB frame if the atom is at rest.

Since the electric dipole transitions[9] occur when the difference $\Delta l \equiv l_{fin} - l_{in}$ is ± 1 and the parity of the system must be the same both in the initial and final states:

> the intrinsic parity of the photon is odd (-1) if parity is conserved in electromagnetic interactions.

The Standard Model provides a rigorous definition of the intrinsic parity for massless particles and assigns an odd parity to the photon, as expected from the empirical derivation above.

5.6.2 The Chinowsky–Steinberger experiment

In 1954, W. Chinowsky and J. Steinberger performed one of the most spectacular derivations of the intrinsic parity of a particle assuming that parity is conserved by strong interactions (Chinowsky and Steinberger, 1954). This result played a central role in the construction of the quark models and the Standard Model. The **Chinowsky–Steinberger experiment** is an experiment designed to study the spectroscopy and decay of "mesonic atoms." These exotic atoms are bound states where one electron is replaced by one negatively charged pion (π^-). At that time, the nature of the pion (a bound state of quarks) was unknown but the mass (139 MeV) and the spin ($s = 0$) were already established through cosmic-ray and accelerator experiments. Mesonic atoms are produced by beams of charged pions created, together with other particles, by the interaction of protons on solid-state targets. The charged pions were produced by a 390 MeV proton accelerator at Columbia University (New York, NY), sign and momentum selected by a magnetic spectrometer, and directed toward a liquid deuterium target.[10] The pion is slowed down by the collision losses until is captured by a deuterium atom. Due to the large pion mass, the orbits of the pion in the mesonic atom are much closer to the nucleus than the electron orbits in a standard atom. This is an ideal situation to study strong interactions, which take place when there is a significant overlap between the wavefunction of the π^- and the one of the nucleus.

The spectroscopy of this atom is qualitatively similar to the one of the hydrogen, except when the atom principal quantum number n is small. Spectroscopy data collected since 1951 (Panofsky *et al.*, 1951) indicate that an excited mesonic atom de-excites quickly (in a few ps) emitting photons until the pion reaches the $n = 7$, $l = 0$ state. At that time, there was no clear explanation of why the lowest state was an $l = 0$ state. This effect was clarified by T.B. Day and colleagues in 1960 and is due to the Stark effect (Bettini, 2014), but its very existence was exploited much earlier to demonstrate that the π^- has negative intrinsic parity. In the Chinowsky–Steinberger experiment (see Fig. 5.2), the arrival of a pion was triggered by a liquid scintillator detector (#2 in the figure) located in the proximity of the deuterium target. Between the triggering detector and the target, a carbon and a polyethylene absorber slowed down the

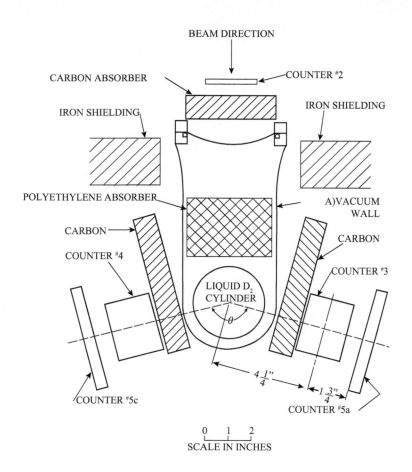

BEAM DIRECTION

COUNTER #2

CARBON ABSORBER

IRON SHIELDING

IRON SHIELDING

POLYETHYLENE ABSORBER

A)VACUUM WALL

CARBON

CARBON

COUNTER #4

COUNTER #3

LIQUID D₂ CYLINDER

θ

$4\frac{1}{4}$"

$1\frac{3}{4}$"

COUNTER #5c

COUNTER #5a

0   1   2

SCALE IN INCHES

Fig. 5.2 Schematics of the Chinowsky–Steinberger experiment. The pions coming from above cross the trigger ("counter #2") and are slowed down by the carbon absorber and the polyethylene absorber. They stop in the liquid deuterium target ("D_2 cylinder") producing two neutrons detected by two liquid scintillators ("counter #3" and "#4"). Additional detectors veto the charged particles entering (not shown in the figure) or exiting the neutron detectors ("counter #5c" and "#5a"). This configuration, in particular, was used to study the angular correlation between the two neutrons given by the angle θ. Reprinted figure with permission from Chinowsky and Steinberger, 1954. Copyright 1954 by the American Physical Society.

pions to maximize the number of stopping π^- in the target. Two additional detectors (the #3 and #4 liquid scintillator detectors) were used to detect the neutrons by the energy released after an elastic scattering with a proton. Additional detectors (#5 in the figure) were used to veto charged particles entering or exiting the neutron detectors. This set-up is well-suited to reconstruct the two neutrons that are produced after the destruction of the mesonic atom provoked by the strong interactions:

$$\pi^- + {}_1^2\text{H} \to n + n. \tag{5.43}$$

As noted by B. Ferretti in 1946, this reaction cannot occur for $l = 0$ (s-wave[11]) unless the pion has a negative intrinsic parity due to a conspiracy between the conservation of angular momentum and the Pauli exclusion principle. In the initial state, the angular momentum is $J = 1$ because the reaction occurs in s-wave ($l_{ini} = 0$), the intrinsic angular momentum (spin) of the pion is zero and the intrinsic angular momentum of the deuteron (i.e. the nucleus of the ${}_1^2\text{H}$ atom) is 1. In the final state, $l_{fin} \equiv l$ is unknown but the intrinsic angular momenta (spin) of the neutrons are 1/2. The conservation of angular momentum implies $J_{fin} = 1$, which is devised by one of the following combinations:

[11]A scattering occurring with $l = 0$ is called an **s-wave** scattering. This word is a remnant of the spectroscopic notation used at the end of 19th century: a scattering with $l = 2, 3, 4$ is still named a **p,d,f-wave**.

$$l = 0, \ s = 1$$
$$l = 1, \ s = 1$$
$$l = 1, \ s = 0$$
$$l = 2, \ s = 1 \qquad (5.44)$$

$s \equiv s_{fin}$ being the total spin of the two-neutron final state ($s = 0$ or 1). Higher angular momenta are forbidden because $J_{fin} > 1$ if $l > 2$.

We show now that the only combination compatible with the Pauli exclusion principle is $l = 1, s = 1$. This happens because the final state is made up of two identical fermions and, therefore, the wavefunction of the system must be antisymmetric. Exchanging the position of the two neutrons in the RF is equivalent to applying simultaneously the parity operator (the space coordinate of the first neutron $\mathbf{x}$ is changed to $-\mathbf{x}$, which is the location of the second neutron in the RF) and an exchange of the two particles in the spin part of the wavefunction.

In non-relativistic QM, the spin and spatial part of the wavefunction factorize.[12] The spin part of the wavefunction of two identical fermions can be computed by the combination rule of the spin:

$$\frac{1}{2} \otimes \frac{1}{2} = 0 \oplus 1. \qquad (5.45)$$

The wavefunction $\alpha(s, s_z)$ can be expressed as a linear combination of the spin-1/2 wavefunction using the Clebsch–Gordan coefficients (see Sec. B.2.1 or (Griffiths, 2018)). If $\alpha_1(s_1, s_{1z})$ is the spin wavefunction of the first neutron with $s = 1/2$ and $s_z = \pm 1/2$, we can write:

$$\alpha(0,0) = \frac{1}{\sqrt{2}} \left[\alpha_1 \left(\frac{1}{2}, \frac{1}{2} \right) \alpha_2 \left(\frac{1}{2}, -\frac{1}{2} \right) - \alpha_1 \left(\frac{1}{2}, -\frac{1}{2} \right) \alpha_2 \left(\frac{1}{2}, \frac{1}{2} \right) \right]$$
$$(5.46)$$

and

$$\alpha(1,1) = \alpha_1 \left(\frac{1}{2}, \frac{1}{2} \right) \alpha_2 \left(\frac{1}{2}, \frac{1}{2} \right)$$
$$\alpha(1,0) = \frac{1}{\sqrt{2}} \left[\alpha_1 \left(\frac{1}{2}, \frac{1}{2} \right) \alpha_2 \left(\frac{1}{2}, -\frac{1}{2} \right) + \alpha_1 \left(\frac{1}{2}, -\frac{1}{2} \right) \alpha_2 \left(\frac{1}{2}, \frac{1}{2} \right) \right]$$
$$\alpha(1,-1) = \alpha_1 \left(\frac{1}{2}, -\frac{1}{2} \right) \alpha_2 \left(\frac{1}{2}, -\frac{1}{2} \right) . \qquad (5.47)$$

More precisely, the wavefunction corresponding to $s = 0$ is proportional to $\alpha(0,0)$ and changes sign swapping the two neutrons: it is an **antisymmetric singlet** (Griffiths, 2018). Note that there are infinite wavefunctions that belong to the antisymmetric singlet: they are all proportional to eqn 5.46. However, since the wavefunction must be normalized to 1, all these wavefunctions differ from eqn 5.46 by at most a unit phase. The wavefunctions corresponding to $s = 1$ are linear combinations of the functions of eqn 5.47 and do not change if we

[12]This is not true in relativistic QM but the speeds of the particles in the mesonic atom are small and we can safely perform a non-relativistic approximation.

exchange the two neutrons: they are **symmetric triplets**. Again, there are infinite wavefunctions that belong to this class. However, all of them can be expressed as linear combinations (plus an arbitrary unit phase) of the three functions of eqn 5.47. In conclusion,

> if we exchange the two neutrons, the final state wavefunction ψ_{nn} changes as:
>
> $$\psi_{nn} = \psi_{spatial}\psi_{spin} \rightarrow (-1)^l(-1)^{s+1}\psi_{nn} \qquad (5.48)$$
>
> and the term $(-1)^{s+1}$ accounts for the change of sign of the spin part of the wavefunction ($+1$ for the triplet, -1 for the singlet).

Since ψ_{nn} must be antisymmetric, the Pauli exclusion principle dictates that $l + s$ must be even. The only even combination of l and s compatible with the conservation of the angular momentum is $l = 1$, $s = 1$ in eqn 5.44.

Let us move now to parity. The parity operator does not affect the spin part of ψ_{nn} but only the space part. Hence,

$$\mathcal{P}\psi_{nn} = (-1)^l \mathcal{P}_n \mathcal{P}_n \psi_{nn} = (-1)^l \psi_{nn} \qquad (5.49)$$

where $\mathcal{P}_n$ is the intrinsic parity of the neutron and $\mathcal{P}_n^2 = 1$, whatever is the actual value of $\mathcal{P}_n$. In fact (see Sec. 5.6.3), the intrinsic parity of neutrons and protons is conventional and equal to $+1$. The parity of the final state system is thus -1. In the initial state, the intrinsic parity of the deuteron is positive because is a bound state of a proton and a neutron in s-wave ($l_{^2_1\mathrm{H}} = 0$). Again,

$$\mathcal{P}\psi_{ini} = (-1)^l_{ini}\mathcal{P}_d \mathcal{P}_{\pi^-}\psi_{ini} = \mathcal{P}_{\pi^-}\psi_{ini}. \qquad (5.50)$$

As a consequence,

> if the parity is conserved in strong interactions, the intrinsic parity of the π^- is -1.

5.6.3   Intrinsic parity of the elementary fermions

The experimental determination of the intrinsic parity is useful to characterize new particles. On the other hand, if the elementary components of the particle and their relative motion are known, the Standard Model can predict the intrinsic parity. For an elementary fermion, the intrinsic parity is conventional because elementary fermions are not bound states of more fundamental particles: parity-conserving interactions can detect only the changes of the intrinsic parities and are not sensitive to their absolute value. The convention employed in the Standard Model is the following:

- all negatively charged elementary leptons and all neutrinos have positive intrinsic parity. This rule fixes the intrinsic parity of all the leptons: the electron (e^-), the muon (μ^-), the tau lepton (τ^-), the electron neutrino (ν_e), the muon neutrino (ν_μ), and the tau neutrino (ν_τ)

- all quarks – up (u), down (d), strange (s), charm (c), beauty (b), and top (t) – have positive intrinsic parity.

Relativistic quantum mechanics provide a general theorem due to P. Dirac that fixes the intrinsic parity of all elementary antiparticles. We will demonstrate this theorem for the electron in Example 5.4:

> the intrinsic parity of an antifermion is opposite to the intrinsic parity of the corresponding fermion. The intrinsic parity of an antiboson is the same as the corresponding boson. Hence, the intrinsic parity of all elementary fermions is $+1$ by convention but the intrinsic parity of all elementary antifermions (antileptons and antiquarks) must be -1.

The **parity of a pair of particles** follows from the rules above. In the RF of the pair, the wavefunction can be factorized as

$$\psi = \psi_{space}(r)Y_{l,m}(\theta,\phi)\psi_{spin}. \tag{5.51}$$

The parity operator changes ψ as:

$$\mathcal{P}\psi = (-1)^l \mathcal{P}_1 \mathcal{P}_2 \tag{5.52}$$

where l is the angular momentum quantum number of the pair and $\mathcal{P}_1$ ($\mathcal{P}_2$) is the intrinsic parity of the first (second) particle. Hence:

- the wavefunction of a pair of particles can be an eigenstate of the parity operator only in the RF. In this frame, **the parity of a pair of particles** is $(-1)^l \mathcal{P}_1 \mathcal{P}_2$.
- the parity of a fermion-antifermion ($F\bar{F}$) pair is $(-1)^{l+1}$ since $\mathcal{P}_F = -\mathcal{P}_{\bar{F}}$.
- the parity of a boson–antiboson ($B\bar{B}$) pair is $(-1)^l$ since $\mathcal{P}_B = \mathcal{P}_{\bar{B}}$.

Please, keep in mind that the parity operator and the **exchange operator** that exchanges two particles in the wavefunction are two different objects. They are the same operator only if we consider the space part of the wavefunction in the RF or, equivalently, if all particles are spinless. If the particles have a spin, the outcomes of the two operators are different because the exchange of the particles affects the spin part of the wavefunction, too. This distinction is noteworthy if the particles are identical. For instance, a pair of spinless particles (e.g. a π^+ and a π^-) can have any possible parity because the wavefunction transforms as:

$$\psi \to (-1)^l \psi \tag{5.53}$$

if we exchange the π^+ with the π^-. There are no special requirements on ψ here since the particles are different and the Pauli exclusion principle is not in force. If the particles are identical (e.g. $\pi^+\pi^+$ or $\pi^-\pi^-$), ψ must be symmetric because the pions are bosons. So, only states with $l = 0, 2, 4\ldots$ are allowed. Since the parity of the $\pi^+\pi^+$ pair is $(-1)^l$, the parity of the system must be $+1$.

To compute the wavefunction after exchanging a pair of particles, you may want to know how the spin part changes. This is given by the QM theorems on Fermi–Dirac and Bose–Einstein statistics (Cohen-Tannoudji *et al.*, 1991):

> if we exchange a pair of fermions, the spin part of the wavefunction changes as $(-1)^{s+1}$. If we exchange a pair of bosons, the spin part of the wavefunction changes as $(-1)^s$.

This neat statement generalizes the results discussed above about the electron spins in the symmetric triplet and antisymmetric singlet (eqns 5.46 and 5.47).

5.6.4   The intrinsic parity of composite states

The intrinsic parity of the protons and the neutrons follows from the previous recipes combined with the results of the quark models of Chap. 8. A proton is a bound state of three quarks: up, up, and down. A neutron is a bound state of up, down, and down. If we label l_{12} the angular momentum of the first two quarks and l_3 the angular momentum of the third quark with respect to the center of mass of the first and second quarks, the intrinsic parity of the proton is:

$$(-1)^{l_{12}+l_3}\mathcal{P}_u\mathcal{P}_u\mathcal{P}_d \tag{5.54}$$

and the intrinsic parity of the neutron is:

$$(-1)^{l_{12}+l_3}\mathcal{P}_u\mathcal{P}_d\mathcal{P}_d. \tag{5.55}$$

As discussed in Chap. 8, protons and neutrons are the ground states of the three-quark bound systems and, therefore, have the lowest angular momenta. As a consequence,

> protons and neutrons have the same intrinsic parity and this parity is $+1$ ($l_{12} = l_3 = 0$).

The deuteron is a bound state of a proton and a neutron and its intrinsic parity must be evaluated in the center of mass (CM) of the two particles. If l_{pn} is the angular momentum of the two-body system in CM, the intrinsic parity of the deuteron is $+1$ only if l_{pn} is zero or even. Note, however, that there is no guarantee a priori that the ground state of a bound nuclear system like the deuteron has $l = 0$. So, the intrinsic parity of the deuteron requires either an experimental determination or a complete nuclear physics calculation. At the time of the Chinowsky–Steinberger experiment, the even parity of the deuteron was already established using these kinds of methods. The same consideration holds for a generic composite system, where there are no simple rules to assess the intrinsic parity.

5.6.5   The parity of the elementary bosons*

The elementary fermions are not the only elementary particles of the Standard Model. As noted in Chap.1, this model employs additional particles to describe the interactions among fermions: the photon (electromagnetic interactions), the gluons (strong interactions), the massive spin-1 bosons (the W^+, W^-, Z^0), and the massive spin-0 boson (the Higgs boson). The Standard Model predicts the intrinsic parity of the photon and the gluons to be -1, and the parity of the Higgs boson to be +1. The intrinsic parity of the $W^\pm$ and Z^0 bosons are of no use because they mediate processes that violate the parity conservation law.

5.7   C-parity

A **C-parity** transforms a particle into its anti-particle and vice-versa. Therefore, this operator $\mathcal{C}$ is deeply intertwined with the concept of anti-matter. The C-parity is an inversion because two applications of $\mathcal{C}$ brings the system to the original state.

5.7.1   Classical C-parity

We can define the C-parity transformations even in classical physics. In this case, the C-parity changes the sign of electric charges and it is not difficult to show that:

> the classical intrinsic C-parity of the electromagnetic field is -1.

This famous result follows from the general solution of the Maxwell equations (Griffiths, 2017) that can be expressed in integral form as a function of the charge and current density ($\rho(\mathbf{x}, t)$ and $\mathbf{j}(\mathbf{x}, t)$), the scalar potential $V(\mathbf{x}, t)$, the vector potential $\mathbf{A}(\mathbf{x}, t)$, and the retarded time $t_R = t - |\mathbf{x}' - \mathbf{x}|/c$. The solution in SI is:

$$V(\mathbf{x}, t) = \frac{1}{4\pi\epsilon_0} \int \frac{\rho(\mathbf{x}', t_R)}{|\mathbf{x}' - \mathbf{x}|} d^3\mathbf{x}' \tag{5.56}$$

$$\mathbf{A}(\mathbf{x}, t) = \frac{\mu_0}{4\pi} \int \frac{\mathbf{j}(\mathbf{x}', t_R)}{|\mathbf{x}' - \mathbf{x}|} d^3\mathbf{x}' \tag{5.57}$$

and the corresponding electric and magnetic fields can be computed from the derivative of the potentials as:

$$\mathbf{E} = -\nabla V - \frac{\partial \mathbf{A}}{\partial t} \tag{5.58}$$

and

$$\mathbf{B} = \nabla \times \mathbf{A}. \tag{5.59}$$

Equations 5.58 and 5.59 are used to derive the **Jefimenko equations**, that is the generalization of the Coulomb law for time-dependent e.m. fields. These (quite cumbersome) equations are just eqns 5.58 and 5.59 after replacing the potentials with eqns 5.56 and 5.57 and performing the derivatives.

A change of sign of the electric charges corresponds to:

$$\rho(\mathbf{x}', t_R) \to -\rho(\mathbf{x}', t_R)$$
$$\mathbf{j}(\mathbf{x}', t_R) \to -\mathbf{j}(\mathbf{x}', t_R) \tag{5.60}$$

so that:

$$\mathbf{E} \to -\mathbf{E}$$
$$\mathbf{B} \to -\mathbf{B}. \tag{5.61}$$

Hence, the classical intrinsic C-parity of the photon is -1.

The Standard Model provides a rigorous definition of the C-parity of the photon and demonstrates that it is an eigenstate of the C-parity operator, whose eigenvalue is equal to -1. Like the parity of the photon, this demonstration is outside the scope of the book and can be found in several QFT textbooks including Peskin and Schroeder, 1995 and Schwartz, 2013.

5.7.2   The C-parity of particles

Very few quantum systems are eigenstate of the C-parity operator because if the particle is electrically charged, it transforms into a different particle, a particle with the opposite sign. Hence, for one-particle systems, only neutral particles are candidates to be a C-parity eigenstate in their RF. In this case, the particle is identical to its antiparticle and the wavefunctions differ just by a unit phase. Understanding when a particle is the same as its antiparticle has been one of the deepest mysteries in particle physics and was solved in the late 1960s. Some particles are clearly their own antiparticles and are eigenstates of the C-parity operator. The most prominent examples are the photon and the π^0. Other particles – for example the neutron — have an antiparticle that is different from the original particle. This is evident from the experimental data. An antineutron $\bar{n}$ decays as:

$$\bar{n} \to \bar{p} \; e^+ \; \nu_e \tag{5.62}$$

and a magnetic spectrometer can easily distinguish a positive proton p from a negative proton $\bar{p}$. This is the dominant ($BR \simeq 100\%$) decay mode of $\bar{n}$ and is completely different from the dominant mode of n:

$$n \to p \; e^- \; \bar{\nu}_e. \tag{5.63}$$

On the contrary, there is no evidence of a particle with the same mass, lifetime, and spin of the π^0, which decays differently from $\pi^0 \to \gamma\gamma$.

We will solve all these puzzles in Chap. 8 with one notable exception, the neutrino. At the time of writing, we do not know yet whether the neutrino corresponds to its own antiparticle. We will discuss this open issue in Chap. 11.

The intrinsic C-parity of the π^0 can be established experimentally as we did for the π^-. The π^0 decays into two photons. Hence, in the π^0 rest frame, the two photons are emitted back to back. The photons are spin-1 particles and the spin can be aligned either parallel or antiparallel to the direction of the momentum (see Fig. 5.3).[13] If we apply the C-parity operator to the final state, the operator changes neither the spin nor the space part of the wavefunction (see also Sec. 5.7.3). Therefore, the C-parity of the final state only depends on the intrinsic C-parity of the photon:

$$\mathcal{C}\psi_{\gamma\gamma} = \mathcal{C}_\gamma \mathcal{C}_\gamma \psi = (\mathcal{C}_\gamma)^2 \psi = \psi. \tag{5.64}$$

The last equality results from $\mathcal{C}$ being an inversion, whose eigenvalues are ± 1. It thus holds whatever the C-parity of the photon is. Since the C-parity is conserved in electromagnetic interactions, the intrinsic C-parity of the π^0 must be $+1$. Like parity, the C-parity of elementary fermions is conventional but if the neutrinos are different from the antineutrinos, there are no elementary fermions that are eigenstates of $\mathcal{C}$.

[13]We rigorously prove this statement in Sec. 6.6.2.

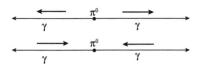

Fig. 5.3 The two relative orientation of the photon spins in the decay of a π^0.

5.7.3   Particle–antiparticle pairs

Particle–antiparticle pairs are neutral systems and are good candidates to be eigenstates of $\mathcal{C}$. The above-mentioned Dirac theorem provides a highly non-trivial result for the case of particle–antiparticle pairs:

> the intrinsic C-parity of a particle-antiparticle pair is $+1$ for boson–antiboson pairs and -1 for fermion–antifermion pairs.

Again, we will demonstrate this theorem for a e^+e^- pair in Sec. 5.9.1. The C-parity is an internal discrete symmetry, so we may wonder whether space and spin play any role in the evaluation of the C-parity of a system. If a change of sign in a multiparticle system is equivalent to a change in the position or exchange of particles, spin and space enter the game. For these systems, the total wavefunction is the product of:

$$\psi = \psi_{space}\psi_{spin}\psi_{intrinsic}. \tag{5.65}$$

The C-parity of the system can be established from the transformation of the space and spin part when we change the sign of the electric charges. Consider for instance a $\pi^+\pi^-$ pair (pair of spinless bosons). As shown in Fig. 5.4, the C-parity operator acts as the exchange operator between the two particles.

If the two particles have no spin, as for the pions, the C-parity operator is equivalent to the parity operator:

$$\mathcal{C}\psi_{\pi^+\pi^-} = \mathcal{P}\psi_{\pi^+\pi^-} = (-1)^l \psi_{\pi^+\pi^-}.$$

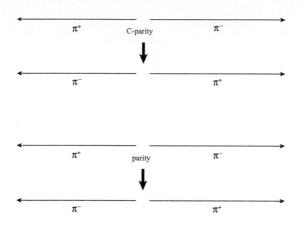

Fig. 5.4 A system of two spinless particles in its RF. In this case, the application of the C-parity operator (exchange of charges) is equivalent to the exchange of the momenta, i.e. to the change of the sign of the momentum of both particles. Hence, the outcome of the application of the C-parity is the same as for the parity.

If the particles have a spin (e.g. an electron–positron pair) the C-parity operator is equivalent to the *exchange* of the two particles (see Fig. 5.5):

$$\mathcal{C}\psi_{e^+e^-} = (-1)^l(-1)^{s+1}\mathcal{C}_{pair}\psi_{e^+e^-} = (-1)^{l+s}\psi_{e^+e^-} \tag{5.66}$$

because the intrinsic C-parity of a fermion-antifermion pair ($\mathcal{C}_{pair}$) is -1.

The most general theorem on the C-parity of a particle–antiparticle pair is the following:

> the C-parity of a particle–antiparticle pair is $(-1)^{l+s}$ for both bosons and fermions.

The demonstration is pretty simple. For a generic pair of particles:

$$\mathcal{C}\psi_{p\bar{p}} = \mathcal{C}\psi_{space}\psi_{spin}\psi_{intrinsic} \tag{5.67}$$

but $\mathcal{C}\psi_{space} = (-1)^l$ because the C-parity, like the parity, exchanges the position of the two particles in space. $\mathcal{C}\psi_{spin} = (-1)^{s+1}$ for fermions and $\mathcal{C}\psi_{spin} = (-1)^s$ for bosons. In addition, $\mathcal{C}\psi_{intrinsic} = +1$ for bosons and -1 for fermions. Hence,

$$\mathcal{C}\psi_{p\bar{p}} = (-1)^{l+s}\psi_{p\bar{p}} \tag{5.68}$$

for both bosons and fermions.

5.8   T-parity and the CPT theorem

5.8.1   Time-reversal

The **T-parity** is an external inversion that changes the sign of time from t to $-t$. In QM, time-reversal is described by an anti-unitary operator $\mathcal{T}$ (see Sec. 5.2) acting as:

$$\mathcal{T}\psi(\mathbf{x}, t) = \psi(\mathbf{x}, -t). \tag{5.69}$$

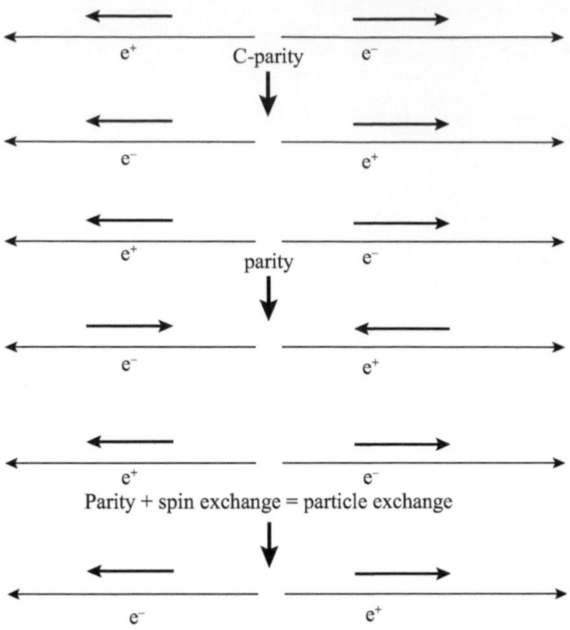

Fig. 5.5 A system of two spin-1/2 particles in its RF. In this case, the application of the C-parity operator (exchange of charges) is equivalent to the exchange of the momenta and the spins. Hence, the outcome of the application of the C-parity is not the same as for the parity but it is equivalent to applying the exchange operator $\psi(\mathbf{x}_1, \mathbf{S}_1, \mathbf{x}_2, \mathbf{S}_2) \to \psi(\mathbf{x}_2, \mathbf{S}_2, \mathbf{x}_1, \mathbf{S}_1)$.

All classical systems are invariant for this transformation. The invariance under time reversal implies that whenever a motion is allowed by a theory, the reversed motion is also allowed and occurs at the same rate forwards and backward in time. This means that the evolution of a system can be seen running from the future to the past if we exchange the boundary conditions so that the final state becomes the initial state. This feature poses severe problems in the definition of the arrow of time because there is not a preferred time direction from past to future at the microscopic level, in striking contradiction with our daily life. In thermodynamics, this problem was solved by L. Boltzmann and J.W. Gibbs relying on statistical mechanics and the concept of entropy (Swendsen, 2020). The same consideration holds in QM moving from a few-body system to a state made of a large number of particles. The main difference is that the conservation of T-parity is *not* a symmetry of the Standard Model. Once more, the bad guys are the weak interactions that show a tiny asymmetry in the scattering amplitudes if the final state particles are time-reversed.

Until 2012, the (overwhelming) experimental evidence for T-violation in weak interactions came from the observation of processes that violate the **charge conjugation and parity (CP) symmetry**. A CP transformation corresponds to the application of the parity and C-parity operators to a microscopic system. We will discuss this evidence in Sec. 13.7. The very existence of CP violation implies the violation of the time-reversal symmetry because all SM processes conserves the **charge conjugation, time reversal, and parity (CPT) symmetry**. This is the transformation obtained applying the parity, the C-parity, and the

T-parity to the wavefunction of a QM system:

$$\mathcal{CPT}\psi_p(\mathbf{x}, t) = \psi_{\bar{p}}(-\mathbf{x}, -t) \tag{5.70}$$

where p is a generic particle. After 2012, we obtained direct experimental evidence of T-parity violation, independently of CP. A renowned example is the comparison of the transition rates of:

$$B^- \to \bar{B}^0 \quad \text{versus} \quad \bar{B}^0 \to B^- \tag{5.71}$$

where the B hadrons are particles containing the b quark and the transitions occur by exchanges of weak mediators (Lees *et al.*, 2012). No differences have ever been observed for strong and electromagnetic interactions, where time reversal is an actual symmetry: the T-parity is an anti-unitary operator commuting with the Hamiltonian of quantum electrodynamics (QED) and quantum chromodynamics (QCD). This property is equivalent to the Boltzmann's **detailed-balance principle**:

> in a multi-body system at equilibrium, each elementary process is in equilibrium with its time-reversed process.

In QM, this principle reads:

$$\sum_{f,i} |\mathcal{M}_{i \to f}|^2 = \sum_{i,f} |\mathcal{M}_{f \to i}|^2 \tag{5.72}$$

where i and f are the initial- and final-state particles. Equation 5.72 can be exploited in many strong and electromagnetic processes. For instance, the cross-section for a $pp \to \pi^+ d$ reaction is nearly[14] the same as $\sigma(\pi^+ d \to pp)$, where d is a deuterium nucleus.

5.8.2 The CPT theorem

As discussed in Chap. 3, the discovery of the antiparticles is considered one of the breakthroughs of the 20th century because antimatter was hinted at neither by SR nor by non-relativistic QM. The nature of antiparticles remained a mystery until 1954. P. Dirac associated these states to negative energy solutions of the Dirac equation (see Sec. 5.9) but the "Dirac hole theory" was plagued by conceptual problems such as the assumption of an infinite negative energy ground state and issues in extending the theory to bosons. Even if the Dirac hole theory and other theories like the Feynman–Stueckelberg interpretation discussed in Chap. 6 are still interesting today, the modern understanding of antiparticles is much more solid. It is based on a theorem demonstrated by Lüders and Pauli in 1954 that links the inversions with the Lorentz invariance. The **CPT theorem** states that:

> the simultaneous application of parity, C-parity, and T-parity (the $\mathcal{CPT}$ operator) is a symmetry in any Lorentz invariant local quantum field theory with a hermitian Hamiltonian.

[14]The differences are only due to the phase space and the spin of the particles and do not depend on the transition matrices. The exact formula reads:

$$\frac{\sigma(pp \to \pi^+ d)}{\sigma(\pi^+ d \to pp)} = \frac{2}{3(2s_\pi + 1)} \frac{p_p^2}{p_\pi^2}$$

where p_p and p_π are the proton and pion momenta and s_π is the spin of the π^+. This equation was used in 1951 to demonstrate that the pions are spinless particles (Bettini, 2014).

For such a broad class of theories – that includes the Standard Model – antiparticles are an inescapable consequence of merging special relativity (Lorentz invariance) with quantum mechanics (local QFT). In particular, any Hamiltonian containing fields that describe elementary charged particles requires additional fields associated with the corresponding antiparticles. These fields are needed because the scattering amplitude of an n-body process must be the same if the scattering occurs from the final state to the initial state (T-parity), the position and momenta of the particles are reversed (parity) and the charge of the particles changes sign (C-parity).

The CPT theorem has been tested for decades and is remarkable that CPT conservation holds even if weak interactions violate separately parity, C-parity, T-parity, and the CP-parity. The conservation of the CPT symmetry substantiates the deep connection between SR, QM, and the existence of antiparticles.

5.9   The Dirac equation

You are probably aware that a general analytical formula giving all the solutions of the polynomial equations $a_0 + a_1 x + a_2 x^2 + a_3 x^3 \ldots + a_n x^n = 0$ exists only for $n \leq 4$. This result was proven in 1824, after 250 years of failed attempts, by N.H. Abel and P. Ruffini and the consequences of this theorem are still an active field of research. Something similar happened for the Schrödinger equation in 1926. The question here is "what is the relativistic generalization of the Schrödinger equation for particles with spin n?" A partial answer was found in 1948 by V. Bargmann and E. Wigner:

> any *free* particle of integer or half-integer spin n has a corresponding equation of motion that is covariant and, therefore, consistent with SR. However, we do not know yet if this is true for all *interacting* particles.

If the particle interacts with some fields, for example with a background field as in many atomic physic problems, inconsistencies can arise at any stage if the spin is sufficiently high: violation of causality, unphysical states, violation of unitarity (see Sec. 12.1), etc. This is still an open field of research. We have happy-ending stories for many theories with interacting particles of spin 0, 1/2, 1, 3/2, and 2 but also many no-go theorems for spins higher than 2 (Weinberg, 2005a; Schwartz, 2013).

Unaware of such a level of complexity, many scientists (O. Klein, W. Gordon, L. De Broglie, V. Fock, and others) noted that the Schrödinger equation is ill-posed in SR because it mixes a first-order time derivative with a second-order space derivative, while a covariant formulation requires space-time derivatives of the same order. The simplest covariant equation for a massless particle is:[15]

$$\Box \psi(\mathrm{x}, t) = \partial^\mu \partial_\mu \psi(x^\mu) = 0. \tag{5.73}$$

[15]We remind readers that the d'Alembert operator $\Box$ is a shortand for $\frac{1}{c^2}\frac{\partial^2}{\partial t^2} - \nabla^2$. In the covariant formalism, it easy to show that $\Box = \partial^\mu \partial_\mu$ where ∂^μ is a shorthand for $\partial/\partial x^\mu$. In NU, $c=1$ and we employ NU in the rest of the chapter.

The massive counterpart of eqn 5.73 is the **Klein–Gordon equation:**

$$(\partial^\mu \partial_\mu - m)\psi(x^\mu) = 0 \tag{5.74}$$

where $\psi(x^\mu) \equiv \psi(\mathbf{x}, t)$ is the particle wavefunction. A plane wave solution of eqn 5.74 is:

$$\psi(\mathbf{x}, t) = Ne^{i\mathbf{p}\cdot\mathbf{x} - Et} \tag{5.75}$$

where N is a normalization constant such that the probability of finding the particle in at least one point in space-time is 1. Replacing eqn 5.75 in eqn 5.74, we get the energy-momentum dispersion relation of SR – $E^2 = m^2 + |\mathbf{p}|^2$ – and a set of solutions corresponding to:

$$E = \pm\sqrt{m^2 + |\mathbf{p}|^2} \, , \tag{5.76}$$

and we have the job done if we happily reject negative energy solutions as unphysical states, as we often do in classical mechanics. Unfortunately, this is forbidden in QM because a set of solutions must form a complete basis for the Hilbert space and you cannot pick up your favorite solutions with impunity.

Example 5.2

Negative solutions correspond to negative probabilities of finding the particle of mass m. This statement follows from the probability density ρ computed for the negative energy solutions of eqn 5.74:

$$\rho = i\left(\psi^* \frac{\partial \psi}{\partial t} - \psi \frac{\partial \psi^*}{\partial t}\right) = 2E|N|^2. \tag{5.77}$$

We encourage the readers to go through Exercise 5.11 and compute the probability current $\mathbf{j} = -i(\psi^* \nabla \psi - \psi \nabla \psi^*)$, too. This current, together with eqn 5.77 that plays the role of j^0, can be written in covariant form as a four-current j^μ. Unlike the charge current of eqn 5.24, you will learn that such "probability-current" j^μ is completely meaningless and brings us to negative probability states.

The difficulties experienced by second-order equations were sidestepped by P. Dirac in 1928 choosing – somewhat arbitrarily for our modern standards – a first-order equation, whose covariant form is:

$$\left(i\gamma^\mu \frac{\partial}{\partial x^\mu} - m\right)\psi = 0. \tag{5.78}$$

Unlike non-relativistic QM, ψ is a novel four-component complex vector made of four wavefunctions and is called a **bispinor:**

$$\psi(x^\mu) = \begin{pmatrix} \psi_1 \\ \psi_2 \\ \psi_3 \\ \psi_4 \end{pmatrix}. \tag{5.79}$$

In eqn 5.78, m is the (rest) mass of the particle and γ^μ are 4×4 matrices. A common representation of these matrices is the **Pauli–Dirac representation**:

$$\gamma^0 = \gamma_0 = \begin{pmatrix} \mathbb{I} & 0 \\ 0 & -\mathbb{I} \end{pmatrix} ; \quad \gamma^i = -\gamma_i = \begin{pmatrix} 0 & \sigma_i \\ -\sigma_i & 0 \end{pmatrix}. \qquad (5.80)$$

Here $\mathbb{I}$ is the 2×2 identity matrix and σ_i are the Pauli matrices (see Sec. B.2.1). More explicitly:

$$\gamma^0 = \begin{pmatrix} 1 & 0 & 0 & 0 \\ 0 & 1 & 0 & 0 \\ 0 & 0 & -1 & 0 \\ 0 & 0 & 0 & -1 \end{pmatrix} \quad \gamma^1 = \begin{pmatrix} 0 & 0 & 0 & 1 \\ 0 & 0 & 1 & 0 \\ 0 & -1 & 0 & 0 \\ -1 & 0 & 0 & 0 \end{pmatrix}$$

$$\gamma^2 = \begin{pmatrix} 0 & 0 & 0 & -i \\ 0 & 0 & i & 0 \\ 0 & i & 0 & 0 \\ -i & 0 & 0 & 0 \end{pmatrix} \quad \gamma^3 = \begin{pmatrix} 0 & 0 & 1 & 0 \\ 0 & 0 & 0 & -1 \\ -1 & 0 & 0 & 0 \\ 0 & 1 & 0 & 0 \end{pmatrix} \qquad (5.81)$$

which are called the **Dirac matrices** in the Pauli–Dirac representation. Other representations are possible and have pros and cons but we will stick to the Pauli–Dirac one in the rest of the book.

Equation 5.78 can be written in a compact form as:

$$(i\gamma^\mu \partial_\mu - m)\,\psi(x^\mu) = 0 \qquad (5.82)$$

and is called the **Dirac equation**.

The Dirac equation is one of the fundamental equation of particle physics but it is a delicate object to handle. First, note that the Dirac matrices are *not* four-vectors but just "scalars": since they are constants, they do not change for any Lorentz transformation. We write them in vector-like form because, when combining the matrices, we can formally treat the $\mu = 0, 1, 2, 3$ index as if it were a four-vector index. For instance,

$$\{\gamma^\mu, \gamma^\nu\} \equiv \gamma^\mu \gamma^\nu + \gamma^\nu \gamma^\mu = 2g^{\mu\nu} \qquad (5.83)$$

and the symbol $\{A, B\} \equiv AB + BA$ is called the **anticommutator** of A and B. Similarly, relations as:

$$(\gamma^0)^2 = \mathbb{I} \quad (\gamma^k)^2 = -\mathbb{I}$$
$$\gamma^\mu \gamma^\nu = -\gamma^\nu \gamma^\mu \ \text{if} \ \mu \neq \nu \qquad (5.84)$$

make the use of the pseudo-index μ extremely convenient to carry on the calculations. The price to pay is that we do not know if eqn 5.78 is really covariant. The covariance of the Dirac equation must be demonstrated explicitly and is quite a job.[16]

---

[16]See e.g. the Appendix B.2 in Thomson, 2013.

5.9.1   Solutions of the Dirac equation

The solutions of the Dirac equation are classified into two groups: the functions $u^{(r)}(\mathbf{p})$ corresponding to the positive energy solutions of the Klein–Gordon (KG) equation and the functions $v^{(r)}(\mathbf{p})$ corresponding to the negative energy solutions (Le Bellac, 1991):

$$u^{(r)}(\mathbf{p}) = \frac{1}{\sqrt{E_p + m}} \begin{pmatrix} (E_p + m)\chi^{(r)} \\ (\boldsymbol{\sigma} \cdot \mathbf{p})\chi^{(r)} \end{pmatrix} \tag{5.85}$$

$$v^{(r)}(\mathbf{p}) = \frac{1}{\sqrt{E_p + m}} \begin{pmatrix} (\boldsymbol{\sigma} \cdot \mathbf{p})\chi^{(r)} \\ (E_p + m)\chi^{(r)} \end{pmatrix}. \tag{5.86}$$

In these formulas, $E_p = \sqrt{m^2 + p^2}$ and $\boldsymbol{\sigma} = (\sigma_1, \sigma_2, \sigma_3)$. r can be either 1 or 2, and

$$\chi^{(1)} = \begin{pmatrix} 1 \\ 0 \end{pmatrix} \; ; \quad \chi^{(2)} = \begin{pmatrix} 0 \\ 1 \end{pmatrix}. \tag{5.87}$$

The general solutions of the Dirac equation are:

$$\psi(x^\mu) = u^{(r)} e^{-ip^\mu x_\mu} \quad \text{[positive energy solutions]}; \tag{5.88}$$

$$\psi(x^\mu) = v^{(r)} e^{ip^\mu x_\mu} \quad \text{[negative energy solutions]}. \tag{5.89}$$

In modern terms, these solutions describe spin $1/2$ particles and antiparticles, respectively. $r = 1, 2$ can be interpreted (see Example 5.3) as **spin indexes** and correspond to $s_z = +1/2$ and $s_z = -1/2$, respectively. Therefore, $|\psi(x)|^2$ for $r = 1$ and positive energy solutions give the probability of finding, for example, an electron with the spin aligned in the direction of motion. In this case, the corresponding negative energy solutions give the probability of finding a positron with the spin aligned in the direction of motion. This equation can be applied also to heavier charged leptons and to neutrinos, if neutrinos are different from antineutrinos.

At first sight, the Dirac equation is not particularly appealing compared with the Klein–Gordon equation: it solves the problem of the negative probability but does not wipe off the negative energy solutions. Also, each **spinor** considered separately, that is, the upper or the lower pair of the wavefunction vector in eqn 5.79 is a solution of the Klein–Gordon equation.

Still, the Dirac equation is packed with terrific surprises: it creates a link between spin and covariance so that you cannot factorize the spatial and spin wavefunction anymore, as we were doing in non-relativistic QM. Even more, it shows (see Example 5.3) that:

> if a particle satisfies the Dirac equation, it is a spin-1/2 fermion and its intrinsic magnetic moment is proportional to the spin **S** according to:
>
> $$\boldsymbol{\mu}_S = g_e \mu_B \mathbf{S} = \frac{e}{m_e} \mathbf{S} \qquad (5.90)$$
>
> (in NU). g_e is called the electron spin **g-factor**. $\mu_B = e/2m_e$ is the Bohr magneton, while e and m_e are the electron charge and mass.

[17] The exact value is slightly higher due to the QED corrections discussed in Chap. 6.

Unlike classical physics where the spin does not exist, the angular momenta have a mechanical origin and $g_e = 1$, the Dirac equation predicts $g_e = 2$, in excellent agreement with experimental data.[17] The correct prediction of g_e and, even more, the discovery of the antiparticles by Anderson established the Dirac equation as "the equation of motion of spin-1/2 massive particles" even if Dirac's interpretation of antimatter was quite naïve.

Example 5.3

The Dirac equation describes a particle with spin. A smart way to establish the particle spin is to write the Hamiltonian that makes the Dirac equation an equation of motion and check if it commutes with the angular momentum operator. It does not. For instance (Thomson, 2013), we can define $\gamma^0 \equiv \beta$, $\gamma^1 = \beta\alpha_x$, $\gamma^2 = \beta\alpha_y$, $\gamma^3 = \beta\alpha_z$. The αs depend on the Pauli matrices as:

$$\alpha_i = \begin{pmatrix} 0 & \sigma_i \\ \sigma_i & 0 \end{pmatrix} \qquad (5.91)$$

with $i = x, y, z$. In this case, the Hamiltonian is $\hat{H} = \boldsymbol{\alpha} \cdot \hat{\mathbf{p}} + \beta m$ and there is no way to have $[\hat{H}, \hat{\mathbf{L}}] = 0$ because $[\hat{H}, \hat{\mathbf{L}}] = [\hat{H}, -i(\mathbf{r} \times \nabla)] = [\boldsymbol{\alpha} \cdot \hat{\mathbf{p}} + \beta m, \hat{\mathbf{r}} \times \hat{\mathbf{p}}] = [\boldsymbol{\alpha} \cdot \hat{\mathbf{p}}, \hat{\mathbf{r}} \times \hat{\mathbf{p}}]$. It is not difficult to evaluate the commutator term-by-term for x, y, and z. All terms commute except for the position-momentum pair in the same direction: $[\hat{x}, \hat{p}_x] = i$, $[\hat{y}, \hat{p}_y] = i$, $[\hat{z}, \hat{p}_z] = i$. Hence, $[\hat{H}, \hat{L}_x] = -i(\boldsymbol{\alpha} \times \hat{\mathbf{p}})_x$ and

$$[\hat{H}, \hat{\mathbf{L}}] = -i(\boldsymbol{\alpha} \times \hat{\mathbf{p}}) \neq 0. \qquad (5.92)$$

Equation 5.92 demonstrates that the orbital angular momentum of a Dirac particle does not account for all the angular momentum available to the particle and is not an integral of motion. The symmetry for rotations can be restored for a single free Dirac particle adding another source of momentum, whose origin is non-mechanical: the **particle spin**. If the particle described by the Dirac equation is a spin-1/2 fermion, the **spin operator** is

$$\hat{\mathbf{S}} = \frac{1}{2} \begin{pmatrix} \boldsymbol{\sigma} & 0 \\ 0 & \boldsymbol{\sigma} \end{pmatrix} \qquad (5.93)$$

or, more explicitly:

$$\hat{S}_x = \frac{1}{2}\begin{pmatrix} 0 & 1 & 0 & 0 \\ 1 & 0 & 0 & 0 \\ 0 & 0 & 0 & 1 \\ 0 & 0 & 1 & 0 \end{pmatrix}, \ \hat{S}_y = \frac{1}{2}\begin{pmatrix} 0 & -i & 0 & 0 \\ i & 0 & 0 & 0 \\ 0 & 0 & 0 & -i \\ 0 & 0 & i & 0 \end{pmatrix},$$

$$\hat{S}_z = \frac{1}{2}\begin{pmatrix} 1 & 0 & 0 & 0 \\ 0 & -1 & 0 & 0 \\ 0 & 0 & 1 & 0 \\ 0 & 0 & 0 & -1 \end{pmatrix}. \tag{5.94}$$

Using the commutation relations among the Pauli matrices we can demonstrate (see Exercises 5.9 and 5.10) that:

---

a free Dirac particle has a total angular momentum $\hat{\mathbf{J}} \equiv \hat{\mathbf{L}} + \hat{\mathbf{S}}$ and $[\hat{H}, \hat{\mathbf{J}}] = 0$, as prescribed by the rotational symmetry of any QM system.

Note that the demonstration holds only if the spin is $1/2$, whose operator is given by eqn 5.93. The Dirac equation thus describes only spin-$1/2$ fermions (an electron, a muon, an electron neutrino, etc.).

Another gem of the Dirac equation is the demonstration that the intrinsic parity of an electron is opposite to the one of the positron. We already stated this theorem in a general form in Sec. 5.6.3.

Example 5.4

If ψ satisfies eqn 5.82, the parity operator must have a very peculiar form:

$$\hat{\mathcal{P}} = \gamma^0 \tag{5.95}$$

This is due to the fact that the Dirac equation can be recast as:

$$i\boldsymbol{\gamma} \cdot \nabla\psi - m\psi = -i\gamma^0 \partial_0 \psi \tag{5.96}$$

when we use the compact notation $\boldsymbol{\gamma} \equiv (\gamma^1, \gamma^2, \gamma^3)$ and:

$$\nabla = \left(\frac{\partial}{\partial x}, \frac{\partial}{\partial y}, \frac{\partial}{\partial z}\right). \tag{5.97}$$

Here, we want $\psi'(x', y', z')$ to fulfill the same equation in the mirrored reference system obtained after the parity transformation:

$$i\boldsymbol{\gamma} \cdot \nabla_{\mathbf{x}'}\psi' - m\psi' = -i\gamma^0 \partial_0 \psi' \tag{5.98}$$

where $\nabla_{\mathbf{x}'}$ is:

$$\nabla_{\mathbf{x}'} \equiv \left(\frac{\partial}{\partial x'}, \frac{\partial}{\partial y'}, \frac{\partial}{\partial z'}\right). \tag{5.99}$$

Since $\psi' = \hat{\mathcal{P}}\psi$ and the parity is an inversion ($\hat{\mathcal{P}}^2 = \mathbb{I}$), then $\psi = \hat{\mathcal{P}}\psi'$. We can now replace ψ with $\hat{\mathcal{P}}\psi'$ in eqn 5.96 and multiply the result by γ^0 from the left to get:

$$i\gamma^0\boldsymbol{\gamma}\hat{\mathcal{P}} \cdot \nabla\psi' - m\gamma^0\hat{\mathcal{P}}\psi' = -i\gamma^0\gamma^0\hat{\mathcal{P}}\partial_0\psi'. \tag{5.100}$$

The parity is a very simple operator and the relation between old and new coordinates are rather obvious:

$$(t, x, y, z,) \rightarrow (t, -x, -y, -z).$$

Therefore,

$$\hat{\mathcal{P}}\frac{\partial}{\partial x}\psi' = -\frac{\partial\psi'}{\partial x'} \;\; ; \;\; \hat{\mathcal{P}}\frac{\partial}{\partial y}\psi' = -\frac{\partial\psi'}{\partial y'} \;\; ; \;\; \hat{\mathcal{P}}\frac{\partial}{\partial z}\psi' = -\frac{\partial\psi'}{\partial z'} \tag{5.101}$$

and eqn 5.100 can be rewritten as:

$$- i\gamma^0\boldsymbol{\gamma}\hat{\mathcal{P}} \cdot \nabla_{\mathbf{x}'}\psi' - m\gamma^0\hat{\mathcal{P}}\psi' = -i\gamma^0\gamma^0\hat{\mathcal{P}}\partial_0\psi'. \tag{5.102}$$

Equation 5.102 is equivalent to:

$$i\boldsymbol{\gamma}\gamma^0\hat{\mathcal{P}} \cdot \nabla_{\mathbf{x}'}\psi' - m\gamma^0\hat{\mathcal{P}}\psi' = -i\gamma^0\gamma^0\hat{\mathcal{P}}\partial_0\psi' \tag{5.103}$$

because $\gamma^0\gamma^k = -\gamma^k\gamma^0$, and we get the sought-for eqn 5.98 if $\gamma^0\hat{\mathcal{P}} \sim \mathbb{I}$. At this point, we do not have too many choices. Since $\hat{\mathcal{P}}^2 = \mathbb{I}$, either $\hat{\mathcal{P}} = \gamma^0$ or $\hat{\mathcal{P}} = -\gamma^0$. If we choose, for instance, $\hat{\mathcal{P}} = \gamma^0$, we can compute the intrinsic parity moving to the RF of the Dirac particle. Setting $E = m$ and $\mathbf{p} = 0$ in eqn 5.85, we get:

$$\hat{\mathcal{P}}\psi' = \hat{\mathcal{P}}u^{(1)} = \gamma^0 u^{(1)} = \sqrt{2m}\gamma^0 \begin{pmatrix} 1 \\ 0 \\ 0 \\ 0 \end{pmatrix} = +u^{(1)}. \tag{5.104}$$

In the same way, $\hat{\mathcal{P}}u^{(2)} = +u^{(2)}$, $\hat{\mathcal{P}}v^{(1)} = -v^{(1)}$, $\hat{\mathcal{P}}v^{(2)} = -v^{(2)}$. So, electrons and positrons have opposite intrinsic parity. Note that we get the same result if we make the other choice ($\hat{\mathcal{P}} = -\gamma^0$): the electron–positron parity is opposite but, now, the intrinsic parity of the electron is negative. The relative parity of fundamental fermions are opposite to their antiparticles but the absolute value (± 1) is conventional and does not affect any observable.

Using the Dirac equation we can also demonstrate that the C-parity of an e^+e^- pair is -1, as claimed in Sec. 5.7.3 and discussed in Exercises 5.12 and 5.13.

5.10   What is a spinor?*

Spinors and bispinors were not invented by Pauli or Dirac but inherited from the pioneering studies of E. Cartan in 1913. Still, they are the Rosetta Stone that connects intrinsic angular momenta with SR, the spin-statistics theorem, and CPT conservation. Even now, they continue to play a major role in theoretical physics and pure mathematics. In this section, we want to understand what is the difference between a wavefunction that describes spin-1/2 particles and a wavefunction that describes a particle with zero or integer spin. The answer is in the difference between a vector and a spinor.

Let us consider rotations in space. The unitary operator that describes the rotations is given by eqn 5.21 and the wavefunction of a spinless particle after a rotation can be computed as $\psi' = \hat{U}\psi$ with:

$$\hat{U}(\boldsymbol{\theta}) = e^{i\boldsymbol{\theta}\cdot\hat{\mathbf{L}}}. \tag{5.105}$$

For a spinless particle, $\hat{\mathbf{L}}$ is the orbital momentum operator of Tab. B.1 expressed in NU. If we rotate the reference system around the z-axis, the wavefunction changes as:

$$\psi' = e^{i\theta_z \hat{L}_z}\psi. \tag{5.106}$$

Here $\theta_z \in [0, 2\pi]$ and $\hat{L}_z$ is the z component of $\hat{\mathbf{L}} = \hat{\mathbf{x}} \times (-i\boldsymbol{\nabla})$. For particles with spin, the generalization is straightforward in non-relativistic QM:

$$\psi' = e^{i\theta_z \hat{J}_z}\psi \tag{5.107}$$

with $\hat{\mathbf{J}} = \hat{\mathbf{L}} + \hat{\mathbf{S}}$.

In all these cases, if $\theta_z = 2\pi$, then $\psi'(\mathbf{x}) = \psi(\mathbf{x})$. This feature holds not only in non-relativistic QM but also when ψ is the solution of the corresponding relativistic equation *if the spin of the particle is an integer number*. To be frank, we would have not expected anything else. A 360° rotation brings the reference frame to its original position and the wavefunction should be the same as before the rotation. In general, it is possible to demonstrate that:

> all relativistic wavefunctions describing a particle with integer spin do not change for a 2π rotation along z, exactly as an ordinary geometrical vector.

This result is consistent with our lay expectations but you will be astonished to learn that it does not hold for particles with half-integer spins. For spin-1/2 fermions without orbital angular momentum, this follows from the Dirac equation. In this case, the rotations are described, again, by an operator like $\exp(i\theta_z J_z)$ but the generator of the rotations along z for the wavefunctions that fulfill the Dirac equation has a very peculiar form (Schwartz, 2013):

$$\hat{U}(\theta_z) = e^{i\theta_z S_{12}} \tag{5.108}$$

where

$$S_{12} \sim [\gamma^1, \gamma^2] = \frac{1}{2}\begin{pmatrix} 1 & 0 & 0 & 0 \\ 0 & -1 & 0 & 0 \\ 0 & 0 & 1 & 0 \\ 0 & 0 & 0 & -1 \end{pmatrix}. \tag{5.109}$$

The operator that rotates a Dirac bispinor along z is thus a 4×4 matrix of the form:

$$e^{i\theta_z S_{12}} = \begin{pmatrix} \exp\{\tfrac{i}{2}\theta_z\} & 0 & 0 & 0 \\ 0 & \exp\{-\tfrac{i}{2}\theta_z\} & 0 & 0 \\ 0 & 0 & \exp\{\tfrac{i}{2}\theta_z\} & 0 \\ 0 & 0 & 0 & \exp\{-\tfrac{i}{2}\theta_z\} \end{pmatrix}.$$

$$(5.110)$$

Disdaining common sense, $e^{i\theta_z S_{12}} = -\mathbb{I}$ for $\theta_z = 2\pi$. The geometrical objects that change sign after a 360° rotation are called **spinors** and:

> all wavefunctions of particles with half-integer spins change sign after a 2π rotation along z and are described by spinors.

This unusual feature is the core of the spin-statistics theorem. For instance, if we have two electrons in the RF, one opposite to the other with respect to the origin and both with the spin along the positive z-axis ($\mathbf{S} \sim +\mathbf{k}$), we can perform a rotation by 180° to exchange the two particles, getting a factor i in front of the wavefunction. Then, we can perform another rotation by 180° to go back to the initial state. The result is:

$$\psi_{12} \to -\psi_{21} \qquad\qquad (5.111)$$

and a particle exchange picks up a factor -1. The generalization of these considerations and the rigorous derivation of all previous statements are a tough climbing in group representation theory and goes well beyond the scope of this book (Streater and Wightman, 2000; Weinberg, 2005a). A rigorous derivation took a long time but we can fairly say that most of the issues were solved by M. Fiertz and W. Pauli in 1940. The **spin-statistics theorem** says that:

> the wavefunction of a system of identical particles with integer spin has the same value when the positions of any two particles are swapped. These particles are called **bosons**. The wavefunction of a system of identical particles with half-integer spin changes sign when two particles are swapped. The wavefunction is thus antisymmetric under particle exchange and the corresponding particles are called **fermions**.

As for the CPT theorem and the origin of the antiparticles, this theorem is a result emerging from the unification of SR and QM and is entangled with the conservation of CPT.

Further reading

Symmetries play an important role in our discipline and physics as a whole. Most of the literature about the tools to study symmetries is written by mathematicians for mathematicians and employ a language that suits their needs but is very challenging for physics undergraduates. Fortunately, there are excellent textbooks written especially for physicists that cover basic and advanced applications. A classical book for particle physicists is Georgi, 1999, which covers mostly the continuous groups of relevance for the SM. A more general approach is followed in Zee, 2016, which requires few prerequisites and is quite student-friendly. Costa and Fogli, 2012 covers all material presented in this chapter plus a detailed treatment of the Lorentz and Poincaré groups of SR and the gauge symmetries. An advanced undergraduate book more oriented toward physics applications is Schwichtenberg, 2018.

Exercises

(5.1) Show how the following classical quantities change after a parity transformation:

$$\mathbf{x}, \quad \mathbf{p}, \quad \mathbf{J} \text{ (angular momentum)}, \quad \mathbf{J} \cdot \mathbf{p}, \quad \mathbf{x} \cdot \mathbf{p}$$

$$y = \frac{1}{2} \log \frac{E + p_L}{E - p_L} \text{ (rapidity)}, \quad \mathbf{J} \times \mathbf{p}$$

and identify the invariant quantities.

(5.2) Demonstrate that eqn 5.23 written in NU implies

$$\partial_\mu j^\mu = \frac{\partial \rho}{dt} + \nabla \cdot \mathbf{j} = 0.$$

(5.3) The η is a spinless neutral particle with $m_\eta = 548$ MeV, parity equal to -1 and C-parity equal to $+1$. Identify the possible decay modes within this list:

$$\eta \to \gamma\gamma$$
$$\eta \to \gamma\gamma\gamma$$
$$\eta \to \pi^0\pi^0$$
$$\eta \to \pi^0\pi^0\pi^0$$
$$\eta \to \pi^0\pi^+\pi^-$$
$$\eta \to \pi^0\gamma.$$

[Note: all pions have the same intrinsic parity]

(5.4) A e^+e^- bound state (the **positronium**) can decay both in $e^+e^- \to \gamma\gamma$ and in $e^+e^- \to \gamma\gamma\gamma$. Is it an evidence of C-parity violation?

(5.5) Is a $\pi^+\pi^-$ system a CP eigenstate?

(5.6) Demonstrate eqn 5.83 using the Pauli–Dirac representation of the gamma matrices.

(5.7) * The first deficit of muon neutrinos produced by a source and propagating at large distances up to a detector ("$\nu_\mu \to \nu_\mu$ oscillations") was observed in 1998. Can we observe a difference between the $\nu_\mu \to \nu_\mu$ and $\bar{\nu}_\mu \to \bar{\nu}_\mu$ oscillation probabilities?

(5.8) Consider the **Weyl representation** of the gamma matrices:

$$\gamma^0 = \begin{pmatrix} 0 & \mathbb{I} \\ \mathbb{I} & 0 \end{pmatrix}; \quad \gamma^i = \begin{pmatrix} 0 & \sigma_i \\ -\sigma_i & 0 \end{pmatrix}. \quad (5.112)$$

Show that these matrices fulfill the same **Clifford algebra**

$$\{\gamma^\mu, \gamma^\nu\} = 2g^{\mu\nu}$$

as the gamma matrices in the Pauli-Dirac representation, i.e. fulfill eqn 5.83.

(5.9) * If we define $\tilde{\mathbf{S}}$ as $2\hat{\mathbf{S}}$ in eqn 5.93, show that $[\alpha_x, \tilde{S}_x] = 0$, $[\beta, \tilde{S}_x] = 0$, $[\alpha_y, \tilde{S}_x] = -2i\alpha_z$, and $[\alpha_z, \tilde{S}_x] = 2i\alpha_y$.

(5.10) * Using the results of Exercise 5.9, show that $[\hat{H}, \hat{\mathbf{S}}] = i\boldsymbol{\alpha} \times \hat{\mathbf{p}}$ and, therefore, $[\hat{H}, \hat{\mathbf{L}} + \hat{\mathbf{S}}] \equiv [\hat{H}, \hat{\mathbf{J}}] = 0$. The conservation of angular momentum in the Dirac equation is, thus, ensured only if we account for the spin.

(5.11) * Compute the probability current $\mathbf{j} = -i(\psi^* \nabla \psi - \psi \nabla \psi^*)$ for the solutions of the Dirac equation.

(5.12) ** The C-parity operator in relativistic quantum mechanics can be expressed as

$$\mathcal{C}\psi = i\gamma^2\psi^*$$

(5.113)

where ψ^* is the complex conjugate of the wavefunction of eqn 5.79 and γ^2 is the second Dirac matrix in the Pauli-Dirac representation. Using the solutions of the Dirac equation, show that eqn 5.113 changes an electron into a positron and fulfills the property of the inversions.

(5.13) * Using the previous result, demonstrate that the C-parity of a e^-e^+ pair is -1.

Electromagnetic interactions

<table>
<tr><td></td><td>6</td></tr>
</table>

6.1 Classical electrodynamics and relativity

Electromagnetic interactions are the only fundamental forces that have a complete quantum description that follows from its classical counterpart. The classical theory of electromagnetism was conceived in 19th century and culminated in the unified theory of electric and magnetic forces by J.C. Maxwell. The Maxwell theory summarizes in a system of eight equations the time evolution of the electromagnetic fields $\mathbf{E}$ and $\mathbf{B}$. In SI, these equations read:

$$\nabla \cdot \mathbf{E} = \frac{\rho}{\epsilon_0} \quad \text{(Gauss law)}$$

$$\nabla \cdot \mathbf{B} = 0$$

$$\nabla \times \mathbf{E} = -\frac{\partial \mathbf{B}}{\partial t} \quad \text{(Faraday's law)}$$

$$\nabla \times \mathbf{B} = \mu_0 \mathbf{j} + \mu_0 \epsilon_0 \frac{\partial \mathbf{E}}{\partial t} \quad \text{(AM's law)}$$

where AM is Ampere's law modified by the Maxwell displacement current (**Ampere–Maxwell's law**). Here, ρ and $\mathbf{j}$ are the charge and current densities, respectively. The interaction of charged particles with the electromagnetic fields is described by an additional law: $\mathbf{F} = q(\mathbf{E} + \mathbf{v} \times \mathbf{B})$ where q is the charge of the particle and $\mathbf{v}$ is the particle velocity. This law is due to Maxwell although the full derivation was provided by H. Lorentz in 1895; it is thus named the **Lorentz force**. The metric systems of units and, in particular, SI are inconvenient in electrodynamics because $\mathbf{E}$ [V/m] and $\mathbf{B}$ [T] are measured in different units. This choice is unfortunate in view of special relativity, where the magnetic field arises from the Lorentz transformation of the electric field generated by a charge. In natural units, electric and magnetic fields have the same dimensions: eV^2. An electric field of 1 eV^2 corresponds to a field of 432.90844×10^3 V/m and a magnetic field of 1 eV^2 corresponds to 1.4440271×10^{-3} T. In natural units, Maxwell's equations get a lighter

A Modern Primer in Particle and Nuclear Physics. Francesco Terranova, Oxford University Press.
© Francesco Terranova (2021). DOI: 10.1093/oso/9780192845245.003.0006

form:

$$\nabla \cdot \mathbf{E} = \rho, \tag{6.1}$$

$$\nabla \cdot \mathbf{B} = 0, \tag{6.2}$$

$$\nabla \times \mathbf{E} = -\frac{\partial \mathbf{B}}{\partial t}, \tag{6.3}$$

$$\nabla \times \mathbf{B} = \mathbf{j} + \frac{\partial \mathbf{E}}{\partial t}. \tag{6.4}$$

We get a simpler form because we have set $k = 1/4\pi$ and $c = 1$ in Sec. 1.5.2. $k = 1/4\pi$ implies $\epsilon_0 = 1$. $c = 1$ gives $\mu_0\epsilon_0 = \mu_0 = 1$ since $c = (\epsilon_0\mu_0)^{-1/2}$ in SI. The Maxwell's equations look quite asymmetric and physicists did not realized immediately that the electric and magnetic fields have the same origin. Special relativity (SR) provides a deep insight into these equations because it gives rules to compute the forces originating from a charged particle in motion.

Static electric fields are produced by charged particles at rest. Since magnetic fields are produced by charged particles in motion ("currents"), the Lorentz transformations change an electric field into a superposition of electric and magnetic fields. If the rest frame of a particle seen in the laboratory frame (LAB) has a velocity $\beta = v/c$ along the x axis, the transformed fields are:

$$
\begin{aligned}
E'_x &= E_x & B'_x &= B_x \\
E'_y &= \gamma(E_y - vB_z) & B'_y &= \gamma(B_y + \frac{v}{c^2}E_z) \\
E'_z &= \gamma(E_z + vB_y) & B'_z &= \gamma(B_z - \frac{v}{c^2}E_y).
\end{aligned}
\tag{6.5}
$$

The Lorentz transformations change the electric and magnetic fields only in the direction perpendicular to the motion: as already noted in eqn 3.18, it is the other way round for the space and momentum components. A purely electrostatic field boosted toward the x-axis generates a magnetic field in the $y - z$ plane.

It is possible to demonstrate that classical electrodynamics is the simplest theory consistent with special relativity when assuming the Coulomb law (Jackson, 1998). In its post-Einsteinian formulation, the Coulomb law in SI reads:

---

the force between two charged particles at rest in a given frame is

$$\mathbf{F} = \frac{1}{4\pi\epsilon_0} \frac{q_1 q_2}{r^2} \hat{\mathbf{r}} \tag{6.6}$$

where q_1 and q_2 are Lorentz-invariant quantities called the **electric charges** of the particles. $\hat{\mathbf{r}}$ is the unit-vector along the line between the two particles and r is the distance between the particles. The force is attractive if the signs of the charges are opposite, and repulsive if charges have the same sign.

The electromagnetism is then a truly relativistic theory (Rindler, 2006), which was discovered before SR. Einstein developed SR building on electromagnetism and on the empirical evidence for a universal speed of the electromagnetic waves in vacuum. Note also that SR solves a striking paradox of Maxwell's theory: the interaction of a charged particle with an electromagnetic (e.m.) field is inconsistent with Newton's laws. In Newton's theory, $\mathbf{F} = m\mathbf{a} = m\ddot{\mathbf{x}}$ and forces are invariant for the Galilean transformations of eqn A.1. On the other hand, the Lorentz force is *not* invariant because it depends on the velocity $\mathbf{v}$ of the particle in a given frame.

Since electromagnetism is a relativistic theory, it can be expressed in covariant form. The electric field cannot be a Lorentz invariant due to eqn 6.5. It cannot be decoupled by the magnetic field, either. The simplest covariant term with at least six components is a rank 2 tensor. This tensor is called the **electromagnetic tensor** $F^{\mu\nu}$ and is defined as:

$$F^{\mu\nu} = \begin{pmatrix} 0 & -E_x/c & -E_y/c & -E_z/c \\ E_x/c & 0 & -B_z & B_y \\ E_y/c & B_z & 0 & -B_x \\ E_z/c & -B_y & B_x & 0 \end{pmatrix} \qquad (6.7)$$

or, in NU,

$$F^{\mu\nu} = \begin{pmatrix} 0 & -E_x & -E_y & -E_z \\ E_x & 0 & -B_z & B_y \\ E_y & B_z & 0 & -B_x \\ E_z & -B_y & B_x & 0 \end{pmatrix}. \qquad (6.8)$$

The NU Maxwell equations in covariant form read:

$$\partial_\mu F^{\mu\nu} \equiv \frac{\partial}{\partial x^\mu} F^{\mu\nu} = j^\nu \qquad (6.9)$$

$$\partial_\mu G^{\mu\nu} \equiv \frac{\partial}{\partial x^\mu} G^{\mu\nu} = 0 \qquad (6.10)$$

where:

$$G^{\mu\nu} = \begin{pmatrix} 0 & -B_x & -B_y & -B_z \\ B_x & 0 & E_z & -E_y \\ B_y & -E_z & 0 & E_x \\ B_z & E_y & -E_x & 0 \end{pmatrix} \qquad (6.11)$$

is called the **dual tensor** and is drawn from $F^{\mu\nu}$ replacing $\mathbf{E} \to \mathbf{B}$ and $\mathbf{B} \to -\mathbf{E}$. As expected, charge densities and currents belong to the same four-vector, the **four-current**:

$$j^\mu \equiv (c\rho, j_x, j_y, j_z) \overset{NU}{=} (\rho, j_x, j_y, j_z). \qquad (6.12)$$

The c and $1/c$ factors result from using different dimensions for $\mathbf{E}$ and $\mathbf{B}$, and we get rid of this nuisance employing natural units ($c = 1$). Equation 6.9 brings to Gauss law for $\mu = 0$ and to Ampere–Maxwell's law for $\mu = 1, 2, 3$. Equation 6.10 gives the $\nabla \cdot \mathbf{B} = 0$ law for $\mu = 0$ and Faraday's law for $\mu = 1, 2, 3$.

6.2   Classical gauge transformations

Electrodynamic processes can be described by electromagnetic fields and their interaction with charged particles. On the other hand, the Maxwell equations introduce some correlations between these fields and reduce the actual degrees of freedom. The scalar and vector potentials $V(\mathbf{x}, t)$ and $\mathbf{A}(\mathbf{x}, t)$ lower the number of independent variables from six to four:

$$E_x, E_y, E_z, B_x, B_y, B_z \rightarrow V, A_x, A_y, A_z \tag{6.13}$$

because

$$\mathbf{E}(\mathbf{x}, t) = -\nabla V(\mathbf{x}, t) - \frac{\partial \mathbf{A}(\mathbf{x}, t)}{\partial t} \tag{6.14}$$

$$\mathbf{B}(\mathbf{x}, t) = \nabla \times \mathbf{A}(\mathbf{x}, t). \tag{6.15}$$

In the covariant formalism, the potentials form a four-vector, the **four-potential**:

$$A^\mu = (cV, A_x, A_y, A_z) \overset{NU}{=} (V, A_x, A_y, A_z) \tag{6.16}$$

so that

$$F^{\mu\nu} = \partial^\mu A^\nu - \partial^\nu A^\mu. \tag{6.17}$$

This is the covariant electromagnetic tensor. In eqn 6.17, ∂^μ indicates a contravariant derivative and is linked to the standard gradient in SI units

$$\partial_\mu \equiv \left(\frac{\partial}{\partial ct}, \frac{\partial}{\partial x}, \frac{\partial}{\partial y}, \frac{\partial}{\partial z} \right) \tag{6.18}$$

by

$$\partial^\mu = g^{\mu\nu} \partial_\nu = \left(\frac{\partial}{\partial ct}, -\frac{\partial}{\partial x}, -\frac{\partial}{\partial y}, -\frac{\partial}{\partial z} \right). \tag{6.19}$$

The scalar and vector potentials are not uniquely determined: this means that there are additional degrees of freedom that allow redefining the potentials without changing the fields. The simplest example is a change of the scalar potential by a global constant b. A transformation $V(\mathbf{x}, t) \rightarrow V(\mathbf{x}, t) + b$ neither affects the electric and magnetic fields nor changes the dynamics of the system and the numerical values of the observables. Classical electrodynamics is invariant for a $V(\mathbf{x}, t) \rightarrow V(\mathbf{x}, t) + b$ transformation.

But what are the most general transformations of the scalar and vector potentials that do not change the electric and magnetic fields? More precisely, what are the most general forms of $b(\mathbf{x}, t)$ and $\mathbf{a}(\mathbf{x}, t)$ so that

$$\mathbf{A}'(\mathbf{x}, t) = \mathbf{A}(\mathbf{x}, t) + \mathbf{a}(\mathbf{x}, t) \tag{6.20}$$

$$V'(\mathbf{x}, t) = V(\mathbf{x}, t) + b(\mathbf{x}, t) \tag{6.21}$$

gives $\mathbf{E}'(\mathbf{x}, t) = \mathbf{E}(\mathbf{x}, t)$ and $\mathbf{B}'(\mathbf{x}, t) = \mathbf{B}(\mathbf{x}, t)$? Since $\mathbf{B}' = \nabla \times \mathbf{A}' = \nabla \times \mathbf{A} + \nabla \times \mathbf{a} = \mathbf{B}$, we have $\nabla \times \mathbf{a} = 0$ and $\mathbf{a}$ must be an irrotational vector field. Irrotational fields are also conservative fields, at least in simply connected domains (Griffiths, 2017). These are the only interesting domains in particle physics (and in nature) because the construction of a disconnected domain like an infinite wire or wall requires an infinite amount of energy. A conservative field can always be expressed by the gradient of a scalar function:

$$\mathbf{a}(\mathbf{x}, t) = \nabla\tilde{\alpha}(\mathbf{x}, t) \tag{6.22}$$

and all these functions depend on time and space. The invariance of the electric field implies:

$$\mathbf{E}' = -\nabla V - \nabla b - \frac{\partial \mathbf{A}}{\partial t} - \frac{\partial \mathbf{a}}{\partial t} = \mathbf{E} \tag{6.23}$$

so that

$$\nabla b + \frac{\partial \mathbf{a}}{\partial t} = \nabla\left(b + \frac{\partial\tilde{\alpha}}{\partial t}\right) = 0. \tag{6.24}$$

A function whose gradient is always zero in space is a function that does not depends on $\mathbf{x}$:

$$\left(b + \frac{\partial\tilde{\alpha}}{\partial t}\right) = \kappa(t). \tag{6.25}$$

It is fortunate that $\tilde{\alpha}$ is an arbitrary function and can be redefined at will. A change of variables like

$$\alpha = \tilde{\alpha} + \int_0^t \kappa(t)dt \tag{6.26}$$

gives us:

$$b = -\frac{\partial\alpha}{\partial t} . \tag{6.27}$$

In conclusion, b is not bound to be a simple additive constant but can be a full-fledged function, provided that is expressed as the time derivative of $\alpha(\mathbf{x}, t)$.

The most general transformations of the scalar and vector potentials that do not change the electric and magnetic fields are, therefore, the **classical gauge transformations**:

$$\mathbf{A}'(\mathbf{x}, t) = \mathbf{A}(\mathbf{x}, t) + \nabla\alpha(\mathbf{x}, t)$$
$$V'(\mathbf{x}, t) = V(\mathbf{x}, t) - \frac{\partial\alpha(\mathbf{x}, t)}{\partial t} \tag{6.28}$$

or, in covariant form,

$$A'^{\mu} = A^{\mu} - \partial^{\mu}\alpha(x^{\mu}). \tag{6.29}$$

The choice of $\alpha(\mathbf{x}, t)$ to be a specific function corresponds to **fixing the gauge** of the electromagnetic field. The word "gauge" is inherited from the early attempts of H. Weyl to construct a unified theory of electromagnetic and gravitational interactions and should not be taken too seriously. A classical gauge transformation corresponds to the freedom of changing the numerical values of the potentials without affecting the physical observables.

There are several ways to fix the gauge for electromagnetism and each has pros and cons. The **Coulomb gauge** defines A^μ fixing:

$$\nabla \cdot \mathbf{A} = 0 \tag{6.30}$$

and is one of the favorite gauges in solid-state physics because it determines uniquely A^μ. Particle physicists use more often the **Lorenz gauge:**[1]

$$\partial_\mu A^\mu = 0 \tag{6.31}$$

because $\partial_\mu A^\mu$ is Lorentz invariant and the equations of motion become very elegant (see Sec. 6.6.2). Unfortunately, it does not define uniquely A^μ and we may need additional constraints (see Exercise 6.1).

6.3   Particle Hamiltonian in an e.m. field*

Handling Maxwell's equations in the Hamiltonian or Lagrangian formalism may be somewhat worrisome because both formalisms are based on potentials instead of fields. All in all, the Hamiltonian or the Lagrangian are just mathematical tools to conveniently write the equations of motion, and we do not expect the equations of motion or the physical observables to depend on the choice of the gauge. This is pretty evident in classical physics where the Lagrangian of a point-like particle in an electromagnetic field is defined as

$$L = \frac{1}{2}m|\mathbf{v}|^2 - q(V - \mathbf{v} \cdot \mathbf{A}). \tag{6.32}$$

The **conjugate momentum** is:

$$\mathbf{p} = \frac{\partial L}{\partial \dot{\mathbf{x}}} = \frac{\partial L}{\partial \mathbf{v}} = m\mathbf{v} + q\mathbf{A}. \tag{6.33}$$

Both the Lagrangian and the conjugate momentum depend on α because a change in the choice of the gauge gives:

$$\mathbf{p} \to m\mathbf{v} + q\mathbf{A} + q\nabla\alpha. \tag{6.34}$$

Similarly, the Hamiltonian is defined as:

$$\mathcal{H} = \mathbf{p} \cdot \dot{\mathbf{x}} - L = \mathbf{p} \cdot \mathbf{v} - L = \frac{1}{2m}\left[\mathbf{p} - q\mathbf{A}\right]^2 + qV \tag{6.35}$$

and is hopelessly gauge-dependent. Genuine physical observables are independent of the choice of the gauge. For instance, the kinetic energy of the particle in Newton's mechanics is:

$$\frac{1}{2}m|\mathbf{v}|^2 = \frac{1}{2m}\left(\mathbf{p} - q\mathbf{A}\right)^2. \tag{6.36}$$

A change of gauge changes the **conjugate momentum p** as $\mathbf{p}' = m\mathbf{v} + q\mathbf{A} + q\nabla\alpha$ and the vector field $\mathbf{A}$ as $\mathbf{A}' = \mathbf{A} + \nabla\alpha$ so that

$$\frac{1}{2m}\left(\mathbf{p}' - q\mathbf{A}'\right)^2 = \frac{1}{2m}\left(\mathbf{p} - q\mathbf{A}\right)^2 = \frac{1}{2}m|\mathbf{v}|^2. \tag{6.37}$$

There is real physics behind the kinetic energy and this quantity is gauge-independent under eqn 6.37, as expected. The same considerations hold for the (physical) momentum of the particle, $m\mathbf{v}$, which should not be confused with the **conjugate momentum p**. In general, any physical observable is independent of the choice of α because the equations of motions of the Hamiltonian formalism:

$$\dot{\mathbf{p}} = -\frac{\partial \mathcal{H}}{\partial \mathbf{x}} = -\nabla \mathcal{H}; \quad \dot{\mathbf{x}} = \mathbf{v} = \frac{\partial \mathcal{H}}{\partial \mathbf{p}} \tag{6.38}$$

are equivalent to the equations of motion of Newton's formalism:

$$\mathbf{F} = m\ddot{\mathbf{x}} = q(\mathbf{E} + \mathbf{v} \times \mathbf{B}) \tag{6.39}$$

and none of the solutions – that is, the particle trajectory $\mathbf{x}(t)$ – depends on α.

6.4   Gauge transformations in non-relativistic quantum mechanics*

Quantum mechanics (QM) is a more complex playground for gauge transformations. Quantum systems with classical counterparts are build upon the canonical quantization rules, which replace the Hamiltonian and the **conjugate variables** ($\mathbf{x}$ and $\mathbf{p}$) with the corresponding hermitian operators. The Hamiltonian that describes the motion of a charged particle in an electromagnetic field is given by eqn 6.35 replacing $\mathbf{p}$ with $\hat{\mathbf{p}} \equiv -i\hbar\nabla$ and $\mathbf{x}$ with the position operator $\hat{\mathbf{x}}$. The position operator is defined as $\hat{\mathbf{x}}\psi(\mathbf{x},t) = \mathbf{x}\psi(\mathbf{x},t)$. It is clear that the **semiclassical Hamiltonian**:

$$\hat{H} = \frac{1}{2m}\left[\hat{\mathbf{p}} - q\mathbf{A}(\hat{\mathbf{x}},t)\right]^2 + qV(\hat{\mathbf{x}},t) \tag{6.40}$$

does depend on the choice of the gauge. Changing the gauge, we get another $\hat{H}'$ operator:

$$\hat{H}' = \frac{1}{2m}\left[\hat{\mathbf{p}} - q\mathbf{A}'(\hat{\mathbf{x}},t)\right]^2 + qV'(\hat{\mathbf{x}},t) \neq \hat{H}. \tag{6.41}$$

Following the same strategy as in classical mechanics, we need to change the variables that describe the state of the particle counterbalancing the change of the Hamiltonian to keep untouched the physical observables. In QM, the state variable is the wavefunction $\psi(\mathbf{x},t)$ and the equation of motion in SI is the Schrödinger equation $\hat{H}\psi = i\hbar\partial\psi/\partial t$. As a consequence, we are seeking a (unitary) transformation that changes ψ for a change of gauge so that:

$$H'\psi' = i\hbar\frac{\partial\psi'}{\partial t} \tag{6.42}$$

when ψ is a solution of the Schrödinger equation. Once more, it is worth stressing that handling with gauge freedom forces us to change both the equation of motion and the wavefunction. Nonetheless, we can still hope to build a theory that is invariant after those changes because the physical observables in QM are the probability distributions, which only depend on $|\psi(\mathbf{x}, t)|$.

It is remarkable (Cohen-Tannoudji *et al.*, 1991) that the **gauge transformation**

$$\psi'(\mathbf{x}, t) \equiv \hat{U}(\mathbf{x}, t)\psi(\mathbf{x}, t) = e^{\frac{i}{\hbar}q\alpha(\hat{\mathbf{x}}, t)}\psi(\mathbf{x}, t) \tag{6.43}$$

fulfills this condition:

$$i\hbar\frac{d}{dt}\psi'(t) = i\hbar\frac{d}{dt}\left[\hat{U}(t)\psi(t)\right] = i\hbar\left[\frac{d}{dt}\hat{U}(t) \cdot \psi(t) + \hat{U}(t) \cdot \frac{d}{dt}\psi(t)\right] =$$

$$-q\left[\frac{\partial}{\partial t}\alpha(\hat{\mathbf{x}}, t)\right] \cdot \hat{U}(t)\psi(t) + \hat{U}(t)\hat{H}(t)\psi(t) =$$

$$\left[-q\frac{\partial}{\partial t}\alpha(\hat{\mathbf{x}}, t) + \tilde{\hat{H}}(t)\right]\psi'(t) \tag{6.44}$$

where $\tilde{\hat{H}}(t)$ is the Hamiltonian $\hat{H}(t)$ transformed by $\hat{U}(t)$:

$$\tilde{\hat{H}}(t) = \hat{U}(t)\hat{H}(t)\hat{U}^\dagger(t). \tag{6.45}$$

By virtue of eqn 6.44, the transformed Schrödinger equation (eqn 6.42) is satisfied if:

$$\hat{H}'(t) = \tilde{\hat{H}}(t) - q\frac{\partial}{\partial t}\alpha(\hat{\mathbf{x}}, t). \tag{6.46}$$

To compute $\tilde{\hat{H}}(t)$ and demonstrate that eqn 6.46 holds, we need to evaluate the transformed values of $\hat{\mathbf{x}}$ and $\hat{\mathbf{p}}$. Since $[\hat{x}_i, \hat{x}_j] = 0$ for $i, j = 1, 2, 3$:

$$\tilde{\hat{x}} \equiv \hat{U}(\hat{\mathbf{x}}, t) \, \hat{\mathbf{x}} \, \hat{U}^\dagger(\hat{\mathbf{x}}, t) = \hat{\mathbf{x}}. \tag{6.47}$$

On the other hand, the commutation rule between $\hat{p}_x$ and a generic function of $\hat{x}$ is (see Exercise 6.2)

$$[\hat{p}_x, f(\hat{x})] = -i\hbar f'(\hat{x}) \tag{6.48}$$

so that

$$[\hat{\mathbf{p}}, \hat{U}] = q\nabla\alpha\hat{U} \tag{6.49}$$

and

$$\hat{U}\hat{\mathbf{p}}\hat{U}^\dagger = \hat{\mathbf{p}}\hat{U}\hat{U}^\dagger - [\hat{\mathbf{p}}, \hat{U}]\hat{U}^\dagger = \hat{\mathbf{p}} - q\nabla\alpha. \tag{6.50}$$

Note also that $\hat{U}$ commutes with $\nabla\alpha$ because $\hat{U}$ and α are sole functions of position operators. Equation 6.50 reads:

$$\tilde{\hat{\mathbf{p}}} = \hat{\mathbf{p}} - q\nabla\alpha(\hat{\mathbf{x}}, t). \tag{6.51}$$

The transformed Hamiltonian is given by:

$$\tilde{\hat{H}} = \frac{1}{2m}\left[\tilde{\hat{\mathbf{p}}} - q\mathbf{A}(\tilde{\hat{\mathbf{x}}}, t)\right]^2 + qV(\tilde{\hat{\mathbf{x}}}, t)$$

$$= \frac{1}{2m}\left[\hat{\mathbf{p}} - q\mathbf{A}(\hat{\mathbf{x}}, t) - q\nabla\alpha(\hat{\mathbf{x}}, t)\right]^2 + qV(\hat{\mathbf{x}}, t). \tag{6.52}$$

We now apply the classical gauge trasformations of eqn 6.28 to the Hamiltionian:

$$\hat{H}'(\hat{\mathbf{x}}, t) = \frac{1}{2m} \left[\hat{\mathbf{p}} - q\mathbf{A}'(\hat{\mathbf{x}}, t) \right]^2 + qV'(\hat{\mathbf{x}}, t) =$$

$$\frac{1}{2m} \left[\hat{\mathbf{p}} - q\mathbf{A}(\hat{\mathbf{x}}, t) - q\nabla\alpha(\hat{\mathbf{x}}, t) \right]^2 + qV(\hat{\mathbf{x}}, t)$$

$$-q\frac{\partial \alpha(\hat{\mathbf{x}}, t)}{\partial t} = \tilde{\hat{H}}(\hat{\mathbf{x}}, t) - q\frac{\partial}{\partial t}\alpha(\hat{\mathbf{x}}, t) \tag{6.53}$$

which fulfills eqn 6.46. Even if following the maths above is quite an effort, the result is amazing:

> in non-relativistic QM, we can derive the interaction of a charged particle with a classical e.m. field (i.e. the **semiclassical Hamiltonian**) changing the wavefunction by the **gauge transformation**:
>
> $$\psi \to \hat{U}\psi = \exp\left(\frac{i}{\hbar} q\alpha(\hat{\mathbf{x}}, t) \right) \psi \tag{6.54}$$
>
> i.e. by a phase that arbitrarily changes in time and space, provided that such phase is the same function appearing in the classical gauge transformation of the e.m. field.

As we will see shortly, this idea is at the heart of the Standard Model of particle physics.

6.5 Gauge theories

What we have learned so far can be summarized as follows: the semiclassical Hamiltonian that describes the interaction of a charged particle with an electromagnetic field is written employing the canonical quantization, that is, a procedure for quantizing a classical theory. The classical theory has fields that are not uniquely defined. Their numerical values depend on a function α of space and time. As a consequence, the Hamiltonian is now time-dependent[2] through $\alpha(\hat{\mathbf{x}}, t)$. If we require the equations of motion to be independent on the choice of $\alpha(\hat{\mathbf{x}}, t)$ (i.e. the choice of the gauge), we end up with wavefunctions whose phase can be arbitrarily defined at any point of the space-time. The keyword here is the arbitrariness of the phase for each point in space-time: we do not want α to be just a constant.

We know already that global phases are immaterial in QM: a wavefunction $\psi(\mathbf{x}, t)$ can be redefined through a **global phase transformation** as

$$\psi'(\mathbf{x}, t) \to e^{i\alpha}\psi(\mathbf{x}, t) \tag{6.55}$$

and, in general,

[2]Time-dependent operators are commonly used in the **Heisenberg picture** of QM. In this case, however, both the operators and the wavefunctions are time-dependent. This framework is called **interaction picture** and is presented in Sec. B.5.

all systems in QM are invariant for a global change of the phase of the wavefunctions. After a phase change, all probabilities, scattering amplitudes, and expectation values of the observables stay unchanged. This property can be classified as a continuous symmetry depending on the real-valued parameter α ($\alpha \in \mathbb{R}$) and it is called a **global U(1) symmetry**.

The name of the symmetry is inherited from the jargon of group representations theory (see Chap. 7) and indicates a transformation by a unitary 1×1 complex-value matrix, that is, a complex number z. Since the matrix is unitary, $UU^\dagger = zz^* = 1$ and $|z| = 1$, which gives $z = e^{i\alpha}$ for $\alpha \in \mathbb{R}$, a contorted but mathematically rigorous way to say a "change of phase."

6.5.1 Gauge symmetries

The quantization of the motion of a charged particle in an electromagnetic field requires a much tighter symmetry than the global $U(1)$ symmetry because we want the phase to change arbitrarily in each point of space-time: it requires a **local symmetry**. If a system fulfills such a strong condition, we say it owns a **gauge symmetry**. The **local U(1) symmetry**:

$$\psi'(\mathbf{x}, t) \to e^{i\alpha(\mathbf{x},t)}\psi(\mathbf{x}, t) \tag{6.56}$$

gives the observer the freedom to redefine the phase of the wavefunction in a different way at each point in time and space, that is to redefine the phase *on local basis*. In electromagnetism, the map that associates the point in space and time to the phase chosen by the observer is an arbitrary function $\alpha(\mathbf{x}, t)$, which depends on the choice of the gauge in the electromagnetic potentials. The wavefunction of the charged particle is now correlated with the electromagnetic potentials because both eqns 6.28 and 6.56 must hold. As a consequence,

the semiclassical theory of electromagnetism possesses a gauge symmetry: the local $U(1)$ symmetry. This symmetry determines the form of the Hamiltonian and, in particular, the interaction term between the charged particle and the classical e.m. fields.

We can broaden this statement by saying that:

a theory whose Hamiltonian has a gauge symmetry that allows for the definition of the interaction terms between particles and fields is called a **gauge theory** and the corresponding gauge symmetry is always a **local symmetry**.

Semiclassical electrodynamics is then a gauge theory based on a (local) $U(1)$ gauge symmetry. The symmetry corresponds to the transformations of eqn 6.56. Unfortunately, the word "symmetry" has a historical origin and is somehow misleading for modern physicists. Unlike the actual symmetries in QM studied in Chap. 5, the "gauge symmetry" operator:

$$\hat{U}(t) = e^{\frac{i}{\hbar}q\alpha(\hat{\mathbf{x}},t)} \overset{NU}{=} e^{iq\alpha(\hat{\mathbf{x}},t)} \tag{6.57}$$

does not commute with the Hamiltonian and both $\hat{U}$ and $\hat{H}$ depend on time.

The corresponding $U(1)$ global symmetry is recovered fixing $\alpha(\mathbf{x},t)$ to a constant value like, for example, $\alpha(\mathbf{0},0)$. This is a real symmetry in QM and fulfils the first Noether theorem.

> Thanks to Noether's theorem, the $U(1)$ global symmetry gives a conserved current: eqn 5.23, that is the local conservation of **electric charge**. The electric charge is then conserved in **quantum electrodynamics** (QED) like in classical electromagnetism.

6.5.2   The gauge principle

The construction of semiclassical electrodynamics is a prominent example of the application of the **gauge principle**, a procedure to generate the interaction term from a free theory (Schwartz, 2013). The gauge principle is even more important in quantum field theories (QFTs). There, the wavefunction of (charged) particles are replaced by fields and there is no asymmetry between the quantum description of the particles and the description of the fields. The gauge principle in QFT is a procedure for generating an interaction term from a free Lagrangian that is symmetric with respect to a local continuous symmetry (the gauge symmetry). In particular, in QED, the free Lagrangian describes the motion of spin-1/2 charged particles through the Dirac Lagrangian. Tightening the global U(1) symmetry to a gauge (i.e. local) symmetry requires the introduction of an interaction term between the Dirac Lagrangian and the field that describes the free propagation of massless spin-1 particles (the Maxwell Lagrangian). This term rules the interaction of charged particles with the electromagnetic fields in a way that is consistent both with SR and the principles of QM.

The gauge principle is the spell under which the entire Standard Model has been created and extended to describe strong and weak interactions. We will rigorously show how this principle can be applied to derive the Standard Model Lagrangian – including QED – in the last chapter of this book.

6.6   Quantum electrodynamics

Despite technical complexity, the QFT extension of semiclassical electro magnetism brings a much deeper insight into electromagnetic interactions. QED is the relativistic QFT of electromagnetism. In 1949, F. Dyson set down its first consistent formulation based on the works of H. Bethe, R. Feynman, J. Schwinger, and S. Tomonaga. The view of an electromagnetic interaction in QED is quite different from non-relativistic QM. In particular, the observables in QED are the **scattering amplitudes** between a set of (non-interacting) initial and final state particles. QED, therefore, provides means to compute the cross-sections for any set of charged particles and scattering processes. In QED, the potentials in the Hamiltonian are replaced by the interaction terms in the Lagrangian. We will explicitly do this in Sec. 13.1.1.

> All electrodynamic processes are described by the scattering of photons and charged particles. The Lorentz-invariant transition matrix $\mathcal{M}$ of eqn 2.125 and the corresponding cross-section (eqn 2.126) drive the outcome and the probability of the process.

The transition amplitude can be computed using perturbative techniques developed by J. Schwinger, E. Stueckelberg, and F. Dyson in the 1940s and culminated in the introduction of the **Feynman diagrams** in 1948. R. Feynman realized that the perturbative series of QED can be rewritten as a sum of diagrams.

> A Feynman diagram is a graphical representation of a term of the perturbative series that contributes to the transition amplitude. The sum of all diagrams corresponds to the whole perturbative series.

In the vertices of these diagrams both the energy and momentum are conserved, but p^2 is not necessarily equal to the mass. Through the **Feynman rules**, each diagram can be read as a formula, which is used to compute the cross-section at any order in perturbation theory.

In this introductory text, we will use the diagrams only as a visualization tool to identify the leading contributions to a given scattering process. Still, the reader should be aware that Feynman diagrams are the main route to interpret quantum phenomena in relativistic QM and play a leading role in theoretical physics from particles up to solid-state physics and gravitation.

A QED scattering process can be described by a Lorentz-invariant matrix expanded as a function of powers of the charge of the electron or, equivalently, of the **fine structure constant** α. The numerical value of α is the same both in SI and in NU because α is dimensionless. In NU, $\alpha \equiv e^2/4\pi \simeq 1/137 \ll 1$ and therefore the perturbative series of a two-body scattering ($a + b \to X$, where X is a generic set of particles) is

$$\mathcal{M} = \alpha \cdot A_1 + \alpha^2 \cdot A_2 + \ldots = \sum_n A_n \alpha^n. \tag{6.58}$$

Equation 6.58 provides a reliable method to compute the scattering amplitudes. For instance, a $2 \to 2$ scattering process ($a + b \to c + d$) in the $x - t$ plane (see Fig. 6.1) is described by two incoming lines from the past and two outcoming lines pointing to the future ($t \to +\infty$). The incoming and outcoming lines represent the initial and final states, respectively.

In the language of Feynman diagrams:

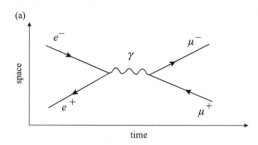

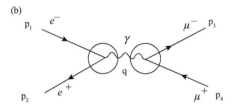

Fig. 6.1 Top: Lowest order diagrams for $e^+e^- \rightarrow \mu^+\mu^-$. Bottom: four-momenta of the off-shell (photon with four-momentum q) and on-shell particles, and location of the two vertices (circles).

- initial- and final-state particles are represented by **on-shell** (or **real**) particles satisfying the SR dispersion relations $p_j^2 = E_j^2 - |\mathbf{p}_j|^2 = m_j^2$. Here, j is any of the particles in the initial or final state. In Fig. 6.1 the on-shell particles are the incoming electron and positron and the outgoing muon and antimuon.

- intermediate particles are called **off-shell** (or **virtual**) particles and do not fulfill the SR dispersion relation. For off-shell particles, $p_j^2 \neq m_j^2$. In Fig. 6.1 the only off-shell particle is the intermediate photon.

- in QED the only interaction vertices among particles are three-line vertices made of two charged particles and a photon. The vertex represents the interaction of the photon with a particle of electric charge z and contributes to $\mathcal{M}$ as $z\sqrt{\alpha}$. These vertices are indicated by two circles in Fig. 6.1 and are zoomed in Fig. 6.2.

- a Feynman diagram for a $I \rightarrow F$ scattering is a connected graph made of I on-shell particles in the initial state and F on-shell particles in the final state. If the diagram has V vertices, it contributes to the perturbative series at the order $\alpha^{V/2}$. The diagram of Fig. 6.1 has $V = 2$ and contributes to the α term of the series.

- charge, energy, and momentum are conserved in each vertex although $p^2 \neq m^2$ for off-shell particles.

Fig. 6.2 The fermion–fermion–photon vertex of QED.

The reader may be puzzled by the role of off-shell particles and how Feynman diagrams are summed up. Off-shell particles naturally arise even in non-relativistic QM in the context of time-dependent perturbation theory (Sakurai and Napolitano, 2017). Particles of mass larger than what is kinematically allowed can be produced for a very short time under the Heisenberg uncertainty principle. The non-zero difference between p^2 and m^2 arises from the intrinsic uncertainty ΔE in a short time Δt.

Fig. 6.3 The most common symbols used when drawing Feynman diagrams. (1) a fermion moving forward in time (up) and space-time (down). (2) An antifermion moving forward in time (up) and space-time (down). (3) A spin 1 boson (e.g. a photon or a Z^0). (4) A spin 0 boson (e.g. the Higgs boson).

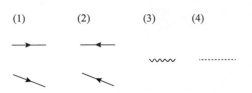

Off-shell particles are also a tool to account for many unobservable intermediate states contributing to the same final state. Similarly, Feynman diagrams are just a graphical representation of a perturbative term and each diagram is a complex number d or, more precisely, a complex function of the initial-state particle momenta and spins. Two diagrams d_a and d_b having the same number of vertices V contribute to the same term of the series: the term of order $\alpha^{V/2}$. As a consequence,

$$\mathcal{M} \sim (d_a + d_b)\alpha^{V/2} \tag{6.59}$$

and

$$\sigma \sim |\mathcal{M}|^2 \sim |d_a + d_b|^2 \alpha^V. \tag{6.60}$$

Equation 6.60 is highly non-trivial because the real and imaginary parts of different Feynman diagrams can interfere if they contribute to the same perturbative order.[3] Again, this is due to the fact that the perturbative series must account for all possible paths between the initial- and final-state particles that are compatible with QM even if they do not represent classical trajectories.

Feynman diagrams are usually written in a two-dimensional plane where space runs in the vertical direction and time runs in the horizontal direction. The symbols used for charged particles, antiparticles, and the photon are shown in Fig. 6.3. Note that charged particles are equipped with arrays that run from past to future for particles and from future to past for antiparticles. In the evaluation of the Feynman diagrams, antiparticles can thus be interpreted as "particles moving backward in time." Even if this interpretation (**Feynman–Stueckelberg interpretation of antiparticles**) has been superseded by the discovery of the CPT theorem, it is still tenable in the domain of perturbative QED.

[3] The interference term (IT) comes from an elementary property of complex numbers in eqn 6.60: $|z_1+z_2|^2 = |z_1|^2 + |z_2|^2 + 2\,\mathrm{Re}\{z_1\bar{z}_2\}$ and IT=Re$\{z_1\bar{z}_2\}$.

6.6.1   The simplest QED process

The simplest Feynman diagram for a $2 \to 2$ scattering is depicted in Fig. 6.1. Since the initial-state particles (e^+e^-) are different from the final ones $(\mu^+\mu^-)$, the only possible diagram with two vertices is an e^+e^- annihilation into a photon that converts into a muon–antimuon pair. The lowest-order diagram has $V = 2$ and contributes to the term of order α in eqn 6.58, that is, to the term A_1. We have already shown in Sec. 3.4.3 that this process cannot occur in vacuum if all particles are on-shell because an $e^+e^- \to \gamma$ annihilation violates four-momentum conservation.

On the other hand, an $e^+e^- \to \mu^+\mu^-$ scattering can occur even in vacuum because the photon is produced off-shell for a time compatible with the Heisenberg uncertainty principle and promptly decays into a $\mu^+\mu^-$ pair. This process is commonly observed in e^+e^- colliders provided that four-momentum conservation holds between initial- and final-state (on-shell) particles. As expected from SR, the kinematic threshold for muon pair production is $\sqrt{s} > 2m_\mu$ where $\sqrt{s}$ is the invariant mass and m_μ the muon rest mass. Once more, these considerations highlight that the Feynman diagram should not be interpreted as a classical trajectory but really as a symbol representing the perturbative series at order α.

The Feynman diagram of Fig. 6.1 also provides the first coefficient of $\mathcal{M}$ in its perturbative expansion:

$$\mathcal{M} = (-e)\langle \mu^+\mu^-|j^\mu|0\rangle \frac{1}{q^2}(-e)\langle 0|j_\mu|e^+e^-\rangle \qquad (6.61)$$

and can be computed analytically in the framework of QFT. Even if the QFT calculations of the cross-sections are not in the scope of this Primer, it is instructive to show the actual formula that corresponds to Fig. 6.1 because it consists of just Dirac spinors, gamma matrices, and the four-momentum exchanged by the photon (Maggiore, 2005):

$$\mathcal{M} = -\frac{e^2}{q^2}g_{\mu\nu}[\bar{v}(p_2)\gamma^\mu u(p_1)][\bar{u}(p_3)\gamma^\nu v(p_4)]. \qquad (6.62)$$

Here, $u(p_1)$ is the spinor[4] of the incoming electron as defined in Sec. 5.9.

$$\bar{v}(p_2) \equiv v(p_2)^\dagger \gamma^0 \qquad (6.63)$$

is the **adjoint spinor** of the incoming positron and is a row vector made up of four components. $\bar{u}(p_3)$ is the adjoint spinor of the outgoing muon and $v(p_4)$ is the spinor of the outgoing antimuon. The adjoint spinors are defined as:

$$\bar{u} \equiv u^\dagger \gamma^0 \quad ; \quad \bar{v} \equiv v^\dagger \gamma^0. \qquad (6.64)$$

The general rule is that, for any QED process, an incoming particle is described by u and an outgoing particle is described by $\bar{u}$. An incoming antiparticle is described by $\bar{v}$ (instead of the particle's u) and an outgoing antiparticle is described by v (see Exercise 6.10). We can restate eqn 6.62 in a compact form introducing j_e^μ, the **electron current** that interacts with the photon and with the muon current j_μ^μ. The upper index of the current is a four-vector index and the lower letter is the particle described by the current (e and μ in this case).

$$j_e^\mu \equiv \bar{v}(p_2)\gamma^\mu u(p_1) \qquad (6.65)$$

$$j_\mu^\mu \equiv \bar{u}(p_3)\gamma^\mu v(p_4) \qquad (6.66)$$

so that

$$\mathcal{M} = -\frac{e^2}{q^2}g_{\mu\nu}j_e^\mu j_\mu^\nu = -\frac{e^2}{q^2}j_e \cdot j_\mu \qquad (6.67)$$

[4]We often use the word spinor as a shorthand of bispinor for the sake of fluency. $u(p_1)$ is a four-component vector, indeed, that is a bispinor.

and

> the QED amplitudes can be expressed as a scalar product of fermion currents multiplied by the **photon propagator**:
>
> $$\frac{g_{\mu\nu}}{q^2}. \tag{6.68}$$
>
> This is the reason why QED is considered the prime example of a **current–current** interaction theory.

On-shell particles fulfil energy and charge conservation so that $q = p_1 + p_2 = p_3 + p_4$. q is called the **four-momentum transfer** because it represents the "kick" of the off-shell particle to the trajectories of the initial-state particles. Note also that $q^2 = (p_1 + p_2)^2 = s$ in this case. This explains the name **s-channel** for the diagram of Fig. 6.1 (see Sec. 6.6.3).

Computing eqn 6.62 may be a hassle for first-order scatterings and becomes a tremendous challenge for higher-order processes. Still, it incorporates the entire dynamics of QED. Once $\mathcal{M}$ is available, it can be plugged in the cross-section formula of eqn 2.128 to get the actual scattering probability. Note, however, that experimental physicists are generally unable to prepare the initial state with well-defined polarizations. A QED cross-section requires summing over all polarizations of the final states and averaging over all polarizations of the initial states. For the diagram of Fig. 6.1:

$$\langle |\mathcal{M}|^2 \rangle = \frac{1}{4} \sum_{A,B} |\mathcal{M}_{AB}|^2 \tag{6.69}$$

where A and B are the polarization of the final states and can be either left-handed LH or right-handed RH.[5] Factor 4 comes from the four possible combinations of polarizations in the initial state: LH-LH, LH-RH, RH-LH, and RH-RH.

Even if the mathematics is similar to relativistic QM, in QED/QFT ψ is no more a wavefunction but the **quantized field** representing the fermion. This makes quite a difference because it allows for the creation and destruction of new particles. For instance, $\langle 0|j_e^\mu|e^+e^-\rangle$ is the expectation value of the j_e current that destroys the e^+e^- initial state. Similarly, $\langle \mu^+\mu^-|j_\mu^\mu|0\rangle$ is the expectation value of the j_μ current that creates a muon–antimuon pair from the vacuum.

Leaving maths aside, there are many features we can get from an inspection of Fig. 6.1. QED belongs to a class of theory where the dynamics is determined by a $j_\mu j^\mu$ (**current–current**) interaction and a massless propagator: **the photon**. j^μ is called a **vector current** because is a (contravariant) vector. Each vertex contributes as $\sqrt{\alpha}$ to the perturbative series. All polarizations contribute to the scattering amplitude both in the initial and final states. QED is the theory of the electric forces. Therefore, all charged particles contribute to the scattering while the only neutral particle of relevance for QED is the photon.

[5]In this context, a LH polarization means that the direction of the spin measured along the direction of the particle is antiparallel to the particle momentum, while for RH polarizations, spin and momentum are parallel. We will discuss LH and RH particles, "helicity," and "chirality" in Sec. 6.6.3 and Chap. 10.

This corresponds to the classical statement that electrodynamics forces are a manifestation of a Lorentz-invariant electric charge. Note that the electric charge is locally conserved in QED because it is conserved in any QED vertex appearing in any Feynman diagram. We will demonstrate in Chap. 10 that QED conserves parity. It also conserves C, T, CP, and CPT parity (Maggiore, 2005). In summary,

> the conservation laws of QED are four-momentum and angular momentum conservation, the conservation of electric charge, and the conservation of the inversions: P, T, C, and their combinations.

The QED vertices of Fig. 6.2 cannot change an electron into a muon or a quark up into a quark charm even if the charge is conserved. **Flavor conservation** is an important property of QED that will be detailed in Chap. 8.

6.6.2 The photon in QED

When we discussed the equation of motion in relativistic QM we ran into trouble because we had to start from a non-relativistic equation (the Schrödinger equation) and guess the Dirac equation before even thinking of quantizing the electron field. We are here in the opposite situation: since classical electrodynamics is a relativistic theory, the equation of motion is already available. For the case of a free photon (no charges or currents so that j^ν in eqn 6.9 is equal to zero) in Lorenz gauge, this is the d'Alembert equation:

$$\Box A^\mu = \partial_\nu \partial^\nu A^\mu = 0 \tag{6.70}$$

whose plane wave solutions for a photon moving along the z axis are:

$$A^\mu = \epsilon^\mu(q) e^{-iq \cdot x} \tag{6.71}$$

where[6] $q^\mu = (\tilde{q}, 0, 0, \tilde{q})$ is the photon four-vector and $q \cdot x = q_\mu x^\mu$. We thus get for free a nice result:

> photons in relativistic quantum mechanics must be massless.

It comes from replacing eqn 6.71 in eqn 6.70:

$$\partial_\nu \partial^\nu A^\mu = -q^2 \epsilon^\mu e^{-iq \cdot x} = 0 \tag{6.73}$$

which holds only if the photon invariant mass q^2 is zero. Besides, the photon has only two degrees of linear (transverse) polarization corresponding to $\epsilon^\mu = (0, 1, 0, 0)$ and $(0, 0, 1, 0)$. Other combinations (circular polarization) are possible but the number of independent polarizations remains two. This result is highly non-trivial because the photon has spin 1 and we would have expected three possible polarizations, $S_z = -1, 0, 1$. We could demonstrate that $S_z = \pm 1$ using the full artillery of group

[6] We write the solutions in complex notations for sake of clarity. The physical fields are real and eqn 6.71 should be interpreted as

$$\epsilon^\mu(q) e^{-iq \cdot x} + c.c. \tag{6.72}$$

where "c.c." stands for "complex conjugate" and provides the real part of the solution.

theory applied to the Lorentz group (Schwartz, 2013). But we can get quite easily the same result by a trick that exploits the gauge transformations.[7] The Lorenz gauge guarantees that there are no independent polarizations along the time coordinate:

$$\partial_\mu A^\mu = 0 \implies \epsilon_\mu q^\mu = 0 \qquad (6.74)$$

and therefore $\epsilon^0 = \epsilon^3$. Then, we employ eqn 6.29 and eqn 6.71 replacing the arbitrary function $\alpha(x^\mu)$ with $iAe^{-iq\cdot x}$, where A is an arbitrary number. We get now:

$$\epsilon^\mu(q) \to \epsilon^\mu(q) + q^\mu. \qquad (6.75)$$

In this way, we are free to redefine $\epsilon^0 = \epsilon^3$ by a translation in Aq^μ:

$$\epsilon^0 \to \epsilon^0 + Aq^0 = \epsilon^0 + A\tilde{q} \qquad (6.76)$$
$$\epsilon^3 \to \epsilon^3 + Aq^3 = \epsilon^3 + A\tilde{q}.$$

We get rid of another polarization state setting $A = -\epsilon^0/\tilde{q}$ so that:

$$\epsilon^0 = \epsilon^3 = 0. \qquad (6.77)$$

This is a lay demonstration of a more general theorem stating that:

> all massless particles in SR have only two spin polarizations: either parallel ($S_z = +1$) or antiparallel ($S_z = -1$) to the direction of motion.

QED needs to quantize the fields and the quantization could compromise some of these features. Fortunately, this is not the case but the demonstration here really needs the full machinery of QFT(Schwartz, 2013). In brief, the photon remains massless by virtue of the gauge principle. If we define the **helicity** of a photon as[8]

$$\mathcal{H} \equiv \frac{\mathbf{S} \cdot \mathbf{q}}{|\mathbf{S}||\mathbf{q}|} \qquad (6.78)$$

then $\mathcal{H}$ is just ± 1. We can identify the former with a photon whose spin is along the direction of motion (right-handed helicity) and the latter with the spin antiparallel to the direction of motion (left-handed helicity).

If the photon were massive, the photon propagator would change, too. Both QFT and QM time-dependent perturbation theory show that a massive photon X would change the scattering amplitude as:

$$\frac{1}{q^2 - m_X^2} \qquad (6.79)$$

even if $q^2 \neq 0$ (off-shell photons). Note that if the mass of this exotic particle were very large, the scattering would become independent of q^2 since $|q^2 - m_X^2| \simeq m_X^2$. Using a powerful accelerator ($\sqrt{s} \simeq m_X$), we would see a clear resonance at $s = m_X^2$.

Unfortunately, the quantization of a massive spin-1 boson is nightmarish because this particle has three degrees of polarization. This is immaterial for QED and strong interactions but is a game changer in the Standard Model. The Higgs mechanism discussed in Sec. 13.1 has been developed to solve this issue.

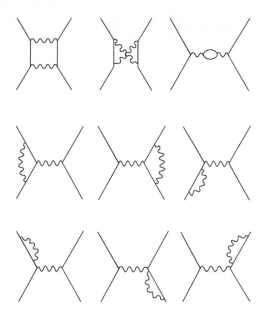

6.6.3 Higher-order diagrams

The second-order contributions to the $e^+e^- \to \mu^+\mu^-$ scattering consist of several diagrams with four vertices, which are depicted in Fig. 6.4:

- **Self-energy** contributions are Feynman diagrams where the off-shell photon creates and annihilates a pair of particles (third diagram in Fig. 6.4).

- **Initial- and final-state corrections** are diagrams where a photon is emitted and absorbed by one of the particles in the initial or final state (sixth to nineth diagrams in Fig. 6.4).

- **Vertex corrections** are due to photons emitted and absorbed by particles belonging to the same vertex (fourth and fifth diagrams of Fig. 6.4).

- **Two-photon exchange** generate diagrams where a photon is emitted by an initial-state particle and absorbed by a final-state particle (first and second diagrams of Fig. 6.4).

Note, in particular, that the self-energy diagrams and, in general, the diagrams where there are closed loops provide information on particles that are not kinematically accessible by slightly changing the values of the scattering amplitude at low energy. We will discuss this important topic in the next section.

Even at first order in perturbation theory, several diagrams can be present if there are identical on-shell particles. A well-known example is the **Bhabha scattering** $e^+e^- \to e^+e^-$. Here, two diagrams contribute to $\mathcal{M}$ at leading order ($V = 2 \implies$ at order α). These diagrams are shown in Fig. 6.5.

The first diagram is similar to the $e^+e^- \to \mu^+\mu^-$ diagram discussed in Sec. 6.6.1. This class of diagrams is called **s-channel** because the four-momentum transferred to the off-shell photon is given by the invariant mass of the system. In s-channel diagrams, s is then the squared four-momentum of the off-shell particle. Loosely speaking, s may be interpreted as the (squared) "invariant mass" of the virtual particle produced by the annihilation of the e^+e^- pair. On the other hand, a **t-channel** diagram (see Fig. 6.5, right) also contributes to the Bhabha scattering at order α. In this diagram, the electron and the positron of the initial state propagate up to the final state but their motion is perturbed by the scattering with the photon. The diagram is named from the **second Mandelstam variable (t)**:

$$t \equiv (p_a - p_c)^2 = (p_b - p_d)^2 \tag{6.80}$$

where p are the four-momenta of the initial- and final-state particles in a $a + b \to c + d$ scattering and p^2 is the usual four-vector norm in the Minkowski space. In the t-channel diagrams, t is the squared four-momentum of the off-shell particle but there is no annihilation of particles with antiparticles.

In some scattering processes, all on-shell particles are identical. The simplest case is the **Möller scattering** $e^-e^- \to e^-e^-$. Unlike the Bhabha scattering, the Feynman formalism must account for the Pauli exclusion principle because the two particles in the final state cannot be distinguished. In particular, in a $a + b \to c + d$ scattering where all particles are identical, we cannot tell whether the particle c comes from a or from b. In non-relativistic QM, this is equivalent to requiring the overall wavefunction to be antisymmetric. In QED, the Pauli principle is accounted for summing the t-channel with an appropriate **u-channel** diagram depicted in Fig. 6.6. The diagram is named from the **third Mandelstam variable (u)**:

$$u \equiv (p_a - p_d)^2 = (p_b - p_c)^2 \tag{6.81}$$

which corresponds to the squared four-momentum of the off-shell photon. This is why we need to add the u-channel diagram only if the final-state particles are identical and we do not have this diagram in the Bhabha scattering.

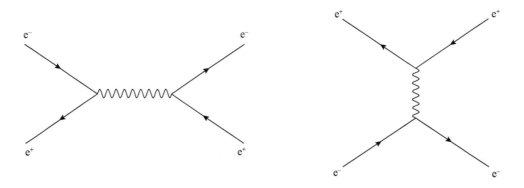

Fig. 6.5 The two lowest-order diagrams contributing to the Bhabha scattering: the s-channel (left) and the t-channel (right).

6.7 Positronium

One of the simplest and most spectacular applications of the Feynman diagrams accounts for the properties of the positronium. The **positronium** is a bound state of an electron and a positron. It can be formed very easily from the thermalization of a positron. A positron in matter reduces its energy both by radiation and ionization losses. At very low speed, it forms an e^+e^- bound state with one of the electrons of the medium and the bound state can decay in two or more photons. The bound state is non-relativistic and the corresponding solution of the Schrödinger equation is identical to the solution of the hydrogen atom except for the fact that the center of mass does not correspond anymore to the position of the positive particle (the proton) but to the position of the two-body reduced mass $\mu = m_e/2$:

$$\frac{1}{\mu} = \frac{1}{m_{e^+}} + \frac{1}{m_{e^-}} = \frac{2}{m_e} . \tag{6.82}$$

C-parity conservation provides a straightforward way to understand when the number of photons in the final state must be even or odd. Since the C-parity of a fermion-antifermion pair is $(-1)^{l+s}$, the states with the lowest angular momenta ($l = 0$) are the $s = 0$ state (**para-positronium**), whose C-parity is $+1$, and the $s = 1$ state (**orto-positronium**), whose C-parity is -1. The C-parity of the photon being -1, the para-positronium decays in an even number of photons and the orto-positronium decays in an odd number of photons.

The leading-order Feynman diagrams for the decay of the para- and orto-positronium are depicted in Fig. 6.7. Since the two (three) photon diagrams have two (three) vertices, we have

$$\mathcal{M}_{para} \sim \alpha \tag{6.83}$$
$$\mathcal{M}_{orto} \sim \alpha^{3/2}. \tag{6.84}$$

The decay width is:

$$\Gamma \sim |\mathcal{M}|^2 \tag{6.85}$$

and

$$\frac{\Gamma_{orto}}{\Gamma_{para}} = \frac{\tau_{para}}{\tau_{orto}} \sim \alpha. \tag{6.86}$$

Therefore, we can predict that the three-photon decay of the positronium is strongly suppressed with respect to the two-photon decay just

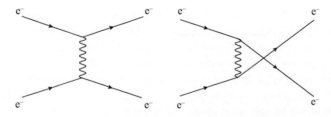

Fig. 6.6 The two lowest-order diagrams contributing to the Möeller scattering: the t-channel (left) and the u-channel (right).

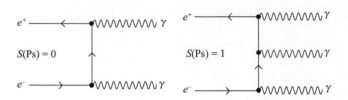

Fig. 6.7 The two lowest-order diagrams contributing to the decay of para-positronium (left) and orto-positronium (right). S(Ps) indicates the spin of the positronium.

inspecting the Feynman diagrams without performing any calculation! In 1949, A. Ore and J.L. Powell carried out the QED calculation of the decay amplitudes at first order in perturbation theory and got:

$$\frac{\tau_{para}}{\tau_{orto}} = \frac{4(\pi^2 - 9)}{9\pi}\alpha \simeq \frac{125 \text{ ps}}{142 \text{ ns}}. \tag{6.87}$$

M. Deutsch observed the positronium for the first time in 1951 and his measurement perfectly confirmed the QED predictions. QED is considered today one of the best-tested theories in the history of science and the safest pillar of the Standard Model.

6.8 The e^+e^- cross-section

The self-energy corrections to e^+e^- scattering offer an intriguing way to seek new particles at accelerators. At leading order, no loops are present in the inner part of the diagram of Fig. 6.1 or in the Bhabha scattering. All the internal lines have a momentum that is completely determined by the external lines and the condition that the incoming and outgoing momenta are equal at each vertex. These diagrams are called **tree-level diagrams**. The total cross-section of a two-body scattering at tree-level is proportional to α^2 (two vertices) and inversely proportional to the flux. Since the Möeller flux factor of eqn 2.129 is $\sim s$ for ultrarelativistic particles (see eqn 2.133),

$$\sigma \sim \frac{\alpha^2}{s} . \tag{6.88}$$

Tree-level QED calculations gives the total cross-section for $e^+e^- \to \mu^+\mu^-$:

$$\sigma = \frac{4}{3}\pi\frac{\alpha^2}{s} \simeq \frac{86.8 \text{ nb}}{s[\text{GeV}^2]}. \tag{6.89}$$

Any deviation from eqn 6.89 suggests the existence of new particles even if the energy of the accelerator used for the e^+e^- collision is not enough to produce the new particles on-shell. Since the early 1960s, experimenters have observed very strong variations in the total e^+e^- cross-section, signaling the production of these particles. Fig. 6.8 shows, in particular, the e^+e^- cross-section into hadrons, that is the bound states of quarks. The presence of a resonance at given energy suggests the production of a new particle with an invariant mass equal to $\sqrt{s}$. None of these states are predicted by QED assuming the electrons (or the muons) to be the only elementary fermions.

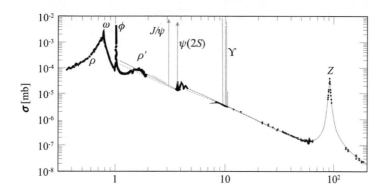

Fig. 6.8 Total $e^+e^- \to$ hadrons cross-section (in mb) as a function of $\sqrt{s}$ in GeV. Reproduced with permission from Zyla *et al.* 2020.

6.9   The running of α

Apart from resonances, the QED cross-sections are expected to increase at high energy due to higher-order Feynman diagrams affecting the fine structure constant. This effect is called the **running of** α.

The **fine structure constant** is a parameter introduced by A. Sommerfeld in 1918 in the framework of the Bohr model of the atom. It is defined as the ratio between the speed of the electron in the first orbit of the Bohr atom and the speed of light:

$$\alpha = \frac{v_1}{c} \overset{SI}{=} \frac{1}{4\pi\epsilon_0}\frac{e^2}{\hbar c} \overset{NU}{=} \frac{e^2}{4\pi} \simeq \frac{1}{137} . \tag{6.90}$$

α also provides the definition of the **charge of the electron in NU**. Unlike SI where the charge is a fundamental unit and is measured in Coulombs, the electric charge is dimensionless in NU, and the absolute value of the electron charge corresponds to:

$$e = \sqrt{4\pi\alpha} \simeq 0.30. \tag{6.91}$$

The fine structure constant plays a central role in perturbative QED as evident from eqn 6.58. A perturbative expansion is meaningful as long as the expansion parameter α is much smaller than 1. Still, QED does not provide a prediction for the value of α. The fine structure constant, that is the strength of electromagnetic interactions, must be determined experimentally by a scattering process and plugged into the QED formulas after summing over the entire perturbative series. This task has been carried out by L. Landau in 1955. In particular, if perturbative methods are valid, it is possible to sum up all Feynman diagrams getting a correction to the photon propagator that reads (Peskin and Schroeder, 1995):

$$\frac{g_{\mu\nu}}{q^2}\frac{e^2}{\left(1 - [\Pi(q^2) - \Pi(q^2 \to 0)]\right)} \tag{6.92}$$

where $\Pi(q^2)$ is an appropriate function of q^2, the (squared) four-momentum exchanged between the particles by the photon propagator. In practice, $\Pi(q^2)$ is computed up to a sufficiently high finite order in perturbation theory.

Intuitively, this is equivalent to saying that the strength of the electromagnetic interactions changes with q^2 and, therefore, α is not a universal constant but, somehow, a function of the energy of the scattering process. Landau was able to show that these corrections can be traced back to self-energy diagrams and the other contributions play no role in changing α. For instance, in a s-channel process like the one of Fig. 6.1 where q^2 is s, the quantum corrections produce a growth of the e.m. strength, that is $\alpha = \alpha(s) > \alpha(s \to 0)$. If we measure α at a scale μ_0^2 where we are sure the theory is perturbative ($\alpha < 1$) and the Feynman diagrams can be safely used, the fine structure constant at another energy scale μ is given by:

$$\alpha(\mu) = \frac{\alpha(\mu_0)}{1 - \frac{\alpha(\mu_0)}{3\pi} \log\left(\frac{\mu^2}{\mu_0^2}\right)} . \tag{6.93}$$

For instance, in an e^+e^- collider like the LEP, producing particles in the s-channel, μ^2 is just s. Equation 6.93 is known as the **Landau formula** and it holds if electrons and positrons are the only elementary fermions. This is an excellent approximation for $\mu \ll m_\mu$ because high-order diagrams with loops are heavily suppressed. Apart from high-energy physics, $\mu \ll m_\mu \simeq 106$ MeV encompasses the vast majority of applications of electrodynamics. In this framework, α is defined at small scales ($\mu = m_e \simeq 0$) using the low-energy limit of the Compton scattering: the Thomson cross-section of eqn 3.41. The most precise determination of α at low energies is $\alpha = 7.2973525664(17) \times 10^{-3} \simeq 1/137$. For $\mu > m_e$, the Landau formula predicts an increase of α. If only electrons existed, perturbative QED would be safe practically at any energy because $\alpha \simeq 1$ at $\mu \simeq 10^{500}$ eV. Perturbative QED is safe also if we consider the full Landau formula:

$$\alpha(\mu) = \frac{\alpha(\mu_0)}{1 - z_f \frac{\alpha(\mu_0)}{3\pi} \log\left(\frac{\mu^2}{\mu_0^2}\right)} \tag{6.94}$$

where z_f is the sum of the squares of the charge was measured experimentally atof all elementary fermions in units of the electron charge. At a center-of-mass energy corresponding to the rest mass of the Z^0 ($s = M_Z^2$ and $M_Z \simeq 91$ GeV), all charged particles except the top quark are kinematically available. These particles are the electron, muon, tau lepton (electric charge: -1), the quark up and charm (electric charge: $+2/3$), the quark down, strange, and beauty (electric charge: $-1/3$). As we will see in Chap. 7, each quark comes in three "colors" so that:

$$z_f(s = M_Z^2) = 3 \cdot (1)^2 + 3 \cdot \left(\frac{2}{3}\right)^2 \cdot 2 + 3 \cdot \left(-\frac{1}{3}\right)^2 \cdot 3 = \frac{20}{3} \simeq 6.67. \tag{6.95}$$

The running of α up to $s = M_Z^2$ was measured experimentally at several colliders using s-channel processes and the results were in perfect agreement with QED. In particular, the experimental data collected at LEP give:

$$\alpha^{-1}(M_Z^2) = 128.936 \pm 0.046 \tag{6.96}$$

that is, α goes from $1/137$ to $1/129$ increasing the center-of-mass energy from m_e to M_Z (five orders of magnitude!). For $s \gg M_Z^2$, also the top quark contributes to z_f. For $z_f = 8$ and $\alpha(\mu^2 = M_Z^2) = 1/129$, we have $\alpha = 1$ at 10^{35} GeV. Again, this is an incredibly large value that makes any QED calculation safe in the purview of perturbation theory. The bad news show up when we come to strong interactions, as we will see in the next chapter.

Exercises

(6.1) Consider a scalar function $f(x^\mu)$ such that $\partial_\mu \partial^\mu f = 0$. Show that even when imposing the Lorenz gauge, the transformation $A'^\mu = A^\mu + \partial^\mu f$ does not change $\mathbf{E}$ and $\mathbf{B}$.

(6.2) Demonstrate eqn 6.48 assuming that f can be derived infinite times in its domain ($f \in \mathcal{C}^\infty$).

(6.3) Is it possible to have a tree level diagram that describes the scattering of two photons: $\gamma + \gamma \to \gamma + \gamma$? Draw the lowest order diagram for this process (**light–light scattering**).

(6.4) Show that the Compton scattering at second order in perturbation theory is described by 17 diagrams.

(6.5) Demonstrate eqn 6.17, i.e. write eqn 6.7 as a function of A^μ and its derivatives to show that $F^{\mu\nu}$ is really a rank-2 tensor of SR.

(6.6) Can we have a process that has both s, t, and u channels at first order (tree-level)?

(6.7) Write a process that has both s-, t-, and u-channels at orders higher than 1.

(6.8) Compute the fine structure constant at $\sqrt{s} = 1$ GeV.

(6.9) * Suppose you are running a e^+e^- accelerator with $\sqrt{s} = 1$ GeV. Can you produce a $\tau^+\tau^-$ pair? Estimate the precision you need in the measured cross-section to get a hint of the existence of the tau lepton by the self-energy contributions.

(6.10) Write the lowest-order Feynman diagram for a $e^+\mu^- \to e^+\mu^-$ scattering and write the corresponding matrix element. [Hint: use the general rule described below eqn 6.64]

(6.11) * Is the $\gamma \to \gamma\gamma$ process kinematically allowed? Show that the process $\gamma \to \gamma\gamma$ is forbidden by the conservation laws of QED.

(6.12) ** Show that even in the case of a hypothetical massive photon (e.g. the Z^0, which is a massive spin-1 particle) the two-photon decay of Exercise 6.11 is forbidden.

(6.13) ** Demonstrate the full **Landau–Yang theorem**: no state with quantum numbers $J^P = 1^\pm, 3^-, 5^-, \ldots$ can decay into two photons.

7 The modern theory of strong interactions

[1] An electron in a closed orbit is accelerated by the centripetal force and experiences the radiation losses of Sec. 3.2.

7.1 Introduction

Physicists started investigating strong interactions at the beginning of the 20th century, well before the raise of quantum mechanics (QM), but we know now that the underlying theory of strong forces has no classical counterpart. Scientists had been suspecting the existence of "strong forces" since 1910. The discovery of Rutherford's atom, which was blatantly inconsistent with classical physics, was a source of rampant speculations. In the Rutherford model, the electrons run through closed orbits without emitting electromagnetic waves.[1] The atom is electrically neutral but the positive $+Ze$ charge of the protons is squeezed into a tiny region of space ("nucleus") within a radius of a few femtometers. Still, there was no evidence of the repulsive forces predicted by electrostatics. After the discovery of the neutron, Wigner demonstrated that the hypothetical force binding the nucleus should have a very short range and fade away just outside the nucleus. Wigner's argument was based on the large binding energy of deuteron ($^2_1$H: 2.2 MeV), triton, that is the nucleus of tritium ($^3_1$H: 8.5 MeV), and the "alpha particle" ($^4_2$He: 28 MeV). The binding energy lies orders of magnitude above the typical energy scale of electromagnetic interactions in atomic physics (eV–keV) and the larger the number of bonds among protons and neutrons, the higher the binding energy. He ended up claiming that an unknown short-range force binds the nucleons (protons and neutrons) close together and the strength of this force increases with the atomic mass number A. The origin of these effects was a conundrum for nearly 60 years.

The modern theory of strong interactions follows straightforwardly by the funding principles of quantum electrodynamics (QED) presented in Chap. 6. It is a gauge theory based on a group slightly more complicated than $U(1)$ but the similarities between the two theories are blurred by non-perturbative effects and are very difficult to spot. This is the reason why it took so long to embed strong interactions into the Standard Model of particle physics.

The theory of strong interactions is called **quantum chromodynamics (QCD)**: it is a gauge theory based on the $SU(3)$ group, and the charges of the strong interactions are three "colors", conventionally called red, blue, and green. The $SU(3)$ group acts on colored particles

A Modern Primer in Particle and Nuclear Physics. Francesco Terranova, Oxford University Press.

© Francesco Terranova (2021). DOI: 10.1093/oso/9780192845245.003.0007

as the QED group $U(1)$ acts on charged particles. Electrically neutral particles do not sense electromagnetic (e.m.) forces. Likewise, particles without colors do not experience strong forces. In atomic physics, a set of neutral atoms may experience faint e.m. forces like the van der Waals forces or the interaction with external magnetic fields due to the magnetic moment. Similarly, a set of quarks like a proton and a neutron can experience fainter forces of strong origin: the forces that keep together the nucleus of an atom.

The road to QCD was steeper than QED because the elementary fermions that sense strong interactions cannot be observed as free states but always form bound states. The hadrons mentioned in Chap. 1 are those bound states. The quarks are hidden in the hadrons and the origin of quark confinement inside hadrons was clarified as recently as in 1973.

Even if the QCD Lagrangian is well known nowadays, drawing predictions from QCD still represents a challenge. A large fraction of strong interaction processes cannot be dealt with by perturbative techniques and defeats the Feynman diagram formalism. Theoreticians are developing non-perturbative techniques to tackle these challenges and QCD remains one of the most lively fields of research in contemporary physics.

7.2 Quarks

As anticipated in Sec. 1.4.2 at the beginning of this book, the quarks are six elementary fermions with half-integer (1/2) spin and fractional electric charge. As a matter of fact, they are the only elementary fermions that sense strong interactions.

The definition of the mass of a quark requires a note of caution because quarks cannot be observed as free particles and their mass is not measured using standard techniques (mass spectrometry, production thresholds, etc.). In Tab. 7.1, we show the best estimate of the **(current) quark masses**, which represent the quark mass terms in the QCD Lagrangian. The reader should be aware of other definitions of masses (e.g. the "constituent quark mass") developed before QCD and employed to describe hadrons in approximate models as the historical quark model of Sec. 8.1. Even within QCD, the current mass is not uniquely defined and any modern measurement must clarify the procedure used for the extraction of the mass from the physical observables.[2]

The word "quark" was coined by M. Gell-Mann in 1963. It is a meaningless term from Joyce's most enigmatic novel *Finnegan's Wake*. That was Gell-Mann's way to warn the readers about the obscure nature of these "particles": at that time, quarks were considered just a formal tool to describe symmetries in strong interactions and were employed to carry out the classification of hadrons and predict some of their properties. Today, we can safely claim that Gell-Mann and his contemporaries were too pessimistic: in the last 40 years, we accumulated overwhelming experimental evidence that quarks are real elementary fermions as are

[2]To be thorough, we report in Tab. 7.1 the current quark masses in the modified minimum-subtraction scheme ($\overline{\text{MS}}$) except for the top quark, where we use the pole mass from the cross-section measurements (Schwartz, 2013, Zyla *et al.*, 2020).

the electrons and the neutrinos, and represent the leading constituents of all hadrons.

Quark	Electric charge	Mass
up (u)	$+2/3$	$2.16^{+0.49}_{-0.26}$ MeV
down (d)	$-1/3$	$4.67^{+0.48}_{-0.17}$ MeV
strange (s)	$-1/3$	93^{+11}_{-5} MeV
charm (c)	$+2/3$	1.27 ± 0.02 GeV
beauty (b)	$-1/3$	$4.18^{+0.03}_{-0.02}$ GeV
top (t)	$+2/3$	172.5 ± 0.7 GeV

Table 7.1 Flavor, electric charge (in units of the proton charge e), and mass of the six quarks. Data from Zyla *et al.*, 2020.

7.3   The charges of the strong interactions

Quarks undergo strong interactions because they are sensed as "charged particles" by the fields of the strong force. This effect mirrors QED, where electrons experience electric forces by their electric charge. In e.m. there is only one type of charge (namely, the "electric charge") and the charge of a particle can be, in principle, any real number. We do not know yet why charges show up as multiples of an elementary charge e but – since Millikan's oil-drop experiment – we know that the available charges are $\pm ne$ where n is an integer number (Cahn and Goldhaber, 2009). The elementary leptons only have $Q = -e$ (the electron, muon, and tau lepton), $Q = +e$ (the corresponding antiparticles), and $Q = 0$ (the neutrinos and their antiparticles, if any). Quarks may have three different types of strong charges: the **blue charge (B)**, the **red charge (R)**, and the **green charge (G)**. The choice of the name of the strong charges ("color": "blue", "red", "green") is as puzzling as the word "quark" and – as the reader may suspect – it has nothing to do with actual colors (wavelength of visible light). Color charges have been introduced in 1964 by O.W. Greenberg and, once more, the name is a cautionary note by the founding fathers of QCD: at that time, color charges were considered theoretical tools since you cannot measure the strong charge of a quark using the experimental techniques employed for electric charges.

Like the electric charge, each color charge might have arbitrary real values but, in practice, charges are a multiple of an elementary unit. We call this elementary unit g for reasons that will be explained later on. g is the strong counterpart of e. A strongly interacting particle can then have $\pm n$ red charge, $\pm m$ green charge, and $\pm p$ blue charge with $n, m, p \in \mathbb{N}$. However, each elementary fermion (the quarks in this case) has a well-defined electric charge and a well-defined strong charge. Both of them are Lorentz-invariant quantities and intrinsic properties of the particle. For all types of quarks, we have only three possibilities: a quark with a blue charge +1 ("blue up quark"), with a red charge +1 ("red up quark"), or with a green charge +1 ("green up quark"). We cannot produce a quark that has simultaneously blue, red, and green charges equal to zero. It simply does not exist, as an electron with an electric charge equal to zero does not exist. Likewise, we cannot produce a quark that has more than a single type of charge different from zero: a simultaneously red (red charge: +1) and blue (blue charge: +1) down quark does not exist, either.

When we discuss the strong charges of a particle we call it collectively the **color** of the particle, which just means the set of strong charges of the particle: if a hypothetical particle is made of three blue down quarks and one red down quark, it has a blue charge = +3 and a red charge = +1. We will show in Sec. 7.6.4 that this particle does not exist or, more precisely, cannot be a hadron.

The antiquarks have the same properties as the quarks but opposite charges. This statement holds both for electric and strong charges. An up antiquark ($\bar{u}$) has an electric charge equal to $-2/3$ in unit of e. An antiquark can only have a blue charge equal to -1 *or* a red charge equal to -1, *or* a green charge equal to -1. An $\bar{u}$ with blue charge equal to -1 is called "an anti-blue (anti)quark" and is indicated as $\bar{u}_{\bar{B}}$. You may guess from Sec. 5.7 that the C-parity operator changes this wavefunction[3] to a blue up quark: u_B.

7.4 Color symmetry

The modern theory of strong interactions is built upon the same gauge principle of QED. The natural formalism of QCD is quantum field theory but, as usual, the most important results can be derived using the formalism of non-relativistic quantum mechanics and the Dirac equation applied to classical relativistic fields.

Let us consider the quark as a point-like particle in QM, whose state is described by a wavefunction. The most general wavefunction of a quark is a superposition of a red quark (i.e. a quark with red charge equal to +1), a blue quark, and a green quark. The QM formalism can be extended to cope with color replacing ψ with a three-vector:[4]

$$\boldsymbol{\psi}(\mathbf{x}, t) = \psi_R(\mathbf{x}, t) \begin{pmatrix} 1 \\ 0 \\ 0 \end{pmatrix} + \psi_G(\mathbf{x}, t) \begin{pmatrix} 0 \\ 1 \\ 0 \end{pmatrix} + \psi_B(\mathbf{x}, t) \begin{pmatrix} 0 \\ 0 \\ 1 \end{pmatrix}. \quad (7.1)$$

$|\psi_R(\mathbf{x}, t)|^2$ is then proportional to the probability of finding a red quark in $\mathbf{x}$ at time t. In the formalism of eqn 7.1, the type of quark (the flavor) described by $\boldsymbol{\psi}(\mathbf{x}, t)$ is well-defined because $\boldsymbol{\psi}(\mathbf{x}, t)$ describes a point-like particle with a mass m. For instance, if the mass is 2.16 MeV (172.5 GeV), the particle is an up (top) quark, as in Tab. 7.1. The strong charge of this quark is not uniquely defined: it may be blue, red,

[3]A hypothetical free quark fulfills the Dirac equation because is an elementary spin-1/2 fermion (eqn 5.78) and, therefore, has the same behavior of a positron under the C-parity operator. In the QFT formulation of strong interactions (QCD), the wavefunctions are replaced by fields but the equation of motion of the fields is again the Dirac equation and the previous result still holds.

[4] There is a subtlety in eqn 7.1 you should not miss. In non-relativistic QM, this wavefunction gives the probability of finding a quark of a given color in a given point in space and time. The formula describes the spatial part of the quark wavefunction and ignores the spin part, which can be factorized. As shown in Chap. 5, in relativistic QM, spin and space do not factorize and $\psi_R(\mathbf{x}, t)$ should be interpreted as a Dirac spinor that describes a red quark. Fortunately, all theorems stated in Chaps 7 and 8 hold in relativistic QM, too, provided we replace $\psi_{R,G,B}$ with the corresponding spinors.

or green. The wavefunction is a superposition of those three possibilities. This is very common in QM. For example, a particle of spin 1/2 may have a wavefunction that is a superposition of a state with $S_z = -1/2$ and $S_z = 1/2$: in this case, it is an entangled state of two S_z spin states. Equation 7.1 represents an entangled state of a quark with a given mass (i.e. a given flavor) with three possible color charges.

> Quantum chromodynamics is built on the principle that the definition of "blue", "red", and "green" is conventional and that no physical observable changes if we redefine a blue state as a linear combination of blue, red, and green states.

Again, this principle mirrors QED, where a redefinition of the phase of the wavefunctions does not affect any physical observable. Under this principle, if we change the definition of blue, $\psi(\mathbf{x}, t)$ transforms into $\psi'(\mathbf{x}, t)$ but the two wavefunctions describe the same physical state. The freedom to redefine the colors of quarks is called **color symmetry** and the transformations that implement the redefinition are unitary operators (see the Wigner's theorem of Sec. 5.2). For three types of charges (blue, red, green), the transformations are provided by complex 3×3 unitary matrices with determinant $+1$. These matrices form a group that is called $SU(3)$, where "S" stands for "special matrices", that is matrices M with det $M = +1$.

Example 7.1

Since $SU(3)$ is a set of matrices and the group operation is the product of matrices, we can demonstrate straightforwardly that this set is a group. The demonstration employs only basic linear algebra (Larson and Falvo, 2009). I remind you that a group is a set G equipped with a binary operation that combines any two elements to form a third element. The set is a **group** if four conditions are satisfied: closure (if g_1 and $g_2 \in G \implies g_1 g_2 \in G$), associativity ($g_1(g_2 g_3) = (g_1 g_2)g_3$), existence of the identity element ($\exists I \in G$ such that $gI = Ig = g$, $\forall g \in G$), and the inverse element ($\forall g \in G$, $\exists g^{-1}$ such that $gg^{-1} = g^{-1}g = I$). These conditions are satisfied by $SU(3)$:

- closure: if $A, B \in G$, AB is unitary because $(AB)^\dagger = B^\dagger A^\dagger$ and $(AB)^\dagger(AB) = \mathbb{I}$. The determinant of AB is $+1$ because for any square matrix A and B, $\det AB = (\det A)(\det B)$. Hence $AB \in SU(3)$.

- associativity: if $A, B, C \in G$, $(AB)C = A(BC)$ for any square matrix.

- identity element: $\mathbb{I}$ is a unit matrix of determinant $+1$ and, hence belongs to G. $\mathbb{I}A = A\mathbb{I} = A$ for any square matrix, including the matrices belonging to $SU(3)$.

- inverse element: if $A \in G$, A is unitary. Therefore, $A^\dagger = A^{-1}$.

$SU(3)$ is then a group. Unlike $U(1)$, it is a **non-abelian** (or **non-commutative**) group because $AB \neq BA$ if $A, B \in SU(3)$.

Using the same technique, we can show that the **SU(N) group** is a group for any integer N. In this case, the set of the group is made of $N \times N$ unitary matrices with unit determinant. None of these groups are abelian except for $N = 1$.

7.5 Representations of continuous groups

QCD is technically more complicated than QED because is based on a non-abelian group. I do not think the reader is already acquainted with $SU(3)$ because this group has few applications in QM. On the other hand, it will be our workhorse both for the description of QCD and for the classification of the hadrons. Given its special role in particle physics, we recap here the basic properties of $SU(3)$ and its representations.

Groups are abstract mathematical objects and it may be extremely difficult to derive theorems by making calculations directly on the members of the group. The theory of representations is aimed at overcoming this barrier. In a representation, each element of a group is mapped into a matrix. The representation should be able to replace the group in all practical needs. In the language of mathematicians:

---

a **representation** of a group on a vector space H of dimension N is a map that associates a linear operator O (i.e. an $N \times N$ matrix) on the vector space to every member of the group. For each matrix:

$$O(g_1 g_2) = O(g_1)O(g_2)$$
$$O(g^{-1}) = O^{-1}(g)$$
$$O(I) = \mathbb{I}$$

where g is a generic element of the group G, I is the identity element of G, and $\mathbb{I}$ is the $N \times N$ identity matrix.

In this case, the matrices preserve all the properties of the group, and handling the matrices is equivalent to handling the element of the groups. The **dimension of the representation** is the dimension of the vector space H. Note that we can build representations in an arbitrary number of dimensions. For instance, the $U(1)$ group of the change of phases in QED is the group of unitary 1×1 matrices, that is, complex numbers of unit modules (e^{ia}). The group is a set of matrices on its own but we can build representations even for vector spaces of dimension larger than 1.

For instance, if $\boldsymbol{\psi}$ is an element of the vector space $\mathbb{C}^2$,

$$\boldsymbol{\psi} = \psi_1 \begin{pmatrix} 1 \\ 0 \end{pmatrix} + \psi_2 \begin{pmatrix} 0 \\ 1 \end{pmatrix} \qquad (7.2)$$

and a two-dimensional representation of $U(1)$ is simply:

$$e^{ia}\boldsymbol{\psi} = \begin{pmatrix} e^{ia}\psi_1 \\ e^{ia}\psi_2 \end{pmatrix} = \begin{pmatrix} e^{ia} & 0 \\ 0 & e^{ia} \end{pmatrix} \begin{pmatrix} \psi_1 \\ \psi_2 \end{pmatrix}. \qquad (7.3)$$

These are quite trivial representations because they *clone* the action of the group in any dimension. We are interested in non-trivial representations like the **irreducible representations (irreps)**. If we build an irrep of G, take an element of the vector space, and apply all possible matrices of the representation (i.e. all possible elements of the group), we will be able to generate all possible elements of the vector space. These representations carry a lot of swag because they can reach any nook of the vector space starting from an arbitrary element of G. The formal statement is less fancy but more precise:

> a representation of a group G on a vector space H of dimension N is irreducible if $\nexists$ any subspace of H that is invariant under the application of G.

Note that the reducible representations always give matrices that can be expressed as block-diagonal (see e.g. the matrix of eqn 7.3). Quite the opposite, there is no way to manipulate an irrep down to a block-diagonal form (Riley *et al.*, 2006).

If $U(a_i)$ $(i = 1 \ldots N)$ is unitary, continuous and smooth on N real parameters, we have already seen in Sec. 5.3 that:

$$U(a_1 \ldots a_N) = e^{i \sum_j a_j T_j} \equiv e^{i a_j T_j}, \qquad (7.4)$$

where T_j are hermitian operators called the **generators** of the group.

The aforementioned $SU(N)$ groups belong to this class and we can show that (Costa and Fogli, 2012):

> the $SU(N)$ group of unitary $N \times N$ matrices with determinant equal to $+1$ has $N^2 - 1$ generators. In particular, **SU(3)** has 8 generators.

Even if we can devise representations of $SU(N)$ in any dimension,[5] the representations in a vector space that has the same dimension as N are somehow special and are called the **fundamental representations**. We will see shortly that the quarks – which have three color charges – transform as the fundamental representation of $SU(3)$.

[5] The reader may be curious to learn whether such an ambitious representation program is successful for any group of interest in physics. A smooth and continuous group can always be represented by matrices. However, only "compact Lie groups" (Costa and Fogli, 2012) can be represented by unitary matrices of finite dimension. $U(1)$, $SU(N)$, the group of rotations in space $SO(3)$, and many other groups used to describe symmetries in physics belong to this class. Unfortunately, the Lorentz group of special relativity (SR) is not compact and there is no way to represent it in such a simple way.

7.6 Quantum chromodynamics

The modern theory of strong interactions is built upon the same gauge principle of QED. The "redundant" degree of freedom of QED was the change of the phase of the wavefunction

$$\psi(\mathbf{x}, t) \rightarrow e^{ia}\psi(\mathbf{x}, t) \tag{7.5}$$

which is a global symmetry of the system describing the dynamic of charged particles. Such a redundancy also appears in the electromagnetic potentials $V(\mathbf{x}, t)$ and $\mathbf{A}(\mathbf{x}, t)$, and corresponds to the classical gauge transformations. As shown in Chap. 6, the e.m. semiclassical Hamiltonian can be derived tightening this global symmetry. If we request $U(1)$ to be a local symmetry, the arbitrary phase gets different values at any point in space-time but, at the same $(\mathbf{x}, t)$ point, the phase convention must be the same both for the wavefunction of the charged particle and the electromagnetic field. Since the local (gauge) condition:

$$\psi(\mathbf{x}, t) \rightarrow e^{ia(\mathbf{x},t)}\psi(\mathbf{x}, t) \tag{7.6}$$

is tighter than eqn 7.5, the Hamiltonian must be changed so that eqn 6.42 holds. In this way, all physical observables remain untouched. The additional terms in the Hamiltonian provide the correct interaction between the electromagnetic field and the charged particle.

The global symmetry of QCD originates from the arbitrariness of the definition of the color charges and transforms the quark wavefunction as:

$$\psi(\mathbf{x}, t) \rightarrow e^{ia_j T_j}\psi(\mathbf{x}, t) \tag{7.7}$$

where T_j are the eight **generators of SU(3)**. This global symmetry is called the **SU(3)$_\mathbf{c}$ color symmetry** because the transformations of $SU(3)$ change the colors of the quarks. Again, the corresponding gauge theory is build upon enforcing the $SU(3)_c$ global symmetry to a local symmetry:

$$\psi(\mathbf{x}, t) \rightarrow e^{ia_j(\mathbf{x},t)T_j}\psi(\mathbf{x}, t) \tag{7.8}$$

so that the definition of what is blue, red, or green can be changed at will at any point of space-time.

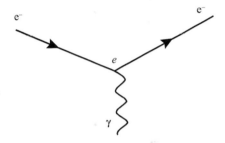

Fig. 7.1 Electron–photon vertex in QED. The e at the vertex corresponds to the electric charge.

> Quantum chromodynamics is a gauge theory based on the $SU(3)_c$ color symmetry. The $SU(3)$ group has 8 generators, which correspond to the mediators of the strong interactions and are called the **gluons**.

Unfortunately, a semi-classical theory of strong interaction cannot be easily defined because the strong fields (gluon fields) have no classical counterpart and the semiclassical approach is of limited use to compute the strong interaction cross-sections. In the following, we explore this theory without relying on a full-fledged QFT formulation. Still, all prominent features of QCD arise from the tools of Chap. 6: the inspection of the Feynman diagrams, the lowest order quark–quark scattering, the properties of the gluons, and the running of the coupling constant.

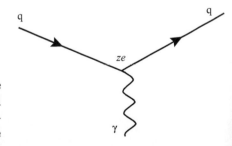

Fig. 7.2 Quark–photon vertex in QED. The charge of the quark in units of e is given by z.

7.6.1   The Feynman diagrams of QCD

The Feynman diagrams of QED are based on the electron–photon vertex of Fig. 7.1 that corresponds to the interaction term of the Lagrangian between the electron and the photon. The only charged particles we considered in Chap. 6 were the electron and the muon but QED works for any particle with an electric charge equal to ze. The quarks have both color and electric charges. They are sensitive to e.m. interactions and the corresponding QED vertices are given in Fig. 7.2. These vertices and, therefore, all QED scattering amplitudes are independent of the color charge of the quarks because the electromagnetic and strong interactions are two different types of interactions and the corresponding theories are completely decoupled. The QED scattering amplitude of a blue up quark is identical to the scattering of a red up quark but the QED amplitudes of a red up quark are *different* from the amplitudes of a red down quark because the electric charge of the two quarks is different. Quarks are sensitive to e.m. interactions but – unless their color charge is neutralized in some way – their strong interactions are overwhelming. The QCD vertex of a quark is shown in Fig. 7.3. **q** is a three-vector of fields[6] because there are three types of color charges in QCD. As a consequence, the strong analog of the fine structure constant must be multiplied by a 3×3 matrix. In this way, the vertex describes the strength of the interaction for all possible scatterings of the quarks, including a blue versus red, a blue versus green, and a green versus red collision. This is immaterial in electromagnetism since only electric charges exist and the matrix cannot be different from 1.

[6]We remind that, in QFT, the quark wavefunction is replaced by a quantized fermionic field. Like the wavefunctions, the values of the fields are complex numbers and $\mathbf{q} \in \mathbb{C}^3$.

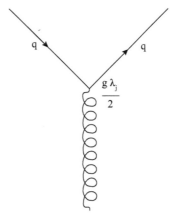

Fig. 7.3 The quark–gluon vertex in QCD. λ_j is the j-th Gell-Mann matrix and is a 3×3 matrix. **q** is a three-component vector of quark fields. g is the unit charge in strong interactions ($\alpha_s = g^2/4\pi$).

This result is not accidental and represents a special case of a theorem derived in 1954 by C.N. Yang and R. Mills that we state without demonstration (Yang and Mills, 1954; Peskin and Schroeder, 1995):

> any $SU(N)$ gauge theory describes an interaction that is produced by $N^2 - 1$ massless mediators with spin equal to 1. These theories are also called **Yang–Mills theories**.

We remind the reader that a massless mediator has only two possible orientations of the spin: $S_z = -1$ and $S_z = +1$. The λ matrices are, therefore, $N \times N$ matrices that couple the fermions to the force mediators and their number must be equal to the generators of $SU(N)$: one for QED and eight for QCD. The most intriguing part of the theorem is that the mediators must be **massless bosons**, that is, spin-1 particles with $m = 0$.

The strength of the QCD interactions is given by:

$$g\, T_j = g\, \frac{\lambda_j}{2} \tag{7.9}$$

or, equivalently,

$$\sqrt{\alpha_s}\, T_j = \sqrt{\alpha_s}\, \frac{\lambda_j}{2} \tag{7.10}$$

where α_s is the fine structure constant of the strong interaction and is called **alpha-strong (α_s)**. We can understand now why we called "g" the unit strong charge in natural units (NU):

> g and e are the unit charge of strong and e.m. interactions, respectively. They are linked to the corresponding fine structure constants α and α_S by:
>
> $$e = \sqrt{4\pi\alpha} \;\; ; \;\; g \equiv g_s = \sqrt{4\pi\alpha_s} \tag{7.11}$$

Note that several authors prefer to write g as g_s ("g-strong") to avoid confusion with the corresponding quantity in the weak interaction theory of Chaps 10 and 12.

The 3×3 matrices are thus the generators of the $SU(3)$ group in the fundamental representation. These matrices are called the **Gell-Mann matrices** since they were employed by M. Gell-Mann in 1964 to classify the quark bound states. Mathematicians are (rightly) outraged by this name because these generators were computed by E. Cartan and H. Weyl in a more general framework at the beginning of the 20th century. The eight **Gell-Mann matrices** are (Gell-Mann, 1962):

$$\lambda_1 = \begin{pmatrix} 0 & 1 & 0 \\ 1 & 0 & 0 \\ 0 & 0 & 0 \end{pmatrix} ; \quad \lambda_2 = \begin{pmatrix} 0 & -i & 0 \\ i & 0 & 0 \\ 0 & 0 & 0 \end{pmatrix} ; \quad \lambda_3 = \begin{pmatrix} 1 & 0 & 0 \\ 0 & -1 & 0 \\ 0 & 0 & 0 \end{pmatrix} ;$$

$$\lambda_4 = \begin{pmatrix} 0 & 0 & 1 \\ 0 & 0 & 0 \\ 1 & 0 & 0 \end{pmatrix} ; \quad \lambda_5 = \begin{pmatrix} 0 & 0 & -i \\ 0 & 0 & 0 \\ i & 0 & 0 \end{pmatrix} ; \quad \lambda_6 = \begin{pmatrix} 0 & 0 & 0 \\ 0 & 0 & 1 \\ 0 & 1 & 0 \end{pmatrix} ;$$

$$\lambda_7 = \begin{pmatrix} 0 & 0 & 0 \\ 0 & 0 & -i \\ 0 & i & 0 \end{pmatrix} ; \quad \lambda_8 = \frac{1}{\sqrt{3}} \begin{pmatrix} 1 & 0 & 0 \\ 0 & 1 & 0 \\ 0 & 0 & -2 \end{pmatrix} .$$

The form of the matrices suggests that some of the gluons couple only with a limited number of color charges. For instance, if we associate colors as in eqn 7.1, there are only two gluons that scatter a red quark into a red quark because:

$$(1 \quad 0 \quad 0)\,\lambda_j \begin{pmatrix} 1 \\ 0 \\ 0 \end{pmatrix} \tag{7.12}$$

is different from zero only if $j = 3$ and $j = 8$. A scattering of a blue quark into a blue quark occurs only through λ_8, that is, through the last gluon.

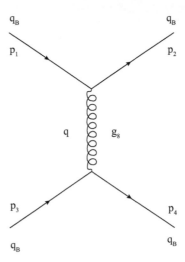

Fig. 7.4 Scattering between two blue quarks. The Gell-Mann matrices ensure that only one gluon (g_8 associated to λ_8) contributes to the amplitude. p_i are the four momenta of the on-shell particles in the initial ($i = 1, 3$) and final ($i = 2, 4$) states. q is the four-momentum of the gluon.

Gluons can change the type of strong charge brought by the quarks: λ_1 and λ_2 scatter a red quark into a green quark, while λ_4 and λ_5 scatter a red quark into a blue quark. Note that charge conservation in QED is evident in the photon vertex of Fig. 7.2 because the electric charge entering the vertex is equal to the one leaving the vertex. In QCD, the vertices may change the type of charge but the total color charge must be conserved. As noted in Sec. 6.5.1, gauge symmetries are not symmetries of QM and do not commute with the Hamiltonian. On the other hand, they are devised starting from a global symmetry that still holds in QED/QCD. For instance, if we choose $\alpha(\mathbf{x}, t) = \alpha(\mathbf{0}, 0)$, the gauge symmetry becomes a simple global symmetry. The global symmetries fulfill the definition of symmetry in QM and follow the Noether theorem of Sec. 5.4.1. As already noted, the global $U(1)$ symmetry provides the conservation of the electric charge in QED.

The global $SU(3)_c$ symmetry ensures the **conservation of color charge** in QCD.

7.6.2   Scattering of quarks

Since the assignment of the type of charge is conventional, the scattering amplitudes must be independent of the color assignment of eqn 7.1.

We might have chosen $(1, 0, 0)$ to be a blue quark instead of a red quark but no physical observable should depend on this choice. Since there is an apparent asymmetry between the number of gluons contributing to a red-to-red scattering (two gluons) versus a blue-to-blue scattering (one gluon), you may want a demonstration that, at the end of the day, the amplitudes are the same.

The tree-level Feynman diagram describing the scattering of a blue quark into a blue quark is shown in Fig. 7.4. The equation associated to this diagram resembles eqn. 6.62 and reads (Schwartz, 2013):

$$\mathcal{M} \sim (\sqrt{\alpha_s})^2 \bar{u}(p_2) \gamma^\mu \frac{\lambda_8}{2} u(p_1) \frac{g_{\mu\nu}}{q^2} \bar{u}(p_4) \gamma^\nu \frac{\lambda_8}{2} u(p_3) \qquad (7.13)$$

where $u(p_1)$ and $u(p_3)$ are the bispinors of the incoming blue quarks, and $u(p_2)$ and $u(p_4)$ are the bispinors of the outgoing blue quarks. As usual (see eqn 6.63):

$$\bar{u}(p_2) \equiv u(p_2)^\dagger \gamma^0$$
$$\bar{u}(p_4) \equiv u(p_4)^\dagger \gamma^0$$

are the adjoint bispinors and $q^2 \equiv q^\mu q_\mu = (p_2 - p_1)^2$ is the four-momentum transferred by the gluon. Since eqn 7.13 depends on two Gell-Mann matrices, it is useful to explicitly write the equation using the Einstein summation rule:

$$\mathcal{M} \sim (\sqrt{\alpha_s})^2 \bar{u}_j(p_2) \gamma^\mu \frac{[\lambda_8]_i^j}{2} u_i(p_1) \frac{g_{\mu\nu}}{q^2} \bar{u}_k(p_4) \gamma^\nu \frac{[\lambda_8]_l^k}{2} u_l(p_3) \qquad (7.14)$$

where i, j, k, and l run from 1 to 3. In eqn 7.14, $[\lambda_8]_i^j$ represents the $\{j, i\}$ element of the 3×3 matrix λ_8.

We do not dare to compute the QCD cross-sections but we can comfortably estimate the relative strength between the blue-to-blue and red-to-red scattering. If we trim all spinors and gamma matrices in eqn 7.14 because they are the same in both processes, the strength of each vertex is (Bettini, 2014):

$$\sqrt{\alpha_s} \begin{pmatrix} 0 & 0 & 1 \end{pmatrix} \frac{\lambda_8}{2} \begin{pmatrix} 0 \\ 0 \\ 1 \end{pmatrix} = \left(-\frac{2}{\sqrt{3}}\right) \frac{1}{2} \sqrt{\alpha_s} = \left(-\frac{1}{\sqrt{3}}\right) \sqrt{\alpha_s} \qquad (7.15)$$

where $(0, 0, 1)$ is the blue quark, as defined in eqn 7.1. The matrix element of the (two-vertex) diagram of Fig. 7.4 is then:

$$\mathcal{M} \sim \left(-\frac{1}{\sqrt{3}}\right)^2 \alpha_s = \frac{1}{3} \alpha_s . \qquad (7.16)$$

Fig. 7.5 shows the corresponding lowest-level diagram for the scattering of a red quark into a red quark. Here, we have two diagrams corresponding to two different gluons: the third and the eighth. In this case, the strength of the vertex of the first diagram is:

$$\sqrt{\alpha_s} \begin{pmatrix} 1 & 0 & 0 \end{pmatrix} \frac{\lambda_3}{2} \begin{pmatrix} 1 \\ 0 \\ 0 \end{pmatrix} = \frac{1}{2} \sqrt{\alpha_s} \qquad (7.17)$$

but the strength in the second diagram is:

$$\sqrt{\alpha_s}\begin{pmatrix}1 & 0 & 0\end{pmatrix}\frac{\lambda_8}{2}\begin{pmatrix}1\\0\\0\end{pmatrix} = \frac{1}{2\sqrt{3}}\sqrt{\alpha_s}\,. \tag{7.18}$$

The matrix element of the red-to-red scattering becomes:

$$\mathcal{M} \sim \left(\frac{1}{2}\right)^2\alpha_s + \left(\frac{1}{2\sqrt{3}}\right)^2\alpha_s = \frac{1}{3}\alpha_s. \tag{7.19}$$

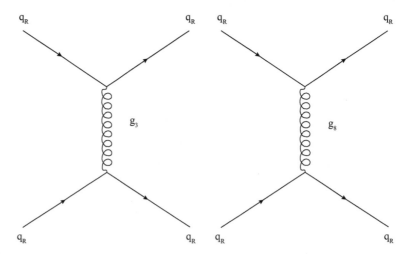

Fig. 7.5 Scattering between two red quarks. In this case, two gluons (g_3 associated to λ_3 and g_8 associated to λ_8) contribute to the amplitude.

The overall strength is, therefore, the same as in the blue-to-blue scattering (see eqn 7.16). But in the red-to-red scattering, the strength is shared between the third and the eighth gluons, and the strength per gluon is halved. This is why eqn 7.19 gives the same result as eqn 7.16.

No magic in here. The amplitude is the same because the $SU(3)_c$ gauge symmetry provides color invariance on a general ground. As disclosed at the beginning of this chapter, QCD is the theory that ensures the invariance of all physical observables by any redefinition of red, green, and blue. This statement brings us to an amazing result:

> we cannot devise an experiment that measures the color of a quark. The very origin of the strong interactions is the arbitrariness of the definition of color and our inability to distinguish quarks of different colors.

This principle is as fundamental as the Pauli exclusion principle, where our inability to distinguish identical fermions leads to totally anti-symmetric wavefunctions.

You may wonder whether the scattering of two blue up quarks is the same as two blue down quarks. The answer is, nearly. The strong interactions are insensitive to the electric charge of the quark because z does not enter the QCD vertices. However, an up quark has a mass different from a down quark and the available phase-space is different, too. Besides, the QED amplitudes are larger for the up quark because the electric charge is bigger. I said "nearly" because all these effects are tiny corrections to the strong interaction matrix element, which is insensitive to the type (flavor) of the quarks and looks at colors only. As a consequence, if we neglect tiny corrections, we can claim that the scattering and interaction of quarks are flavor-independent. More rigorously:

> if quarks have equal masses (**degenerate quarks**) and we neglect electromagnetic interactions, no QCD observable depends on the flavor of the quarks.

This property is called **flavor symmetry** and lays the foundations for the quark models presented in Chap. 8.

7.6.3 The gluon*

Gluons are[7] the strong counterparts of the QED photon and share with the photon several properties. Each gluon is massless and has a spin equal to 1. Therefore, the only polarizations available along the direction of motion are left-handed ($S_z = -1$) and right-handed ($S_z = +1$) — see Sec. 6.6.2. The propagator of the gluon depends on the gauge fixing, as for the photon. In Chap. 6 we worked in the Lorenz gauge ($\partial_\mu A^\mu = 0$) at the price of accepting a leftover of the gauge freedom. We could have chosen alternative gauge fixings that completely determine the e.m. potentials (e.g. the Coulomb gauge) but we gave away these options to preserve the explicit Lorentz invariance of all formulas. The same problem appears in QCD at a much higher level of complexity.

The most general form of the gluon propagator is given by (see e.g. Schwartz, 2013):

$$\delta^{ab} i \frac{-g^{\mu\nu} + (1-\xi)p^\mu p^\nu / p^2}{p^2} \tag{7.20}$$

where ξ is a gauge-fixing constant. The δ^{ab} term (see Fig. 7.6) ensures that a gluon cannot change its type inside a Feynman diagram:[8] a given gluon contributes to the amplitude only if the Gell-Mann matrix interacting with the fermions in the initial state (a) is the same matrix of the final state (b). This does not mean that the gluon cannot change the colors of the quarks but a color change is allowed only if it is allowed by the corresponding Gell-Mann matrix λ_a. To simplify eqn 7.20, the easiest – and most common – way is to resort to the **Feynman gauge** $\xi = 1$. This choice of the gauge preserves the Lorentz invariance and simplifies remarkably the gluon propagator. However, the price to pay is much higher than QED. A QCD amplitude computed in the Feynman gauge must account for the remnants of the gauge freedom introducing non-physical states and the corresponding Feynman diagrams. These phantoms are (deservedly) called **Fadeev–Popov ghosts**. They have their own propagator (see Fig. 7.7):

[7]The name was coined by M. Gell-Mann in 1962 and calls to mind the "glue" that sticks the quarks in the bound states.

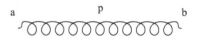

Fig. 7.6 Gluon propagator. a and b are the gluon indexes ($a, b = 1, \ldots 8$). p is the gluon four-momentum.

[8]For the sake of thoroughness, both the photon propagator (eqn 6.68) and the gluon propagator have $p^2 + i\epsilon$ instead of p^2 in the denominator and sidestep a pole if the massless particle is on-shell ($p^2 = 0$). This term is important to compute the Feynman diagrams in QED and QCD, and have a physical interpretation, too, since it preserves causality in relativistic QM (Peskin and Schroeder, 1995; Schwartz, 2013).

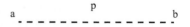

Fig. 7.7 The propagator of the unphysical Fadeev–Popov ghost. Note that the same symbol (the dashed line) is used in other theories to indicate physical scalar particles, too.

$$\delta^{ab}\frac{i}{p^2} \qquad (7.21)$$

and mimic a sort of spin-0 particle that does not obey the Bose–Einstein statistics. Ghosts are essential to get meaningful results in QCD if we employ the Feynman gauge but by no means represent physical particles.

The QCD analog of the electromagnetic tensor:

$$F_{\mu\nu} \equiv \partial_\mu A_\nu - \partial_\nu A_\mu \qquad (7.22)$$

is given by the **strong interaction tensor**:

$$G^a_{\mu\nu} \equiv \partial_\mu A^a_\nu - \partial_\nu A^a_\mu - g f^{abc}[A^b_\mu, A^c_\nu]. \qquad (7.23)$$

In this formula $g = \sqrt{4\pi\alpha_s}$ and a, b, c are the indices that run through the number of generators of $SU(3)$: $a = 1, 2, \ldots, 8$. The **structure constants** f^{abc} are real numbers that can be computed by the commutation relations of the Gell-Mann matrices:

$$[\lambda^a, \lambda^b] = i f^{abc}\lambda^c. \qquad (7.24)$$

The structure constants enter the tensor of the gluon fields because $SU(3)$ is a non-abelian group and produce another striking feature of QCD:

> unlike photons, a gluon can interact with another gluon and the Feynman diagrams of QCD include **three** and **four gluon vertices**, like the ones shown in Fig. 7.8. In general, any gauge theory based on non-abelian $SU(N)$ groups exhibits self-interactions among the gauge fields.

Despite so many differences, the QCD gluons are massless like the photon, in agreement with the Yang–Mills theorem cited in Sec. 7.6.1. Computing QCD amplitudes is a very complex task because self-interactions and ghosts enormously increase the number of diagrams. But, as long as the theory is perturbative and $\alpha_s < 1$, the Feynman diagram formalism works pretty well.

7.6.4 Asymptotic freedom

Until the late 1960s, there was very little hope of explaining strong interactions using a gauge theory[9] because the properties of strong interactions looked completely different than QED. Strong interactions are short-range forces. Both photons and gluons are massless particles and, hence, they should produce long-range forces as the Coulomb force in electromagnetism. A breakthrough in the understanding of strong interactions occurred in 1973 when D. Gross, H.D. Politzer, and F. Wilczek discovered asymptotic freedom in QCD. In Sec. 6.9, we noted that the QED fine structure constant is small and, even if its value slightly increases with energy, it remains well below 1 at any realistic energy.

[9]The mood of the times was brightly summarized by L. Landau who declared: "It is well known that theoretical physics is at present almost helpless in dealing with the problem of strong interactions. ... We are driven to the conclusion that the Hamiltonian method for strong interactions is dead and must be buried, although of course with deserved honour." (Ter Haar, 2013).

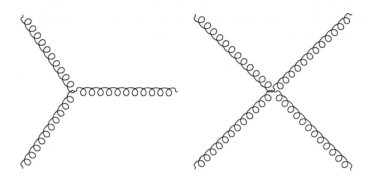

Fig. 7.8 Self interactions of gluons in QCD: three gluon vertex (left) and four gluon vertex (right).

As a consequence, QED amplitudes can be computed using perturbative techniques (e.g. the Feynman diagrams) at any energy scale. In 1973, there were no measurements of α_s, whatsoever. Assuming QCD to be perturbative at least in one (unknown) energy scale μ_0, Gross, Politzer, and Wilczek computed the running of α_s to another scale μ following Landau's approach described in Sec. 6.9. The result - the **Gross–Politzer–Wilczek formula** - is:

$$\alpha_s(\mu) \simeq \frac{\alpha_s(\mu_0)}{1 + \left(\frac{11N_c - 2F}{3}\right)\frac{\alpha_s(\mu_0)}{4\pi}\log\frac{\mu^2}{\mu_0^2}} \ , \qquad (7.25)$$

where the approximation is due to the evaluation of the formula at finite order in perturbation theory. In this formula (Gross and Wilczek, 1973; Politzer, 1973), μ is the new scale[10] where we want to compute α_S, N_c are the number of types of strong charges (the $N_c = 3$ colors of QCD), and F is the number of flavors ($F = 6$). In turn,

$$\alpha_s(\mu) \simeq \frac{\alpha_s(\mu_0)}{1 + \frac{7}{4\pi}\alpha_s(\mu_0)\log\frac{\mu^2}{\mu_0^2}} \qquad (7.26)$$

This formula provides the running of α_s with energy and is the QCD lookalike of eqn 6.93. Comparing eqn 7.26 with eqn 6.93, we note a puzzling plus sign in the denominator:

> unlike QED, the value of α_s decreases with energy and $\alpha_s \to 0$ for $\mu \to \infty$. The strength of QCD interactions decreases at high energy and the quarks become non-interacting (free particles) in the $\mu \to \infty$ limit. This property is called **asymptotic freedom**.

In the 20 years that followed this seminal result, several scattering experiments provided evidence that strong interactions become fainter and fainter when we increase μ. These findings gave an early experimental support to the asymptotic freedom. The perturbative nature of QCD, however, becomes crystal clear at very large energies (see Fig. 7.9). The most precise measurement of α_s to date is for $\mu^2 = s = M_Z^2 \simeq (91.2 \text{ GeV})^2$ and amounts to:

$$\alpha_s(M_Z^2) = 0.1179 \pm 0.0010 \simeq \frac{1}{9} \ . \qquad (7.27)$$

[10]In an e^+e^- collider, if the total cross-section is dominated by the s-channel, μ can be safely considered $\sqrt{s}$. For other channels or processes involving multiple diagrams and particles (e.g. a collision of composite particles like protons in a hadron collider), μ is the Lorentz-invariant four momentum transfer among the elementary particles involved in the scattering. In most cases, the experimentalist only have an approximate estimate of μ. This is the reason why the measurement of α_s from e^+e^- colliders at the peak of the Z^0 is considered a "standard candle" for the determination of $\alpha(\mu_0 = M_Z)$. Note that the production of top quarks is kinematically forbidden at the Z^0 peak (91.2 GeV) because the top is too heavy, and the effective number of flavors is 5 instead of 6.

You may note that even at this energy, α_s is still much stronger than the electromagnetic fine structure constant:

$$\alpha(\mu = m_e \simeq 0) \simeq \frac{1}{137} \; ;$$

$$\alpha(\mu = M_Z) \simeq \frac{1}{128} \; .$$

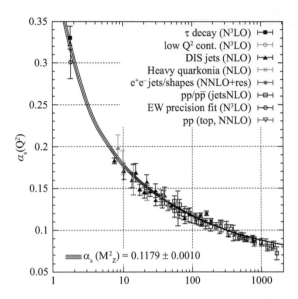

Fig. 7.9 The value of α_s as a function of the four-momentum transferred in the scattering ($Q^2 \simeq \mu^2$) in GeV. The continuous line is the Gross–Politzer–Wilczek formula corrected for the number of available flavors at a given energy (see footnote 10). Experimental measurements and their errors are shown by dots. Reproduced with permission from Zyla *et al.*, 2020.

M_Z is a wise choice for μ because $\alpha_s \simeq 1/9 \ll 1$ and we are safely inside the perturbative regime of QCD. Through eqn 7.25, we can compute the value of the energy where α_s is of order 1 and the QCD perturbative series of no use.

> QCD enters the non-perturbative regime for energies smaller that $\Lambda_{\mathrm{QCD}} \simeq 300$ MeV. At these energies, the scattering amplitudes cannot be computed at any order using the Feynman diagrams.

Equation 7.25 indicates that strong interactions cannot be dealt with by Feynman diagrams at the energies available to the experimentalists when QED was established. This is the reason for the delay between the formulation of QED and QCD. In particular, the entire domain of nuclear physics and radioactivity is within the scope of **non-perturbative QCD**.

Asymptotic freedom provides an elegant explanation of why quarks cannot be observed as free particles. Fig. 7.10 shows the lowest-order Feynman diagram producing, for example, a u_R and $\bar{u}_{\bar{R}}$ pair by the annihilation of an electron–positron pair at energies $\gg \Lambda_{\mathrm{QCD}}$. If Fig. 7.10 were a QED process, the quarks would fly apart emitting radiation photons due to the scattering with other charged particles in the medium. Since the emission probability is proportional to the fine structure constant and α is nearly the same at any energy, a charged particle

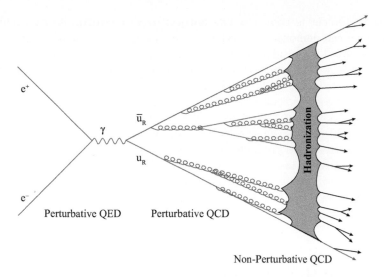

Fig. 7.10 An electron–positron pair produces a photon that generates a quark–antiquark pair. In this case, the type of quark is "up" and the color is "red." The region where $\alpha_s < 1$ is labeled "perturbative QCD." The hadronization phase where quarks create bound states with zero color charges (hadrons) occurs in the non-perturbative regime ($\alpha_s > 1$). Redrawn from Buskulic *et al.*, 1995.

emits a constant amount of photons per unit length. This is true even if the energy is reduced after each photon emission. If a quark emits a gluon (the strong counterpart of radiation losses), the quark energy is reduced because the gluon carries away a fraction of the initial energy. After the gluon emission, α_s becomes larger than the initial α_s and, hence, the probability of a second emission will be enhanced. This gluon emission cascade stops when the initial quark loses most of its energy and the final state particles – gluons or additional quarks produced by $g \to q\bar{q}$ processes – combine to reach a final state where particles can fly apart. As soon as the quark reaches an energy lower than Λ_{QCD}, the Feynman diagrams are of no use to describe the gluon–quark shower because the perturbative expansion does not hold. This stage is called the **hadronization** phase. At this stage, we can only guess that the final-state particles will be bound states of zero total color charge because these are the only states that do not undergo strong interactions. However, the path that brings quarks to the color-neutral final state cannot be traced by perturbative QCD. Since the late 1970s, we have at our disposal non-perturbative techniques to solve the equations of motion of quarks below Λ_{QCD}. These methods demonstrate that:

> in any QCD scattering process, the N_f final states are made of quark and gluon bound states, whose wavefunction is a $SU(3)_c$ singlet.

$SU(3)_c$ singlets are wavefunctions with zero total color charge that are invariant (but a unit phase ϕ) for any transformation of $SU(3)_c$:

$$\psi_{final} = \prod_{k=1}^{N_f} \psi_k \tag{7.28}$$

and

$$e^{ia_j T_j}\psi_k = e^{i\phi_k}\psi_k \quad \forall\, k = 1, \ldots, N_f. \tag{7.29}$$

[11]The mass gap problem is the fourth Millenium Problem of the Clay Mathematics Institute and, unlike the Poincaré conjecture (problem n.1), has not been solved, yet. It is considered of great mathematical interest because the demonstration could disclose analytical methods to solve non-perturbative theories like QCD, quantum gravity, non-abelian gauge theories and the like.

This statement is known as the **conjecture of confinement** because has not been demonstrated yet in axiomatic quantum field theory. This lacking proof is considered a major issue in mathematical physics, where the problem is known as the **mass-gap problem**.[11] Nonetheless, the numerical evidence in favor of the conjecture is overwhelming and, so far, quark confinement has always been corroborated by experiments. Free quarks, if exist, could be detected quite easily because they have a fractional electric charge. A free quark would have a peculiar ionization pattern and could be identified, for example, by measuring the dE/dx ionization losses inside a detector. These particles were searched for decades in dedicated and general-purpose experiments at energies ranging from $E \simeq m_u$ to $\sqrt{s} = 14$ TeV (LHC). Despite a few early thrills, free fractional-charge particles have never been observed.

7.7   Do quarks exist?*

Quarks were introduced as a classification tool for hadrons: hypothetical particles with a fractional electric charge that bring a hypothetical color charge. The color charge definition (red, green, blue) is discretionary and eqn 7.25 implies that free quarks cannot be observed a priori. This is enough to raise skepticism about the existence of quarks as elementary particles in the sense used in Sec. 1.2. To silence the doubters, we need physical observables that are uniquely linked to the point-like constituents of the hadrons and processes that test the perturbative calculations of QCD.

The first evidence in favor of quarks as elementary fermions came from electron–proton and electron–neutron scattering experiments ($e^- N \rightarrow e^- X$), where $N = p, n$ is a nucleon (i.e. a proton or a neutron) and X a set of hadrons. These experiments investigated the structure of the nucleons with a perfectly known probe, the photon.

The experimenters who performed these pioneering studies in the 1960s were puzzled by the results because – despite all Gell-Mann's caveats – they had a quite naive notion of quarks. For instance, they considered the proton as a composite state of three quarks, seen as charged incompressible "balls," whose mass was of the order of $m_p/3$ minus some binding energy.[12] They were designing the experiment like soldiers that want to see what is inside a sack by shooting it with a machine gun using electrons instead of bullets. If the bag contains only soft material like styrofoam, the bullets cross the bag with small and smooth trajectory changes. If there are three iron balls inside the styrofoam, some bullets bounce at large angles when they hit the iron, like in Rutherford's experiment. The distribution of the scattering angles depends on the number, mass, and size of the iron balls and the determination of these parameters was the goal of the first SLAC experiments. Conversely, data were supporting neither the styrofoam nor the iron ball hypothesis but draw the attention of several physicists, including J.D. Bjorken and R. Feynman.

[12]To be fair, this view was not so simple minded. Gell-Mann did not know the actual quark masses and, in the historical quark model, the quark masses were inflated to account for the large proton mass. We will see in Sec. 8.1 that the inflated mass – called the constituent quark mass – has no place in the modern theory of hadrons because the hadron mass is dominated by non-perturbative interactions among a large number of gluons and quark–antiquark pairs.

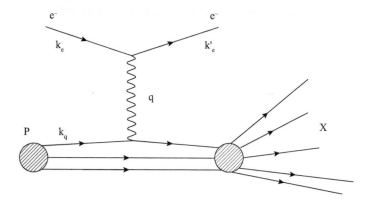

Fig. 7.11 Deep inelastic scattering of an electron interacting with a proton. P is the four-momentum of the proton. $k_q = xP$ is the four-momentum of the parton (e.g. an up quark), q is the photon four-momentum flowing from the electron to the proton. This energy transfer breaks apart the proton creating a shower of hadrons (X). The initial and final four-momenta of the electron are k_e and k'_e, respectively.

A $e^- N \to e^- X$ scattering may occur with a large momentum transfer from the electron to the proton so that the proton is broken and produces a set of hadrons X in the final state. These collisions are called **deep inelastic scatterings** (DIS). In classical physics, DIS corresponds to such a hard collision that the rigid body hit by the probe breaks apart, unveiling the details of the inside. A scattering of an electron onto a hypothetical constituent of the proton is depicted in Fig. 7.11. The initial and final four-momenta of the electrons are labeled k_e and k'_e, respectively. The four-momentum lost by the electron is:

$$k_e - k'_e \equiv q. \tag{7.30}$$

q is called the four-momentum transfer because it corresponds to the four-momentum flowing from the electron to the proton. If we want to see the inner constituents with good resolution, we must look at high momentum transfer events and hope that the De Broglie wavelength of the probe (the photon in this case) is comparable to the size of the quarks.

Since a DIS is described by a t-channel, at high energy (k_e and $k'_e \gg m_e \simeq 0$), q^2 must be space-like. Let us demonstrate this.

Example 7.2

k_e and k'_e are the four-vectors of the initial- and final-state electrons. Then:

$$(k'_e - k_e)^2 = k_e^2 + k'^2_e - 2k_e \cdot k'_e \tag{7.31}$$

If the electron is ultra-relativistic, its energy $E \gg m_e$. Assuming, without loss of generality, that the electron moves along the z-axis of the laboratory frame (LAB),

$$k_e \simeq (E, 0, 0, E)$$
$$k'_e \simeq (E', 0, E' \sin\theta, E' \cos\theta)$$

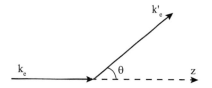

Fig. 7.12 Scattering of an electron to a nucleon (p or n). The final (initial) state electron has a four-momentum k'_e (k_e) and the scattering angle in LAB is labeled θ.

where θ is the scattering angle of Fig. 7.12. Therefore,

$$k_e \cdot k_e' = EE'(1 - \cos\theta) > 0 \tag{7.32}$$

and

$$(k_e - k_e')^2 = k_e^2 + k_e'^2 - 2k_e \cdot k_e' \simeq -2k_e \cdot k_e' < 0 \tag{7.33}$$

because k_e^2 and $k_e'^2$ are the electron masses and can be neglected. Therefore, eqn 7.33 gives a negative number, as expected. Lorentz invariance ensures that q^2 is negative in any reference frame – including LAB – and the event is space-like, as in any t-channel.

It is customary to introduce a new variable that is always positive: $Q^2 = -q^2$. Q^2 can be interpreted as the four-momentum transferred to the proton by the photon, that is, by the e.m. probe.

Unfortunately, we do not know the four-momentum k_p of the "balls" inside the proton and we can only claim that is a fraction of the whole proton four-momentum P: $k_p = xP$, where $0 \leq x \leq 1$. Since DIS are produced by head-on collisions of ultra-relativistic protons and electrons, we can safely neglect the transverse momentum of the proton elementary constituents. Nonetheless, we have no idea of the probability to find the four-momentum of the hypothetical constituent (k_q) between xP and $(x + dx)P$ inside a proton. We call the corresponding probability density function $f_q(x)$ the **parton distribution function** (see Fig. 7.31). R. Feynman introduced the **partons** in 1969 to indicate the elementary constituents of the nucleon. The hypothesis behind the parton model is that:

> the proton is made of many constituents (partons), whose mass is negligible compared with the proton mass.

Is this statement supported by the SLAC data? The focus of Feynman's analyses were events that can probe the dimension and mass of a parton carrying a fraction x of P: events with large Q^2 but a finite Q/x ratio. If the partons are light particles, we can compute the $e^- q \rightarrow e^- q$ cross-section using QED. The result follows from the first order Feynman diagram describing the t-channel scattering of two massless particles (Larkosky, 2019):

$$\frac{d\sigma}{d\hat{t}} = \frac{2\pi\alpha^2 Q_f^2}{\hat{s}^2}\left(\frac{\hat{s}^2 + \hat{u}^2}{\hat{t}^2}\right). \tag{7.34}$$

This equation holds in the hypothesis that the partons are spin-1/2 particles like the electrons because the corresponding amplitude is:

$$\frac{1}{4}\sum_{\text{spin}}|\mathcal{M}|^2 = 2e^4 Q_i^2 \left(\frac{\hat{s}^2 + \hat{u}^2}{\hat{t}^2}\right) \tag{7.35}$$

[13]The hat on top of the variables is useful to avoid confusion with Mandelstam's variables in the *proton*-electron scattering.

where Q_i is the charge of the i-th parton, and $\hat{s}, \hat{t}, \hat{u}$ are the Mandelstam variables for the parton-electron scattering.[13] You can try to demonstrate eqn 7.34 using the crossing theorem (see Exercise 7.9)but you cannot miss Feynman's great move: he attacked a non-perturbative QCD problem, recasting it as a first-order QED perturbative calculation.

We now write the Mandelstam variables as a function of kinematic observables neglecting the electron and proton masses.

$$\hat{s} = (k_e + k_q)^2 = (k_e + xP)^2 \simeq 2xk_e \cdot P \simeq xs$$
$$\hat{t} = (k_e - k'_e)^2 = q^2 = -Q^2$$
$$\hat{u} = -\hat{t} - \hat{s} = Q^2 - xs$$

where k_q is the parton four-momentum and s is the squared inviariant mass (first Mandelstam variable) of the electron-proton scattering. The last equality comes from $\hat{s} + \hat{t} + \hat{u} = 0$ for massless particles, which is demonstrated in Exercise 7.4.

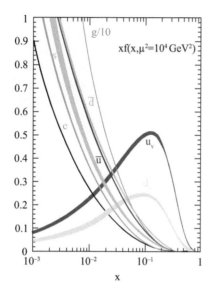

Fig. 7.13 Parton distribution function in a nucleon expressed as $xf(x, Q^2)$. The partons corresponding to the valence quarks are labeled u_v and d_v. The gluon contribution ($g/10$) is reduced by a factor of 10 to ease the reading. The functions are computed for $\mu^2 \simeq Q^2 = 10^4$ GeV^2. The width of the lines indicates the estimated errors. Reproduced with permission from Zyla et al., 2020.

Under the assumptions of the parton model – partons are nearly massless particles with spin-1/2 – and using the previous formulas for $\hat{s}$, $\hat{t}$, and $\hat{u}$, eqn 7.34 reads:

$$\frac{d\sigma}{dQ^2} = 2\pi\alpha^2 Q_q^2 \left[\frac{1 + \left(1 - \frac{Q^2}{xs}\right)^2}{Q^4} \right] . \tag{7.36}$$

Since we do not know the momentum fraction of the interacting parton x, we must resort to the parton distribution functions and write:

$$\frac{d^2\sigma}{dQ^2 dx} = 2\pi\alpha^2 Q_q^2 f_q(x) \left[\frac{1 + \left(1 - \frac{Q^2}{xs}\right)^2}{Q^4} \right] . \tag{7.37}$$

This is the electron–parton scattering cross-section in Feynman's parton model. Equation 7.37 seems quite a miserable result because depends on

x, which cannot be measured. This is false, indeed. Since x can be linked to the scattering angles:

$$k'_q + k'_e = k_e + xP \implies k'_q = (k_e - k'_e) + xP = q + xP. \tag{7.38}$$

The partons are nearly massless,

$$(k'_q)^2 \simeq 0 = (q + xP)^2 = q^2 + 2xq \cdot P \tag{7.39}$$

and the parton momentum fraction is:

$$x = \frac{Q^2}{2q \cdot P} = \frac{k_e \cdot k'_e}{(k_e - k'_e) \cdot P}. \tag{7.40}$$

Please do not forget that we are considering a $e^- q \to e^- q$ scattering. This is the scattering between an electron and a parton q, and the final particles are on-shell. Otherwise, we could not assume $(k'_e)^2 \simeq 0$. A well-designed experiment[14] can measure P by the proton energy. q and $Q^2 = -q^2$ are given by eqn 7.30. We then get x by the kinematics of the electrons and we can merrily forget the messy hadronic system X.

Since a proton is made of several partons, the master formula of the **Feynman parton model** for an electron–proton scattering is:

$$\frac{d^2\sigma}{dx\,dy} = 2\pi\alpha^2 s \left[\sum_q Q_q^2 x f_q(x) \right] \frac{1 + (1-y)^2}{Q^4}$$

$$= 2\pi\alpha^2 s F_2(x) \frac{1 + (1-y)^2}{Q^4} \tag{7.41}$$

where

$$y \equiv \frac{Q^2}{xs}. \tag{7.42}$$

Equation 7.41 is just eqn 7.37 expressed as a function of y and summed over all partons inside the proton. Unlike eqn. 7.37, which holds for a single parton, eqn 7.41 is the actual cross-section measured at SLAC looking at the electron-proton DIS scattering. $F_2(x)$ is called the F_2 **structure function** and, according to this model, is a sole function of x. This feature is called **Bjorken's scaling** because was conjectured by Bjorken in 1969: the proton looks the same for the photon probe no matter how hard the proton is struck. More rigorously, the structure function in the limit $Q^2 \to +\infty$ depends only on the finite Q^2/x value. Then, the proton structure function does not depend on the strength of the hard scattering.[15]

The essence of Feynman's model is encoded in the $[1 + (1 - y)^2]$ term. This term describes the scattering of (massless) electrons with a point-like massless fermion. Feynman and Bjorken thus demonstrated that:

[14] For a detailed review of the pioneering SLAC-MIT experiments and their successors see (Taylor, 1991). The scientists that designed and led them – J. Friedmann, H.W. Kendall, and R.E. Taylor – were awarded the 1990 Nobel Prize for Physics for the first empirical evidence of the existence of quarks.

[15] Bjorken's scaling is an approximation that holds up to Q^2 corrections. These corrections are mild and follow a logarithmic trend in the experimental data while most QFT predict a strong violation of the scaling. This is not true in asymptotically free theories, where the corrections are logarithmic. We know now that the only QFT that show asymptotic freedom in the four-dimensional Minkowski space-time are gauge theories based on non-abelian groups.

> the proton or any other hadron is made of many elementary constituents called partons. The **partons** are point-like particles with spin-1/2 and can be considered massless if the scattering is hard ($Q^2 \to \infty$). Each of these particles brings a fraction x of the total hadron momentum and the corresponding x distribution (the **structure function**) depends on x, only.

Going back to the machine gun firing the bag, we can claim neither that the bag is full of styrofoam nor that three heavy iron balls are located in the styrofoam. The outcome of the gunfire is that the sack contains styrofoam and a large number of tiny marbles, whose mass is negligible compared with the mass of the material inside the sack.[16]

Are these tiny marbles Gell-Mann's quarks? First, they are spin-1/2 fermions because eqns. 7.34 and 7.41 hold only for these particles. Besides, the structure functions can be measured experimentally and plugged into eqn 7.41 to get information about the parton charges. Finally, we can sum all the x of the partons to check whether they sum to 1. These tests were performed in the 1970s and 1980s and the results are summarized below:

[16]I draw this example after a conversation with C. Oleari, a renowned expert of perturbative QCD. The original example is too gory to be published but, if you are interested, you may get the full version from him.

> the parton model is fully compatible with QCD because a proton is not made of three massive quarks but is a complex system. It has three low-mass **valence quarks** and a **sea** of gluons and quark–antiquark pairs. The charge and spin of the partons are the same as the quarks. Since a quark–antiquark charge sums to zero and gluons are neutral particles, the sea charge is zero and the net charge of the hadron is determined by the valence quarks, only. The valence quarks of a proton and a neutron are *uud* and *udd*, respectively. The sum of the x of all quarks is $\sim 0.5P$ and the rest of the momentum is brought by the gluons.

Therefore, the identification of partons with quarks is fully appropriate and there is no need to introduce the constituent mass and the like: the only mass that counts is the QCD quark mass ("current mass"), quoted in Tab. 7.1 and at the beginning of this book.

7.8 Jets and QCD predictions

If QCD is the correct theory of strong interactions, the core of the proton must be richer than a semiclassical state of three point-like particles, as seen in Sec. 7.7. At the dawn of QCD, physicists did not exploit the hadronic system, trying to get hints of the main QCD features only using isolated particles – generally leptons – in deep inelastic scattering. In this way, the theory was surely perturbative because the momentum transfer was higher than Λ_{QCD} and the hadronization regime was sidestepped. These techniques were important to understand the meaning of Bjorken's scaling, the evolution of the parton distribution functions, and to gain evidence of asymptotic freedom. Later on, the study of the post-hadronization final state became central for the development of the theory and it is the focus of this section.

7.8.1   Spatial distribution of hadrons

The qualitative discussion that brought us to Fig. 7.10 is in contradiction with the early views of strong interactions. Following Wigner's considerations in Sec. 7.1, the strong forces should be turned on when the wavefunctions of two particles significantly overlap. If the energy of two particles is very high and the two particles collide so that their wavefunctions occupy nearly the same volume, the strong interactions should produce a wealth of hadrons, somehow uniformly distributed in space, and the number of hadrons per unit surface should be proportional to the center-of-mass (CM) energy. This does not happen at all at $\sqrt{s} > \Lambda_{\text{QCD}}$ because the gluon cascade tends to be collinear to the final state quarks and the directions of the hadrons are clustered near the direction of flight of those quarks. These streams of collimated hadrons are called **jets** and are very common in collider experiments (see Sec. 4.6.2). In a sense, the jet properties (number, direction, energy, number of particles per unit surface, total transverse momentum, etc.) are the most prominent observables to test the predictions of QCD.

Perturbative QCD provides a simple tree-level formula for the $e^+e^- \to q\bar{q}g$ process, that is the emission of a gluon together with a quark–antiquark pair:

$$\frac{d\sigma}{dx_1 dx_2}(e^+e^- \to q\bar{q}g) = \sigma_0 \left(3 \sum Q_q^2\right) \frac{2\alpha_s}{3\pi} \frac{x_1^2 + x_2^2}{(1-x_1)(1-x_2)}. \quad (7.43)$$

The x variables are the momentum fraction brought by the quark (x_1) and the antiquark (x_2). The gluon momentum fraction (x_3) is then fixed by four-momentum conservation. Q_q is the quark electric charge and 3 is the number of colors. σ_0 is the QED cross-section for $e^+e^- \to \mu^+\mu^-$, that is eqn 6.89. The weird features of QCD show up in the singularities that arise when any of the x goes to 1. The singularity testifies for the transition from the perturbative to the non-perturbative regime. If the entire momentum is taken by a single quark, the other particles run into the non-perturbative regime in a very early stage of the gluon cascade and eqn 7.43 becomes unreliable. On the other hand, in most cases, the energy is fairly shared among the quark, the antiquark, and the gluon, and three well-separated jets appear in the detector. In practice, if the transverse momentum of the gluon is significantly larger than 1 GeV, the jet associated with the gluon can be easily detected and the scattering produces a **three-jet event**. Otherwise, the two collinear jets look like a single object, and the gluon is incorporated into the leading jet. The scattering then produces a fake two-jet event. This is the reason why:

> the largest the transverse momenta of the quarks (or the gluon), the cleanest their association with the jets in the detector.

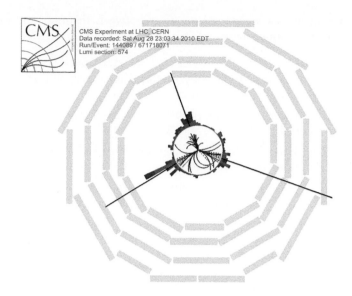

In general, if a QCD experimental observable can be consistently predicted by perturbative QCD, the observable is said to fulfill **infrared and collinear (IRC) safety**. All modern tests of QCD predictions in the $\mu \gg \Lambda_{\mathrm{QCD}}$ regime are based on IRC observables (Muta, 2010).

7.8.2   The spin of the gluon

The appearance of **two-jet event** at colliders was observed for the first time in 1975. The experiments performed at SLAC and running at $\sqrt{s} = 7.5$ GeV observed well-defined pairs of jets that could be associated with the production of quark pairs. Three-jet events were observed at DESY in 1979 as high-x coplanar streams of particles. The three-jet event rate was in good agreement with the prediction of eqn 7.43 and the information available on $f_q(x, Q^2)$ at that time. A three-jet event from the CMS experiment at the LHC is shown in Fig. 7.14.

One of the most spectacular applications of IRC safe observables was the use of **thrust** to determine the spin of the gluon. The thrust τ is the maximum among x_1, x_2, x_3 in a three-jet events. We can compute the cross-section for the thrust taking eqn 7.43 and summing over all phase space points that provide a given value of τ (Larkosky, 2019). A collider detector can measure separately the distribution of x_1, x_2, and x_3, whose value strongly depends on the spin of the gluon. Experimental data accumulated at colliders and, in particular, at DESY in Hamburg in the late 1980s, proved conclusively that the gluon is a spin-1 particle.

7.8.3   Modern techniques

Even if the non-perturbative regime of QCD is still largely unknown, the study of QCD predictions at $\mu > \Lambda_{\mathrm{QCD}}$ (**perturbative QCD**) has reached an unprecedented level of sophistication. Nowadays, a QCD analysis is performed as a cooperative effort between theorists and experimentalists in the following way. The researchers:

- compute the rate and kinematics of final state particles at parton level (quarks and gluons) by solving the corresponding QCD Feynman diagrams at a given order in perturbation theory;
- implement the corresponding probability distribution functions in a Monte Carlo generator (see Sec. 3.11);
- match the parton kinematics with semi-empirical hadronization models to obtain the hadrons expected in the final state;
- simulate the detector response and the jet reconstruction algorithms as described in Chaps 3 and 4 to build a realistic model of the detector data;
- compare IRC safe observables between data and Monte Carlo to validate the QCD predictions.

We will further discuss QCD techniques for high-energy physics (i.e. much above Λ_{QCD}) in Chap. 13, together with the role played by QCD in the discovery of the Higgs boson.

Exercises

(7.1) Show that a two-quark bound state is not a physical state in QCD.

(7.2) Demonstrate that the scattering amplitude of two green quarks is the same as for two red quarks.

(7.3) If Λ_{QCD} were close to the atomic energy scale (1 eV-1 keV), would you be able to read this book?

(7.4) Consider a $a + b \to c + d$ scattering. Show that at high energy $s + t + u = 0$ where s, t, u are the three Mandelstam variables defined in Chap. 6. Recast $s + t + u$ as a function of the CM energy and the scattering angle.

(7.5) * Suppose that only four quark exists. Compute the corresponding value of Λ_{QCD} and the value of μ where $\alpha_s \to +\infty$. Discuss the reliability of these types of calculations.

(7.6) The lowest order (i.e. tree-level) cross-section of $e^+e^- \to \mu^+\mu^-$ in QED can be written as

$$\frac{d\sigma}{d\cos\theta} = \frac{\pi\alpha^2}{2s}(1 + \cos^2\theta). \qquad (7.44)$$

Derive the corresponding formula for $e^+e^- \to q\bar{q}$.

(7.7) * Demonstrate that the scattering angle θ of eqn 7.44 can be written as

$$\cos\theta = \frac{t-u}{s} \qquad (7.45)$$

$$1 + \cos^2\theta = 2\frac{t^2 + u^2}{s^2}. \qquad (7.46)$$

(7.8) * Using the results of the two previous exercises show that:

$$\frac{d\sigma}{dt} = \frac{6\pi\alpha^2 Q_q^2}{s^2} \cdot \frac{u^2 + t^2}{s^2} \qquad (7.47)$$

for a $e^+e^- \to q\bar{q}$ scattering.

(7.9) * In a $a + b \to c + d$ process, a **crossing symmetry** is a transformation that moves the initial state fermion to the final state transforming it into an antiparticle and vice versa. A theorem by M. Gell-Mann and M.L. Goldberger (Gell-Mann and Goldberger, 1954) shows that the cross-section of the crossed process can be obtained from the original process rotating by 90° the Feynman diagram and exchanging the time axis with the space axis. For a $a(p_1)+b(p_2) \to c(p_3)+d(p_4)$ process, this is equivalent to replacing $p_2 \to -p_4$ and $p_4 \to -p_2$, where p_i are the four-momenta of the particles. Using this theorem derive eqn 7.34

(7.10) * Show that the u/s ratio for the partons in DIS is experimentally observable because

$$\frac{u}{s} = -\frac{k'_e \cdot P}{k_e \cdot P}. \qquad (7.48)$$

(7.11) * Compute, if possible, the energy where α and α_s have the same value.

(7.12) ** A jet is made of four charged pions and two π^0. Each particle has an energy of 15 GeV. Estimate the energy resolution of the jet with and without a tracking system, assuming that the detector has an e.m. and hadronic calorimeter with standard performance (see Chaps 3 and 4).

(7.13) * The differential cross-section for deep inelastic scattering between an electron and a photon is different from zero and has a form similar to eqn 7.41:

$$\frac{d^2\sigma}{dxdQ^2} = \frac{2\pi\alpha^2}{xQ^4} \left[(1 + (1-y)^2)F_2^\gamma(x, Q^2) \right.$$
$$\left. -y^2 F_L^\gamma(x, Q^2) \right] \qquad (7.49)$$

where $F_2^\gamma(x, Q^2)$ and $F_L^\gamma(x, Q^2)$ are the photon structure functions. Why is this cross-section different from zero? Can you observe Bjorken's scaling in photons, too?

(7.14) ** Design an experiment to observe free quarks if the confinement conjecture breaks down at 100 GeV. Describe the type of detectors and the resolutions that are needed.

8 Flavor symmetries and the quark models

[1]Nowadays, the terms "mesons" (in Greek, particles of intermediate mass) and "baryons" (heavy particles) are meaningless because they were coined before the discovery of heavy quarks like c or b. A meson made of heavy quarks (e.g. a $b\bar{b}$ bound state) has a mass bigger than baryons made, for instance, of u and d quarks (protons, neutrons, etc.).

8.1 Hadrons

The vast majority of particles discovered to date are bound states of quarks and gluons. This is a consequence of confinement since only multi-quark bound states are candidate color singlets: an isolated quark cannot be a singlet because it has a net color charge different from zero and cannot be invariant for $SU(3)_c$ transformations. The same consideration holds for a pair of quarks or a single gluon. But, for instance, a quark–antiquark pair, three bound quarks, or three bound antiquarks can fulfill the confinement condition and are good candidates to describe the physical states in strong interactions. The set of all color singlets are called **hadrons** although this term is often used to indicate the set of all **mesons** (quark–antiquark pairs), **baryons** (three-quark bound states), and **antibaryons** (three-antiquark bound states). The word "hadron" is inherited from ancient Greek and means stout: in this context, a hadron is a particle that interacts strongly.[1] Nearly all quantum chromodynamic (QCD) final states are either mesons or baryons. If we forgive the exotic states covered in Sec. 8.9, the set of the hadrons is just the union of all possible mesons and baryons.

You may wonder whether it is possible to find all QCD final states using QCD itself. To get the full list, we should solve the theory in the non-perturbative regime and identify all bound states, like in atomic physics. The total energy of the bound state would provide the mass of the corresponding hadron. QCD also provides the spin, intrinsic parity, C-parity, and T-parity of the bound state, if any. This is a task that particle physicists have just undertaken using non-perturbative tools and, in complex systems like the nuclei, it is still out of reach. On the other hand, since 1964 we have had at our disposal a rather precise classification method that works for hadrons. As expected from Chap. 5, this method is based on symmetries and, in particular, on the **flavor symmetry** of QCD.

> If all quarks have the same mass (degenerate quarks), the QCD scattering matrices do not depend on the type of quarks in the initial and final state. As a consequence, the transformations among the types (flavors) of quarks commute with the QCD Hamiltonian, whatever it is.

A Modern Primer in Particle and Nuclear Physics. Francesco Terranova, Oxford University Press.
© Francesco Terranova (2021). DOI: 10.1093/oso/9780192845245.003.0008

We already demonstrated this statement in Chap. 7, noting that the quark–gluon and gluon–gluon vertices only depend on the color charges. This flavor symmetry holds in quantum electrodynamics (QED), too.

The flavor symmetry is the pillar on which the time-honored **static quark models** are based. The models are built on a (quite rough) approximation that originates from the aforementioned symmetry:

> in the limit of degenerate quarks, i.e. quarks with equal masses, no QCD observable depends on the flavor of the quarks. As a consequence, any redefinition of the type of quark does not change the dynamics of the system.

This statement is equivalent to the previous one because the observables depend on the matrix element and the phase-space. The latter is the same for any flavor if the quark masses are degenerate. The flavor symmetry is an approximation of reality because $m_u \ll m_t$ by five orders of magnitude.

If quarks were really degenerate in mass, the non-relativistc wavefunction of the bound system would be:[2]

$$\psi(\mathbf{x}, t) = \psi_{space}\psi_{spin}\psi_{color}\psi_{flavor}. \tag{8.1}$$

Note that final states are considered stationary states and, hence, $|\psi(\mathbf{x}, t)|$ is equal to $|\psi(\mathbf{x})|$. The term ψ_{flavor} provides the probability of finding a given quark in a given flavor. For the wavefunction of a single quark, we have:

$$\psi(\mathbf{x}) = \psi_u(\mathbf{x}) \begin{pmatrix} 1 \\ 0 \\ 0 \\ 0 \\ 0 \\ 0 \end{pmatrix} + \psi_d(\mathbf{x}) \begin{pmatrix} 0 \\ 1 \\ 0 \\ 0 \\ 0 \\ 0 \end{pmatrix} + \psi_s(\mathbf{x}) \begin{pmatrix} 0 \\ 0 \\ 1 \\ 0 \\ 0 \\ 0 \end{pmatrix}$$

$$+ \psi_c(\mathbf{x}) \begin{pmatrix} 0 \\ 0 \\ 0 \\ 1 \\ 0 \\ 0 \end{pmatrix} + \psi_b(\mathbf{x}) \begin{pmatrix} 0 \\ 0 \\ 0 \\ 0 \\ 1 \\ 0 \end{pmatrix} + \psi_t(\mathbf{x}) \begin{pmatrix} 0 \\ 0 \\ 0 \\ 0 \\ 0 \\ 1 \end{pmatrix}. \tag{8.2}$$

Here, $|\psi_j(\mathbf{x})|^2$ should be interpreted as the probability density of finding that quark in the "j" flavor state.

Several notes of caution are in order. Equation 8.2 makes no sense if quarks are non-degenerate because particles with different masses cannot be considered different states of the same particle. Even worse, you cannot demand the amplitudes to be invariant for transformations among flavors because every time you change the type of quark, you also change the available phase space.

[2]We use the non-relativistic approximation that factorizes the spin and space part of the wavefunction for sake of simplicity and because the bound states are often non-relativistic. The same results hold in the full relativistic treatment.

The "historical quark models" addressed this issue in a naïve but effective way. Consider, for instance, just the up- and down-quark. The mass difference between them is large because $m_d \simeq 2m_u \simeq 4$ MeV but both quarks are much lighter than the lightest mesons, the pions, whose masses range from 135 MeV to 139 MeV. It means that the mass of the pion does not depend on the actual mass of the quarks but on something else. Lacking QCD, the founding fathers of the quark models ignored this "something else" and inflated the mass of the quarks to make them comparable with the pion mass. They called these inflated masses the **constituent quark mass**. In this way, they used the fundamental (but yet unknown) flavor symmetry of QCD to classify the hadrons and the ancillary hypotheses on the inflated quark mass to estimate the mass of the hadrons. Going from two to three quarks, the issues get more complicated because $m_s \gg m_d$ and inflating the masses is not enough to describe the energy of the hadrons even if the classification, the predictions on the particle spin, parity, and C-parity were in perfect agreement with the experiments. After a few years, scientists realized that all the advantages of the static quark model stemmed from the use of the flavor symmetry and all the disadvantages from the ancillary hypotheses on the constituent mass and other ad-hoc statements introduced to reproduce the hadron properties.

In this book, we will study the quark models in their modern interpretation, leaving aside the ancillary hypotheses.

> An F-quark model is an approximate classification of the hadrons. The classification is performed under the hypothesis that only F quark flavors exist and that the masses of these quarks are all the same (degenerate).

We will explain why the quark models work so well in Sec. 8.5. In the meanwhile, we take note that nearly all hadrons we observe on earth (protons and neutrons) are made of up and down quarks and we start our classification with $F = 2$.

This makes sense because, if the energy available is not enough to produce a heavy quark, such a type of quark is unavailable. The most striking examples are **nuclear reactions**, where the energy available is $\ll m_s$. Nuclear reactions produce only up or down quarks, which are accidentally much lighter than the strange quark and have a mass much smaller than any hadron. Within this (excellent!) approximation, the up and down quarks can be safely considered *massless*[3] and, hence, *degenerate*. Again, since the QCD vertices of Fig. 7.3 do not distinguish between up and down quarks, the scattering amplitude must be independent of any redefinition of what is "up" or "down." This approximate form of the QCD flavor symmetry is called the **two-quark approximation** and the corresponding symmetry of the Hamiltonian is called the **isospin symmetry**. The formal definition of the isospin symmetry is:

---

[3] The approximation of massless quarks is called the **chiral approximation** and is briefly discussed in Chap. 13.

> in the limit $F = 2$ and $m_u = m_d$, the transformations of ψ_{flavor} that redefine the flavor of quarks, i.e. the transformations
>
> $$\psi_{flavor} \to e^{ia_j T_j} \psi_{flavor} \qquad (8.3)$$
>
> are a symmetry of the strong interaction Hamiltonian. These transformations are represented by unitary operators U belonging to the $SU(2)$ group and U commutes with the Hamiltonian.

This symmetry is the cornerstone of the **two-quark model** of hadrons and nuclear physics, too. In particle physics, it provides a powerful classification method of the hadrons made of up and down quarks.

8.2 Classification of stationary states in atomic physics

Before developing the two-quark model, we recap how a symmetry of the Hamiltonian can be employed to classify all stationary states of a non-relativistic quantum system without solving the corresponding Schrödinger equation. A well-known example is the **rotation symmetry** in atomic physics that originates the **composition rules** for the angular momentum.[4] Consider two point-like particles inside a central scalar field (e.g. two electrons with an angular momentum j_1 and j_2). The wavefunction of each particle can be expressed as $|j_1, j_{z1}\rangle$ and $|j_2, j_{z2}\rangle$. The two-particle wavefunction is the combination of the two states and is given by the **Clebsch–Gordan formula**:

$$|j, j_z, j_1, j_2\rangle = \sum_{j_{z1}, j_{z2}} |j_1, j_{z1}, j_2, j_{z2}\rangle \, \langle j_1, j_{z1}, j_2, j_{z2}, |j_1, j_2, j, j_z\rangle. \qquad (8.4)$$

$\langle j_1, j_{z1}, j_2, j_{z2}, |j_1, j_2, j, j_z\rangle$ are real numbers (the **Clebsch–Gordan coefficients**), whose value only depends on the group of the rotations. This is the $SO(3)$ group: 3×3 real-valued orthogonal matrices. Using the tools of representation theory, it is possible to show that the $SO(3)$ group is (locally) isomorphic to $SU(2)$ and, in the following, we will consider the two groups as if they were identical. For two particles with $j = 1/2$, this formula corresponds just to eqns 5.46 and 5.47: one wavefunction with $j = 0$ (singlet) and three wavefunctions with $j = 1$ (triplet).

As noted in Sec. 5.4, a rotation of the axes of the reference frame by an orthogonal matrix R transforms the wavefunction ψ into $U(\theta)\psi = e^{-i\boldsymbol{\theta} \cdot \mathbf{L}}\psi$. We can now apply all possible operators U by scanning all possible values of $\boldsymbol{\theta}$ and see what happens to the wavefunctions:

- the singlet wavefunction $|0, 0, 1/2, 1/2\rangle$ does not change except for a unit phase.[5] The singlet $|0, 0, 1/2, 1/2\rangle$ ($j = 0$ state) is then invariant for rotations in space.

- a function belonging to the triplet changes in such a way that we will always be able to express any transformed function as a linear

[4]For the readers not acquainted with quantum mechanics (QM), the composition rules of the angular momentum are summarized in Appendix B and derived in Griffiths, 2018.

[5]Please keep in mind that in Chap. 5 we denoted $|0, 0, 1/2, 1/2\rangle$ as $\alpha(0, 0)$.

combination of $|1, -1, 1/2, 1/2\rangle$, $|1, 0, 1/2, 1/2\rangle$, and $|1, 1, 1/2, 1/2\rangle$ ($j = 1$ states). Rotations in space over any of the triplet wavefunctions then generate a three-dimensional vector space.

These classes cover all the wavefunctions we can create from two $j = 1/2$ particles in a central potential. Since the rotations are a symmetry of the Hamiltonian, it can be shown (see Exercise 8.12) that:

> states with the same value of j have the same energy and are called **degenerate states**.

This famous result can be derived in many ways but in Exercise 8.12 we obtain it without solving the Schrödinger equation of the atom. You can guess now why we recalled the technique used to classify stationary states in atomic physics: we are planning to do the same for the quarks replacing the friendly Schrödinger equation with the weird QCD formulas. And we will not even try to solve them!

8.2.1   A glimpse of group representation theory*

The changes of a hydrogen wavefunction due to rotations are a good playground to practice with group representation theory. How does Sec. 8.2 sound to mathematicians? $U(\theta) \equiv e^{-i\theta \cdot \mathbf{L}}$ is a unitary operator and represents one element of the rotation group $SO(3)$. Since $SO(3)$ is a smooth, continuous group, we can go from one element of the group to the other by changing θ. Suppose that U belongs to an irreducible representation (irrep), as defined in Sec. 7.5. In this case, we can be sure that applying $U(\theta)$ to a wavefunction by spanning all values of θ produces a set of wavefunctions that constitutes a vector space. U applied to a singlet is just a one-dimensional representation of $SO(3)$ because the vector space is just $|0, 0, 1/2, 1/2\rangle$ multiplied by a complex number, that is, a 1-dimensional vector space, whose base is $|0, 0, 1/2, 1/2\rangle$. If we apply U to a function of the triplet (e.g. $|1, 1, 1/2, 1/2\rangle$) and run over θ, we end up with a 3-dimensional vector space, whose base is $|1, -1, 1/2, 1/2\rangle$, $|1, 0, 1/2, 1/2\rangle$, and $|1, 1, 1/2, 1/2\rangle$. Every wavefunction we find during the run can be expressed as a linear combination of $|1, -1, 1/2, 1/2\rangle$, $|1, 0, 1/2, 1/2\rangle$, and $|1, 1, 1/2, 1/2\rangle$. U applied to the triplet functions belongs then to the 3-dimensional irrep of $SO(3)$. In conclusion, the physics of Sec. 8.2 signal to mathematicians that:

> if two states have two different energies, they belong to the vector spaces of two different irreducible representations (irreps) of $SO(3) \simeq SU(2)$. The catalog of all irreps fixes the set of states with different energies. If two states have the same energy, they are degenerate states belonging to the same vector space, and the number of degenerate states for a given irrep is given by the dimension of the vector space.

Example 8.1

For a wavefunction of total angular momentum j the number of degenerate states or, if you prefer, the dimension of the irrep is $2j + 1$. This is the formal reason why a triplet $(j = 1)$ has three degenerate states while a singlet $(j = 0)$ cannot have more than one degenerate state. The only way to break the degeneracy, that is, to produce states with small differences in energy starting from degenerate states, is to break the underlying symmetry. In atomic physics, this is done, for example, adding an electric or magnetic field that creates a preferential axis and, thus, breaks the rotation invariance of the Hamiltonian. Atomic physicists employ this technique to produce the Stark and Zeeman effect, respectively.

The classification described for singlets and triplets can be extended to an arbitrary value of j by using eqn 8.4. All possible wavefunctions generated from the combination of two identical particles with angular momentum j_1 and j_2 can be classified using a quantum number j that runs from $|j_1 - j_2|$ to $j_1 + j_2$. Each value of j provides a class of functions, whose linear combinations generate a vector space of dimension $2j + 1$. As expected,

> in the language of group representation theory, the angular momentum j classifies all irreducible representations of $SU(2)$ in the vector space V of the wavefunctions ψ. The dimension of V is $2j + 1$. If ψ_1 and ψ_2 belong to V, they are eigenstates of the Hamiltonian with the same eigenvalue, i.e. they are degenerate states.

8.3   Baryons in the two-quark model

If we account only two quarks to assemble the hadrons as bound states, all possible wavefunctions can be classified looking at how they transform under $SU(2)$. This classification provides states with the same energy, that is, *hadrons with the same mass*, without solving the QCD equations of motion.

We start our classifications from baryons, that is, the bound state of three quarks. Since these quarks come in two types ("up" and "down") the ψ_{flavor} part of the wavefunction of eqn 8.1 belong to the $j = 1/2$ vector space. Following a historical notation, we label the quantum number associated with flavor as I and we call it **isospin**. The name was chosen in 1932 by W. Heisenberg[6] and is very unfortunate: the reader should be aware that isospin has nothing to do with spin. It shares with real spin (intrinsic angular momentum) only the combination rule of eqn 8.4. In the two-quark model, we describe the wavefunction ψ_{flavor} of the quarks by the third component of the isospin and identify the up

[6]You may be puzzled by this name, coined 32 years before the discovery of the quarks and chosen by a physicist that ignored their existence. I will disclose this minor conundrum in Chap.9.

and down quarks with these components:

$$|I = 1/2, I_z = 1/2\rangle : \quad \text{up quark} \tag{8.5}$$

$$|I = 1/2, I_z = -1/2\rangle : \text{down quark.} \tag{8.6}$$

A baryon is a bound state of three quarks. If we combine the first and second quark, we get:

$$\frac{1}{2} \otimes \frac{1}{2} = 0 \oplus 1 \tag{8.7}$$

that is, a singlet and a triplet wavefunctions. The triplet wavefunctions[7] can be combined with the wavefunction of the third quark giving:

$$1 \otimes \frac{1}{2} = \frac{1}{2} \oplus \frac{3}{2}. \tag{8.8}$$

Baryons in the two-quark model can be classified into a two-dimensional ($2I + 1$ for $I = 1/2$) and a four-dimensional ($2I + 1$ for $I = 3/2$) vector space and we expect six different particles: two with low mass ($I = 1/2$) and four with higher masses ($I = 3/2$). Particles with the same isospin are "degenerate states" and must have the same mass.

The wavefunction of these particles is given by the Clebsch–Gordan formula of eqn 8.4 because $SU(2)$ is virtually identical to $SO(3)$:

$$\left|\frac{3}{2}, \frac{3}{2}, 1, \frac{1}{2}\right\rangle = \left|1, 1, \frac{1}{2}, \frac{1}{2}\right\rangle \tag{8.9}$$

$$\left|\frac{3}{2}, \frac{1}{2}, 1, \frac{1}{2}\right\rangle = \sqrt{\frac{2}{3}}\left|1, 0, \frac{1}{2}, \frac{1}{2}\right\rangle + \sqrt{\frac{1}{3}}\left|1, 1, \frac{1}{2}, -\frac{1}{2}\right\rangle \tag{8.10}$$

$$\left|\frac{3}{2}, -\frac{1}{2}, 1, \frac{1}{2}\right\rangle = \sqrt{\frac{2}{3}}\left|1, 0, \frac{1}{2}, -\frac{1}{2}\right\rangle + \sqrt{\frac{1}{3}}\left|1, -1, \frac{1}{2}, \frac{1}{2}\right\rangle \tag{8.11}$$

$$\left|\frac{3}{2}, -\frac{3}{2}, 1, \frac{1}{2}\right\rangle = \left|1, -1, \frac{1}{2}, -\frac{1}{2}\right\rangle \tag{8.12}$$

and

$$\left|\frac{1}{2}, \frac{1}{2}, 1, \frac{1}{2}\right\rangle = -\sqrt{\frac{1}{3}}\left|1, 0, \frac{1}{2}, \frac{1}{2}\right\rangle + \sqrt{\frac{2}{3}}\left|1, 1, \frac{1}{2}, -\frac{1}{2}\right\rangle \tag{8.13}$$

$$\left|\frac{1}{2}, -\frac{1}{2}, 1, \frac{1}{2}\right\rangle = -\sqrt{\frac{2}{3}}\left|1, -1, \frac{1}{2}, \frac{1}{2}\right\rangle + \sqrt{\frac{1}{3}}\left|1, 0, \frac{1}{2}, -\frac{1}{2}\right\rangle. \tag{8.14}$$

The value of I_z provides the number of up and down quark in the bound state because I_z is an additive quantum number, like a conventional angular momentum. The **electric charge** of the particle is given by the sum of the electric charge of the quarks. Hence, the state $I_z = +3/2$ will have three up quarks and:

$$Q = +\frac{2}{3}e + \frac{2}{3}e + \frac{2}{3}e = +2e. \tag{8.15}$$

This particle has twice the electric charge of the proton. The $I = 3/2$ particles are called Deltas ($\Delta^-, \Delta^0, \Delta^+, \Delta^{++}$) and have the same mass: 1232 MeV. The $I = 1/2$ particles are the proton and neutron and have (nearly) the same mass: 938 and 939 MeV, respectively. Equations 8.13 and 8.14 show that the proton is a bound state of uud quarks and the neutron of udd. You can read it in the right-hand side of eqn 8.13 because a state with $I = 1$ and $I_z = 0$ is a pair of quarks with opposite isospin (ud), while a state with $I = 1/2$ and $I_z = 1/2$ is just u. The term $\left|1, 0, \frac{1}{2}, \frac{1}{2}\right\rangle$ in eqn 8.13 is thus made of two up quark and one down quark. Likewise, $\left|1, 1, \frac{1}{2}, -\frac{1}{2}\right\rangle$ is a uu pair accompanied by a down quark.[8]

The mass of the proton and the neutron are not the same because the two-quark model is based on approximations: the mass of the up quark is lighter than m_d and the two particles have different electric charges so that QED effects may also play a role. The determination of the difference of mass between the neutron and the proton is beyond the scope of the quark model based on the flavor symmetry. It can be addressed only by non-perturbative QCD techniques and, in particular, by **lattice QCD** (Gattringer and Lang, 2009). The proton–neutron mass difference was computed using the full QCD+QED Lagrangian in 2015 and results are in excellent agreement with the experimental measurements (Borsanyi *et al.*, 2015).

The quark models not only predict the number of particles with degenerate mass but also their spin. The total wavefunction of a baryon is

$$\psi_{space}\psi_{spin}\psi_{color}\psi_{flavor} \tag{8.16}$$

and ψ_{spin} can be derived by combining three particles (quarks) with $1/2$ spin. Note that in this case, we are speaking about the *real* spin, the intrinsic angular momentum of the quarks. Again, the combination of the spin wavefunction gives $1/2$ and $3/2$ states. The proton and neutron must be spin-$1/2$ particles and the Δ must be spin-$3/2$ particles. A spin different from these values would falsify the two-quark model.

8.3.1 Physical states

If we combine three quarks of two flavors, we have in principle 2^3 possible states but we considered up to now only six particles. In fact, only a fraction of all possible states corresponds to physical states. This is a subtle manifestation of the Pauli exclusion principle. Since we are considering massless quarks, an up quark is just a down quark in a different flavor state. A baryon wavefunction is a combination of identical fermions in different space, spin, color, and flavor states and must obey the Pauli exclusion principle. As a consequence,

> the baryon wavefunction must be antisymmetric for the exchange of two quarks.

Confinement requires ψ_{color} to be a color singlet. The theory of representation of $SU(3)$ (see also Sec. 8.6) provides the only combination of red, green, and blue wavefunctions that is invariant by $SU(3)$ transformations:

[8] A common source of confusion among novices is the difference between a particle made of uud and the corresponding wavefunction. The proton is a baryon made of uud. We may be tempted to write it as $|uud\rangle$ but, even if you do it to shorten the formulas, please remember that its wavefunction in the two-quark model is given by eqn 8.13. This equation is not equivalent to $|u\rangle |u\rangle |d\rangle$ or "$|uud\rangle$," unless you specify what $|uud\rangle$ really means.

$$\psi_{color} = \frac{1}{\sqrt{6}} \left(|RBG\rangle - |BRG\rangle + |BGR\rangle - |GBR\rangle + |GRB\rangle - |RGB\rangle \right).$$
(8.17)

$|RGB\rangle$ is a ψ_{color} such that the first quark is red (it has a red color charge equal to $+1$), the second is green, and the third is blue. If we exchange two quarks in eqn 8.17:

$$\psi_{color} \to (-1)\psi_{color}$$
(8.18)

and the function is **totally antisymmetric**.

The space component of the wavefunction ψ_{space} describes the probability of finding the quarks in space. This was already discussed in Chap. 5 when dealing with the intrinsic parity of the baryon. For the case under consideration (low-lying bound states) all internal angular momenta are zero and:

$$\psi_{space} \to (+1)\psi_{space}$$
(8.19)

if we exchange two quarks. Note that a bound state of three point-like particles in a central potential can be described in QM by the $Y_{l,m}(\theta, \phi)$ Laplace spherical harmonics of eqn 5.37. Spatially excited bound states can be classified using the same quantum number of atomic physics: the principal quantum number n that generates "radially excited" baryons and the angular quantum number l. The quark model does not provide firm guidance on the energy of these excited states, as we will see in Sec. 8.6.1. Since $\psi = \psi_{space}\psi_{spin}\psi_{color}\psi_{flavor}$ must be antisymmetric:

> the Pauli exclusion principle dictates $\psi_{spin}\psi_{flavor}$ to be symmetric for the exchange of any quark pair in any low-lying $(n = 0, l = 0)$ baryon.

The theory of representation of $SU(2)$ gives us the tools to evaluate how ψ_{spin} and ψ_{flavor} change if we exchange two quarks. In the two-quark model, flavor is modeled by two $I = 1/2$ particles and the spin of the quark is $1/2$. Using eqn 8.4 to combine first two $1/2$ states ($1/2 \otimes 1/2 = 0 \oplus 1$) and, then, these states with the third particle, we get:

$$\frac{1}{2} \otimes \frac{1}{2} \otimes \frac{1}{2} = \frac{3}{2}_{(S)} \oplus \frac{1}{2}_{(M12)} \oplus \frac{1}{2}_{(M23)}.$$
(8.20)

The subscript S indicates that the wavefunction is fully symmetric for the exchange of any of the quark pairs. This can be explicitly demonstrated using eqn 8.4, as shown in Exercise 8.8. ψ_{spin} with spin $3/2$ and ψ_{flavor} with isospin $I = 3/2$ are symmetric. As a consequence,

$$\psi = \psi_{space}\psi_{spin}\psi_{color}\psi_{flavor} = Y_{00}\psi_{spin}^{S=3/2}\psi_{color}^{singlet}\psi_{flavor}^{I=3/2}$$
(8.21)

is antisymmetric and obeys the Pauli exclusion principle. Hence,

> the four wavefunctions with $I = 3/2$ represent physical states (particles) with spin $3/2$. These particles are the Δ particles described by eqs 8.9-8.12.

The two wavefunctions with $S = 1/2, S_z = \pm 1/2$ (or $I = 1/2, I_z = \pm 1/2$) are not eigenstates of the exchange operator. If we swap the first and second quark in the first (second) $1/2_{M12}$ wavefunction, we will get the same (opposite) wavefunction because $M12$ stands for "symmetric (anti-symmetric) for the exchange of the first and second particle." However, if we exchange the second and third particle we get a different function. These states are called **mixed symmetry** states. We can build fully symmetric wavefunctions from linear combinations of the $M12$ and $M23$ functions. In QM, this technique is called **symmetrization** and it gives two additional physical states to the quark model:

$$\psi_{spin}\psi_{flavor} = \left(\psi_{spin}^{S=1/2,M12}\psi_{flavor}^{I=1/2,M12} + \psi_{spin}^{S=1/2,M23}\psi_{flavor}^{I=1/2,M23}\right.$$

$$\left. +\psi_{spin}^{S=1/2,M13}\psi_{flavor}^{I=1/2,M13}\right) \tag{8.22}$$

where $\psi_{spin}^{S=1/2,M12}\,\psi_{flavor}^{I=1/2,M12}$ is a combination of $M12$ wavefunctions fully symmetric for the exchange of the first and second quark, etc. Note that we can define $\psi_{spin}^{S=1/2,M13}$ as a linear combination of the $M12$ and $M13$ wavefunctions because:

$$\frac{1}{2}_{(M13)} = \frac{1}{2}_{(M12)} + \frac{1}{2}_{(M23)}. \tag{8.23}$$

Similar considerations hold for $\psi_{flavor}^{I=1/2,M13}$. We encourage the reader to go through the Clebsch–Gordan formula of eqn 8.4 to verify the identities above (see Exercises 8.16 and 8.17).

The physical states are two particles with spin $1/2$ and $I = 1/2$. The former has $I_z = -1/2$ and represents a udd bound state (neutron), the latter has $I_z = 1/2$ and is a uud bound state (proton). The $\psi_{spin}\,\psi_{flavor}$ proton wavefunction can be written as:

$$\psi_{spin}\psi_{flavor} = \sqrt{\frac{1}{18}}\,[|u\uparrow u\downarrow d\uparrow\rangle + |u\downarrow u\uparrow d\uparrow\rangle - 2|u\uparrow u\uparrow d\downarrow\rangle$$

$$+ |u\uparrow d\uparrow u\downarrow\rangle + |u\downarrow d\uparrow u\uparrow\rangle - 2|u\uparrow d\downarrow u\uparrow\rangle$$

$$+ |d\uparrow u\uparrow u\downarrow\rangle + |d\uparrow u\downarrow u\uparrow\rangle - 2|d\downarrow u\uparrow u\uparrow\rangle] \tag{8.24}$$

where $|u\uparrow u\downarrow d\uparrow\rangle$ is a wavefunction representing the first quark with flavor u and $S_z = +1/2$, the second quark with flavor u and $S_z = -1/2$, the third quark with flavor d and $S_z = +1/2$. Equation 8.3.1 provides the full wavefunction of the proton in the two-quark model. Again, the quark model predicts not only the quark content of the particles but also their spin. No other symmetric combinations of spin and isospin can be built from the combinations of three quarks. Hence,

> for low-lying baryons ($l = 0$), the total number of physical states that obey the Pauli exclusion principle is 6, i.e. it is lower than the 2^3 combinatorial expectation.

The two quark model provides a detailed description of mesons, too: we will discuss the mesons in the two- and three-quark model in Sec. 8.7.

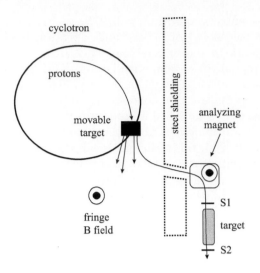

Fig. 8.1 Simplified layout of the Anderson–Fermi experiment at the University of Chicago. The protons from the cyclotron hit the movable target producing hadrons. A π^+ is steered toward the shielding groove by the fringe magnetic field of the cyclotron. The direction of the field is perpendicular to the figure and shown in the circle. The π^+ momentum is measured by the "analyzing magnet" that bends the pion toward the target. The survival rate of pions is given by the rates of the S1 and S2 scintillators. The original layout is available in Anderson, 1982.

8.4 The discovery of the $\Delta(1232)$

The two-quark model was formulated by M. Gell-Mann and G. Zweig in 1964, together with the three-quark model discussed in Sec. 8.6. At that time, the Δ states were already known together with their mass and spin. Physicists were aware of tens of other particles that we now interpret as excited states of the proton, neutron, Δ, or particles with s quarks. Most of these particles were discovered in the 1950s challenging the very concept of "elementary particle." The discovery of the low-lying Δ states marked the beginning of this proliferation era. About 15 years later, however, these candidate elementary particles were understood to be bound states of quarks.

The Δ was discovered in 1952 by H.L. Anderson and E. Fermi in a seminal experiment with pions. Pions were produced from a 450 MeV proton accelerator at the University of Chicago by the scattering of protons with a movable beryllium target. The final state particles (hadrons) were mostly pions with a contamination of protons and neutrons.

The accelerator employed there was a cyclotron, which has a rather strong (> 0.1 T) magnetic fringe field in the surrounding of the accelerator. The experimentalists were shielded from neutrons by an iron wall located between the control room and the accelerator (see Fig. 8.1). Fermi mapped the fringe field during the commissioning phase of the accelerator to identify where positive and negative pions would have impacted on the wall. He also installed the target on a remote-controllable trolley to get the beam intensity by measuring the heat released by the charged particles, that is, by temperature monitoring. In this way, they were able to maximize nearly in real time the flux of positive (negative) pions reaching the upstream (downstream) part of the iron wall.

As reported by H.L. Anderson, "Fermi calculated the trajectories from the map of the cyclotron magnetic field and slots were cut in the steel shield that separated the cyclotron from the experimental room

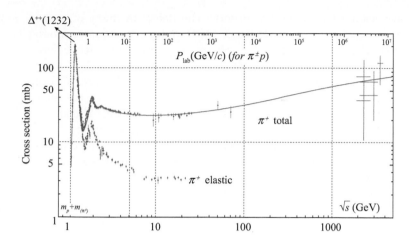

Fig. 8.2 Total cross-section for the $\pi^+ p \to X$ scattering as a function of the invariant mass $\sqrt{s}$ (bottom axis) or the pion momentum in the LAB frame (top horizontal axis). Reproduced with permission from Zyla *et al.*, 2020.

according to his prescriptions. The negative pions emitted in the forward direction came out of the cyclotron through a thin window in the vacuum chamber. Positive pions came out if they were emitted in the backward direction. The positive pion beams were of lower intensity but they came out readily when the magnetic field of the cyclotron was reversed" (Anderson, 1982). Behind the wall Anderson installed a magnetic spectrometer and a few scintillator counters to measure the sign and momentum of the pions. The total cross-section was measured by counting the number of absorbed pions in a liquid hydrogen target using eqn 2.120.

The cross-section as a function of the momentum of the π^+ in the laboratory (LAB) frame ($p_\pi \equiv |\mathbf{p}|_{\pi^+}$) and of the invariant mass $\sqrt{s}$ of the system is shown in Fig. 8.2. The **Anderson–Fermi experiment** is a fixed target ($a + b \to c + d$; here: $\pi^+ p \to \pi^+ p$) experiment where the Δ particle is not seen in a direct manner but through the observation of a huge peak in the cross-section lineshape, i.e. in σ as a function of p_π or $\sqrt{s}$.

Since the maximum of the peak is at $p_\pi \simeq 300$ MeV and, hence, the kinetic energy $T_a \simeq 191$ MeV, the energy of the pion is $E_a = T_a + m_\pi \simeq 330$ MeV. The rest mass of the particle produced by the $\pi^+ p$ scattering is the invariant mass in the rest frame:

$$M = \sqrt{2m_b E_a + m_a^2 + m_b^2} \simeq 1232 \text{ MeV.} \qquad (8.25)$$

The lifetime of the particle measured in its rest frame is the inverse of the width of the corresponding Breit–Wigner distribution (Sec 2.3): $\tau \simeq 6 \times 10^{-24}$ s, corresponding to $\Gamma = 110$ MeV. The data of the Anderson–Fermi experiment can be interpreted as the creation of a short-lived state with an electric charge equal to $+2e$ and $M = 1232$ MeV. This state promptly decays in a $p\pi^+$ state. Such a short lifetime is not surprising since the decay is caused by strong interactions: due to the strong coupling, the decay amplitude and width are much larger than electromagnetic interactions and the lifetime is extremely short. This is also the reason why we cannot observe the decay in flight of a Δ or any other state that decays through QCD processes. Even assuming that the particle is boosted in the laboratory frame, the path before decay is $\gamma c\tau = \gamma \cdot 3 \times 10^8 \cdot 6 \times 10^{-24} \simeq \gamma \cdot 1.8$ fm. For a 1 TeV/c momentum particle in the lab frame ($\gamma \simeq 830$), the traveled distance before decay is smaller than the size of an atom! The observation of **resonances** in the cross-section lineshape is, therefore, the method of choice for the identification of particles that decay by strong interactions.

The state observed by Anderson and Fermi is now called $\Delta^{++}(1232)$. For particles that decay by strong interactions the mass is part of the name and is put in parenthesis after the letter that classifies the state in the corresponding quark model. Since protons and neutrons are degenerate states in the quark models, they are labeled N ("nucleons"). The proton is then N and its excited states are N(1440), N(1529), etc. All these states decay nearly immediately into protons and pions by strong interactions. Using the same experimental apparatus, Anderson and Fermi also observed the $\Delta^0(1232)$ from π^-p interactions. The Δ^0 has the same mass of the Δ^{++}, as expected by the two-quark model. Further experiments established the existence of the Δ^- and Δ^+ and measured the spin to be 3/2. The data on strongly interacting resonances were an essential input for the development of the quark models.

8.5   The origin of hadron masses*

The success of the two-quark model is astonishing in the light of QCD and must hide a more profound explanation. A QCD proton is a complex system where the three quarks used to classify the hadron seem to play a minor role. These quarks are called **valence quarks** because they determine the eye-catching features of the hadron: the electric charge and the spin. On the other hand, the model overlooks the electric charge because it deals with strong interactions only. Similarly, the mass of the hadron is completely dominated by quark–gluon or gluon–gluon interactions because the quark mass is much smaller than the hadron mass. The hadron mass has, therefore, a dynamic origin that has nothing to do with the rest mass of the valence quark.

Even if a complete explanation of this conundrum requires mastering quantum field theory (QFT) and spontaneously broken symmetries (Sec. 13.1.2), the intuitive reason is straightforward. A quark with a net color charge is a strongly interacting particle that continuously produces soft gluons or quarks. As discussed in Sec. 7.6.4, the quark-induced soft cascade can be quenched only by neutralizing the color charge. A null color charge results from adding, for instance, an antiquark with opposite color charge (a meson) or several quarks, whose color charges sum up to zero (a baryon). On the other hand, QM prevents a complete overlap by the Heisenberg uncertainty principle. As a consequence, there will be always a mismatch among the quark wavefunctions, which prevents the complete neutralization of color charges. A small mismatch causes a strong force, which originates a hadron mass (i.e. the binding energy of the bound state) much larger than the valence quark (Wilczek, 2006). The same phenomenon occurs in atomic physics: in a hydrogen atom, the electron should collapse into the proton but the size of the electron wavefunction is such that the probability of finding the electron exactly inside the proton is small. The stationary states of the hydrogen atoms are a compromise between the electromagnetic forces and the Heisenberg uncertainty principle, and the binding energy of the ground state is different from zero.

This qualitative statement is deeply rooted in QCD and has been confirmed by non-perturbative calculations, which can compute the hadron masses with good precision.

Finally, if the mass of the valence quarks are immaterial, why are valence quarks so important to predict the spin? The sea of quarks and gluons surrounding the valence quarks has an electric charge that must sum to zero, otherwise we would observe a violation of charge that is forbidden by QED. What if the sea of quarks and gluons inside the hadron has an angular momentum that sums to zero? Since the quark mass is tiny, the flavor symmetry becomes a real symmetry of QCD and there are no ways to combine the wavefunctions of the valence quarks other than the combinations of Sec. 8.3. These combinations fix the spin of the hadrons. Unfortunately, we have no evidence that the angular momenta of the other particles sum to zero, and experimental data suggest that the valence quark contribution is marginal. So, the quark model should not work for the spin of the proton. This problem is called the **proton spin crisis**. It remains – at least partially – an open question in our field (Myhrer and Thomas, 2010).

8.6 The three-quark model

How far can we increase the value of F in the quark models? A modern answer to this question is given by lattice QCD. If we set $F = 3$ and $m_u = m_d = m_s$, the computed proton mass differs from the real mass by just 10% (Wilczek, 2006). We are still in a comfort zone. The three quark model is a generalization of the $F = 2$ model described in Sec. 8.3 and provides a powerful classification tool of baryons and mesons with u, d, and s quarks. However, since $m_s \gg m_u$ the corresponding flavor symmetry is quite a rough approximation.[9] As a consequence, the model provides the correct number of physical state and their spin but the predictions on degenerate energy eigenstates (hadron with equal masses) are wrong by a factor of m_s/Λ_{QCD} or more. Particles with bound s quarks are predicted to have the same mass of particles made of u and d quarks, which is in striking contradiction with experimental data. The **three-quark model** was proposed for the first time by Gell-Mann and Zweig in 1964 and is based on the following assumption:

[9]A more rigorous treatment based on the renormalization group demonstrates that the size of the errors on the binding energy increases as the ratio between the heaviest quark and Λ_{QCD}. So, mass predictions are completely unreliable for $F > 3$. This is the reason why u, d, s are called "light quarks" and c, b, t are considered heavy quarks (Weinberg, 2005a).

in the limit $F = 3$ and $m_u = m_d = m_s$, the transformations of ψ_{flavor} that redefine the flavor of quarks, i.e. the transformations

$$\psi_{flavor} \rightarrow e^{ia_j T_j} \psi_{flavor} \qquad (8.26)$$

are a symmetry of the strong interaction Hamiltonian. These transformations are represented by unitary operators U belonging to the $SU(3)$ group and U commute with the QCD Hamiltonian.

From the technical point of view, the three-quark model poses new challenges. We cannot resort anymore to the angular momentum composition rules to construct the hadron wavefunctions from the quark ones. This comes from the fact that, unlike $SU(2)$, $SU(3)$ is not isomorphic to $SO(3)$. The theory of representation of Lie groups provides, anyway, an analog of the Clebsch–Gordan formula for $SU(3)$:

$$|y, \gamma, i^2, i^z, c^1, c^2\rangle$$
$$= \sum_{y_1, y_2} \sum_{i_1^2, i_2^2} \sum_{i_1^z, i_2^z} \langle y_1, y_2, i_1^2, i_2^2, i_1^z, i_2^z | y, \gamma, i^2, i^z, c^1, c^2 \rangle |y_1, y_2, i_1^2, i_2^2, i_1^z, i_2^z\rangle$$

$$(8.27)$$

where the $y_1, y_2, i_1^2, i_2^2, i_1^z, i_2^z, y, \gamma, i^2, i^z, c^1, c^2$ are a generalization of the operators that define the angular momentum (j_1, j_2, j) and their additive component (j_{1z}, j_{2z}, j_z) (Hall, 2015). The generators of $SU(3)$ have already been discussed in eqn 7.7. It is worth noting that even if the group describing the color symmetry is mathematically the same as the one of the $F = 3$ flavor symmetry, the physics behind them is completely different.

The $SU(3)_c$ color symmetry of QCD is an exact global symmetry of strong interactions that is promoted to a local (gauge) symmetry to build the QCD Lagrangian. The $SU(3)$ flavor symmetry – sometimes labeled $SU(3)_f$ – is a (very) approximate symmetry of the Hamiltonian of QCD and plays no fundamental role in particle physics.

Since there is no "angular momentum" in the classifications of $SU(3)$, it is customary to use a more general notation to label the combination of wavefunctions. Instead of using j, we group the wavefunctions by the dimension of the associated vector space V as $\{2j+1\}$. In this notation, the $SU(2)$ combination rule $1/2 \otimes 1/2 = 0 \oplus 1$ reads $\{2\} \otimes \{2\} = \{1\} \oplus \{3\}$. A quark wavefunction is changed by a general $SU(3)$ transformation as

$$\psi \rightarrow e^{ia_j T_j} \psi \qquad (8.28)$$

where $j = 1, \ldots, 8$ and T_j are the Gell-Mann matrices of Chap. 7. In the – somewhat arcane – language of representation theory,

quarks transform as the $\{3\}$ representations of $SU(3)$.

Using[10] the generalized Clebsch–Gordan formula of eqn 8.27, it is possible to demonstrate that:

$$\{3\} \otimes \{3\} \otimes \{3\} = \{1\}_A \oplus \{8\}_{M12} \oplus \{8\}_{M23} \oplus \{10\}_S. \qquad (8.29)$$

Equation 8.29 provides the classification of all possible flavor wavefunctions ψ_{flavor} for a baryon. The subscripts A and S indicate that the wavefunctions are antisymmetric and symmetric for the exchange of a quark pair. Again, $M12$ indicates that the wavefunction does not change (except for a $\pm$ sign) for the exchange of the first and second quark but *does* change if we try to swap the third quark with one of the others.

If the Pauli exclusion principle were not in force, we would have 27 low-lying ($n = 0, l = 0$ in ψ_{space}) baryons. These states correspond to the 3^3 possible combinations of three types of quarks in a baryon made of three quarks. Like in the two-quark model, we have to pin down how many of these combinations fulfill the Pauli exclusion principle, assuming that u, d, and s are identical particles.

There is no doubt that that $\{10\}_S$ are physical states: as for the two-quark model, $\psi_{space}\psi_{color}$ is antisymmetric, $\psi_{spin}^{S=3/2}$ is symmetric and $\psi_{flavor} = \{10\}_S$ is symmetric, too. Hence,

$$\psi = \psi_{space}\psi_{color}\psi_{spin}^{S=3/2}\psi_{flavor} \qquad (8.30)$$

is totally antisymmetric and fulfills the Pauli exclusion principle. These particles belong to the **baryon decuplet**. Their spin is $3/2$ and the particles are shown in Fig. 8.3. The number of up and down quark is indicated by the value of I_z: $I_z = 0$ for s, $I_z = +1/2$ for u, and $I_z = -1/2$ for d. Once more, this is a quite unfortunate notation: in the three quark model, isospin plays no fundamental role and should be considered as a tame label for the type of quark. The number of strange quarks is shown in the vertical axis.[11] The spin-3/2 particles belonging to the $\{10\}_S$ decuplet is, clearly, 10. The masses of the particles are shown on the right-hand side of the figure and the vertical axis indicates the number of strange quarks contained inside the particles (see also Fig. 8.4).

The ψ_{flavor} wavefunction is shown in Fig. 8.4. Here $ddd \equiv |ddd\rangle$ is the wavefunction where the flavor of the first, second, and third quark is "down." Instead of using the general formula of eqn 8.26, we can derive these wavefunctions considering *all possible symmetric combination of three quarks*. This is done in Exercise 8.15. In particular, there is only one possible totally symmetric combination of three strange quarks: $|s, s, s\rangle$. The particle corresponding to $|s, s, s\rangle$ is the first particle predicted by the quark model (Gell-Mann and Ne'eman, 1961) not yet known at that time. It is a particle with spin-3/2 and has the same charge as the electron. Gell-Mann called it Ω^- ("the last particle") and was experimentally discovered in Brookhaven in 1964. The discovery of the Ω^- gave a tremendous boost to the quark models (and Gell-Mann's fame).

No low-lying physical state corresponds to $\psi_{flavor} = \{1\}_A$ because of eqn 8.20; in this case, the spin wavefunction cannot compensate

[10]If you have gone through Sec. 8.2.1, note the $\{3\}$ irreps of $SU(3)$ are just 3×3 matrices acting on a vector space, whose basis consists of quark wavefunctions. Hence, the quarks are the basis of the vector space that changes through the fundamental representations of $SU(3)$, as defined in Sec. 7.5.

[11]In fact, the isospin in $SU(3)$ might be defined – if needed – because $SU(2)$ is a subgroup of $SU(3)$. In Fig. 8.3, we could avoid it replacing I with the corresponding set of up and down quarks (see Exercise 8.10) in the horizontal axis of Fig. 8.3. The use of isospin in the classification of particles in the three-quark model is just a habit and we might have dropped it with impunity.

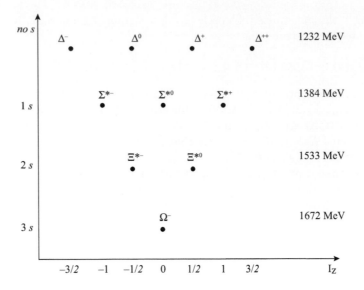

Fig. 8.3 The spin 3/2 particles belonging to the $\{10\}_S$ decuplet. The masses of the particles are shown on the right. The vertical axis indicates the number of strange quarks contained in the particles. The horizontal axis shows the number of u and d quarks (see text).

for the change of sign in ψ_{flavor}. The $\psi_{flavor} = \{1\}_A$ singlet flavor wavefunction is:

$$\psi_{flavor} = \frac{1}{\sqrt{6}} \left(|usd\rangle - |sud\rangle + |sdu\rangle - |dsu\rangle + |dus\rangle - |uds\rangle \right). \quad (8.31)$$

Equation 8.31 has the same form as eqn 8.17 (just replace "red" with "up", "green" with "down", and "blue" with "strange") because from the mathematical point of view, it represents the singlet of the same $SU(3)$ group (just replace "color" with "flavor"). However,

> the antisymmetric form of ψ_{color} for baryons is a fundamental physics law that comes from confinement while ψ_{flavor} has no special meaning. There is no shame in lacking the physical state that corresponds to the flavor singlet.

As for the two-quark model, the product of the $\{8\}$ flavor wavefunction and the $S = 1/2$ wavefunction can be symmetrized and gives eight physical states, the spin-1/2 **baryon octet**:

$$\psi_{spin}\psi_{flavor} = \frac{1}{\sqrt{18}} \left(\psi_{spin}^{S=1/2, M12} \psi_{flavor}^{\{8\}_{M12}} + \psi_{spin}^{S=1/2, M23} \psi_{flavor}^{\{8\}_{M23}} \right.$$
$$\left. + \psi_{spin}^{S=1/2, M13} \psi_{flavor}^{\{8\}_{M13}} \right). \quad (8.32)$$

Again, $\{8\}_{M13} = \{8\}_{M12} + \{8\}_{M23}$ and $\psi_{spin}^{S=1/2, M13} = \psi_{spin}^{S=1/2, M12} + \psi_{spin}^{S=1/2, M23}$. The states of the baryon octet are all physical states and correspond to spin-1/2 baryons. Their names and quark content are shown in Fig. 8.5 and Fig. 8.6, respectively. Group representation theory ensures that no other symmetric combination exists except for the octet and the decuplet. The Pauli exclusion principle thus reduces the number of possible states from the maximum number of combinations of

The flavor wavefunctions of the {10}$_S$ decuplet. The vertical axis corresponds to the number of strange quarks in the bound state. The horizontal axis shows the number of u and d quarks (see text).

three types of quarks in three positions ($3^3 = 27$) to 18. Clearly, this statement holds as long as the space part of the wavefunction has $l = 0$ (s-waves). **Excited baryons** where $\psi_{flavor}\psi_{spin}$ is antisymmetric do exist if l is odd because $\psi_{space}\psi_{spin}\psi_{flavor}\psi_{color}$ becomes antisymmetric. We expect their bound energy (mass of the baryon) to lie above the s-wave states and experimental data support this prediction.

The quark model offers a very elegant explanation of one of the deepest mysteries of particle physics in the 1950s: why some particles are identical to antiparticles and others are not.

> If the particle is a baryon, its wavefunction is never an eigenstate of the C-parity operator, even if the baryon is electrically neutral, because the corresponding antibaryon is composed of $\bar{q}_1\bar{q}_2\bar{q}_3$ antiquarks.

The neutron is different from the antineutron $\bar{n}$ since the neutron is not an elementary particle but a bound state of udd and its antiparticle is a bound state of $\bar{u}\bar{d}\bar{d}$. The C-parity of mesons is less trivial and will be discussed in Sec. 8.7.

8.6.1   Parity and excited states

The parity of a baryon depends on the space part of the wavefunction. An s-wave has positive parity because $\mathcal{P}Y_{00} = Y_{00}$ where $\mathcal{P}$ is the parity operator and Y_{00} is the $l = 0$ harmonic function of Sec. 5.6.1. Naïvely, one can describe the relative motion of the three quarks employing classical mechanics: in this case, we can define the relative angular momentum of the first two quarks as l_{12} and the angular momentum of the third quark with respect to the center of mass of the 1–2 pair as l_3. Since the intrinsic parity of a quark is $+1$, the parity of a baryon is

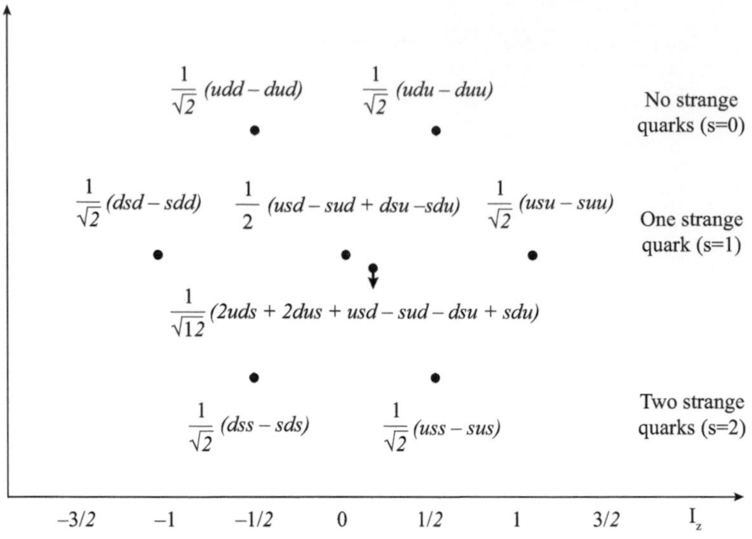

Fig. 8.5 The flavor wavefunctions of $\{8\}_{M12}$ states. The vertical axis corresponds to the number of strange quarks in the bound state. The horizontal axis shows the number of u and d quarks (see text).

$$(-1)^{l_{12}}(-1)^{l_3} P_{q_1} P_{q_2} P_{q_3} = (-1)^{l_{12}+l_3} \tag{8.33}$$

and

$$(-1)^{l_{12}}(-1)^{l_3} P_{\bar{q}_1} P_{\bar{q}_2} P_{\bar{q}_3} = (-1)^{l_{12}+l_3+1} \tag{8.34}$$

for an antibaryon ($P_{\bar{q}} = -1$). For s-waves the quark model predicts the parity of the proton to be $+1$. This is a quite trivial prescription that comes from the conventional assignment $P_u = P_d = +1$. On the other hand, the model prescribes positive parity also for the Λ (see Fig. 8.6). This prediction is less trivial because the Λ particle decays by weak interactions and can be used to investigate parity violation in weak forces.

The reader should not forget that the three-quark model is based on the crude approximation of degenerate quark masses. This is evident in the poor quality of the predictions on the baryon masses and excited states. For instance, the three-quark model predicts the mass of the decuplet to be identical for each particle, which is definitely false. The mass of the Σ^0 is 252 MeV heavier than the proton. This is certainly due to the strange quark but its connection with the s quark mass is non-trivial. Even if many semi-empirical rules were devised in the past to describe the mass pattern of particles with s quarks, we now compute these patterns using lattice QCD and, since 2008 (Durr *et al.*, 2008), these results are in excellent agreement with the experimental data. State-of-the-art calculations indicate that the flavor symmetry employed in quark models is really an approximation of the QCD dynamics and that QCD is the fundamental theory of strong interaction even in its non-perturbative regime.

Similar considerations hold for the excited states. The first excited state of the nucleon ($N = p, n$) has a mass of 1440 MeV and is called $N(1440)$. It decays quickly in a nucleon emitting also a pion. Surprisingly, this state (called "Roper resonance") is highly excited: it

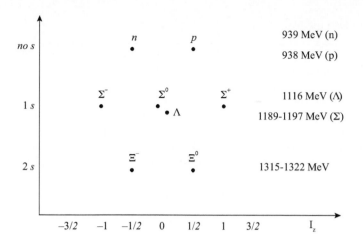

Fig. 8.6 The particles of the spin 1/2 octet. They correspond to the symmetrized wavefunctions of eqn 8.32. The vertical axis gives the number of strange quarks in the bound state. The horizontal axis shows the number of u and d quarks (see text). The mass range of the Σ (Ξ) shows the mass of the lighter and heavier Σ (Ξ) particle.

is a radially excited state with $n = 2$ and $l = 0$. The "low-lying" $n = 1$, $l = 1$ states $N(1520)$ and $N(1535)$ have a mass larger than the Roper resonance. The reason most likely resided in the subtleties of non-perturbative QCD but it is not understood yet.

8.7   Mesons in the two- and three-quark models

Mesons are bound states of a quark–antiquark pair. The combination rules are a straightforward extension of the rules for the baryons with two important exceptions:

- the antiquark is always a different particle than the quark even in the approximation $m_u \simeq m_d \simeq m_s$ and, therefore, the Pauli exclusion principle is not in force. Meson wavefunctions do not need to be antisymmetric.

- an antiquark wavefunction changes for a transformation of a smooth continuous group in a different way than for a quark.

Relativistic quantum mechanics fixes the transformation rule of an antiparticle under a continuous and smooth group:

$$\bar{\psi} \to U^* \bar{\psi} = e^{-ia_j T_j^*} \bar{\psi}. \tag{8.35}$$

In the language of representation theory, if $\{D\}$ is an m-dimensional representation of a group in a (m-dim) vector space V, the **conjugate representation** is called $\{\bar{D}\}$ and acts on the elements of the vector space as:

$$v \to U^* v = e^{-ia_j T_j^*} v. \tag{8.36}$$

As a consequence,[12]

> in the three-quark model, the antiquarks transform as the fundamental conjugate representation of $SU(3)_f$.

For particles that transform under conjugate representations, we should restart from scratch the construction of the Clebsch–Gordan formulas and compute how the particle and antiparticle wavefunctions combine to form n-quark, m-antiquark bound states. Fortunately (see Example 8.2), the combination rules are very similar to quarks both in $SU(2)$ and in $SU(3)$.

A meson wavefunction can be described as

$$\psi = \psi_{space}\psi_{spin}\psi_{color}\psi_{flavor}. \tag{8.37}$$

The space part is particularly simple because describes the relative motion of two particles in their center-of-mass frame: $\psi_{space} = Y_{l,m}(\theta, \phi)$. The spin is the combination of two $1/2$ spins: $1/2 \otimes 1/2 = 0 \oplus 1$ giving a singlet and a triplet.

The combination rule of $SU(3)$ for a pair of quark and antiquark (i.e. the formula corresponding to eqn 8.29) is:

$$\{3\} \otimes \{\bar{3}\} = \{1\} \oplus \{8\}. \tag{8.38}$$

We can use this formula for two tasks. First, to evaluate ψ_{color}. The confinement conjecture states that ψ_{color} must be a singlet under the transformation of $SU(3)_c$. It is fortunate that eqn 8.38 includes one singlet, which is the (only) combination of a red, green, or blue quark with an antiquark that is invariant for $SU(3)$ transformations. If the color singlet did not exist, no mesons could be formed. This combination is:

$$\psi_{color} = \frac{1}{\sqrt{3}}\left[|B\bar{B}\rangle + |R\bar{R}\rangle + |G\bar{G}\rangle\right] \tag{8.39}$$

where B $(\bar{B})$ is the blue quark (antiblue antiquark) wavefunction, that is, it gives the probability of finding a blue quark in the meson. To avoid any source of confusion, it is worth recalling, once more, that a blue quark has a blue charge equal to $+1$. A "blue antiquark" or, equivalently, an "antiblue antiquark" is an antiquark with a blue charge equal to -1.

Finally, in the three-quark model, eqn 8.38 can be used to compute ψ_{flavor}. The singlet wavefunction is:

$$\psi_{flavor} = \frac{1}{\sqrt{3}}\left[|u\bar{u}\rangle + |d\bar{d}\rangle + |s\bar{s}\rangle\right] \tag{8.40}$$

while the eight wavefunctions belonging to the octet are depicted in Fig. 8.7.

Since the Pauli exclusion principle does not apply here, all combinations are physical states. The low-lying mesons ($l = 0$) are thus:

$$\psi = Y_{00}\left(\frac{1}{\sqrt{3}}\left[|R\bar{R}\rangle + |R\bar{R}\rangle + |G\bar{G}\rangle\right]\right)\psi_{spin}\psi_{flavor} \tag{8.41}$$

that is,

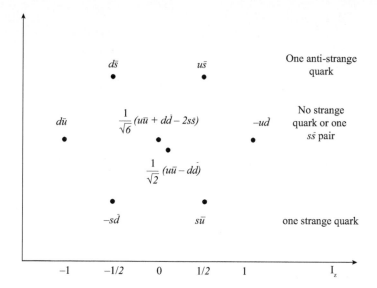

Fig. 8.7 The flavor wavefunctions of the {8} octet for quark-antiquark pairs (mesons). The vertical axis gives the number of strange quarks in the bound state. The horizontal axis shows the number of u and d quarks or their antiparticles expressed as I_z (see text).

- one particle with spin zero corresponding to $\psi_{spin} = \psi_{spin}^{S=0}$ and $\psi_{flavor} = \frac{1}{\sqrt{3}} \left[|u\bar{u}\rangle + |d\bar{d}\rangle + |s\bar{s}\rangle \right]$. This particle is called η_0 (see Sec. 8.7.2).

- one particle with spin 1 corresponding to $\psi_{spin} = \psi_{spin}^{S=1}$ and $\psi_{flavor} = \frac{1}{\sqrt{3}} \left[|u\bar{u}\rangle + |d\bar{d}\rangle + |s\bar{s}\rangle \right]$. This particle is called ϕ_0 (see Sec. 8.7.2).

- eight particles with spin zero corresponding to $\psi_{spin} = \psi_{spin}^{S=0}$ and $\psi_{flavor} = \{8\}$. These particles are the pions, the kaons, and the η_8. They are shown in Fig. 8.8. As for the baryons, the vertical axis indicates the number of strange (or antistrange) quarks in the wavefunction. In the horizontal axis, the number of up- and down-type quarks can be inferred by the sum of their I_z components. Note that for mesons, we have to account also for antiquarks, which – in the two quark model – have the same isospin as the quark but opposite values of I_z. Since u has $I_z = +1/2$, $\bar{u}$ has $I_z = -1/2$. For the down quark, d has $I_z = -1/2$ and $\bar{d}$ has $I_z = 1/2$.

- eight particles with spin 1 corresponding to $\psi_{spin} = \psi_{spin}^{S=1}$ and $\psi_{flavor} = \{8\}$. These particles (see Fig. 8.9) are the ρ, the K^* mesons, and the ϕ_8.

All in all, the three-quark model predicts the existence of 18 mesons at $l = 0$. Half of them (9 particles) are spinless and the other 9 particles have spin equal to 1.

Mesons, of course, can be in excited states as for baryons. They can be described in the two-quark model, as well, and the **two-quark model mesons** are a subset of the aforementioned particles. In this model, the flavor wavefunctions can be derived from the isospin composition rule: ψ_{flavor} is given by $1/2 \otimes 1/2 = 0 \oplus 1$. We then have only eight particles:

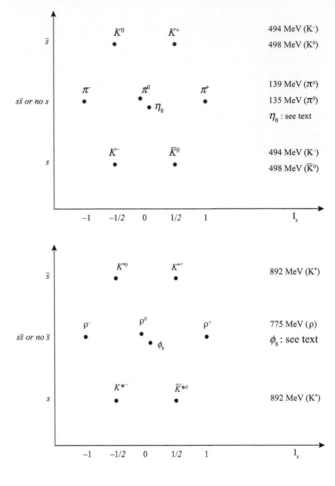

Fig. 8.8 The mesons of the spin 0 octet. The vertical axis gives the number of strange or anti strange quarks. The horizontal axis shows the number of u and d (anti)quarks expressed as I_z (see text).

Fig. 8.9 The mesons of the spin 1 octet. The vertical axis gives the number of strange or anti strange quarks. The horizontal axis shows the number of u and d (anti)quarks expressed as I_z (see text).

the triplets are the pions and the ρ mesons, the singlets are the η and ω particle.

8.7.1 Discrete symmetries of mesons

As for the baryons, the quark model is an excellent classification tool and provides non-trivial information on spin, parity, and C-parity. A meson is a particle–antiparticle pair and, therefore, the intrinsic parity of a meson is:

$$(-1)^l P_q P_{\bar{q}} = (-1)^{l+1}. \tag{8.42}$$

A neutral meson can be an eigenstate of the C-parity and this property can be established by transforming the quark into the corresponding antiquark in ψ_{flavor}. If it is an eigenstate, the corresponding eigenvalue is:

$$\text{C parity} = (-1)^{l+s}. \tag{8.43}$$

This result follows from the general particle–antiparticle pair theorem of Sec. 5.7.3. For the low-lying mesons ($l = 0$), $P = -1$ and $C = +1$

for spinless mesons, and $P = -1$ and $C = -1$ for mesons with spin 1. A spinless mesons with $l = 0$ and $C = -1$ has never been observed, in agreement with the quark model.

8.7.2 Limitations of the quark model*

The quark model fails when we consider degenerate states, that is, states with equal masses. In the three-quark model, pions and kaons have the same mass, which is not what we observe in the experiments. This is, again, due to the crude $m_u = m_d = m_s$ approximation. There are even subtler effects that cannot be predicted by the quark models. Experimental data indicate that the physical states corresponding to η_0 and η_8 are in fact a superposition of the η_0 and η_8 wavefunctions. The same effect holds for the $S = 1$ states ϕ_0 and ϕ_8. The actual states are:

$$
\begin{aligned}
|\eta\rangle &= \cos\theta_\eta |\eta_8\rangle + \sin\theta_\eta |\eta_0\rangle \\
|\eta'\rangle &= -\sin\theta_\eta |\eta_8\rangle + \cos\theta_\eta |\eta_0\rangle
\end{aligned}
\tag{8.44}
$$

where $\theta_\eta \simeq 11°$ and,

$$
\begin{aligned}
|\omega\rangle &= \cos\theta_\phi |\phi_0\rangle + \sin\theta_\phi |\phi_8\rangle \\
|\phi\rangle &= -\sin\theta_\phi |\phi_0\rangle + \cos\theta_\phi |\phi_8\rangle
\end{aligned}
\tag{8.45}
$$

where $\theta_\phi \simeq 35°$. All these effects are very intriguing. For instance, $\theta_\phi \simeq 35°$ implies that $\psi_{flavor}(\phi) \simeq |s\bar{s}\rangle$, that is, the ϕ particle ($\phi(1020)$) is a simple bound state of an s quark and its antiquark. Unfortunately, the value of the parameters θ_η and θ_ϕ are beyond the capabilities of the quark models. Even if modern developments provide further insight, they go well beyond the scope of this book.

Example 8.2

For the two quark model, the flavor of quarks transform as $\begin{pmatrix} u \\ d \end{pmatrix} \rightarrow U \begin{pmatrix} u \\ d \end{pmatrix} = e^{-\frac{i}{2}a_j\sigma_j} \begin{pmatrix} u \\ d \end{pmatrix}$. The flavor of antiquarks transform as the conjugate representation of SU(2): $\begin{pmatrix} \bar{u} \\ \bar{d} \end{pmatrix} \rightarrow U^* \begin{pmatrix} \bar{u} \\ \bar{d} \end{pmatrix} = e^{-\frac{i}{2}a_j\sigma_j^*} \begin{pmatrix} \bar{u} \\ \bar{d} \end{pmatrix}$. It is possible to use the composition rules of quarks even for antiquarks just replacing the vector $\begin{pmatrix} \bar{u} \\ \bar{d} \end{pmatrix}$ with

$$
\begin{pmatrix} \tilde{u} \\ \tilde{d} \end{pmatrix} = i\sigma_2 \begin{pmatrix} \bar{u} \\ \bar{d} \end{pmatrix} = \begin{pmatrix} \bar{d} \\ -\bar{u} \end{pmatrix}
\tag{8.46}
$$

where $\sigma_2 = \begin{pmatrix} 0 & -i \\ i & 0 \end{pmatrix}$ is the second Pauli matrix (Carlsmith, 2012). This is due to the fact that $(i\sigma_2)\boldsymbol{\sigma}^*(i\sigma_2)^{-1} = -\boldsymbol{\sigma}$ where $\boldsymbol{\sigma}$ is the vector of the three Pauli matrices. Using this identity after replacing the exponential of a matrix with its Taylor series, we have

$$(i\sigma_2)e^{-i\frac{\boldsymbol{a}\cdot\boldsymbol{\sigma}^*}{2}}(i\sigma_2)^{-1} = e^{i\frac{\boldsymbol{a}\cdot\boldsymbol{\sigma}}{2}} = e^{i\frac{a_j\sigma_j}{2}}. \tag{8.47}$$

If we transform $i\sigma_2\begin{pmatrix}\bar{u}\\\bar{d}\end{pmatrix}$ by the (conjugate) transformations of $SU(2)$, we have

$$i\sigma_2\begin{pmatrix}\bar{u}\\\bar{d}\end{pmatrix} \rightarrow i\sigma_2 e^{-ia_j\frac{\sigma_j^*}{2}}\begin{pmatrix}\bar{u}\\\bar{d}\end{pmatrix} = i\sigma_2 e^{-ia_j\frac{\sigma_j^*}{2}}(i\sigma_2)^{-1}(i\sigma_2)\begin{pmatrix}\bar{u}\\\bar{d}\end{pmatrix}$$

$$= e^{ia_j\frac{\sigma_j}{2}}(i\sigma_2)\begin{pmatrix}\bar{u}\\\bar{d}\end{pmatrix}. \tag{8.48}$$

As a consequence $i\sigma_2\begin{pmatrix}\bar{u}\\\bar{d}\end{pmatrix}$ transforms as $\begin{pmatrix}u\\d\end{pmatrix}$. From the practical point of view, the classification of mesons in the two quark model can be done using the pseudo-variable $\begin{pmatrix}\tilde{u}\\\tilde{d}\end{pmatrix}$ and replacing $\tilde{u}$ with $\bar{d}$ and $\tilde{d}$ with $-\bar{u}$ at the last stage.

8.8   The heritage of the past

The quark model is one of the major achievements of particle physics in the 1960s and the results gained by the use of the flavor symmetry have been confirmed by QCD. The understanding of hadrons as bound states of quarks required several decades of research and it is not surprising that some obsolete notations are still in use in the literature and textbooks. The most remarkable case concerns the strange quark. Since strong and electromagnetic interactions are flavor-independent, there is no Feynman diagram in QCD/QED that can transform a strange quark into a down quark. In general,

> flavor-changing interactions are forbidden both in QCD and in QED.

When the lightest particles with strange quarks were discovered (e.g. the kaon mesons), particle physicists realized that their decay lifetime was much longer than the typical lifetime of strong or electromagnetic interactions. If only strong and e.m. interaction existed, kaons would be stable particles. Weak interactions (see Chaps. 10 and 13) provide tree-level flavor-changing interactions and cause the decay of the kaon but since these interactions are weak, the decay lifetime is very long. Particle physicists in the 1950s responded to these challenges by introducing a new quantum number called **strangeness** and claiming that strangeness was a conserved quantum number in strong and electromagnetic interactions. In this language, a particle with a quark s has a strangeness equal to -1 and a particle with an antiquark $\bar{s}$ has a strangeness equal to $+1$. Hence a K^+ has strangeness $= +1$ because it is a $u\bar{s}$ state and the ϕ particle ($s\bar{s}$) has strangeness $= 0$. This is the reason why the decuplet and octet of Figs. 8.3, 8.6, and 8.8 are often expressed in the strangeness versus isospin plane. In this book, we use a more up-to-date style, drawing planes where the number of strange quarks is in the vertical axis and the number of u and d quarks, expressed by their isospin, is in the horizontal axis. In principle, we can introduce similar numbers for the charm quark ("charm quantum number" or simply "charm") or the b quark. For b, this number was called "bottomness" since in the past the b quark was also called "bottom." A particle with a charm quark has a charm number equal to $+1$ and a particle with b quark has a bottomness number $= -1$. In general, quarks with negative electric charges have negative "quark quantum numbers" (strangeness, bottomness) and quarks with positive electric charges have positive "quark quantum numbers" (charm).

In the good old days, people attempted to introduce "topness" as well, that is, the top quark quantum number. We know now that the theory of top quark bound states (the six-quark model) has no room in particle physics. The top quark is so heavy and its phase space so large that decays before the hadronization takes place. So, there are no top quark bound states even if we consider particles with tiny lifetimes.

There is, however, a quantum number built on these principles, which is very interesting and still in use. It is called the **baryon number** and is computed assigning a baryon number equal to $+1/3$ to any quark and a baryon number equal to $-1/3$ to any antiquark. In this way, a baryon has a baryon number equal to $1/3 + 1/3 + 1/3 = 1$ and an antibaryon has a baryon number equal to -1. As we will see in Sec. 10.4.1, the baryon number is conserved by all fundamental interactions.

The baryon number enters a popular formula proposed in 1956 that played a central role in the development of the quark model. The **Gell-Mann Nishijima formula** is a simple relationship between the electric charge of a bound state and the isospin, the baryon number, and the quark quantum number of the particles composing the bound state:

$$Q = I_z + \frac{B + S + \ldots}{2} \tag{8.49}$$

where I_z is the third component of the isospin ($-1/2$ for a down quark, $+1/2$ for an up quark, $+1/2$ for an antidown, $-1/2$ for an antiup), S is the strangeness (-1 for s, $+1$ for $\bar{s}$) and $\ldots$ indicates additional quark quantum numbers: bottomness for b quarks and charm for c quarks.

8.9   Generalization of the quark model

The three-quark model described Sec. 8.2 can be generalized following two paths. First, we can build an n-quark model from the approximation $m_u \simeq m_d \ldots \simeq m_n$. The four-quark model neglects the mass of the charm quark and is based on an approximation that is even cruder than the three-quark model. The modern classification of bound states is based on the **five-quark model** and the corresponding $SU(5)_f$ group. This is just a classification model based on the approximation $m_u \simeq m_d \simeq m_s \simeq m_c \simeq m_b$. It is, anyway, useful to predict the number of physical states expected after the hadronization process, and the spin, parity, and C-parity of the bound states. As mentioned above, the six-quark model is of no practical use due to the huge mass of the top quark and its decay before hadronization.

Bound states of four (**tetraquark**) and five quarks (**pentaquarks**) are possible, too, since the color contribution of their wavefunction can be a color singlet. The experimental observation of these states has been extremely hard-won because it is difficult to distinguish a genuine tetraquark from a bound state of two mesons (a two-meson molecule). Similarly, a pentaquark can be easily confused with a baryon–meson molecule. We gained very convincing evidence of the existence of tetraquarks and pentaquarks between 2013 and 2018 by the LHCb experiment at the Large Hadron Collider (LHC). Evidence for these "exotic" states was gained also at the BES-III collider in Beijing, and by the BELLE experiment at KEK in Japan. On the other hand, it is still not clear why bound states of two gluons (**glueball**) have not been observed and QCD results are not mature yet (Ochs, 2013).

Further reading

In this chapter, we presented the quark models starting from the symmetries of QCD and exploiting many techniques of group representation theory. If you are interested in the historical development of the field and the "particle proliferation era," I suggest L.M. Brown 1990. The enthusiasm of finding, once more, an underlying simplicity below an apparent complexity is evoked in a masterpiece of popular science written by the founding father of the quark model: Gell-Mann, 1994. A complete derivation of the properties of continuous smooth groups and their applications to strong interactions are available in Zee, 2016 and Georgi, 1999. On the physical side, the quark model is the bedrock of hadron spectroscopy, which has not been discussed in this book but is covered in many graduate-level textbooks such as Bettini, 2014 and Amsler, 2018.

Exercises

(8.1) Compute all possible electric charges of the hadrons.

(8.2) What are the spins of the mesons and the baryons?

(8.3) Identify at least one meson heavier than a baryon [Hint: to cross-check your results you can resort to the PDG repository cited in Exercise 2.5].

(8.4) Employing the three-quark model, compute the parity, C-parity, and spin of the $\phi(1020)$

(8.5) * Demonstrate that the $\phi(1020) \rightarrow \pi^0\pi^0$ decay cannot occur.

(8.6) Why did Fermi use liquid hydrogen for the study of the $\Delta(1232)$?

(8.7) * In the Anderson–Fermi experiment, the authors observed only the absorption coefficient of the pions, i.e. the disappearance probability of the incoming π^+. Why did they claim that the only possible decay of the $\Delta^{++}(1232)$ was in a π^+p state?

(8.8) * In the two-quark model, show that the 3/2 states of eqn 8.20 are fully symmetric using the Clebsch–Gordan formula.

(8.9) Consider an excited state of the proton (e.g. the $N(1529)$ state). How does this particle reach the ground state? Do you consider more likely a radiative decay like $N(1592) \rightarrow N(939) + \gamma$ or a pion emission like $N(1592) \rightarrow N(939) + \pi^0$?

(8.10) Redraw Fig. 8.3 without relying on the concept of isospin.

(8.11) * If a quantum state is a linear combination of two degenerate states, show that this state is also an eigenstate of the Hamiltonian and has the same energy as the others.

(8.12) * Demonstrate the following theorem: if $\mathbf{S}$ is a unitary operator describing a symmetry and $|\alpha\rangle$ is an eigenstate of $\mathbf{S}$, then $|\alpha\rangle$ and $\mathbf{S}|\alpha\rangle$ are degenerate states of the Hamiltonian.

(8.13) In the previous exercise, what happens if the degenerate states are not linearly independent?

(8.14) * Using eqn 8.4, demonstrate that the ψ_{spin} of a 3/2-spin baryon is symmetric.

(8.15) * Compute the wavefunctions of Fig. 8.4 just seeking for all symmetric combinations of u, d, and s wavefunctions.

(8.16) * Show that the combination of a state with isospin 0 with a state with isospin 1/2 does not produce a symmetric wavefunction. Does the result hold for spin wavefunctions, too?

(8.17) ** Demonstrate eqn 8.20.

From QCD to nuclear physics

9

9.1 The discovery of the neutron

The first observation of the neutron was the milestone that clarified the difference between strong and weak interactions. It marks the inception in **nuclear physics** – 40 years before quantum chromodynamics (QCD). In response to a puzzle observed by W. Bothe, H. Becker, I. Curie-Joliot, and F. Joliot, a few physicists, including E. Rutherford, E. Majorana, and J. Chadwick, speculated about a new neutral particle with a mass comparable to the mass of the proton. The puzzle was solved by Chadwick and N. Feather in 1932. Like Curie-Joliot, they employed a radioactive source to produce $_4^4$He nuclei, also called "α particles," to disintegrate light nuclei. Unlike their French colleagues, Chadwick and Feather developed an efficient detector for protons based on an ionization chamber and an α source. The layout of the experiment is shown in Fig. 9.1. Protons were produced by unidentified radiation emerging from a paraffin wax, which is a material extremely rich in protons. The unknown radiation was the outcome of the interaction of the α particles with a low Z element, beryllium.

Chadwick noted that the energy of the proton could exceed 5 MeV. The kinetic energy spectrum was measured by the range of protons in the ionization chamber employing the Bethe formula in eqn 3.19:

$$R = \int_{T_0}^0 \left[\frac{dE}{dx}\right]^{-1} dE \tag{9.1}$$

that gives the initial kinetic energy T_0 by measuring R. The experiment aimed to show that this new type of radiation could not be attributed to photons. Chadwick's findings supported the claim of Rutherford that the explanation based on photons was untenable. If the unknown radiation were a flux of photons that hits the protons by Compton scattering, the energy of the photons should be enormous. The maximum proton energy corresponding to a head-on scattering is given by eqn 3.82, replacing m_e with m_p. The proton kinetic energy in natural units (NU) is then $T = (m_p^2 + |\mathbf{p}_p|^2)^{1/2} - m_p$ and the maximum value that can be attained is:

$$T_{max} = \frac{2E^2}{m_p + 2E} \ . \tag{9.2}$$

A Modern Primer in Particle and Nuclear Physics. Francesco Terranova, Oxford University Press.
© Francesco Terranova (2021). DOI: 10.1093/oso/9780192845245.003.0009

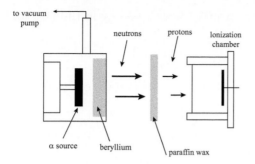

Fig. 9.1 Layout of the first run of the Chadwick experiment. The source of α particles is a sample of polonium. Neutrons are produced mainly by $^9\text{Be} + {}^4\text{He} \rightarrow {}^{12}\text{C} + n$ and hit the paraffin wax. The ionization chamber detects the outgoing protons produced by the $n + p \rightarrow n + p$ elastic scattering. The paraffin was replaced by other targets in the subsequent runs (see text).

To have such a Compton edge for protons with $T_{max} \simeq 5$ MeV, we need photons with 51 MeV energy. These energetic photons are only produced by accelerators or cosmic rays but never appear in radioactive decays or low-energy nuclear reactions like α-beryllium scattering. To test alternative hypotheses, Chadwick performed an experiment replacing the paraffin wax with several materials of different Z. In particular, he employed gaseous targets where molecular and chemical effects can be neglected.

The Chadwick–Rutherford hypothesis was that the unidentified radiations were a heavy neutral particle with a mass similar to m_p. Since in the scatterings of Chadwick's experiment, all particles are non-relativistic, the kinematics follow the classical energy and momentum conservation laws. Fig. 9.2 shows a neutron–proton scattering in the laboratory frame (LAB), where the proton is at rest in the initial state and is scattered with an angle ϕ^{LAB}.

In LAB,

$$(m_n + m_p)v_{CM} = p_{m_n} + p_{m_p} = m_n v_0 \tag{9.3}$$

where $\mathbf{v}_{CM} = v_{CM}\mathbf{k}$ is the initial center-of-mass velocity and v_0 is the initial neutron velocity in LAB. Hence,

$$v_{CM}\mathbf{k} = \frac{mv_0}{m_n + m_p}\mathbf{k} \tag{9.4}$$

is the velocity of the center of mass, whose direction is along the z-axis. The neutron velocity in the rest frame (RF) of the center of mass is, therefore,

$$v_n^{RF} = \frac{m_p}{m_p + m_n}v_0 \tag{9.5}$$

as expected by a two-body scattering in classical physics. Similarly,

$$v_p^{RF} = -\frac{m_n}{m_p + m_n}v_0 \ . \tag{9.6}$$

To compute the momentum in LAB, that is, in the $x-z$ plane of Fig. 9.2, we can either apply the Galilean transformations of eqn A.3 or, equivalently, perform the vector sum $\mathbf{v}_n{}^{LAB} = \mathbf{v}_n{}^{RF} + \mathbf{v}^{CM}$ using the cosine rule (see Example 9.2). In the approximation $m_n \simeq m_p$,

$$v_n^{RF} = \frac{m_n}{m_n + m_p}v_0 \simeq \frac{1}{2}v_0. \tag{9.7}$$

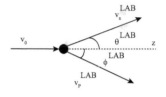

Fig. 9.2 Scattering of a neutron with velocity v_0 on a proton at rest in the LAB frame. The $\mathbf{k}$ unit-vector is along the z-axis.

The ratio of the neutron kinetic energy in the final and initial state in LAB is:

$$\frac{T^{fin}}{T^{ini}} = \left(\frac{v_n^{LAB}}{v_0}\right)^2 = \frac{(1 + \cos\theta^{RF})}{2} . \tag{9.8}$$

In general, for an atom with mass number A (see Example 9.2),

$$\frac{T^{fin}}{T^{ini}} = \left(\frac{v_n^{LAB}}{v_0}\right)^2 = \frac{A^2 + 1 + 2A\cos\theta^{RF}}{(A+1)^2} . \tag{9.9}$$

The Chadwick–Rutherford hypothesis implies that the maximum and minimum energy that the presumptive neutral particle can transfer to a nucleus depends on A, only. The maximum kinetic energy for the proton corresponds to the minimum of T^{fin}/T^{ini} for the neutron:

$$\left.\frac{T^{fin}}{T^{ini}}\right|_{min} = \left(\frac{A-1}{A+1}\right)^2 . \tag{9.10}$$

For a neutron–proton scattering, the minimum T_{fin} is zero. It corresponds to a head-on scattering ($\theta^{RF} = \pi$) of two equal-mass particles where the projectile is stopped in LAB and all the momentum is transferred to the proton in the forward direction. Chadwick and Feather performed the experiment with protons (paraffin wax), helium ($A = 4$), lithium ($A = 7$), nitrogen ($A = 14$), and other targets, measuring the maximum kinetic energy of the protons by the range in the ionization chamber. The outcome of these measurements strongly supported eqn 9.10.

The very fact that this new particle (the **neutron**) was produced by breaking nuclei with α particles suggests that neutrons are a part of the nuclei.[1] The empirical evidence obtained by Chadwick also supported $m_n \simeq m_p$, which implies that the nuclei are bound states made of Z protons and $A - Z$ neutrons. This new nuclear model replaced the hypothesis of a nucleus made of Z protons and $A - Z$ electrons, which was already in a striking contradiction with atomic spectroscopy data, thanks to the work of R. Kronig, L. Ornstein, and F. Rasetti.

[1] The term **subnuclear physics** as an obsolete synonym of particle physics originates from here.

9.2 Nuclei

It is tempting to say that a nucleus is an ensemble of Z protons and $A - Z$ neutrons. This statement seems in contradiction with QCD because neutrons and protons are color singlets and cannot experience strong interactions if their wavefunctions do not overlap significantly. Without any overlap, electrostatic repulsive forces would destabilize the nuclei not only in the old Rutherford model but even in the most up-to-date quantum mechanics (QM) treatment. We saw in Sec. 8.5 that a mismatch in the overlapping wavefunctions is unavoidable under the Heisenberg uncertainty principle and produces the nuclear forces that keep the nucleons together. We will discuss the QCD origin of nuclear forces in Sec. 9.6.

We know already from the **isospin symmetry** of Sec. 8.1 that a quark scattering does not depend on the quark flavor in the limit of mass-degenerate quarks. Protons and neutrons are thus identical particles in QCD, as long as the $m_u = m_d$ approximation holds. They are called **nucleons** and labeled N.

Due to its time-honored history, nuclear physics has its own terminology we should know:

- A **nucleus** is a bound state of Z protons and $A - Z$ neutrons. It is indicated as ^A_ZName or ^A_ZName$_N$ where Name is the name of the element in the Periodic Table. This notation is redundant because the position of Name in the Periodic Table already provides Z, while the number of neutrons N is just $A - Z$. Chemists and nuclear physicists drop out the redundant information and write ^{A}Name like ^{2}H, ^{7}Li, ^{235}U, etc. For instance, ^{2}H is a nucleus with $A = 2$ and $Z = 1$. $Z = 1$ corresponds to the "hydrogen" element in the first row of the Periodic Table. Therefore, ^{2}H is a proton–neutron bound state, the **deuteron**[2].

- Two nuclei with the same Z but a different number of neutrons are called nuclear **isotopes**. The hydrogen, deuterium, and tritium nuclei are isotopes of the same element of the Periodic Table and are labeled as H $\equiv$ ^{1}H, ^{2}H, and ^{3}H. The **natural abundance** is the probability to find an isotope in the earth compared to all other isotopes of the same element. It is expressed as the percentage of atoms with a given isotope found in a representative sample of the earth's composition.

- **Isobars** are nuclei of different chemical elements that have the same number of nucleons: ^{40}Ar, ^{40}K, and ^{40}Ca are a chain of isobars. You may wonder whether isobars – which means "equal-weight nuclei" – have actually the same weight. This is true only if we assume $m_n = m_p$ and neglect the binding energy of the bound state.

- Two nuclei with the same number of neutrons but different Z are called **isotones**. Two light isotones are helium-4 (^{4}He) and the tritium nucleus (^{3}H), where $N = 2$. Note that the former is stable and has a natural abundance close to 100%. The latter is extremely rare and decays by weak interactions in about 12 years. To the best of my knowledge, the word isotone has no special meaning. It is probably a word pun where the "p" of "isotope" is replaced by a "n" to signify the equality in the number of neutrons instead of protons (Z).

- Two nuclei with the same A and Z but in different excited states are called **isomers** if the excited-state lifetime is sufficiently long. The word was introduced by F. Soddy in 1917. This term is usually restricted to isomers with half-lives of 10^{-9} seconds or longer. Otherwise, they are called excited states with prompt decays. Both types reach the ground state emitting photons and are labeled with

[2]In this case, deuteron indicates the nucleus and deuterium the corresponding atom. Most of the names of the nuclei are no more in use, like the "protium", the nucleus of hydrogen (i.e. the proton), or the "triton," the nucleus of tritium. A good practice is to state explicitly the atom or nucleus we are speaking of ("the tritium nucleus" instead of triton).

an asterisk; for example, ^{57}Fe* is a meta-stable state of ^{57}Fe with a lifetime of 141 ns.

This notation is useful but should be handled with care to avoid confusion when dealing simultaneously with atomic and nuclear physics concepts. For instance, we prefer the word **nuclide** when we mean a class of nuclei, that is, a specific nuclear species. The difference between nucleus and nuclide is then like the difference between "atom" and "element."

There are subtleties when we define quantities that appear in the Periodic Table, too. The **atomic number** Z is the number of protons in the nuclide and defines the element in the Periodic Table. Luckily, there is no ambiguity here. The **mass number** A of a nuclear species is the sum of protons and neutrons in a nuclide and, sometimes, is called the **atomic mass number**. A is an integer and it is *not* the number that appears as the atomic weight in the Periodic Table of the elements. The mass number is approximately equal to the **atomic mass** or, more precisely, to the **isotopic mass**: this is the mass of an atom with a given number of protons and neutrons, in other words the mass of an atom made of a given isotope and its electrons. Unlike the dimensionless number A, the atomic mass is expressed in kg or, more commonly, in atomic mass units (u). Note that the atomic mass is not simply the sum of the masses of the constituents of the atom ($Zm_e + Zm_p + (A-Z)m_n$) because the sum must be corrected for the binding energy of nucleons and electrons. The electron binding energy can be safely neglected, but the binding energy of the nucleons may produce a significant **mass defect**. For instance, ^{35}Cl has a mass number $A = 35$ but the atomic mass (isotopic mass) is 34.96885 u. We can also define the corresponding dimensionless quantity expressing the atomic mass in units of u. This is the **relative atomic mass** and is labeled A_r. A_r for ^{35}Cl is 34.96885 and is dimensionless.

The "mass" quoted in the Periodic Table is the **standard atomic weight**, also called the **standard relative atomic mass**. For an atom, it is the arithmetic mean of the relative atomic masses of all isotopes of that element weighted by each isotope's abundance on earth. The standard atomic weight is dimensionless, but we can get it in kg multiplying this quantity by u.

Most of the readers are acquainted with the concept of **(unified) atomic mass unit**. This quantity is indicated as u (formerly *amu* or Dalton) and corresponds to 1/12 of the mass of an unbound atom of ^{12}C. From the considerations above, we expect $u < m_p \simeq m_n$ because of the nuclear binding energy and the presence of the $Z = 6$ electrons. Thus,

$$u \simeq m_N - \frac{E_{bn}}{12} - \frac{E_{be}}{12} + \frac{6m_e}{12} \qquad (9.11)$$

where m_N is the average mass of the nucleon $(m_p + m_n)/2$, E_{bn} and E_{be} are the binding energy of the nucleus and the electrons, and m_e is the electron mass. The average value of E_{bn} is about 8 MeV but there are many nuclides that make exceptions (see Sec. 9.4.1). The value of u is:

$$u = 1.66053906660(50) \times 10^{-27} \text{ kg (SI)}$$
$$= 931.49410242(28) \text{ MeV/c}^2 \text{ (NU).} \qquad (9.12)$$

As usual, we need the definition of mole and Avogadro number to go from a microscopic atomic mass to a macroscopic amount of material. Since November 2018, a **mole** of a substance made, for example, of unbound atoms is simply a set of $6.02214076 \times 10^{23}$ atoms. This number – the **Avogadro number** N_A – was chosen so that the mass of one mole of a substance in grams is numerically equal to the average mass of one atom of the compound expressed in u. This rule can be applied not only to unbound atoms (the only case of interest in this book) but also to molecules, metals, and complex compounds such as polymers of indeterminate molecular size.

u is pretty close to m_p but not identical to it. Since the mass in u of an atom is numerically close, but not exactly equal, to the number of nucleons A contained in the nucleus, the molar mass (grams per mole) of a substance made, for example, of a single unbound atom, is $\simeq$ Am_pN_A. The molar mass of unbounded hydrogen atoms corresponds to $\simeq 1$ gram but the actual value is slightly larger (1.008 g/mole). Similarly, the molar mass of a hydrogen molecule H_2 is 2.016 g/mole (twice the atomic molar mass of H). Then, unless an exceptional precision is requested, the contributions of the electron binding energy can be neglected and most of the difference with respect to the naïve calculation is due to the presence of the isotopes and the nuclear mass deficit (nuclear binding energy). Anyway, nuclear and particle physicists should be aware of these subtleties. Some degree of pedantry is necessary, for example for precise cross-section calculations, which depend on the target mass, and to avoid misunderstandings with atomic scientists and chemists.

9.3 NN states and the deuteron

The deuteron, the nucleus of the deuterium atom, is a bound state of a proton and a neutron. It is the simplest nuclear bound state and plays a role similar to the hydrogen atom in atomic physics. There are two ways to study the deuteron. The first one is based on lattice QCD where the interacting particles are the six quarks bounded in two color singlets (p and n). The second one is purely empirical and is based on the experimental evaluation of the average potential $V(r)$ among the nucleons, which is used to solve the corresponding Schrödinger equation. It goes without saying that the second approach was the only technique available up to 1973, while the lattice QCD approach is still in its infancy.

Figure 9.3 (left) shows a lattice QCD calculation of the deuteron potential compared with modern semi-empirical potentials extracted

Fig. 9.3 Left: Effective NN mean potential computed by lattice QCD. The errors estimate the numerical uncertainty and the systematics due to various approximations. Right: Three examples of modern semi-empirical NN potentials. The dashed vertical lines show the range where light meson exchange dominates. Redrawn under CC-BY-03 from Ishii *et al.*, 2012 (Copyright 2012 Elsevier) and with permission from Ishii *et al.*, 2007 (Copyright 2007 American Physical Society).

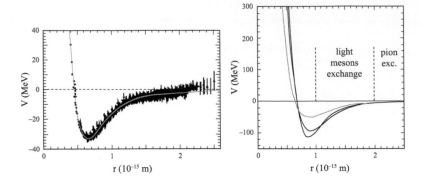

from experimental data. As expected, if the distance between the two nucleons is small, the potential is strongly repulsive because the superposition of the wavefunctions results in a local violation of the confinement conjecture (neutrality of all color charges), that is the nucleon experiences a net color force that restores the minimum distance among the two nucleons. This **repulsive core** is well reproduced by lattice QCD despite the approximations made in the numerical calculations. The behavior at intermediate distances is still beyond the capabilities of lattice QCD. In this range, scattering experiments can be interpreted as an exchange of a color singlet of moderate mass – mostly a light meson – between the neutron and the proton. If the distance is even larger, only the lightest mesons contribute to the force. This phenomenon resembles what happens in perturbative quantum electrodynamics (QED):

> a Feynman diagram where the off-shell particle is massive ($M \neq 0$) and $\sqrt{s}$ is far from M (i.e. far from the resonant production of the particle M) suppresses the scattering amplitudes as M^{-2}

but, unfortunately, it occurs in the non-perturbative regime and the demonstration is highly non-trivial. Since the lightest meson is the pion, the large distance dynamics of a NN state is dominated by the **pion exchange**. A common method that originates from the seminal works of Y. Yukawa in 1930 is to build a quantum field theory (QFT) where the pion plays a role similar to the photon in QED but with a mass different from zero and compute the $p + n \rightarrow p + n$ amplitude using Feynman diagrams with an off-shell π exchange. This approach provides a reasonable approximation of the potential at large distances. Some interesting results can be obtained even in the intermediate range of distances adding heavier mesons to the pions.

The solution of the Schrödinger equation for the potential of Fig. 9.3 (right) shows stationary states both with discrete energies (bound states) and in the continuum (states with free nucleons). The deuteron can be produced as a stable bound state with a binding energy $E_b = 2.224$ MeV. This number corresponds to the minimum energy needed to reach a continuum state and break the deuteron. It plays the same role as the first ionization energy of the hydrogen (13.6 eV) in atomic physics. These values set the energy scales for atomic (1–1000 eV) and nuclear-physics processes (1–10 MeV). Using the same techniques, we can demonstrate that:

> no stable bound states exist for $n - n$ and $p - p$ systems.

In particular, the **diproton** ($p - p$ system) is highly unstable due to the spin-dependent part of the nuclear potentials and the Pauli exclusion principle, which conspire to lower the size of E_b. Some evidence of the **dineutron** ($n - n$ states) was gained in 2012 at the Univ. of Michigan and interpreted as an extremely short-lived state produced by nuclear reactions. These states are somehow in between a real bound state and a system of two co-propagating loosely interacting particles. By no means can they be interpreted as states similar to the positronium or a hypothetical "neutronium" sitting in the $Z = 0$ column of the Periodic Table. It is worth noting that the diproton would be stable if strong interactions were a bit larger at this energy. Since the diproton production would change substantially the evolution of the early universe, this argument is one of the favorite examples to support the **anthropic principle** (Carter, 1974; Barrow and Tipler, 1988), the principle stating that the data we collect on the universe is biased by our very existence because a universe unable to host intelligent life can exist (and maybe exists) but cannot be observed. Citing Carter, "we must be prepared to take account of the fact that our location in the universe is necessarily privileged to the extent of being compatible with our existence as observers."

9.3.1   Spin and parity of the deuteron

As anticipated in Sec. 5.6.2, the total angular momentum of the deuteron results from the combination of the nucleon spins (1/2 each) and the orbital angular momentum. The ground state has $l = 0$ and – from the solution of the corresponding Schrödinger equation – deuteron bound states exist only if the spin of the proton and neutron are parallel. Hence, the total momentum is equal to 1. This result is testified by atom spectroscopy through the study of the deuterium's hyperfine structure. It is customary to label the angular momentum of the electrons and the nuclide as $\mathbf{J}$ and $\mathbf{I}$, respectively.[3] The total angular momentum of the atom is then $\mathbf{F} = \mathbf{J} + \mathbf{I}$, while the total angular momentum of the nuclide (the **nuclear spin**) is just $\mathbf{I}$. Since the hyperfine structure of the atoms depends on $\mathbf{F}$, spectroscopy data provide information on $\mathbf{I}$, too.

The **intrinsic parity** of the deuteron follows from the usual factorization of the wavefunction for a two-body system and is $(-1)^l P_n P_p = (-1)^l$ because the intrinsic parities of the nucleons are both $+1$. The deuteron bound state has, then, positive intrinsic parity. The angular

[3] $\mathbf{I}$ should not be confused with the isospin of Chap. 8. Isospin in nuclear physics is always indicated as $\mathbf{T}$ and we will use the nuclear physics notation here and in Chap. 11 to ease the comparison with literature.

momentum + parity information is commonly summarized as I^P. For the deuteron $I^P = 1^+$.

9.4   Multi-body nuclides

Like atoms, going from a two-body to a multi-body system ($Z > 2$) poses major challenges. They are particularly severe in nuclear physics, where the two-body problem is already non-integrable, and QCD runs low on reliable predictions. We believe that, sooner or later, QCD will play a role in multi-body nuclear physics but, to date, most of the information is drawn from semi-empirical models and symmetries.

Nuclear physicists developed a plethora of semi-empirical models, each with strengths and weaknesses, over more than a century of research. The classical models are the Fermi gas model to estimate the momentum distribution of the nucleons, the liquid drop model to substantiate the empirical mass formulas for nuclides, the collective model to predict the magnetic and electric moments of the nucleus, and the shell model. In this chapter, we will discuss only the shell model, which is inspired by atomic physics and fully compatible with QM. It provides predictions on the spin and parities of most nuclides and gives a proper setting to discuss nuclear decays in Chap. 11, links to QCD, and other modern approaches.

9.4.1   The shell model

Treating an ensemble of tightly packed nucleons as an atom around a central potential seems to be at least far-fetched, if not completely useless. Still, in the 1930s, E. Wigner, M. Goeppert-Mayer, and J.H.D. Jensen noted some striking similarities between the binding energy of the atoms and the ones of the nuclides. Stability in atoms is achieved when the outer shell is filled with electrons and the atoms with the tightest bonds are typically noble gases where all low-lying shells are occupied. Peaks of stability were observed also in the **Chart of Nuclides**[4] (IAEA, 2020) shown in Fig. 9.4. The tightest bounds appear regularly in the form of special values of Z and N. These values are called **magic numbers**: 2, 8, 20, 28, 50, 82, and 126. The existence of magic numbers was associated with the complete filling of a **nuclear shell**. Nuclides, where both Z and N belong to the set of magic numbers, are called **doubly magic** and are particularly tightly bounded. The postulates of the **shell model** are:

- all nucleons are preexisting inside the nuclide. Each nucleon generates a force field that affects the dynamics of all other nucleons

- the motion of a nucleon is determined by an effective mean potential generated by all other nucleons

- each nucleon can occupy a given energy level inside the nuclide in compliance with the Pauli exclusion principle

[4]The Chart of Nuclides or "Segrè chart" is the plot describing the nuclides in the Z versus $N \equiv A - Z$ plane (or vice versa). In nuclear physics, it plays a role similar to the Periodic Table of the Elements. The plot shows that the most stable nuclides have $Z/A \simeq 1/2$ if Z is sufficiently low, while high-Z stable nuclides are richer in neutrons ($Z/A < 1/2$). A strong imbalance between neutrons and protons generally produces very short-living isotopes.

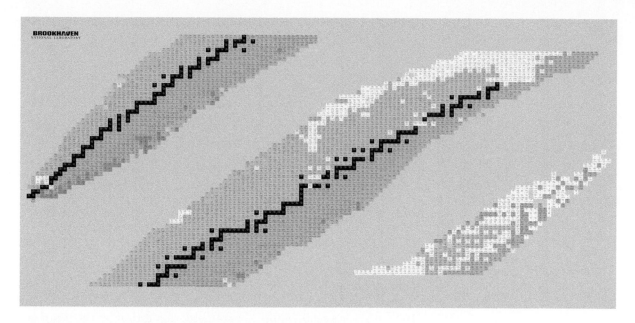

Fig. 9.4 The Chart of Nuclides. The chart shows all known nuclides in the Z versus $N \equiv A - Z$ plane. To ease the reading, the plot is split into three areas corresponding to different N ranges. The nuclides with the lowest Z and N are shown in the upper-left corner. Each box represents a nuclide. The colors (in grayscale) indicate the location of stable (black) and unstable nuclides that decay in different modes: α decay (white), β^+ decay or electron capture (dark gray), and β^- decay (light gray, typically on top of the black boxes). These decay modes are explained in Chap. 11. Image available in the public domain – Courtesy of the US Department of Energy and BNL.

- fully occupied **shells** (a set of discrete energy levels defined by some quantum numbers) correspond to very stable nuclei, that is, nuclei where the binding energy is large.

This formulation sidesteps the problem of the orbits: if all low-lying states are occupied by nucleons, a scattering among two nucleons can produce an effect only if the nucleon is moved to the upper **valence level**. If the energy transferred by an N–N scattering is smaller than the energy gap among levels, there is no way to relocate the nucleon to the valence level. As a consequence, a nucleon can orbit even having an arbitrarily large number of scatterings, but the probability to find it inside its level remains constant in time. In this way, the shell model reproduces the concept of orbiting fermions without resorting to the large volumes needed in atomic physics.

9.4.2 Mean potential of the shell model

The most natural choice for the effective potential would be a central potential, whose shape is fitted to reproduce all magic numbers. Unfortunately, such potential does not exist. The closest approximation is the **Saxon–Wood potential** V_{SW} defined as:

$$V_{SW}(r) \equiv -\frac{V_0}{1 + e^{(r-R)/a}} \qquad (9.13)$$

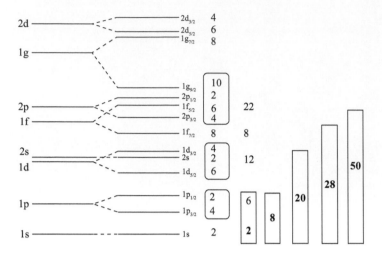

Fig. 9.5 Levels available to a single nucleon from the shell model with both $V_{SW}(r)$ and $V_{LS}(r)$. The boxes on the right indicate the shells. The rectangles show the number of protons (or neutrons) that are needed to completely fill these shells. The numbers inside (in bold) correspond to the magic numbers.

where V_0, R and a are free parameters. Note that R is a sort of classical radius of the nucleus and is expected to grow as $r_0 A^{1/3}$ with $r_0 = 1.2 \times 10^{-15}$ m. r_0 is a semi-empirical parameter that estimates the mean radius of a nucleon. The Saxon–Wood potential reproduces the magic numbers up to 20 but fails at higher values. The breakthrough in model building occurred in 1949 when Goeppert-Mayer and Jensen introduced another ingredient inspired by atomic physics, a **spin–orbit interaction** term proportional to $\mathbf{L} \cdot \mathbf{S}$ that reads:

$$V = V_{SW}(r) + V_{LS}(r)\mathbf{L} \cdot \mathbf{S}. \tag{9.14}$$

Even if the formalism of spin–orbit interactions is very similar to electromagnetic interactions, its origin is deeply rooted in non-perturbative QCD and only partially understood to date. A nuclide can, therefore, be described by:

$$\mathbf{L} \cdot \mathbf{S} = \frac{1}{2} \left(\mathbf{I}^2 - \mathbf{L}^2 - \mathbf{S}^2 \right) \tag{9.15}$$

where $\mathbf{L}$ and $\mathbf{S}$ are the angular momentum and spin of a *single* nucleon. The expectation value in NU is thus:

$$\langle \mathbf{L} \cdot \mathbf{S} \rangle = \frac{1}{2} \left[i(i+1) - l(l+1) - s(s+1) \right] \tag{9.16}$$

like in atomic physics, and $s = 1/2$. Then,

$$\langle \mathbf{L} \cdot \mathbf{S} \rangle = \frac{1}{2} \left[i(i+1) - l(l+1) - s(s+1) \right] = \begin{cases} \frac{l}{2}, & \text{if } i = l + \frac{1}{2} \\ -\frac{(l+1)}{2}, & \text{if } i = l - \frac{1}{2} \end{cases}. \tag{9.17}$$

The energy splitting among levels increases with l because

$$\langle \mathbf{L} \cdot \mathbf{S} \rangle|_{i=l+1/2} - \langle \mathbf{L} \cdot \mathbf{S} \rangle|_{i=l-1/2} = \frac{1}{2}(2l + 1) \tag{9.18}$$

but we can revert the ordering choosing empirically $V_{LS}(r) < 0$. We can then use the standard spectroscopic notation nl_i to classify the nuclides using the principal quantum number n, the orbital quantum number

l, and the total angular-momentum quantum number i. The i quantum number is a function of l and the nucleon spin number s. In spectroscopic notation, $l = 0, 1, 2, 3$ correspond to s, p, d, f states. Hence, a nucleon in the $n = 4, l = 1$ level with spin $1/2$ occupies either the $4p_{3/2}$ or the $4p_{1/2}$ level. Fig. 9.5 demonstrates how an appropriate choice of $V_{LS}(r)$ is able to reproduce all magic numbers for a single nucleon with the effective potential of eqn 9.14.

9.4.3 Classification of nuclides in the shell model

To apply the shell model to a real nuclide, we first need to fill the lowest levels with all protons accounting for the Pauli exclusion principle. Hence, a nl level can host at most two protons: one in $nl_{l+1/2}$ and the other in $nl_{l-1/2}$. We must repeat the same procedure filling the *same* levels with neutrons. The degeneracy of each level, that is the maximum number of protons or neutrons that can host, is $2i + 1$. A nuclide like $^{15}_{8}O_7$ has:

Protons:	Neutrons:
$1s_{1/2}$ full with two p	$1s_{1/2}$ full with two n
$1p_{3/2}$ full with four p	$1p_{3/2}$ full with four n
$1p_{1/2}$ full with two p	$1p_{1/2}$ partially filled with one n

and most of its properties depends on the partially filled shell $p_{1/2}$. The neutron in the partially filled shell is the **valence nucleon** of ^{15}O.

9.4.4 Spin and parity

The shell model predicts that the physical properties of the nuclei are mostly determined by the outer shells. As a consequence, ^{15}O is expected to have spin $1/2$ because $I = 1/2$ for the last occupied level. Since a p-state has $L = 1$, the nuclide has an odd parity ($I^P = 1/2^-$). These predictions are in agreement with experimental data. Similarly, $^{17}_{8}O_9$ (see Fig. 9.6) is dominated by a partially filled $1d_{5/2}$ level that hosts just one neutron. The nucleon spin is then $I = 5/2$ but now the parity is even ($I^P = 5/2^+$).

The shell model can determine the **spin-parity** of all ground states of nuclei. Shells that are completely filled must have a total angular momentum equal to zero because the degeneracy of the shell (i.e. the number of levels per shell) is $2I + 1$, which is always odd. Hence, for each I_z, we have in the same level a nucleon with $I'_z = -I_z$ such that:

$$\sum_{k=0}^{2I+1} (I_z)_k = 0. \tag{9.19}$$

All doubly magic nuclides are, therefore, spinless, and this prediction is confirmed by experimental data. On the other hand, the shell model is unable to compute the spin of all other nuclides. Still, the experimentalists demonstrated since long that all nuclides with even-N and

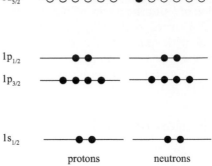

Fig. 9.6 Filling of the shells of ^{17}O for protons ($Z = 8$) and neutrons ($N = 9$). The black dots correspond to filled levels and the empty dots stand for empty levels. The spin-parity of the nuclide is determined by the last unpaired neutron in the $1d_{5/2}$ level ($l = 2$, $I = 5/2$). The spin-parity of the nuclide is thus $\frac{5}{2}^{+}$. Redrawn from Krane, 1988.

even-Z are spinless, even if the last shell is only partially filled. We can supplement the shell model with the **pairing hypothesis**:

> all pairs made of a neutron and a proton in a given subshell always couple to give a combined angular momentum equal to zero, even when the subshell is not filled. This rule, however, holds for the ground state, only.

The most striking consequence is that the nuclear spin only depends on the last uncoupled nucleon. Nuclides with odd-A have just one unpaired neutron, whose level determines the net nuclear spin of the nuclide (see Fig. 9.5). Unfortunately, even with the pairing hypothesis, we cannot predict the spin of nuclides with even-A but both odd-Z and odd-N because we have no information on the angular momentum orientation of the unpaired nucleons. In this ill-fated occurrence, the pairing hypothesis provides at least the parity of the nuclide:

> the total parity of a nuclide is the product of the parities of the last proton and the last neutron.

This prediction has been successfully tested by nuclear experiments, too. Despite these successes, the shell model remains a semi-empirical model. It is disconnected from QCD and relies on *ad-hoc* assumptions like the nuclear spin–orbit interaction and the pairing hypothesis. It also fails in predicting the nuclear magnetic moments (Martin, 2009) and brings no information on nuclear mass.

9.4.5   Island of stability

The shell model predicts additional magic numbers as 184, 258, etc. The largest magic number ever observed is 126 for neutrons and 82 for protons. Large numbers cannot be produced experimentally for values much greater than $Z = 110$. The element with the largest Z, however, has $Z = 118$ and $N = 177$ and was discovered in 2002 in Russia. This element, called oganessum, has a lifetime of about 1 ms and is not far from a $N = 184$ nuclide or even a hypothetical doubly magic $Z = 126$

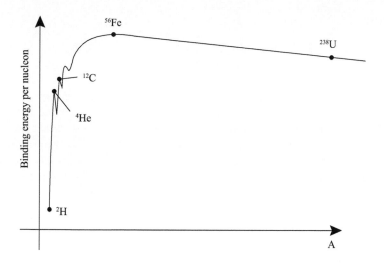

Fig. 9.7 Binding energy per nucleon as a function of A up to uranium ($A = 238$). Along the curve that interpolates the values, we point out a few special nuclides (^{4}He, ^{12}C, etc.), whose binding energy is particularly large. The largest energy corresponds to ^{56}Fe.

and $N = 184$ nuclide. There is a consensus that around these values the Chart of Nuclides should unveil an **island of stability**, that is, a set of nuclides with a lifetime significant higher than most trans-uranic elements ($Z > 92$). At present, however, this is beyond our experimental capabilities. Besides, since the shell model remains a semi-empirical model with several free parameters (V_{LS}, V_{SW}, etc.) and hypotheses, any extrapolation at high A-values is rather questionable.

9.5 The mass of the nuclides

The semi-empirical models provide some hints on the actual mass of the nuclides but no deep explanation of them because the models are disconnected from the findings of QCD. In this section, we will discuss these hints and their most celebrated result: the semi-empirical mass formula. A more profound approach emerging from QCD is presented in Sec. 9.6.

9.5.1 The SEMF

Nuclear physicists measured the binding energy of many nuclides immediately after the discovery of the neutron, and, by the end of the second World War, this catalog comprised hundreds of them. Figure 9.7 shows the binding energy per nucleon ($B(Z, A)/A$) as a function of the mass number A of the nuclide. The smoothing of Fig. 9.7 can be described by a semi-empirical formula inspired by the liquid drop model and written in 1935 by C.F. von Weizsäcker. This formula is called the **SEMF** – semi-empirical mass formula – or the **(Bethe)–Weizsäcker formula** and describes the binding energy per nucleon quite well in a broad range of A. The SEMF is a formula that provides the atomic mass $m_A(Z, A)$:

$$m_A(Z, A) = Z(m_p + m_e) + (A - Z)m_n - B(Z, A) \qquad (9.20)$$

where

$$B(Z, A) \equiv a_v f_1(A) + a_s f_2(A) + a_c f_3(A, Z) \qquad (9.21)$$
$$+ a_a f_4(A, Z) + a_p f_5(A, Z) \qquad (9.22)$$

is the binding energy. For a bound state, the binding energy must be positive because the mass of the nuclide must be smaller than the mass of its constituents. Note that the SEMF gives an *atomic mass* and not a nuclear mass, and is expressed in units of u. It then accounts for the electrons and their binding energies, too. Fortunately, in nearly all cases of interest, the binding energy of the electrons can be neglected. The relationship between atomic and nuclear mass is then trivial:

$$m_N = m_A - Z m_e + \sum_{i=1}^{Z} B_i^e \simeq m_A - Z m_e. \qquad (9.23)$$

Table 5 shows the values of the a's terms, which must be fitted to reproduce Fig. 9.7, and the functions $f_1, \ldots f_5$ together with their physical interpretation[5].

[5] The interpretation of the terms of the SEMF is quite arbitrary and is based on semi-empirical corrections inspired by the liquid drop model.

The term $f_1(A) = a_v A$ is called the **volume term** and comes directly from the liquid drop model. In this model, the nucleus is a droplet of an incompressible fluid and, therefore, its energy is, first of all, dominated by the volume of the fluid. Assuming the same dimension for protons and neutrons, the volume is proportional to A. The second term $f_2(A) = -a_s A^{2/3}$ is a heritage of the liquid drop model, too. Nucleons located at the surface of the drop are not fully surrounded by other nucleons and experience a smaller force. This is why the **surface term** is negative and proportional to the surface of a sphere:

$$a_a \sim 4\pi r^2 = 4\pi(\sqrt[3]{A})^2 \sim A^{2/3}. \qquad (9.24)$$

The third term has a classical electrostatic origin and is called the **Coulomb term**. It arises from the repulsion of the protons packed inside the nucleus in between the neutrons. If we consider a uniform charge distribution (Martin, 2009), whose radius is proportional to $A^{1/3}$, the electrostatic potential energy is $\sim -Z(Z-1)/A^{1/3}$ and

$$f_3(A, Z) = -a_c \frac{Z(Z-1)}{A^{1/3}} \simeq -a_c \frac{Z^2}{A^{1/3}}. \qquad (9.25)$$

The approximation on the right of eqn 9.25 holds only for large values of Z. In the case of the Coulomb term, f_3 depends both on A and Z.

The last two terms of the SEMF are directly inspired by QM and the shell model. The Pauli exclusion principle favors nuclides where $N \equiv A - Z \simeq Z$, that is, nuclides with the same number of neutrons and protons. This principle is encoded into the SEMF by the **asymmetry term** that penalizes nuclides with a large imbalance between protons and neutrons. For the sake of simplicity, we suppose that the energy difference between two nucleons in adjacent shells is constant and amounts[6] to δ. If we remove a proton from one level (e.g. level $1p_{1/2}$ in Fig. 9.6 left) and put a neutron in the upper level (level $1d_{5/2}$ in the right part of Fig. 9.6),

[6] See Fig. 9.6 but assume that all levels have the same vertical distance δ.

the energy increases by δ. The move corresponds to changing (A, Z, N) to $(A, Z-1, N+1)$ and $N-Z$ is now equal to 2. It increases the energy of the nuclide by δ and tends to destabilize the system. In Sec. 9.5.2, we will see that repeating this move many times, the value of δ is reduced and the increase of energy corresponds to the penalty term[7]:

$$f_4(A, Z) = -a_a \frac{(A - 2Z)^2}{A}. \qquad (9.26)$$

The fifth term of the SEMF originates from the pairing hypothesis of the shell model. Nucleons, whose wavefunctions are highly overlapped because of the short distance, tend to couple producing a tight bond. In particular, when both Z and N are odd, the two nucleons located in the outermost shell create a pair, even if their species is different (a proton and a neutron), producing a spin-0 singlet. We can further reduce the binding energy transforming a proton into a neutron or vice versa, so that nuclides with odd-Z and odd-N are disfavored. An empirical parametrization of this effect is given by the **pairing term** of the SEMF that reads:

$f_5(A, Z) = +a_p A^{-1/2}$ for Z–even and N–even
$f_5(A, Z) = \ \ 0$ \qquad for Z–even and N–odd or viceversa
$f_5(A, Z) = -a_p A^{-1/2}$ for Z–odd and N–odd

Note that the functional form of f_5 is purely empirical and the choice of $\pm a_p A^{-1/2}$ is driven by the quality of the fit of Fig. 9.7. A form like $\pm \tilde{a}_p A^{-3/4}$ also provides a good fit (Krane, 1988) and may be employed, especially in particular ranges of A.

[7]We use the notation of Martin, 2009 with updated values of the SEMF coefficients instead of the popular notation of Krane, 1988. The two notations differ by a factor of four in the value of a_a but have the same physical meaning.

Term	Value (MeV/c^2)	Function	Interpretation
a_v	15.8	$a_v A$	volume term (a)
a_s	17.8	$-a_s A^{2/3}$	surface term (b)
a_c	0.71	$-a_c Z(Z-1)A^{-1/3}$	Coulomb correction (c)
a_a	94.8	$-a_a(Z - A/2)^2/A$	asymmetry term (d)
a_p	11.2	$\pm a_p A^{-1/2}$ or 0	pairing term (e)

Table 9.1 The empirical coefficients of the SEMF, their values in MeV/c^2, the functional form of $f_1 \ldots f_5$, and their physical interpretation. (a) the attractive force per nucleon depends on the number of nucleons surrounding it. (b) nucleons close to the surface are affected by a smaller number of surrounding nucleons and the correction is proportional to the surface of the nucleus. (c) electrostatic forces repel protons reducing the stability of the nuclide. (d) nuclides are more stable if $N = Z$ ($A - 2Z = 0$) but this stability line moves toward $N > Z$ for large A (e) due to the pairing hypothesis of the shell model, even–even nuclides are more stable (+ sign) than odd–odd nuclides (− sign). Pairing effects are negligible for even–odd or odd–even nuclei ($a_p = 0$). See text for details.

9.5.2 Fermi gas model and the pairing term of the SEMF

E. Fermi attacked the problem of predicting Fig. 9.7 in a different way. He considered the nucleus as an ensemble of two uncorrelated fermion species (p and n) in a finite volume: the volume of the nuclide. Each species fills the available energy levels from bottom to top according to the Pauli exclusion principle, reaching the highest level that corresponds to a (non relativistic) momentum $p_F = \sqrt{2ME_F}$. E_F is the maximum energy reached by a nucleon and is called the **Fermi energy**. M is the mass of the nucleons. The Fermi–Dirac statistics provide the density of available states within p and $p + dp$. In SI, it reads:

$$n(p)dp = \frac{4\pi V}{(2\pi\hbar)^3}p^2 dp \; . \tag{9.27}$$

Integrating eqn 9.27 from 0 to p_F we can compute the number of neutrons and protons:

$$N = \frac{V(p_F^n)^3}{3\pi^2\hbar^3} \; ; \quad Z = \frac{V(p_F^p)^3}{3\pi^2\hbar^3} \; . \tag{9.28}$$

V can be evaluated experimentally measuring the radius R of the nuclide by scattering experiments and

$$V = \frac{4}{3}\pi R^3 \simeq \frac{4}{3}\pi R_0^3 A \tag{9.29}$$

where R_0 is the average nucleon radius and amounts to 1.21×10^{-15} m (1.21 fm). We can then estimate the **Fermi momentum** p_F as:

$$p_F \simeq p_n \simeq p_p = \frac{\hbar}{R_0}\left(\frac{9\pi}{8}\right)^{1/3} \simeq 250 \text{ MeV}/c. \tag{9.30}$$

This is an astonishing result. If this **Fermi gas model** is a faithful representation of the nucleus, a particle scattered by a nucleus will not see the constituents at rest, but the nucleons will be moving randomly with momentum $0 < p < p_T$ and energy $0 < E < E_F$. Since the 1970s, we have impressive experimental evidence of this prediction gained by the systematic study of neutrino–nucleus scattering. The existence of the **Fermi motion** produces an intrinsic spread of the kinematics of the outgoing particles and is today one of the leading systematics of neutrino cross-section measurements (Sajjad Athar and Singh, 2020).

The Fermi gas model provides a simple explanation of the pairing term in the SEMF (Martin, 2009). The total kinetic energy of the nucleus is the sum of the average energy of the protons and neutrons:

$$E_{kin} = N\langle E_{kin}^n\rangle + Z\langle E_{kin}^p\rangle. \tag{9.31}$$

$\langle E_{kin}^n\rangle$ and $\langle E_{kin}^p\rangle$ can be computed integrating the energy over the the probability density function of eqn 9.27:

$$\langle E_{kin}\rangle = \frac{\int_0^{p_F} E_{kin}(p)n(p)dp}{\int_0^{p_F} n(p)dp} = \frac{\int_0^{p_F} E_{kin}p^2 dp}{\int_0^{p_F} p^2 dp} \; . \tag{9.32}$$

The result is:

$$E_{kin} = \frac{3\hbar^2}{10MR_0^2} \left(\frac{9\pi}{4}\right)^{2/3} \left(\frac{N^{5/3} + Z^{5/3}}{A^{2/3}}\right) \qquad (9.33)$$

as long as the radii of the proton and neutron are the same. This function has a minimum at $N = Z$, as expected. We can now move around a neighbourhood of the minimum using as a perturbation the difference between N and Z. The minimum of E_{kin} is recovered for $\epsilon \equiv N - Z \to 0$. N is thus $(A+\epsilon)/2$ and Z is $(A-\epsilon)/2$. If we perform a Taylor expansion of eqn 9.33 around ϵ/A, as in Exercise 9.14, we find:

$$E_{kin} = \frac{3\hbar^2}{10MR_0^2} \left(\frac{9\pi}{8}\right)^{2/3} \left(A + \frac{5}{9}\frac{(N-Z)^2}{A} + \dots\right). \qquad (9.34)$$

The first term is equivalent to the volume term of the SEMF, while the second term is proportional to:

$$\frac{(N-Z)^2}{A} = \frac{(A-2Z)^2}{A} \qquad (9.35)$$

and gives $f_5(A, Z)$ within a factor of ~ 2.

9.6 The QCD origin of nuclear mass

Empirical models are extremely useful to estimate the mass of nuclides in a given region of A and for a vast portfolio of applications. However, they do not attack the mass problem in the most natural way: relying either on ab-initio QCD calculations or using the symmetries of QCD, as done in quark models. Modern nuclear physics is addressing this challenge with more appropriate tools since we are aware of the fundamental theory that leads to the formation of nuclei.

9.6.1 Isospin in nuclear physics

The pioneer of this approach was E. Wigner, who followed the modern path well before the discovery of QCD. Wigner benefited from a finding of Heisenberg already mentioned in Sec. 8.3. Heisenberg noted that the scattering of neutrons and of protons have the same amplitudes, except for a tiny correction due to the electromagnetic forces. He suspected an underlying symmetry and a corresponding conserved quantity. He called the quantity "isospin," which is a contraction of **isobaric spin**. In Heisenberg's view, this symmetry exchanges neutrons with protons and, therefore, creates isobars. If we assign $I_z = +1/2$ to the proton as we did for the up quark and $I_z = -1/2$ to the neutron as we did for the down quark, we can build "isomultiplets" using the same composition rules of the spin. At that time, "isospin" or "isobaric spin" was a fascinating name: an internal symmetry that acts like an external one (rotations in space), and the name chosen by Heisenberg is still in use. For us, it is an unfortunate choice because isospin recalls the name "spin," which has nothing to do with the flavor symmetry of QCD. Nevertheless, this story explains why the isospin symmetry was introduced 30 years before the discovery of the flavor symmetry and quarks.

Wigner was aware of Heisenberg's isospin symmetry and used systematically the isospin for the classification of the nuclei in a way that resembles what Gell-Mann and Zweig did for the classification of hadrons.

The isomultiplets are then a set of nuclides that can be assembled combining each nucleon with isospin $1/2$ and $I_z = +1/2$ (proton) or $I_z = -1/2$ (neutron) using the $SU(2)_f$ composition rules. As you already know, these rules are formally identical to the composition of angular momenta. Members of the same isomultiplets are states with the same mass number A but different protons, that is, they are isobars. As expected from QCD, members of the same isomultiplets are practically the same particle for what concerns strong interactions, even if their chemical properties are completely different because different Z's bring the element in different columns of the Periodic Table. This is not surprising: the chemical properties of an element depend on electromagnetic forces, which are neglected by the isospin-based classification.

More generally, any nuclide has a flavor part of the wavefunction ψ_F that is defined by the $SU(2)$ composition rule of the isospin of each nuclide. The rule is given by eqn 8.4. T_z is additive and

$$T_z = \sum_{k=1}^{A} T_z^k \tag{9.36}$$

where k runs through all nucleons inside the nuclide. On the other hand, the value of T follows the usual combination rules of angular momentum for A particles of $T = 1/2$. In this way,

> nuclides can be gathered in **isomultiples** or isobaric analog states (IAS) with a well-defined value of T and with T_z ranging from $-T$ to $+T$. The IAS multiplicity is $2T + 1$.

9.6.2 Isomultiplets and the mass of nuclei

A spectacular application of IAS was discovered by Wigner in 1957 and can be used to estimate the mass of the nuclei neglecting e.m. forces and all effects that violate the isospin symmetry. It follows directly from the Wigner–Eckart theorem of QM (Cohen-Tannoudji *et al.*, 1991). Consider a set of nuclides with total isospin T and T_z plus additional quantum numbers α. The binding energy of a nuclide comprises all interactions among nucleons. Its value $E_{\alpha,T}$ is given by:

$$H \left| T, T_z, \alpha \right\rangle = E_{\alpha,T} \left| T, T_z \right\rangle. \tag{9.37}$$

Note that neither the strong Hamiltonian H nor $E_{\alpha,T}$ depend on T_z because H conserves the isospin symmetry and the states with the same T_z are degenerate like the states of Sec. 8.2. Wigner, who was completely unaware of the reasons why nucleons interact inside nuclei, just posited that the interactions were due to the sum of generic two-body forces between the nucleons. Covariance in special relativity (SR) requires that all forces can be expressed as a sum of scalars, vectors, and tensors, as demonstrated in Sec. 2.3.1. Besides, these forces are a perturbation of the mass of the nucleus because $Zm_p + (A - Z)m_n$ is much larger than the binding energy.

As a consequence, the Hamiltonian can be written at second order in perturbation theory as:

$$H = H^0 + H' = H^0 + (H^{(1)} + H^{(2)}) \qquad (9.38)$$

where H' is the perturbation Hamiltonian and $H^{(1)}$ and $H^{(2)}$ are the first and second term of its Taylor expansion. The Wigner-Eckart theorem states that, for a generic operator that can be expressed as the sum of scalars, vectors, and tensors and, in particular, for the Hamiltonian H':

$$\langle j, m | H' | j', m' \rangle = \langle j, m, j', m' \rangle \langle j | H' | j' \rangle \qquad (9.39)$$

where $\langle j, m, j', m' \rangle$ are the Clebsch–Gordan coefficients and the **reduced matrix** $\langle j | H' | j' \rangle$ does not depend any more on m and m'. Playing this game with the isospin T and T_z instead of j and j_z, we get (Benenson and Kashy, 1979)

$$\langle \alpha, T, T_z | H' | \alpha, T, T_z \rangle = \langle \alpha, T | H^{(0)} | \alpha, T \rangle + \frac{T_z}{T(T+1)} \langle \alpha, T | H^{(1)} | \alpha, T \rangle$$
$$+ \frac{3T_z^2 - T(T-1)}{[(2T-1)T(T+1)(2T+3)]^{1/2}} \langle \alpha, T | H^{(2)} | \alpha, T \rangle. \qquad (9.40)$$

Equation 9.40 gives the celebrated **isobaric multiplet mass equation** (IMME):

$$\langle \alpha, T, T_z | H | \alpha, T, T_z \rangle = a + bT_z + cT_z^2. \qquad (9.41)$$

The coefficients a and b can be measured experimentally to determine all masses of the IAS members. Modern techniques based on multibody simulations also provide quantitative statements on these coefficients that can be compared with experimental data.

9.6.3 Modern techniques*

Protons and neutrons are complex bound systems where the rest mass of the valence quarks plays a minor role The big players are gluons and soft particles arising from the large coupling strength ($\alpha_s > 1$) of non-perturbative QCD. We could imagine that, once protons and neutrons get closer, these interactions become the leading forces in the nuclei, too. Fortunately, this is not the case.[8] Equations 9.40 and 9.41 work because the interactions among nucleons are much smaller than the interactions inside nucleons. As a consequence, nucleon–nucleon interactions can be treated using perturbative techniques. For instance, we can describe these interactions at "large" ($10^{-13} - 10^{-14}$ m)distance by

[8]In very special conditions, two nuclei can merge with such a large four-momentum transfer to penetrate the inner core of the potential and create a melting pot of gluons, valence, and sea quarks (see Sec. 7.7). This state of matter is called **quark–gluon plasma** and is currently under study at the LHC and in celestial bodies (Yagi *et al.*, 2008).

the exchange of pions and, in general, we can safely assume that the core of each nucleon will be unaffected by other nucleons because the QCD potential is strongly repulsive at 10^{-15} m. But what is the origin of this impassable core? A nucleon is not structure-less, but to gain evidence of its inner structure, we need – at least – to excite its internal energy levels ($\Delta(1232)$, $N(1440)$, etc.). The production of these states requires hundreds of MeV. This is not surprising because touching the structure of a nucleon means to perturb the cancellation of color charges carried out by the valence quarks. These energy scales are not in the domain of nuclear reactions, which require a handful of MeV. Again, this is the nuclear counterpart of a well-known property of atomic physics. You cannot see the nucleus as a structured object because its excited states are not reachable by the energies of atomic transitions (1–1000 eV).[9] In nuclear physics, the protons and neutrons inside the nucleus look like structure-less elementary particles. Treating nucleons like elementary particles remarkably eases the description of nuclei. First, we can replace quarks with hadrons at large distances and develop models where interactions occur by the exchange of light mesons or heavier particles, as in Fig. 9.3 (right). This replacement plays a central role in **chiral perturbation theories**, which exploit the additional symmetries of QCD arising when light quarks are considered massless (16 conserved Noether currents). Similarly, low A nuclei can be considered a few-body problem, treated using chiral theories, quantum Monte Carlo techniques, or the Fadeev equations (Donnelly *et al.*, 2017). Large A nuclei (many-body systems) are the hardest systems to handle. They are treated by introducing a mean field produced by the whole set of nucleons and employing the Hartree–Fock or other approximations to solve the corresponding equations of motion. This is a cutting-edge field where lattice QCD will likely make the difference in the forthcoming years.

[9]All these examples fall into the category of **quantum censorship**: we cannot appreciate the inner core of a quantum system if the size of the perturbations is too small. (Wilczek, 2012).

9.7 Kinematics of nuclear reactions

Nuclear scattering and reactions occur only if all conservation laws are satisfied, moving from the initial to the final state. Kinematic constraints are easier to be implemented than in Chap. 2 since the vast majority of nuclear reactions occur at non-relativistic velocities and, therefore, SR enters only in the mass-to-energy transformation. Nuclear physicists gained the first empirical evidence of the Einstein energy–mass relation by reactions like:

$$p + {}^{7}\text{Li} \rightarrow {}^{4}\text{He} + {}^{4}\text{He}. \tag{9.42}$$

Note that ${}^{4}\text{He}$ is a doubly magic nuclide and is expected to be tightly bound. ${}^{4}\text{He}$ is so difficult to be broken ($E_b = 28.3$ MeV) that appears nearly as a particle like the pion or the proton. For historical reasons, it is still called the $\boldsymbol{\alpha}$ **particle**.

A standard two-body nuclear reaction is a scattering of the form:

$$a + X \rightarrow b + Y \tag{9.43}$$

where a projectile a hits a nuclide to produce two final state bodies: b and Y. In most nuclear experiments, the projectile is much lighter than the nucleus, and the final state contains one light, tightly bound particle or nuclide. If $m_a \ll m_X$ and $m_b \ll m_Y$, we can use a neater notation equivalent to eqn 9.43:

$$X(a, b)Y. \tag{9.44}$$

This is specially useful when a is a proton, neutron, photon, or α. We can group the corresponding class of processes as (n, γ) – photo-production – reactions or (α, n) – neutron-production – reactions.

A **nuclear reaction** must fulfill energy, momentum and angular momentum conservation, the conservation of parity, C-parity, T-parity, baryonic number, and isospin. We do not need to justify these laws because they follow directly from QCD. Note, however, that the isospin symmetry and the corresponding unitary operator $\hat{T}$ is an approximation and therefore $[\hat{T}, \hat{H}] = 0$ only in the limit $m_u = m_d \ll \Lambda_{QCD}$. Energy conservation can be written in NU as:

$$m_X + T_X + m_a + T_a = m_Y + T_Y + m_b + T_b \tag{9.45}$$

where T_X and T_Y are the kinetic energy of X and Y. In the non-relativistic limit, $T_X \equiv \frac{1}{2} m_X v_X^2$ and this expression is appropriate in all cases of interest in this chapter.

The mass that can be used to feed the reaction is given by the mass difference between the initial and final states. This is called the **Q-value** of the reaction:

$$Q \equiv (m_X + m_a - m_Y - m_b). \tag{9.46}$$

The Q-value represents the amount of energy absorbed or released during the reaction and plays a role similar to enthalpy in thermodynamics. In eqn 9.45, the Q-value is also the change of kinetic energy in the final state:

$$Q = (T_Y + T_b - T_X - T_a). \tag{9.47}$$

We must be careful when we compute T_a and T_b because, in some cases, these particles might be relativistic. In a photo-production reaction, b is a photon and $T_b = E_\gamma - m_\gamma = |\mathbf{p}_\gamma|$.

The Q-value classifies nuclear reactions as the enthalpy classifies thermodynamic processes:

- $Q > 0$ corresponds to an **exothermic reaction**. The mass of the nuclides, which also depends on the binding energy, is partially converted into kinetic energy of the final products.

- $Q < 0$ corresponds to an **endothermic reaction**. Here, the kinetic energy of the initial products must be converted into the final state nuclide mass, which includes its binding energy. That is the most common way to create heavy particles in particle physics and is used for similar purposes in nuclear physics.

Momentum conservation gives three kinematic constraints. If the scattering occurs in the $x-z$ plane (we set $p_y = 0$ for all particles without loss

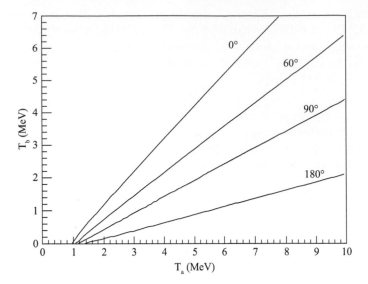

Fig. 9.8 T_b versus T_a in MeV for the ^{3}H(p,n) ^{3}He reaction. The lines correspond to different values of the scattering angle θ of b in LAB (see eqs 9.48 and 9.49).

of generality) and θ, ϕ are the scattering angles of b and Y, respectively:

$$p_a = p_b \cos\theta + p_Y \cos\phi. \tag{9.48}$$

$$0 = p_b \sin\theta - p_Y \sin\phi. \tag{9.49}$$

The energy and momentum conservation constrain the possible final states but does not close the kinematics of the event. In this notation, the four unknowns are θ, ϕ, p_b, p_Y. The constraints are only three, eqs. 9.45, 9.48, and 9.49.

In most nuclear experiments, we reconstruct only the light final-state products. Hence, ϕ and p_Y remain unknown. Solving eqs. 9.45, 9.48, and 9.49 for ϕ and p_Y we get:

$$T_b^{1/2} = (m_Y - m_b)^{-1} \left[(m_a m_b T_a)^{1/2} \cos\theta \tag{9.50} \right.$$
$$\left. \pm \left\{ m_a m_b T_a \cos^2\theta + (m_Y + m_b)[m_Y Q + (m_Y - m_a)T_a] \right\}^{1/2} \right].$$

This formula is puzzling because it predicts two solutions for each scattering process but do not forget that kinematics does not uniquely define all variables, as seen in Sec. 2.4. Fortunately, the kinetic energy is uniquely defined for nearly all values of T_a. For a ^{3}H(p,n) ^{3}He scattering, Fig. 9.8 shows that the second solution is unphysical in most of the parameter space and, in particular, for T_a slightly higher than its minimum value. Here, the measurement of θ at fixed T_a uniquely defines the energy of b but two solutions are possible close to the reaction threshold. This is visible in Fig. 9.9, which is the expanded view of Fig. 9.8 for a scattering angle of $0°$.

Each nuclear reaction has an absolute minimum at:

$$T_{min} = (-Q)\frac{m_Y + m_b}{m_Y + m_b - m_a} \qquad \text{if } Q < 0 \tag{9.51}$$

$$T_{min} = 0 \qquad \text{if } Q > 0 \tag{9.52}$$

which corresponds to the **threshold kinetic energy**. The threshold is reached at $\theta = 0$ because this configuration minimizes the amount of transverse energy of the final state products. Exothermic reactions occur at any energy of the projectile and T_b is always single-valued (Krane, 1988).

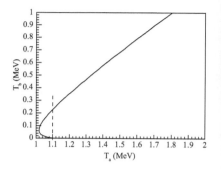

Fig. 9.9 Expanded view of the $0°$ curve in Fig. 9.8. In this region, eqn 9.7 has two solutions (see e.g. the dashed line) and T_b is not uniquely defined by kinematics.

9.8   Fission reactions

Fission reactions were discovered in 1938 by O. Hahn and F. Strassmann. They are by far the most famous nuclear reactions and shaped the science, history, and society of the 20th century. **Fission** is the disruption of a nucleus, which results in the creation of lighter nuclei and neutrons. In most cases, the initial state nucleus breaks into two nucei plus, possibly, a few neutrons. The word "fission" is borrowed from life sciences and means splitting into two pieces. Fission is unavoidable in the Chart of Nuclides because the binding energy per nucleon increases with A (see Fig. 9.7), reaches the absolute maximum in a doubly magic isotope of iron (^{56}Fe), and decreases monotonically at high A producing loosely bound nuclei. The tendency of a nucleus to break apart thus increases with A for $A > 56$. The fission can be catalyzed by an incoming neutron. In this section, we are only concerned with **neutron-induced fission**, while other types of fission will be discussed in Chap. 11. For instance, **spontaneous fission** occurs as a standard decay process in metastable nuclei.

A neutron impinging on a high-A nucleus can be captured creating a nuceus with $A + 1$. The new $A + 1$ nuceus is generally in an excited state. If the energy of this state is higher than the potential barrier where the nucleons are confined, the nucleus breaks apart. Since high-A nuclei are richer in neutrons than protons, the breaking process results in a few (typically two) tightly bound fragments and some free neutrons. To understand more quantitatively the induced fission process, we can use the findings of the shell model.

A nuclide like ^{235}U can absorb neutrons even if their kinetic energy T is close to zero because the neutron has no electric charge and penetrates the nucleus of the uranium without sensing the electromagnetic forces. ^{235}U is an even–odd nuclide: it has an even number of protons and an odd number of neutrons. ^{236}U is even–even and, according to the shell model, is more tightly bound. As a consequence, the reaction ^{235}U $+ n \rightarrow$ ^{236}U is exothermic and the energy released is the difference between the binding energy of ^{235}U and ^{236}U. The energy amounts to 6.5 MeV and is larger than the 5 MeV potential barrier. This process does not occur in threshold-less mode in ^{238}U because of the higher potential barrier. Hence,

> the neutron induced fission of ^{235}U is threshold-less, while the neutron induced fission of ^{238}U has a finite threshold of about 1.2 MeV.

The fission of ^{238}U requires a minimum neutron energy because ^{238}U is even–even and ^{239}U is more loosely bound than ^{238}U. The binding energy of the last neutron in ^{239}U, which is the energy that can be released to commence fission, is just 4.8 MeV, while the potential barrier has a height of 6.0 MeV. Therefore, the incoming neutron must have $T > 6.0 - 4.8 = 1.2$ MeV and the reaction ^{238}U$+n \rightarrow {}^{239}$U is endothermic. In real life, fission in ^{239}U is a relevant process only if the neutron energy is significantly larger than this threshold.

Induced fission for energy production is employed using fissile materials like ^{233}U, ^{235}U, ^{239}Pu, ^{241}Pu, which are even–even and, therefore, threshold-less. We need neutrons with $T \simeq 1 - 2$ MeV to get a fission in ^{232}Th, ^{238}U, ^{240}Pu, and ^{242}Pu. The workhorse of nuclear reactors and nuclear weapons is uranium, which is rather abundant on earth. However, natural uranium is mostly composed by ^{238}U and the natural abundance of ^{235}U is just 0.7%.

9.8.1   Fission of uranium

If we bombard a natural sample of uranium with neutrons, the fission cross-section is completely dominated by ^{235}U at low energies (see Fig. 9.10). For this isotope $\sigma_{tot} \simeq \sigma_{fission}$ and other processes as ^{235}U$(n,\gamma)^{236}$U do not contribute by more than 16%. The 0.1 eV–1 keV region is very complicated due to the occurrence of many resonances. These resonances lead to fission only in ^{235}U while for ^{238}U we mainly observe ^{238}U$(n,\gamma)^{239}$U photoproduction. Above this energy, σ_{tot} and $\sigma_{fission}$ show again a smooth behavior and $\sigma_{fission}$ for ^{235}U is still of the order of a few barns. For ^{238}U, $\sigma_{fission}$ is sizable just for $T_n \gg 2$ MeV (1 barn at 10 MeV). The fragments produced by the ^{235}U fission cover a wide range of nuclei that carry away an average energy per fission of about 180 MeV. The neutrons that are produced together with the fis-

Fig. 9.10 Total (σ_{total}) and fission ($\sigma_{fission}$) cross-section (in barn) as a function of the incoming neutron kinetic energy T (in MeV) for ^{235}U. Data from Nuclear Data Center, 2017.

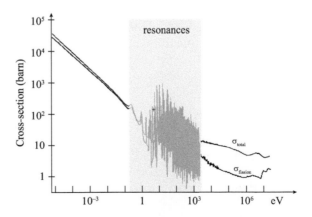

sion bring ~2.6% of the energy and the mean number of these **prompt neutrons** is $n \simeq 2.5$. Like fragments, the number of neutrons and the energy distribution is very broad: $0 < n < 6$ and 0.5–4 MeV (average energy: 2.5 MeV). The neutron sample is further enriched by radioactive beta decay (see Sec. 11.4) producing **delayed neutrons** that account for 13% of the fission energy. Radioactive decays also produce a large neutrino flux. Since the neutrino cross-sections are 18 orders of magnitude smaller than the typical fission cross-sections at the MeV scale, the energy carried by neutrinos is lost in any realistic device and does not contribute to the heat produced by nuclear reactors or weapons.

A **nuclear reactor** exploits fission to produce a large amount of energy. To achieve this goal, the fission rate must be kept under control and, possibly, stay constant until the exhaustion of the fuel. Since every fission produces 2.5 neutrons, after the first fission we have 2.5 neutrons and one fission event. In the next step, each neutron produces one fission and we end up with $1 + 2.5 = 3.5$ total fission events and $2.5 \times 2.5 = 6.25$ neutrons propagating inside the reactor. The rate of events, therefore, increases exponentially at each step.

The actual rate is determined by the **effective multiplication factor k** defined as the number of neutrons produced after $m + 1$ steps divided by the number of neutrons produced after m steps. The case $k = 1$ corresponds to the condition of **criticality**: the fission chain reaction is self-sustaining and the power (energy per unit time) is nearly constant. The reactor is **supercritical** if $k > 1$ and causes an increase of power over time until the fuel is completely exhausted or the reactor container melts down and the fission material is ejected outside the plant (nuclear accident). If $k < 1$ the reactor is **subcritical** and the power decreases over time.

Example 9.1

Reactor monitoring is essential for safety reasons and to avoid misuse of its waste for the fabrication of nuclear weapons. Monitoring, however, requires access to the reactor premises and the access may not be possible if the control agency is not allowed by the owner of the reactor. Still, modern neutrino detectors can record a small fraction of the neutrinos produced by a nuclear reactor (Cribier, 2011; Kim, 2016). The detection probability decreases with the distance of the detector from the reactor because the neutrino flux decreases as L^{-2}. The reason is purely geometrical: the flux of any free-streaming radiation (in this case, the neutrinos) from a point source decreases as the area of the sphere between the source and the detector. We will see in Chap. 10 that direct detection of neutrinos has been achieved thanks to the enormous flux produced by nuclear reactors. For instance, ^{235}U produces 1.92 $\bar{\nu}_e$ per fission with an energy greater than 1.8 MeV. We can then use an exothermic reaction like $\bar{\nu}_e + p \rightarrow n + e^+$ to record the neutrinos by looking at the positron.[10] The cross-section of this reaction averaged over the neutrino energy spectrum is 3.42×10^{-43} cm^2 but the flux of a 1 GW nuclear reactor is $\simeq 10^{20}$ $\bar{\nu}_e$/s. Therefore, the detection of the $\bar{\nu}_e$ is not a desperate task even at a large distance (see Exercise 9.15).

In principle, high-efficiency neutrino detectors could be used for reactor diagnostics and control. In particular, they can monitor some basic features of the state of the reactor (whether it is on or off, the isotopic composition of the fuel versus time, etc.) even at large distances from the plant, and several research groups are currently involved in reactor monitoring with neutrinos, for example, for non-proliferation studies.

9.8.2 Nuclear reactors

A nuclear reactor produces heat generated by a controlled fission chain. The heat is used to increase the temperature of a fluid (water in most cases) and transformed into electric power by turbines. The basic components of a nuclear reactor are:

- **Fuel**: the most common fuel for power reactors is natural or enriched uranium, that is a uranium sample whose ^{235}U abundance is increased up to 2–3%. Special reactors employ ^{239}Pu or ^{233}U or a mixture of them.

- **Moderator**: it slows down fast neutrons to reduce resonant absorption and capture in ^{238}U and make the neutrons available for the fission of ^{235}U. The moderation process is detailed in Example 9.2. If the fuel is natural uranium, the moderator must compensate for the low abundance of ^{235}U minimizing the neutron losses. Water does not help because of the relatively high cross-section of deuteron production $(n + p \to d + \gamma)$, which depletes the neutron flux. **Heavy water** is a water molecule where hydrogen is replaced by deuterium (^{2}H$_2$O or, equivalently, d$_2$O). This moderator can be used instead of water and further reduces neutron losses. Even better, carbon in the form of **graphite** does the job and is much cheaper than heavy water. Graphite's drawback is that the reactors require a larger amount of material and graphite brings some additional safety risks (corrosion of the moderator, loss of coolant, etc.). On the other hand, if the fuel is enriched and neutron loss is not an issue, **natural water** is by a great deal the best option.

- **Control rods** provide feedback to stabilize the power of the reactor and operate in a safe condition near criticality. They are usually made of cadmium, which is a powerful neutron absorber. The speed at which a rod must be moved is given by the size of the neutron increase for small variations of k. The number of neutrons at time t is given by:

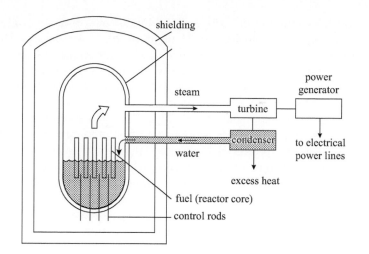

Fig. 9.11 Simplified layout of a boiling water reactor. The graphite bars moderate the neutrons. The reactor power (i.e. the amount of energy produced in the core per unit time) can be fine-tuned at the per-mill level in k by the control rods. The water boils to the top of the core and the steam moves the electric power generator (turbine). In the return line, the water cools the core and the heat exchange between the system and the environment remains constant.

$$N(t) = N(0) \exp \left[\frac{n(k-1)t}{[\delta n - n(k-1)]t_d} \right] \qquad (9.53)$$

where $n = 2.5$ neutrons per fission, δn is the number of delayed neutrons per fission, and t_d is the delay time. Common values are $\delta n \simeq 0.02$ and $t_d \simeq 13$ s. A per-mill variation of k causes an increase in the neutron flux of factor 2 in 60 s, which is a safe margin for any feedback system based on cadmium rods.

The **reactor core** is the volume of the nuclear reactor containing the fuel components and moderators. The heat is generated inside the core. A scheme of a classical boiling water nuclear reactor is depicted in Fig. 9.11. The boiling water drives the turbine and acts as a coolant for the reactor core. A safer option is a homogeneous heavy water reactor, where a loss of coolant brings to a loss of moderator, too. In this way, the reactor becomes subcritical well before melting the core.

Power reactors are facilities producing several gigawatt-hours (GWh) of thermal power, whose maintenance cost is very small compared to the initial investment. As a consequence, they are facilities designed to operate for decades. At the time of writing, nuclear reactors generate about 10% of the entire energy produced by humankind and there are about 450 plants in operation. Despite several major advantages (cost per GWh, efficiency, constant availability, tunability of the power production rate, limited emission of greenhouse gases or CO_2, etc.), there are three major issues that reactor science has to face. First, safety. Accidents in nuclear reactors not only jeopardize the life of the operators but the health of entire regions because they can inject into the environment an impressive amount of long-living radioactive isotopes causing illness and poisoning of natural resources. Major accidents are extremely rare, especially for the newest reactor generations, and most of them occurred in old plants. The first serious accident took place in an army research reactor in 1961 (the SL-1 accident). The readers are surely aware of Chernobyl's disaster caused by a high-power channel-type reactor (the oldest type still in operation) in 1989 and the Fukushima disaster originated from a boiling-water Generation II plant in 2011. Accelerator Driven Subcritical Reactors (ADS) may solve this issue because the reactor is intrinsically subcritical ($k < 1$) and the missing neutrons are provided by a high-current particle accelerator serving a fixed target for neutron production. Still, no ADS is as yet in operation and several other options are under development (Generation IV). The second issue is the disposal of nuclear waste because the spent fuel of any reactor is rich in long-living isotopes produced by fission. At present, we have no other choice than safe storage of this material, which will be active and dangerous for hundreds of years. Again, high-current particle accelerators may help with the transmutation of the most dangerous isotopes into short-living ones, but this approach remains speculative. Note also that, just after a reactor shutdown, the remaining fuel produces heat due to radioactive decays. Such a heat source was also a major concern in the Fukushima accident. Finally, nuclear reactors can be misused to produce nuclear weapons, as discussed in Sec. 9.8.3.

Example 9.2

Moderators are proton-rich materials used to slow down fast neutrons without absorbing them. The slowdown process is a series of $n - p$ scatterings. From Fig. 9.2 and the cosine rule, we can compute the neutron scattering velocity in LAB and in RF assuming $m_n = m_p$.

$$(v_n^{LAB})^2 = (v_n^{RF})^2 + (v_p^{RF})^2 - 2v_n^{RF}v_p^{RF}\cos(\pi - \theta^{RF}) \tag{9.54}$$

$$(v_n^{RF})^2 = (v_n^{LAB})^2 + (v_p^{RF})^2 - 2v_n^{LAB}v_p^{RF}\cos\theta^{LAB}. \tag{9.55}$$

Recalling eqn 9.7, the proton velocity in RF is:

$$v_p^{RF} \simeq \frac{1}{1 + A}v_0 = \frac{v_0}{2} \tag{9.56}$$

for $A = 1$ (proton). The scattering angle in LAB can be derived inserting eqs 9.5 and 9.56 in 9.54. Inserting v_n^{LAB} in eqn 9.55, we get:

$$\cos\theta^{LAB} = \frac{\cos\theta^{RF} + 1}{\sqrt{2(1 + \cos\theta^{RF})}} \ . \tag{9.57}$$

Writing eqn 9.57 for a general neutron-nucleus scattering, as in Chadwick's experiment, we get both:

$$\cos\theta^{LAB} = \frac{A\cos\theta^{RF} + 1}{\sqrt{A^2 + 1 + 2A\cos\theta^{RF}}} \tag{9.58}$$

and eqn 9.9. This equation demonstrates that the energy of the scattered neutron is in the range:

$$\left(\frac{A - 1}{A + 1}\right)^2 E_0 < E < E_0 \tag{9.59}$$

where E_0 is the initial kinetic energy of the neutron. The first term of the inequality goes to zero if $A = 1$ and substantiates the use of hydrogen-rich compounds as moderators.

If the neutron energy is less than ~ 15 MeV, the scattering is isotropic because occurs in s-wave ($l = 0$). The scattering probability is then:

$$dP = \frac{d\Omega}{4\pi} = 2\pi \sin\theta^{RF}\,\frac{d\theta^{RF}}{4\pi} = \frac{1}{2}\sin\theta^{RF}\,d\theta^{RF} \tag{9.60}$$

From eqn 9.9, we get

$$\frac{dE}{E_0} = 2\frac{A}{(A+1)^2}\sin\theta^{RF}\,d\theta^{RF} \tag{9.61}$$

and, therefore,

$$\frac{dP}{dE} = \frac{(A+1)^2}{4AE_0} \ . \tag{9.62}$$

This is an important result because it shows that:

> after one scattering, the energy distribution of a monoenergetic beam of neutrons is constant in the range from $\left(\frac{A-1}{A+1}\right)^2 E_0$ to E_0.

This result can be used to compute the energy distribution of the neutron after two scatterings and can be extended up to the n-th scattering. The general formula for a $A = 1$ (hydrogen moderator) was computed by E. Condon and G. Breit in 1936. It is

$$\frac{dP_n}{dE} = \frac{1}{E_0(n-1)!}\left(\log\frac{E_0}{E}\right)^{n-1} \tag{9.63}$$

where P_n is the probability at the n-th scattering. The **lethargy change** is defined as the logarithmic change of energy after a scattering:

$$u \equiv \log E_0 - \log E = \log\frac{E_0}{E} \tag{9.64}$$

and can be used to compute how many scatterings are needed to thermalize a fast neutron. From eqn 9.9,

$$u(\theta^{RF}) = \log\frac{(A+1)^2}{A^2+1+2A\cos\theta^{RF}} \ . \tag{9.65}$$

Integrating the lethargy change over the solid angle, we get the mean lethargy change in a scattering:

$$\langle u \rangle = \int \frac{d\Omega}{4\pi}u(\theta^{RF}) = \frac{1}{2}\int \log\frac{(1+A)^2}{A^2+1+2A\cos\theta^{RF}}(-\sin\theta^{CM})\,d\theta^{RF}$$

$$= 1 + \frac{(A-1)^2}{2A}\log\frac{A-1}{A+1} \ . \tag{9.66}$$

Equation 9.66 shows that the average lethargy change per scattering is constant and does not depend on E_0: a result that remarkably simplifies the design of a moderator. The moderator brings the initial energy E_0 of the neutron to a final value E_f in n_f collisions, where:

$$n_f = \frac{u}{\langle u \rangle} = \frac{1}{\langle u \rangle} \log \frac{E_0}{E_f} \ . \tag{9.67}$$

In nuclear reactors, we need to slow down MeV-scale neutrons to thermal energy (25 meV). A water moderator achieves this goal for 1 MeV neutrons after $n_f = 17.5$ scatterings, while a graphite moderator based on ^{12}C needs $n_f = 111$ collisions.

9.8.3 Fission-based nuclear weapons

A nuclear weapon is a much simpler device than a nuclear reactor because it releases its enthalpy very quickly triggering an explosion. This **detonation** increases the entropy of the system up to destroying the weapon, which does not have any control feedback. This is why these devices (improperly called "atomic bombs") were employed for military applications soon after the discovery of nuclear fission.

The main physics challenge of a nuclear weapon is to deliver a system that remains subcritical even in hazardous situations (manipulation, transportation, and drop on the target) and detonates only in the proximity of the target. The detonation corresponds to increasing k well above criticality ($k > 2$ in most cases) in a few ms or less.

This task can be accomplished by changing abruptly the density of a critical fuel mass by a factor of 2 to 3. The approach followed during the Second World War by the scientists of the Manhattan project (L. Groves, R. Oppenheimer, and many others) was the **gun-type fission assembly** (see Fig. 9.12, top). Here, two subcritical masses are positioned in a barrel and one is launched onto the other to achieve supercriticality. This technique is simple but requires a large amount of fissile material because the mass is not compressed. Besides, the insertion time to reach supercriticality is relatively slow (a few ms). This is a serious drawback: if the spontaneous fission rate is too large, the detonation can be triggered while the highly supercritical mass is still in the formation stage. ^{239}Pu, for instance, has the lightest critical mass and is the ideal fuel for nuclear weapons. On the other hand, ^{239}Pu has a rate of spontaneous fission[11] of about 10 fission/s/kg. At this rate, the creation of a highly supercritical mass that does not detonate during formation is very unlikely. Therefore, the first nuclear bomb was based on large quantities of ^{235}U instead of small quantities of ^{239}Pu. To assemble Little Boy – the first nuclear bomb used in warfare and dropped on Hiroshima – the scientists of the Manhattan Project used 50 kg of uranium enriched up to 85% of ^{235}U.

Fat Man – the second bomb dropped on Nagasaki three days later – was based on a different technique that quickly superseded the gun-type mechanism: the **implosion-based method** depicted in Fig. 9.12

[11] Like neutron-induced fission, natural fission is accompanied by several fast neutrons that contribute to the neutron budget of the device.

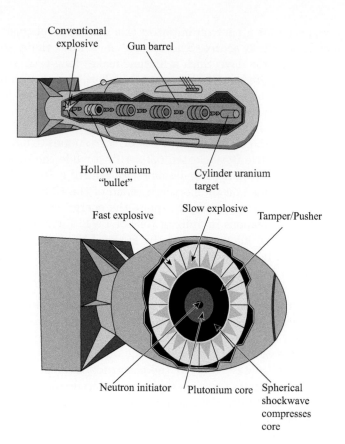

Fig. 9.12 Comparison between the gun-type (Little Boy) and implosion-type (Fat Man) formation schemes for fission-based nuclear weapons. In the first case (top), the conventional explosive is used to shoot the hollow uranium cylinders toward the main uranium target, as if they were bullets. In the second case (bottom), the conventional explosive located in the outer layer of a sphere compresses both the ^{238}U pusher and the plutonium core. Detonation is triggered by the neutron initiator located inside the core. Redrawn from Wikimedia, 2014b under CC-BY-SA.

(bottom). Fat Man is spherical and the subcritical fuel is located in the core of the sphere. A blanket of conventional explosive is mounted just below the surface of the sphere. The space between the core and the explosive is filled with a layer of dense material. At detonation, the explosive compresses both the layer and the core. The increase of density brings the fuel to $k > 2$ and this process is accelerated by the layer, too. Fat Man used a ^{238}U layer acting both as a pusher to increase the pressure inside the core and as a reflector for the neutrons produced by the core. In this way, $k > 2$ can be achieved with an even smaller mass of fuel. Unlike Little Boy, supercriticality is achieved only for a few tens of μs, and spontaneous fission may not be enough to trigger the chain reaction and detonate the bomb. This issue was solved by adding a **neutron initiator** to the core of the bomb, that is, a source of α particles that gets closer to a small beryllium target at the time of the implosion and produces neutrons by:

$$\alpha + {}^9\text{Be} \rightarrow {}^8\text{B}^* + \alpha + n. \tag{9.68}$$

This reaction creates low energy neutrons, but the cross-section is smaller than the reaction used in Chadwick's experiment (see Sec. 9.1):

$$\alpha + {}^9\text{Be} \rightarrow {}^{12}\text{C}^* + n. \tag{9.69}$$

Beyond the need for a neutron initiator, this weapon is plagued by the finite lifetime of the α source. The lifetime of ^{210}Po – the α source of Fat Man – is just 138 days. Such a lifetime makes long-term storage of nuclear weapons quite impractical. Since the 1950s, the neutron initiator has been replaced by combining fission and fusion processes in a new class of weapons: the thermonuclear weapons described in Sec. 11.8.2.

The detonation of Fat Man fissioned about 1 kg of the 6.2 kg of plutonium located in the core. The energy released corresponds to 8.8×10^{13} J. For comparison, Little Boy released 6.3×10^{13} J while the most powerful fission bomb ever built (Tsar-bomb) released 210×10^{15} J in a 1961 nuclear test performed on the northern island of the USSR archipelago Novaya Zemlya. It is customary to quote the energy released by these non-conventional weapons in units of tons of TNT. Little Boy provided the equivalent of 16 kilotons (16×10^6 kg) of TNT killing more than 100,000 people either a few minutes after detonation by the X-ray blast and heat or, in less than four months, by the fallout of the fission products.

The reader may be surprised by the fact that nuclear fission weapons – based on a technology invented more than 70 years ago – are not so widespread and no nuclear attack has been performed since 1945. The non-proliferation of nuclear weapons is the outcome of a very demanding effort. The best way to avoid the construction of new weapons is to prevent the production and accumulation of fissile materials and forbid the development or use of isotope enrichment technologies for these nuclides. ^{239}Pu is extremely rare on earth and is generally produced by nuclear reactors. Hence, reactor monitoring is the priority of the International Atomic Energy Agency (IAEA) and the Nuclear Non-Proliferation Treaty (1970). Similarly, careful tracking of uranium ores and enrichment tools is a critical task for these organizations. Despite these efforts, there are currently (at least) nine states that developed nuclear weapons and just a percentage of this armory could destroy the whole planet in minutes. Even today – decades after the Cold War – non-proliferation remains a grand challenge for governments, our society, and nuclear scientists.

Further reading

Nuclear physics is a discipline on its own. It is deeply connected to QCD, but over the years nuclear physicists have developed specific methods to study nuclear properties and scattering, like the techniques sketched in Sec. 9.6.3. A classic, although sometimes obsolete, introduction to this field for undergraduates is Krane, 1988. A more up-to-date review is available in Bertulani, 2007. If the reader wants to explore the links to QCD and a modern set of techniques, especially for cross-section calculations, an excellent textbook at the graduate level is Donnelly et al., 2017. This book requires some knowledge of QFT. An introduction emphasizing experimental methods is provided in Hans, 2011. The science and engineering of nuclear reactors are presented in Shultis and Faw, 2019. A comprehensive review of nuclear engineering can be found in Murray and Holbert, 2019.

Exercises

(9.1) In Chadwick's detector, compute the ratio between the maximum proton kinetic energy in lithium, iron, and water.

(9.2) Classify the following nuclides identifying the sets of isotopes, isobars, isotons, and isomers: ^{1}H, ^{2}H, ^{3}H, ^{3}He, ^{4}He, ^{4}Li, ^{57}Co, ^{57}Fe, and ^{57}Fe*.

(9.3) A proton with a kinetic energy of 50 MeV hits a ^{208}Pb nucleus in a head-on collision. What is the smallest distance to the surface of the target that the proton can reach?

(9.4) Compute the mass of ^{37}Cl using the SEMF and the difference with the actual value from the Chart of Nuclides (IAEA, 2020). Do the same for ^{56}Fe. Is the difference larger than ^{37}Cl?

(9.5) * A starting point for the liquid drop model is to consider the nucleus as an uniformly charged sphere of radius R. A more sophisticated approximation is a density based on the Fermi function and given by:

$$\rho(r) = \rho_0 \left[1 + \exp\left\{ \frac{R - r}{a} \right\} \right]^{-1} \quad (9.70)$$

where R, a and ρ_0 are semi-empirical parameters. Compute the mean-squared charge radius $\langle r^2 \rangle$ both for the uniform sphere model and the model based on eqn 9.70.

(9.6) Compute the approximate density of a neutron star, i.e., a star made of tightly packed neutrons. What are the forces that make a neutron star a stable celestial object?

(9.7) Compute the spin-parity of the ground state of ^{7}Li in the shell model.

(9.8) * Identify the most stable nuclides in the shell model and compare the results with the actual lifetime in the Chart of Nuclides (IAEA, 2020).

(9.9) * Show that ^{14}C, ^{14}N, and ^{14}O are isobaric analog states (IAS). Is there a difference among their ground-state mass? What is the origin of the differences?

(9.10) How can we compute the **doubly differential cross-section**:

$$\frac{d\sigma}{dE_b d\Omega} \quad (9.71)$$

for the process of eqn 9.43 if we cannot measure E_b?

(9.11) ** The general scattering theory of QM provides the total cross-section for a $a + X \rightarrow b + Y$ reaction (Krane, 1988; Weinberg, 2012). The cross-section is expressed as a sum of partial waves according to the value of l ($l = 0$ is the s-wave, $l = 1$ the p-wave, $l = 2$ the d-wave, etc.) and reads:

$$\sigma_{tot} \equiv \sigma_r + \sigma_{el} = \sum_{l=0}^{+\infty} \pi \lambdabar^2 (2l + 1)(1 - |\eta_l|^2)$$

$$+ \sum_{l=0}^{+\infty} \pi \lambdabar^2 (2l + 1)|1 - \eta_l|^2 \quad (9.72)$$

where η_l accounts for the change in the l partial wave due to the scattering. σ_{el}, σ_r, and σ_{tot}, are the elastic, inelastic ("reaction"), and total cross-sections, respectively, and $\lambdabar \equiv \lambda/2\pi$ is the **reduced De Broglie wavelength**. Show that the total cross-section expanded in **partial waves** is:

$$\sigma_{tot} = \sum_{l=0}^{+\infty} \pi \lambdabar^2 (2l + 1)(1 - \text{Re}\{\eta_l\}) . \quad (9.73)$$

Is it possible to have an elastic scattering with $\sigma_r = 0$ or an inelastic scattering with $\sigma_{el} = 0$?

(9.12) ** If the target in a $a + X \rightarrow b + Y$ reaction is modeled as a black disk absorber, the only partial waves that contribute to the cross-section and are, therefore, absorbed are $l \leq R/\lambdabar$. Demonstrate that, unlike the classical case studied in Sec. 2.7, the total cross-section is twice the geometrical area:

$$\sigma_{tot} = 2\pi (R + \lambdabar)^2 . \quad (9.74)$$

Why? [Hint: for total absorbtion $\eta_l = 0$.]

(9.13) Using the SEMF and neglecting the pairing term, compute the energy produced by the fission of ^{235}U in the $n + {}^{235}\text{U} \rightarrow {}^{92}\text{Rb} + {}^{140}\text{Cs} + 3\,\text{n}$ reaction.

(9.14) * Demonstrate eqn 9.34.

(9.15) * A neutrino detector whose target is made of 1 tonne of water is located 100 m far from a 1.5 GW reactor core. Assuming an average detection efficiency for the inverse beta decay of 20%, how many $\bar{\nu}_e$ are detected per day? Neglect the scattering on oxygen.

(9.16) * Compute the number of scattering needed by a ^{4}He moderator to reduce the neutron energy from 1 MeV to 1/40 eV (**thermal energy**). Why cannot we use ^{4}He moderators instead of graphite in nuclear reactors?

(9.17) * The relative abundance of ^{238}U/^{235}U is 7.2×10^{-3} and the energy produced in a reactor core using natural uranium is due to ^{235}U and amounts to 200 MeV per fission. If the reactor produces 1 GW power, how much plutonium will be created per year if all neutrons captured by ^{238}U generate ^{239}Pu. Note that the capture cross-section $n + {}^{238}\text{U} \to {}^{239}\text{Pu}$ averaged on the neutron energies is about 1 barn and the fission cross-section of ^{235}U is $\simeq 600$ barn.

(9.18) * If the ^{239}Pu in the previous exercise is extracted by the core with an efficiency of 1%, can you build a nuclear weapon? Estimate the power of the weapon assuming the same performance as Fat Boy.

Weak interactions

10

10.1 Spotting weak processes

Weak interactions are nowadays embedded into the Standard Model and described by a unified theory of "electroweak" forces developed in the late 1960s. Such unification is the culmination of more than 40 years of research driven by a set of outstanding experimental findings. The deepest of such findings was the observation that weak interactions do not conserve parity. Weak interactions are the only forces that violate parity, C-parity, and T-parity. Besides, the electroweak Lagrangian is not invariant for combinations of C and P transformations (CP parity). The electroweak theory is a local theory consistent with the axioms of quantum mechanics (QM) and special relativity (SR), and, therefore, weak interactions must violate T-parity to conserve CPT (see Sec. 5.8). Among all known forces, weak interactions are the forces with the smallest number of symmetries and posed tremendous challenges to the establishment of the Standard Model.

Before the second World War, there was still confusion between strong and weak interactions: both phenomena were discovered during the investigation of the nuclear decays discussed in the next chapter. The spontaneous emission of electrons (**beta decays**) provided the first evidence of natural radioactivity (Bequerel, 1896) and the corresponding emission of positrons was observed by I. Curie-Joliot and F. Joliot in 1934. Unlike the nuclear reactions discussed in Sec. 9.7, beta decays represented a conundrum because the sum of the energy of the electron and the final state nucleus was systematically lower than the initial energy. Likewise, beta decays seemingly violate both momentum and angular momentum conservation. Year after year, weak interactions arose like a freak in the zoo of nuclear physics.

Following a very speculative hypothesis put forward by Pauli,[1] E. Fermi developed a successful theory of weak interactions based on a new type of neutral particles that we now call **neutrinos**. Neutrinos are spin-1/2 particles with no electric charge, no QCD vertices, and a mass much smaller than the mass of the electron. They are elementary leptons that experience only weak and gravitational forces. The Fermi theory provided excellent predictions on the lifetime of the β-decays and an estimate of the neutrino cross-sections with matter, paving the way to their experimental observation. The existence of these ghostly particles was conclusively established when the first nuclear reactors came into play as a powerful source of electron-antineutrinos ($\bar{\nu}_e$). F. Reines and

[1]In 1931, Pauli introduced a virtually unobservable final-state particle created in beta decays to preserve the energy–momentum conservation laws at the microscopic level.

A Modern Primer in Particle and Nuclear Physics. Francesco Terranova, Oxford University Press.
© Francesco Terranova (2021). DOI: 10.1093/oso/9780192845245.003.0010

C.L. Cowan observed for the first time neutrino interactions ($\bar{\nu}_e + p \to e^+ + n$) in 1956, in the proximity of the Savannah River (South Carolina) power plant. As a bonus, they showed once and for all that weak interactions do exist and cannot be led back to strong or electromagnetic phenomena.

The neutrino is the ideal probe to investigate weak forces because this particle interacts with ordinary matter only by weak processes and the size of gravitational effects is negligible in any scattering of interest in particle physics.

10.2   Classification of weak interactions

Since the electron and the neutrinos are stable particles, the lightest particle that can decay is the muon. Quantum chromodynamics (QCD) and quantum electrodynamics (QED) decays are not an option because of flavor conservation (see Sec. 8.8). Weak interactions offer a way out:

$$\mu^- \to e^- \nu_\mu \bar{\nu}_e. \tag{10.1}$$

The muon decay is a three-body decay where two types of neutrinos are emitted: a muon-type neutrino and an electron-type antineutrino. Since neutrinos are undetected, the observer records an apparent lack of energy and momentum, which are carried away by the νs. The energy shared among the particles amounts to the difference between the mass of the muon and the mass of the electrons ($\simeq 100$ MeV). Even if the muon is the lightest particle that decays by weak interactions, there are nuclear processes where the available energy is much less than 100 MeV. For these processes one can hope to see a sharp drop of the electron energy spectrum slightly before the maximum allowed value: the drop would show that the neutrinos have a finite mass m_ν so that the available energy is reduced by m_ν. For instance, ^{3}H – tritium, the bound state of two neutrons and one proton – decays as:

$$^3\text{H} \to ^3\text{He } e^- \bar{\nu}_e . \tag{10.2}$$

The available energy for massless neutrinos is just 18.6 keV and corresponds to the Q-value of the reaction. Even if physicists have been searching for the drop in the endpoint of the electron energy produced by tritium decays for more than 70 years, this effect has not been observed to date and evidence for $m_\nu \neq 0$ was gained in 1998 through other means (see Secs 10.4.3 and 13.6). Contemporary physicists classify weak interactions into three groups:

- **Leptonic processes**: processes where both initial and final state particles are leptons. Renowned examples are the decay of the muon or the scattering of neutrinos:

$$\nu_e e^- \to \nu_e e^- \text{(elastic scattering)}. \tag{10.3}$$

Another noteworthy example is the inelastic scattering of neutrinos, where the neutrino is transformed into a charged lepton:

$$\nu_\mu e^- \to \nu_e \mu^- \text{(inelastic scattering)}. \qquad (10.4)$$

As we will see in Sec. 10.3, inelastic scattering is also used to classify the neutrino flavor.

- **Semileptonic processes**: these are processes where the initial and final state include both leptons and hadrons. A classical example is the decay of the neutron:

$$n \to p\, e^- \bar{\nu}_e . \qquad (10.5)$$

This decay can be interpreted in the context of the quark model as the flavor-changing transition:

$$d \to u\, e^-\, \bar{\nu}_e . \qquad (10.6)$$

Nuclear beta decays belong to the category of semileptonic decays provided that the initial and final state contain nuclei, In β^- **decays**, a neutron transforms into a proton that stays bound to the nucleus. Hence, a nucleus (Z, A) with Z protons and $A - Z$ neutrons transforms into

$$(Z, A) \to (Z + 1, A)\, e^- \bar{\nu}_e \quad [\beta^- \text{ decay}] \qquad (10.7)$$
$$(Z, A) \to (Z - 1, A)\, e^+ \nu_e \quad [\beta^+ \text{ decay}] . \qquad (10.8)$$

The β^+ **decay** of eqn 10.8 produces a positron. This process was observed in 1934, two years after Anderson's discovery of antimatter.

- **Hadronic processes**: in these processes only hadrons are present in the initial and final state. A classical example is the decay of the Λ baryon:

$$\Lambda \to p\, \pi^- . \qquad (10.9)$$

Equation 10.9 cannot be due to strong interactions because it corresponds to a flavor-changing transition among quarks:

$$s \to u\bar{u}d . \qquad (10.10)$$

We can now answer the question that opens this chapter: how can we spot weak processes? It is apparent from this classification scheme that weak interactions can be easily recognized because either involve the production or scattering of neutrinos or produce a change in the flavor of quarks. **Flavor-changing processes**, which are allowed neither by QED nor by QCD, are a prominent feature of weak forces.

10.3   Neutrino flavors

Neutrinos were considered massless particles until 1998, and even today, precise measurements of neutrino masses are beyond the state-of-the-art of experimental physics. Unlike an electron or a muon, we cannot say if a neutrino is ν_e or ν_μ by measuring its mass. Still, in 1962, L.M. Lederman, M. Schwartz, and J. Steinberger were able to demonstrate that neutrinos come in different flavors, like quarks. They showed that neutrinos that are produced in association with a muon (e.g. from a $\pi^+ \to \mu^+ \nu_\mu$ decay) always produce muons in the final state. Experimental data support the statement that, for two generic sets of X and Y hadrons,

$$\nu_\mu X \to \mu^- Y \tag{10.11}$$

is an allowed transition, while

$$\nu_\mu X \to e^- Y \tag{10.12}$$

is forbidden. In the original 1962 experiment, this principle was established by the lack of electrons produced by the scattering of a ν_μ beam in a large mass detector. This rule is very well tested nowadays, although it had to be reformulated more subtly after the discovery of neutrino oscillations (see Sec. 10.4.3 and 13.6). It took decades to gain direct evidence of the existence of neutrinos but this is not a surprise: since weak interactions are much fainter than electromagnetic interactions, the typical neutrino cross-section in the MeV-GeV energy range goes from 10^{-42} to 10^{-38} cm^2. Some ultra-rare interactions as the coherent neutrino scattering in nuclei were established only in recent times. **Coherent scattering**, in particular, was observed at the Oak Ridge Spallation Neutron Source in 2017.

The discovery of the tau charged lepton (Pearl, 1977) warned physicists of the existence of the corresponding tau neutrino (ν_τ) and the study of τ decays provided strong indirect evidence for it. Once more, the direct observation of a ν_τ came much later: the

$$\nu_\tau X \to \tau^- Y \tag{10.13}$$

process was observed for the first time at Fermilab in 2000 by the DONUT Collaboration.

10.4   Conservation laws and symmetries of weak interactions

Despite early thrills caused by the energy and momentum taken by neutrinos, after the discovery of neutrinos scientists realized that weak interactions fulfill the standard energy, momentum, and angular momentum conservation. As a consequence, its Lagrangian is invariant for space-time translations, and rotations in space. The theory of weak interactions is completely decoupled from QCD and, therefore, the color of the quarks is immaterial in weak processes. Electric charge is conserved, too. Unlike color, there is a deep connection between electric and weak forces that will be discussed together with the electroweak theory in Sec. 12.2.2.

The weak force Lagrangian is not invariant for inversions: parity, C-parity, and T-parity. Similarly, weak processes may change the flavor of quarks and leptons provided that three conservation rules are satisfied: the conservation of electric charge, baryon number, and lepton number.

10.4.1 Baryon number

As anticipated in Sec. 8.8, we can assign an additive quantum number to quarks called the **baryon number** (B). All quarks have $B = +1/3$ and all antiquarks have $B = -1/3$. B is called the baryon number because mesons are quark–antiquark pairs and, therefore,

$$B_{\text{meson}} = +\frac{1}{3} - \frac{1}{3} = 0 \tag{10.14}$$

while

$$B_{\text{baryon}} = +\frac{1}{3} + \frac{1}{3} + \frac{1}{3} = +1$$
$$B_{\text{antibaryon}} = -\frac{1}{3} - \frac{1}{3} - \frac{1}{3} = -1\,.$$

Intuitively, the baryon number is a quark counting number where quarks count as positive terms and antiquarks as negative terms. We know already that the baryon number is conserved in QED and QCD both in the perturbative and non-perturbative regime, and we have overwhelming experimental evidence that the total baryon number of a system is conserved in weak processes, too. Still, the Standard Model adds a subtlety:

> the baryon number is conserved in strong and electromagnetic scatterings and in all *perturbative* weak scatterings.

This subtlety is immaterial in this Primer but you can imagine that the non-perturbative regime of weak forces was in action just after the Big Bang. A baryon number violation may leave a remnant in our world even if does not occur anymore.[2]

10.4.2 Lepton number

We can assign an additive quantum number L to leptons, as well:

$$L = +1 \ \text{ for } e^-, \ \mu^-, \ \tau^-, \ \nu_e, \ \nu_\mu, \ \nu_\tau$$
$$L = -1 \ \text{ for } e^+, \ \mu^+, \ \tau^+, \ \bar{\nu}_e, \ \bar{\nu}_\mu, \ \bar{\nu}_\tau\,.$$

The total **lepton number** in an initial state is the sum of the number of leptons that are present either as free particles or in bound states (e.g. the positronium). Lepton number conservation implies that:

[2]Violation of the baryon number is one of the necessary conditions (**Sakharov conditions**) to produce a universe where matter is much more abundant than anti-matter. Why is our universe made of matter instead of anti-matter is completely unclear. Such a simple question is a prominent issue in contemporary physics.

> the total lepton number of the initial state is conserved in perturbative weak interactions and in any other particle physics process (QED/QCD).

Again, we raise caveats on what may have been happened just after the Big Bang, where L can be violated by non-perturbative weak effects. There are several extensions of the Standard Model that speculate on combined L and B violations in the early universe to produce the matter–antimatter imbalance we observe today. None of them has been corroborated by experimental data as yet.

Example 10.1

Many light particles would be stable in QED/QCD due to flavor conservation. A transition to a lighter state may arise from a change of flavor, which occurs in weak processes if lepton and baryon numbers are conserved. The most spectacular example is the decay of the neutron, which is a stable bound state in QCD since it belongs to the low-lying ($l = 0$) baryon octet. Baryon number conservation prevents many decay into leptons or mesons as $n \to \nu_e \bar{\nu}_e$, $n \to e^+ e^-$, $n \to \pi^+ \pi^-$, or even $n \to \gamma\gamma$. Since the mass of the neutron is slightly higher than the mass of the proton, the only available decay is then $n \to p + $ leptons. Charge and energy conservation implies that the final state must contain at least one electron: $n \to p + e^- + $ neutrinos. Finally, lepton number conservation requires at least one antineutrino and forbids $n \to p + e^- + \nu_e$ (or ν_μ, ν_τ). But what is the flavor of this antineutrino? We will see below that the only possible flavor is $\bar{\nu}_e$. As a consequence, neutrons can only decay in $n \to p\, e^-\, \bar{\nu}_e$ at lowest order in perturbation theory. Other high-order modes are strongly suppressed. For example, the second-order process $n \to p\, e^-\, \bar{\nu}_e \gamma$ has a $BR \simeq 9 \times 10^{-3}$.

Table 10.1 Assignment of the lepton family numbers. These numbers are all zero for quarks and elementary bosons.

Part.	L	L_e	L_μ	L_τ
e^-	1	1	0	0
μ^-	1	0	1	0
τ^-	1	0	0	1
ν_e	1	1	0	0
ν_μ	1	0	1	0
ν_τ	1	0	0	1
e^+	-1	-1	0	0
μ^+	-1	0	-1	0
τ^+	-1	0	0	-1
$\bar{\nu}_e$	-1	-1	0	0
$\bar{\nu}_\mu$	-1	0	-1	0
$\bar{\nu}_\tau$	-1	0	0	-1

10.4.3   Lepton family number

The experiments on neutrinos demonstrate that an electron neutrino must be produced always in association with an electron. The same law holds for muon (tau) neutrinos that are always produced in association with a muon (tau lepton).

This law suggests a further additive quantum number called **lepton family number** (or **lepton flavor number**). The electron family number L_e is $+1$ for electrons and electron neutrinos and -1 for positrons and electron antineutrinos. Similarly, the muon (tau) family number L_μ (L_τ) is $+1$ for muons (taus) and muon (tau) neutrinos and -1 for antimuons (antitau) and muon (tau) antineutrinos, as shown in Tab. 10.1. If the lepton family numbers L_e, L_μ, and L_τ are separately conserved, $n \to p\, e^-\, \bar{\nu}_\mu$ or $n \to p\, e^-\, \bar{\nu}_\tau$ is forbidden even if it conserves the baryon and lepton number.

Lepton flavors for neutrinos are subtler than for quarks or charged leptons because they are not defined using the mass of the particle. Any observer can distinguish a muon from an electron because of its rest mass even if the charge is the same. For neutrinos, the flavor is an intrinsic feature of the particle, that is, the capability to produce charged leptons in the final state: an electron neutrino is *the only particle* that can produce an electron through an inelastic scattering as $\nu_e X \to e^- Y$.

Our understanding of neutrino flavor changed dramatically in 1998 when neutrino oscillation experiments demonstrated that neutrinos have a (tiny) mass of the order of tens of meV (1 meV = 0.001 eV). When a neutrino with a given flavor is produced with a momentum $\mathbf{p}$, its wavefunction is a linear superposition of the three mass eigenstates m_1, m_2, and m_3. The mass eigenstates propagate in space at different velocities

$$v = \frac{|\mathbf{p}|}{E} = \frac{|\mathbf{p}|}{(m_i^2 + |\mathbf{p}|^2)^{1/2}} \qquad (10.15)$$

so that an observer located at a distance L from the source detects a different linear combination of m_1, m_2, and m_3. In turn, an observer far from the source has a non-zero probability of observing a ν_μ instead of a ν_e. This "spontaneous" flavor transition is called **neutrino oscillation** and is discussed in detail in Sec. 13.6. If a neutrino oscillation occurs, we may detect processes like:

$$\pi^+ \to \mu^+ \nu_\mu \xrightarrow{\text{L large}} \mu^+ \nu_e. \qquad (10.16)$$

From the experimental point of view, the observer will report the appearance of ν_e at a distance L observing, for instance, a scattering like:

$$\nu_e \, X \to e^- \, Y. \qquad (10.17)$$

The conservation of the lepton family number is valid only at the time when the neutrino is created or if neutrino oscillations can be neglected, for example because L is too small. Coming back to the neutron decay (Example 10.1), the only allowed decay of the neutron at $t = 0$ is

$$n \to p \, e^- \, \bar{\nu}_e \qquad (10.18)$$

although neutrino oscillations can change the flavor of $\bar{\nu}_e$ in $\bar{\nu}_\mu$ or $\bar{\nu}_\tau$ at later times.

Example 10.2

The muon is a stable particle in QED because the lepton family number conservation forbids $\mu \to e\gamma$. This process may occur through higher-order processes involving neutrino oscillations but the corresponding branching ratio is extremely small ($\ll 10^{-50}$). If the lepton family number is conserved, we need at least two neutrinos in the final state to compensate for the initial-state muon and final-state electron. The leading decay mode of the muon can thus only be

$$\mu^- \to e^- \bar{\nu}_e \nu_\mu \,. \tag{10.19}$$

Other higher order modes like $\mu \to e^- \bar{\nu}_e \nu_\mu \gamma$ are, again, strongly suppressed ($BR \simeq 6 \times 10^{-8}$).

Weak interactions highly reduce the number of stable particles. To the best of our knowledge, the only **stable particles** are the proton, the photon, the electron, and the neutrinos. Neutrons can be stabilized only in bound states where the available energy is smaller than all possible final states. These bound states are the stable nuclei we studied in Chap. 9. As noted in Sec, 10.4.1, baryon and lepton conservation do not have the status of an exact conservation law – like, for example, the conservation of electric charge or the angular momentum – because they may be violated in special conditions. It comes as no surprise that many extensions of the Standard Model predict decay modes where the baryon number is not conserved and, therefore, produce unstable protons. Proton decays, however, have never been observed.

10.5 Parity violation

The choice of a right-handed or left-handed system is a matter of convention as it is, for instance, the choice of the origin of a reference frame. Changing the origin corresponds to performing a translation and all fundamental interactions are invariant for translations in space. We have already shown that such invariance results in the conservation of momentum. So – citing Pauli – "how can a fundamental force of nature be sensitive to an arbitrary choice?" We will see shortly that nature turned out to be subtler than Pauli.

10.5.1 Chiral theories

Let us consider a particle that has no rest mass, spin-1, and that interacts with a new hypothetical force different from QED and QCD. We may think about it as an exotic photon that does not obey the laws of QED. An observer in a given frame can measure the motion and the relative orientation of the spin S with respect to the direction of motion.

For a spin 1 massless particle, a measurement of S_z along the axis of motion results only in $S_z = +1$ or $S_z = -1$ in NU (natural units), as seen in Sec. 6.6.2. In classical electrodynamics, $S_z = +1$ corresponds to a circularly polarized light wave: the "right-handed polarization." $S_z = -1$ corresponds to a "left-handed polarization". If this hypothetical particle is measured as left-handed, that is, the spin is antiparallel to the direction of motion (see Fig. 10.1 top), it will appear as left-handed to any other inertial observer. The left-handedness of this hypothetical photon is then a Lorentz-invariant quantity. This is a consequence of the fact that the particle is massless: if an observer wants to see this particle as right-handed, he must run faster than the particle in its direction of motion (see Fig. 10.1). In this case, he sees in its own rest-frame (Fig. 10.1 bottom) the sign of the momentum reversed and measures the spin parallel to the direction of motion. But since no observer can travel faster than light, a left-handed massless particle will be left-handed for any observer.

We can define the **helicity** of a particle as the relative orientation of spin and momentum $\mathcal{H} = \mathbf{p} \cdot \mathbf{J}/(|\mathbf{p}||\mathbf{J}|)$ — see Sec. 6.6.2 — and claim that:

> the helicity of a massless particle is Lorentz invariant; it has the same value for any observer and can be considered an intrinsic property of the particle.

Fig. 10.1 Top: in the laboratory frame (LAB), both a massive particle and a man move from the left to the right with speed v and v', respectively ($v < v'$). An observer in LAB (not shown in the figure) measures the particle helicity as LH because the spin direction (thick arrow) is opposite to the momentum direction (thin arrow). Bottom: In the rest frame (RF) of the man, the particle moves from the right to the left but the spin is unchanged. The helicity in RF is thus RH. Image credit: V. Terranova.

In a sense, the helicity of a massless particle is as fundamental as its electric charge. In the early 1950s, Tsung-Dao Lee and Chen-Ning Yang noted that this idea could be used to build **chiral theories**. A chiral theory is a theory where the strength of the interaction depends on the left-handedness (or right-handedness) of the particles. For instance, a possible chiral theory is a theory where this hypothetical particle interacts with matter only if it is left-handed (LH), while no interactions are allowed if it is right-handed (RH). It is easy to show that the theory violates parity conservation. A parity transformation changes $\mathbf{p} \to -\mathbf{p}$ and $\mathbf{J} \to \mathbf{J}$ so that the helicity changes the sign: $\mathcal{H} \to -\mathcal{H}$. Since the interaction probabilities change going from LH to RH particles, the transition probabilities of a system will be different in the two cases. Chiral theories like this one are an outcome of special relativity –which provides the maximum speed– and quantum mechanics –which provides intrinsic angular momenta. They defeated Pauli's common sense: you cannot change the orientation of the reference frame with impunity because the change is equivalent to changing an intrinsic property of a particle.

10.5.2 Chirality

Helicity and chirality are fundamental to understand weak interactions and originate from the Dirac equation discussed in Sec. 5.9. We already introduced the definition of helicity for a hypothetical massless particle. We do it now for any particle: the relative orientation of the spin of a particle with respect to the direction of motion is called **helicity** and is defined as:

$$\mathcal{H} = \frac{\mathbf{p} \cdot \mathbf{J}}{|\mathbf{p}||\mathbf{J}|} \tag{10.20}$$

where $\mathbf{p}$ and $\mathbf{J}$ are the momentum and angular momentum of the particle. We remind the reader that we define the helicity dividing the scalar product by $|\mathbf{p}|$ $|\mathbf{J}|$ but other definitions are in use too (see Sec 6.6.2). If the angular momentum is along the direction of the particle, then $\mathcal{H} = +1$ and the helicity of the particle is called **right-handed** (RH). If the direction of the angular momentum is opposite to the direction of motion of the particle, $\mathcal{H} = -1$ and the helicity of the particle is called **left-handed** (LH). We already noted that the helicity of a particle is not a Lorentz-invariant quantity. An inertial observer moving in the same direction of the particle but with a higher speed will see the momentum reversed and the spin unchanged (see Fig. 10.1). This observer disagrees on the helicity of the particle measured by an observer moving with a speed slower than the speed of the particle. As a consequence, in a given frame it is possible to express any wavefunction as a linear combination of LH and RH helicity states.

> The **chirality** of a particle is defined as the eigenvalue of ψ, if the particle wavefunction ψ is an eigenstate of the **chirality operator**:
>
> $$\gamma^5 \equiv i\gamma^0\gamma^1\gamma^2\gamma^3 \tag{10.21}$$
>
> where the gamma matrices are defined by eqn 5.81.

In the Pauli–Dirac representation,

$$\gamma^5 \equiv i\gamma^0\gamma^1\gamma^2\gamma^3 = \begin{pmatrix} 0 & \mathbb{I} \\ \mathbb{I} & 0 \end{pmatrix} = \begin{pmatrix} 0 & 0 & 1 & 0 \\ 0 & 0 & 0 & 1 \\ 1 & 0 & 0 & 0 \\ 0 & 1 & 0 & 0 \end{pmatrix}. \tag{10.22}$$

Note that the form of γ^5 depends on the choice of the matrix representation and has the following properties. It behaves like an inversion if applied twice:

$$\gamma^5\gamma^5 = (\gamma^5)^2 = \mathbb{I}, \tag{10.23}$$

it is self-adjoint:

$$\gamma^{5\dagger} = \gamma^5, \tag{10.24}$$

and anticommutes with any γ^μ:

$$\gamma^5\gamma^\mu = -\gamma^\mu\gamma^5. \tag{10.25}$$

In particular,

$$\psi_L = \frac{1}{2}(1 - \gamma^5)\psi \qquad (10.26)$$

is a **LH chiral state** because $\gamma^5\psi_L = -\psi_L$. Therefore, ψ_L is an eigenstate of γ^5, whose eigenvalue is -1.

$$\psi_R = \frac{1}{2}(1 + \gamma^5)\psi \qquad (10.27)$$

is a **RH chiral states** because $\gamma^5\psi_R = \psi_R$.

From a mathematical point of view,

$$P_L \equiv \frac{1}{2}\left(1 - \gamma^5\right) \qquad (10.28)$$

is a **LH chirality projector**, an operator that projects the wavefunction into its LH component. The corresponding **RH chirality projector** is:

$$P_R \equiv \frac{1}{2}\left(1 + \gamma^5\right) \qquad (10.29)$$

and selects the RH component of ψ. P_L and P_R do the opposite for anti-particles: P_L (P_R) projects an antiparticle in the $+1$ (-1) chirality state.

The Dirac equation correlates the LH and RH helicity states[3] because

$$(\gamma^\mu p_\mu - m)\,u = (E\gamma_0 - \mathbf{p}\cdot\boldsymbol{\gamma} - m)u = 0 \qquad (10.30)$$

can be recast as:

$$\begin{pmatrix} (E-m)\mathbb{I} & -\mathbf{p}\cdot\boldsymbol{\sigma} \\ \mathbf{p}\cdot\boldsymbol{\sigma} & -(E+m)\mathbb{I} \end{pmatrix} \begin{pmatrix} u_A \\ u_B \end{pmatrix} = \begin{pmatrix} \mathbf{0} \\ \mathbf{0} \end{pmatrix} \qquad (10.31)$$

so that

$$u_A = \frac{\mathbf{p}\cdot\boldsymbol{\sigma}}{E-m}u_B; \quad u_B = \frac{\mathbf{p}\cdot\boldsymbol{\sigma}}{E+m}u_A. \qquad (10.32)$$

where $\boldsymbol{\sigma}$ is the three-vector of the pauli matrics and $\mathbb{I}$ is the 2 x 2 identity matrix. The LH chirality component is then:

$$\psi_L = \frac{1}{2}(1-\gamma^5)\psi = \frac{1}{2}\begin{pmatrix} u_A - u_B \\ u_B - u_A \end{pmatrix} \qquad (10.33)$$

and a chirality eigenstate mixes up LH and RH helicity components. On the other hand, if z is the direction of motion of the fermion the upper component of eqn 10.33 is

$$\frac{1}{2}(u_A - u_B) = \frac{1}{2}\left(1 - \frac{\mathbf{p}\cdot\boldsymbol{\sigma}}{E+m}\right)u_A$$

$$= \frac{1}{2}\left(1 - \frac{p_z}{E+m}\right)u_A^{(1)} + \frac{1}{2}\left(1 + \frac{p_z}{E+m}\right)u_A^{(2)} \xrightarrow{m\to 0} u_A^{(2)}. \qquad (10.34)$$

The same result holds for the lower component of eqn 10.33. LH chirality states for fermions are thus LH helicity states in the massless limit,

[3]For the sake of simplicity, we consider a fermion running along the z-axis. so that $p_x = 0$ and $p_y = 0$. The corresponding wavefunction is given by eqn 5.85 and the (four-component) bispinor u can be written as the sum of the RH ($r = 1$) and LH ($r = 2$) helicity states:

$$u = \begin{pmatrix} u_A \\ u_B \end{pmatrix}$$

$$= \sqrt{E+m}\begin{pmatrix} \chi^r \\ \frac{\boldsymbol{\sigma}\cdot\mathbf{p}}{E+m}\chi^r \end{pmatrix}$$

where u_A and u_B are two-component row vectors (spinors). More explicitly,

$$u = \begin{pmatrix} \sqrt{E+m} \\ 0 \\ \frac{p_z}{\sqrt{E+m}} \\ 0 \end{pmatrix} + \begin{pmatrix} 0 \\ \sqrt{E+m} \\ 0 \\ \frac{-p_z}{\sqrt{E+m}} \end{pmatrix}$$

$$\equiv \begin{pmatrix} u_A^{(1)} \\ u_A^{(2)} \\ u_B^{(1)} \\ u_B^{(2)} \end{pmatrix}$$

that is, the LH chirality corresponds to a $u^{(2)}$ state where the spin is aligned in the direction opposite to the motion. In the same way, we can show that for massless antifermions the LH chirality component is the RH helicity state (see Exercise 10.13). If the particle is massive, the chirality components are polluted by "wrong" helicity components. Their contribution, however, is suppressed as:

$$1 - \frac{p_z}{E+m} \xrightarrow{E \gg m} 1 - \beta. \tag{10.35}$$

As a consequence,

> the chirality is equivalent to the helicity if the mass of the particle is negligible with respect to the particle energy. In this ultra-relativistic limit, the probability of observing a LH chirality fermion as a RH helicity fermion is suppressed as $\simeq (1 - \beta)$.

10.5.3 The experiment of Chien-Shiung Wu

The fundamental nature of parity conservation was challenged in the early 1950s due to an anomaly in the decay of the K mesons ("$\theta - \tau$ puzzle"). Following the suggestion of Lee and Yang, Chien-Shiung Wu provided the first evidence of parity violation in beta decays in 1957. Soon after, a particle physics experiment by R.L. Garwin, L.M. Lederman, and M. Weinrich corroborated her results.

The **experiment of C.S. Wu** is based on a very low temperature cryostat developed at the US National Bureau of Standards. It employs a well known decay of ^{60}Co:

$$^{60}\text{Co} \ (J = 5) \rightarrow {}^{60}\text{Ni}^{**} \ (J = 4) \ e^- \bar{\nu}_e \rightarrow$$
$$^{60}\text{Ni}^* \ (J = 2) \ \gamma \rightarrow {}^{60}\text{Ni} \ (J = 0) \ \gamma. \tag{10.36}$$

The first decay is a semileptonic weak process with the emission of an electron and an antineutrino. Since the nickel is produced in an excited state, the decay is followed by two electromagnetic transitions emitting a photon at 1173 and 1332 keV, respectively. If it were possible to align the spin of ^{60}Co in a well-defined direction using, for example, a magnetic field, the distribution of the electron would provide an observable sensitive to parity violation: the scalar product $\mathbf{p} \cdot \mathbf{J}$, where $\mathbf{p}$ is the electron momentum and $\mathbf{J}$ the spin of ^{60}Co. If parity is conserved, the probability of emitting an electron at an angle θ cannot depend on $\mathbf{p} \cdot \mathbf{J}$ because parity transforms $\mathbf{p} \cdot \mathbf{J}$ into $-\mathbf{p} \cdot \mathbf{J}$. In particular, the number $N(\theta)$ of electrons emitted per unit time at an angle θ with respect to the direction of $\mathbf{J}$ is such that:

$$N(\theta = 0°) = N(\theta = 180°) \tag{10.37}$$

because:

$$\frac{\mathbf{p} \cdot \mathbf{J}}{|\mathbf{p}||\mathbf{J}|} = \cos \theta \tag{10.38}$$

and $\cos 0° \neq \cos 180°$ On the other hand, if RH chirality particles do not interact with matter, as speculated by Lee and Young, there is quite a good chance of observing $N(\theta = 0°) \neq N(\theta = 180°)$. The size of the inequality could also provide further insight into the underlying chiral theory.

Designing an experiment based on the magnetic alignment of ^{60}Co is very difficult because the thermal motion randomizes the orientation of the angular momentum with respect to the magnetic field. The statistical distribution of the aligned subsample follows a Boltzmann distribution so that the degree of alignment ("polarization," P_{ol}) is proportional to

$$P_{ol} \sim \exp\left(-\frac{\boldsymbol{\mu} \cdot \mathbf{B}}{kT}\right). \tag{10.39}$$

Here k is the Boltzmann constant and $\boldsymbol{\mu}$ is the dipole magnetic moment of the nucleus, which is of the order of the nuclear magneton $\mu_N = e\hbar/2m_p$ (in SI units). In particular, $|\boldsymbol{\mu}| \simeq 3.8\mu_N$ for ^{60}Co. Since the nuclear magneton is $m_p/m_e \simeq 1836$ times smaller than the Bohr magneton, achieving polarization of nuclei is much more difficult than for atoms. The laboratory of Wu was equipped with one of the few cryostats in the world able to reach a temperature of 10 mK for a few minutes. This cryostat employed a cooling technique ("adiabatic magnetization of paramagnetic salts") detailed in Sec. 10.5.5 for the readers who are fascinated by the ingenuity of quantum thermodynamics.

10.5.4 Experimental evidence of P-violation

The low-temperature measurements of Wu were based on a polarization technique developed by C.J. Gorter and M.E. Rose in 1948. The apparatus consisted of a cobalt source installed inside the cooling paramagnetic salt (Fig. 10.2). The magnetic field required to polarize the ^{60}Co atoms was generated by a solenoid inserted along with the outermost shield of the cryostat just after cooling. The electrons emitted on top (green line in Fig. 10.2) were recorded by a scintillator located just above the source and the light produced by the scintillator was transported on top of the cryostat by a plastic light guide (lucite) connected to a photomultiplier (PMT in Fig. 10.2). Unlike the electrons, the photons produced in the decay (dashed black lines in Fig. 10.2) escape the cryostat and are detected by two sodium-iodide (NaI) scintillators located approximately at $\theta = 0°$ and $\theta = 90°$. The study of the photons played an important role in the experiment. Photon emission is a QED process and conserves parity. Hence, $N_\gamma(0°) = N_\gamma(180°)$. However, the angular distribution of the photons is not uniform and can be computed analytically in QED (see Sec. 11.6). The distribution is peaked at $\theta = 90°$ but the peak fades away if the temperature is too high because the alignment of the nuclei with the magnetic field is lost. Hence, the measurement of the number of photons at $\theta = 0°$ and $\theta = 90°$ provides a direct measurement of P_{ol}.

Fig. 10.2 Layout of Wu's experiment. The photomultiplier (PMT) collects the light that comes from electrons impinging into the scintillator (not visible in the figure) installed on top of the source (red circle). The lucite light guide (LG) brings the light outside the cryostat and is connected to the PMT. In the figure, two photons from the de-excitation of ^{60}Ni are depicted as black dashed arrows. An electron from the beta decay of ^{60}Co is shown as a green arrow. The spin of the ^{60}Co (vertical red arrow) is polarized according to the direction of the current in the coils. NaI[0] and NaI[90] collect photons emitted at about $\sim 0°$ and 90° with respect to the direction of the magnetic field produced by the coils. The magnet producing the cooling field of the CMN is removed as soon as the temperature reaches 10 mK, just before the start of the measurement. Reprinted figure with permission from Wu *et al.* 1957. Copyright 1957 by the American Physical Society.

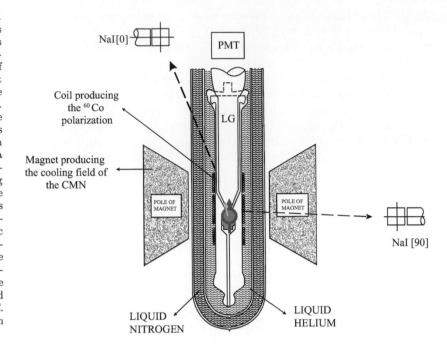

Unlike Gorter and Rose, Wu also measured the electron distribution from the beta decay using the scintillator located inside the cryostat. Wu and her collaborators achieved $P_{ol} \simeq 60\%$, which was sufficient to establish a clear asymmetry between the electrons produced at $\theta = 0°$ and $\theta = 180°$ (see Fig. 10.4). Instead of rotating the scintillator, the rate of down-moving electrons was measured reversing the sign of the magnetic field.

If we neglect the finite polarization, and we assume $P_{ol} = 1$, the expected distribution of the electrons can be parametrized as:

$$N_e(\theta) \simeq 1 + a\frac{\mathbf{p} \cdot \mathbf{J}}{|\mathbf{p}||\mathbf{J}|} = 1 + a\cos\theta \tag{10.40}$$

where a is an unknown parameter depending on the underlying theory. Accounting for finite polarization, the distribution is given by:

$$N_e(\theta) \simeq 1 + aP_{ol}\beta_e\cos\theta = 1 + aP_{ol}\frac{\mathbf{p} \cdot \mathbf{J}}{E|\mathbf{J}|}. \tag{10.41}$$

The presence of the velocity of the electron β_e in the formula is because we are assuming the underlying theory to be a chiral theory. Since the experiment measures relative orientations between momenta and spin (helicity), eqn. 10.40 holds only for massless electrons. β_e in eqn. 10.41 accounts for the probability to observe electrons whose helicity provides a faithful measurement of the chirality. We computed this probability in eqn 10.35.

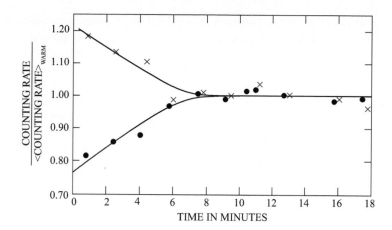

Fig. 10.4 (Upper curve with crosses) Rate of electrons detected in the PMT when the magnetic field is in the downward direction, i.e. the electrons are emitted in the direction opposite to the spin at an angle $\theta = 180°$. (Lower curve with dots) Rate of electrons detected in the PMT when the magnetic field is in the upward direction, i.e. the electrons are emitted in the same direction as the spin at an angle $\theta = 0°$. The rate is expressed as the electron counting rate divided by the electron counting rate at warm. After a few minutes, the paramagnetic salt warms up and $P_{ol} \to 0$: the electron emission is isotropic and the ratio of the rates $\to 1$. Reprinted figure with permission from Wu *et al.* 1957. Copyright 1957 by the American Physical Society.

In the first round of experiments of Wu, $P_{ol} \simeq 0.6$ and $\beta_e \simeq 0.8$. Even if the polarization achieved by the cryostat was not particularly high, it was sufficient to show that parity is violated in the distribution of the electrons (see Fig. 10.4), independently of any theory assumption. Further rounds achieved polarizations $> 90\%$ and the fit of a turned out to be compatible with $a = -1$. It means that, in this process, the electrons are emitted more often in the direction opposite to the direction of the spin of the nucleus. The conservation of angular momentum (see Fig. 10.3) suggests that electrons produced by weak interactions have a thing for being left-handed. This was the seed that blossomed into the modern theory of weak forces.

10.5.5 Cooling at 10 mK*

Chien-Shiung Wu took advantage of a technology breakthrough that occurred in 1933 when W.F. Giauque and D.P. MacDougall provided a method for cooling solid-state targets below 1 K. This technique is called **adiabatic demagnetization of paramagnetic salts**. The cooling agent is a paramagnetic salt whose entropy at $T \simeq 1$ K or lower is dominated by the total angular momentum J of the ions in the crystal. If the J of the ions are not polarized, the entropy of the salt is given by:

$$S = R \log(2J + 1) \tag{10.42}$$

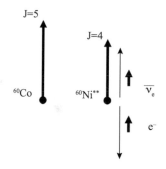

Fig. 10.3 (left) A ^{60}Co nucleus with $J = 5$ oriented along the vertical axis (i.e. the magnetic field). (right) The decay products ^{60}Ni** ($J = 4$), e^- ($J = 1/2$) and $\bar{\nu}_e$ ($J = 1/2$) when the electron is emitted in the direction opposite to the magnetic field. In this configuration, the electron helicity is LH and the antineutrino helicity is RH.

where R is the ideal gas constant and $2J+1$ represents all possible orientations of J. If we apply a magnetic field, J starts to be polarized and the value of S decreases because the system is more ordered. This entropy change can be employed to generate cooling power (Pobell, 2007). In practice, the salt is precooled at the lowest temperature achievable by standard techniques. Wu's cryostat had two shieldings cooled respectively at 77 K (liquid nitrogen) and 1.8 K (pumped liquid helium) and the salt (cerium magnesium nitrate, CMN) was thermalized at the pumped liquid helium temperature. The cooling cycle for a CMN is depicted in Fig. 10.5. The initial temperature T_i of point "A" in Fig. 10.5 is about 1 K. The cycle starts when a magnetic field is applied to produce an isothermal magnetization from $B = 0$ to $B = B_i$ (point "B"). During this field ramp, the total angular momenta J are oriented and the entropy decreases. To keep a constant temperature, the CMN must be in equilibrium with the precooling bath (the pumped liquid helium in Wu's cryostat). Later on, the salt is thermally isolated from the bath and is adiabatically demagnetized reducing the field from B_i to $B_r \simeq 10^{-3}$ Tesla (point "C"). From B to C, the entropy must be constant if the demagnetization rate is slow because the transformation is adiabatic and reversible. As a consequence, the entropy growth of eqn 10.42 must be compensated by a temperature decrease: the CMN cools down to a few mK. Once the magnetic field reaches B_f, the temperature would be constant only in an ideal isolated system. In fact, in a few minutes, the salt warms up along the entropy curve at the final demagnetization field (B_r=const) from T_f to T_i due to unavoidable heat leaks with the rest of the cryostat. This technique provides the cooling power needed to perform the physics measurement described in Sec. 10.5.4 ($T \simeq 10$ mK) but the time available for the measurement is just a few minutes. This is the reason why, although adiabatic demagnetization can attain even lower temperatures than Wu's cryostat (2–3 mK), it is no more used in low-temperature physics and has been replaced by modern ^{3}He-^{4}He dilution refrigerators (Pobell, 2007).

10.6   The chirality of neutrinos

Unlike electrons, the measurement of the helicity of the neutrino is practically equivalent to a measurement of its chirality. Detecting neutrinos, however, is a major experimental challenge and there is no direct way to test the orientation of their spin. Still, their chirality was established one year after the discovery of parity violation. The experiment that determined the neutrino chirality was performed by M. Goldhaber, L. Grodzins, and A.W. Sunyar in 1958 and stands as a masterpiece in experimental physics.

The chirality of neutrinos was determined through a very peculiar decay chain, where a neutrino is emitted with an energy similar to the energy of the subsequent gamma de-excitation. The nucleus used in this case was an isotope of the europium, ^{152}Eu.

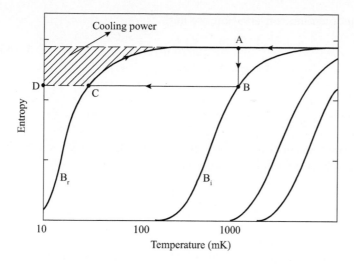

Fig. 10.5 Cooling cycle of a paramagnetic salt (CMN) with an angular momentum $J = 1/2$. The plot shows the molar entropy in arbitrary units as a function of the temperature for different values of the magnetization field (B_r, B_i). The cooling power of the salt at the end of the cycle is given by the shaded area. Redrawn from Betts 1989.

10.6.1   Electron capture and photon re-absorption

Many nuclei are unstable because atomic electrons can interact weakly with the nucleus producing a neutrino in the final state. This interaction was discovered by G. Wick and L. Alvarez in the 1930s and corresponds to:

$$(A, Z) + e^- \rightarrow (A, Z - 1) + \nu_e. \tag{10.43}$$

Equation 10.44 describes the spontaneous transmutation of a proton into a neutron in bound states ($e^- u \rightarrow \nu_e d$). This process is called **electron capture** and, unlike standard beta decays, is a two-body process, and the neutrino is emitted at fixed energy. ^{152}Eu is a $J = 0$ nucleus that transforms into an excited state of samarium (^{152}Sm*, $J = 1$) emitting a ν_e at 911 keV. The electron is located in the innermost atomic shell (K-shell) and, therefore, has a large probability to overlap with the wavefunction of the nucleus. This process was studied in detail by Grodzins in 1958, who noted that the emitted photon from:

$$^{152}\text{Sm}^*(J = 1) \rightarrow {}^{152}\text{Sm}(J = 0) + \gamma \tag{10.44}$$

has an energy of 963 keV. If the ^{152}Sm* were produced at rest, the photon could not be re-absorbed by another ^{152}Sm nucleus. The energy of the photon is not enough to induce a ^{152}Sm + $\gamma \rightarrow {}^{152}$Sm* transition because part of the energy of the de-excitation is lost in the recoil of the ^{152}Sm nucleus. The recoil energy can be computed in a frame where the ^{152}Sm* is at rest so that:

$$p^\mu_{ini} = (M_i, 0, 0, 0)$$
$$p^\mu_{fin} = (E_f, 0, 0, p_f) = \left(\sqrt{M_f^2 + p_f^2}, 0, 0, p_f\right)$$
$$p^\mu_\gamma = (p, 0, 0, p)$$

where M is the mass of ^{152}Sm*, M_f is the mass of ^{152}Sm, p is the photon momentum, and p_f is the recoil momentum of ^{152}Sm. Applying four-momentum conservation, we get:

$$|p| = |p_f| \tag{10.45}$$

and

$$p = \frac{M_i^2 - M_f^2}{2M_i} = \frac{(M_i - M_f)(M_i + M_f)}{2M_i} \simeq M_i - M_f. \tag{10.46}$$

Since the mass of the nucleus is huge ($\simeq 152$ GeV), the recoiling nucleus is non-relativistic and its energy is:

$$E_f^{kin} = \frac{p_f^2}{2M_f} \simeq 3 \text{ eV}. \tag{10.47}$$

Even if 3 eV is a small quantity, it is sufficient to inhibit the re-absorption of the photon, which lacks twice this value(6 ev) . You may wonder whether either the thermal motion or the Heisenberg uncertainty principle can compensate for the missing 3 eV. They do not. The intrinsic width of the decay due to the uncertainty principle is $\simeq h/\tau$ and contributes to a broadening of 20 meV. It is then immaterial for re-absorption. The thermal motion results into a Doppler broadening, whose full width at half maximum (FWHM) is:

$$\Delta p \simeq \sqrt{\frac{8(\log 2)\text{kT}}{M_f}} p \simeq 10^{-6} p \simeq 0.9 \text{ eV} \tag{10.48}$$

at room temperature (kT=26 meV). This value is too small to compensate for the recoil and, therefore, the only source of re-absorption can be the Doppler shift due to the production of ^{152}Eu in flight.

10.6.2   The Goldhaber experiment

The detailed study performed by Grodznis on the ^{152}Eu decay chain brought to the design of an experiment where all information on the helicity of the neutrino produced in the electron capture is transferred to the most energetic photon produced by ^{152}Sm*. In the electron capture of ^{152}Eu, the excited samarium nucleus is recoiling with an energy given by:

$$E_{\text{Sm}^*}^{kin} \simeq \frac{p_\nu^2}{2M_f} \tag{10.49}$$

where p_ν is the momentum of ν_e. This formula can be derived following the same line of reasoning of Eq. 10.47 by replacing the recoiling ^{152}Sm with the recoiling ^{152}Sm*, which is produced after the interaction of the electron with the ^{152}Eu. Since $p_\nu \simeq 911$ keV, the speed of the non-relativistic recoiling nucleus is:

$$\beta_{\text{Sm}^*} = \sqrt{\frac{2E_{\text{Sm}^*}^{kin}}{M_i}} \simeq 5.8 \times 10^{-6}. \tag{10.50}$$

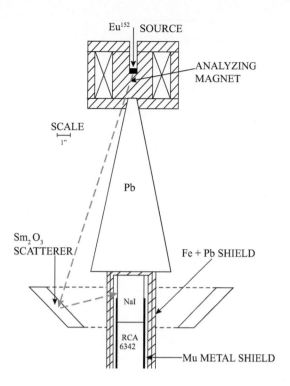

SCALE
⊢—⊣
1"

Fig. 10.6 The layout of the neutrino helicity experiment. The source is located on the top, inside a magnetized iron cradle. The most energetic photons that are not shielded by the iron reach a slab of samarium (Sm_2O_3) and are re-emitted isotropically. They can thus reach the NaI scintillator located at the bottom read out by a photomultiplier (RCA6342). One of these photons is depicted as a dashed arrow. The lead cone prevents the photons to reach the scintillator directly from the source. The coil on the top is used to generate the magnetic field in the iron, whose sign depends on the flow of the current. Reprinted figure with permission from Goldhaber *et al.* 1958. Copyright 1958 by the American Physical Society.

If a photon is emitted in the same direction of flight of Sm*, it will be Doppler shifted toward higher frequencies (blue shifted). The relativistic Doppler formula (see Sec. A.6) provides the size of the shift:

$$p' = \gamma(1+\beta)p = \sqrt{\frac{1+\beta}{1-\beta}}\, p \qquad (10.51)$$

where β is $\beta_{Sm^*} \simeq 5.8 \times 10^{-6}$. The photons emitted along the Sm* are so energetic that can be re-absorbed by another Sm atom because:

$$\Delta p = p\gamma(1+\beta) - p \simeq 5.3 \text{ eV} \qquad (10.52)$$

and the momentum shift exceeds 6 eV if we account for the thermal broadening (eqn 10.49). Photons emitted in the opposite direction are red-shifted and cannot be re-absorbed.

The **Goldhaber–Grodznis–Sunyar experiment** (Fig. 10.6) consists of a ^{152}Eu source located on top of a magnetized iron cradle. The iron support acts as a helicity selector for the photons. If the magnetic field is such that the spins of the electrons in the iron are aligned in the direction opposite to the photon spin (see Fig. 10.7), the photoelectric cross-section substantially increases because the photon can be absorbed inducing a spin-flip of the electron. The iron, in this case, strongly reduces the amount of RH photons. If the magnetic field is reversed, the iron filters out the LH photons. The transmitted photons (see Fig. 10.6) reach a slab of samarium where they can be re-absorbed and re-emitted

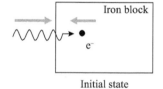

Initial state

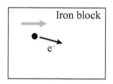

Final state after the photoelectric effect

Fig. 10.7 A photon whose spin is antiparallel to the magnetic field and, hence, to the spin of the electrons in the iron is absorbed by photoelectric effect because the cross-section of the process is enhanced by spin-flip. If the orientation of the photon and electron spin is the same, the corresponding cross-section is suppressed and the photon can cross the iron block.

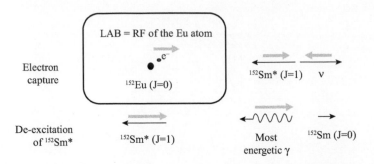

Fig. 10.8 The figure demonstrates that the helicity of the neutrino is identical to the helicity of the most energetic photon. (Top figure)The initial state in the LAB frame is shown inside the black box. The only source of angular momentum is the spin of the electron, which is depicted as a light-gray arrow. The electron capture produces LH (RH) ^{152}Sm* if the neutrino is LH (RH). In the figure the neutrino is considered LH. (Bottom figure) The most energetic photon is produced in the direction of flight of the ^{152}Sm*. This photon brings the entire angular momentum of the system because $J(^{152}$Sm$) = 0$; it is then LH (or RH) if ^{152}Sm* is LH (or RH).

isotropically. Only photons re-emitted toward the inner part of the apparatus can be recorded by an NaI scintillator. Since re-absorption selects the most energetic photons, the magnetic field selects the helicity of these photons. But what is the link between the helicity of the most energetic photon and the corresponding neutrino?

As depicted in Fig. 10.8, if the ^{152}Eu is at rest in the lab frame close up[4] and, if neutrinos are LH, the Sm* will be emitted in the opposite direction with respect to the neutrino, and angular momentum conservation implies the Sm* to be LH, too. When the ^{152}Sm* reaches the ^{152}Sm ground state, the photon emitted in the direction of flight of the Sm* brings the angular momentum since Sm has $J = 0$. Conservation of angular momentum implies that the photon is LH. We can repeat the same argument starting from the hypothesis that neutrinos are RH. In this case, we will have a RH Sm* recoiling against the neutrino and the photon emitted along the same line of flight of Sm* will be RH. In conclusion:

> in the Goldhaber experiment, the helicity (and, therefore, the chirality) of the most energetic photon is the same as the helicity (and, therefore, the chirality) of the neutrino.

The observation of photons in the NaI for a given orientation of the field provided strong experimental evidence that:

> neutrinos produced by weak interactions are left-handed chirality particles.

[4]This is an excellent approximation because the thermal motion is negligible at room temperature. At 300 K, kT is just 26 meV.

10.7   The V-A theory

After such a long look at the astounding ideas of Wu, Goldhaber, and their collaborators, it is now time to wrap up their experimental findings and show that the speculations of Lee and Yang give an adequate description of weak forces.

The Goldhaber experiment showed that only LH chirality neutrinos are produced in weak interactions. Since the chirality projector of eqn 10.28 applied to an antiparticle selects the RH component of the wavefunction, only RH chirality antineutrinos are produced by these forces. For massive particles, the statement is difficult to be tested experimentally because we generally observe helicities, not chiralities. Still, Wu's experiment shows that the electrons are *preferentially* emitted with LH helicity, which is consistent with the production of LH chirality electrons because electrons are much heavier than neutrinos. All in all, nature has chosen the most extreme chiral theory conceived by Lee and Yang:

> only LH chirality particles and RH chirality anti-particles interact by weak forces. A RH chirality particle or a LH chirality antiparticle are particles that do not interact with matter in the underlying theory of weak forces.

This theory – named the **V-A theory** – was developed by several authors and, in particular, by R. Marshak, E.C.G. Sudarshan, R. Feynman, and M. Gell-Mann starting from 1957. This time, the choice of the name is fortunate and captures the essence of the theory.

10.7.1   Parity in vector and axial couplings*

The Dirac equation provides a straightforward way to show that QED *does* conserve parity. The QED vertices are given by eqn 6.65 for a s-channel, where an incoming electron (u) annihilates with an incoming positron $\bar{v}$. Similarly, for a t-channel where we have only an incoming particle (u) and an outgoing particle $(\bar{u})$, the vertex is given by:

$$j_e^\mu \equiv \bar{u}(p_2)\gamma^\mu u(p_1). \tag{10.53}$$

We have already shown that the parity operator in relativistic QM is given by γ^0 (eqn 5.95). Spinors and adjoint spinors transform as (Thomson, 2013):

$$\mathcal{P}u = \gamma^0 u \tag{10.54}$$

$$\mathcal{P}\bar{v} = (\mathcal{P}v)^\dagger \gamma^0 = v^\dagger \gamma^{0\dagger} \gamma^0 = v^\dagger \gamma^0 \gamma^0 = \bar{v}\gamma^0 \tag{10.55}$$

because $\gamma^{0\dagger} = \gamma^0$ (see eqn 5.81). We can then show (Exercise 10.14) that:

$$\mathcal{P}j_e^0 = j_e^0 \tag{10.56}$$

$$\mathcal{P}j_e^i = -j_e^i \tag{10.57}$$

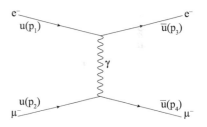

Fig. 10.9 The lowest order Feynman diagram for a $e^-\mu^- \to e^-\mu^-$ scattering in QED. The only contribution is a t-channel. The (bi)spinors of the incoming and outgoing particles are indicated next to the particle name, together with the corresponding four-momenta.

where $i=1,2,3$. The QED matrix element corresponding to the t-channel of Fig. 10.9 is (see eqns 10.53 and 6.67):

$$\mathcal{M} = -\frac{e^2}{q^2}g_{\mu\nu}j_e^\mu j^\nu = -\frac{e^2}{q^2}j_e \cdot j_\mu. \tag{10.58}$$

This quantity is invariant for parity transformations because:

$$\mathcal{P}(j_e \cdot j_\mu) = \mathcal{P}(j_e^0 j_e^0 - j_\mu^0 j_\mu^0) = j_e^0 j_e^0 - (-j_\mu^i)(-j_\mu^i) = j_e \cdot j_\mu. \tag{10.59}$$

> The properties of the gamma matrices provide a simple and rigorous demonstration that QED conserves parity.

The same result holds for QCD, too. What happens if the chirality operator enters the game? We can define an **axial current**[5]:

$$j_A^\mu = \bar{u}\gamma^\mu\gamma^5 u \tag{10.60}$$

together with the **vector current** of eqn 10.53:

$$j_V^\mu = \bar{u}\gamma^\mu u \tag{10.61}$$

and – as in QED – we can show (see Exercise 10.15) that:

$$\mathcal{P}j_A^0 = -j_A^0 \tag{10.62}$$
$$\mathcal{P}j_A^i = j_e^i \tag{10.63}$$

using the γ^5 properties described in Sec. 10.5.2. Then,

$$j_A \cdot j_A \tag{10.64}$$

conserves parity, too. But if we mix up vector and axial currents, parity conservation is gone:

$$\mathcal{P}j_V \cdot j_A = -j_V \cdot j_A \tag{10.65}$$

and we enter the domain of chiral theories.

10.7.2 From the Fermi to the V-A theory

The vector current:

$$j_V^\mu = \bar{u}\gamma^\mu u \tag{10.66}$$

and the axial current:

$$j_A^\mu = \bar{u}\gamma^\mu\gamma^5 u \tag{10.67}$$

are the basic ingredients of any chiral theory because give access to parity violation in a way that is consistent with SR and QM. Consider, for instance, the inelastic scattering:

$$e^-\nu_\mu \to \mu^-\nu_e. \tag{10.68}$$

In the original Fermi theory, there was no room for parity violation because the weak interactions were described by hypothetical four-line

[5] We drop the four-momenta and the type of particles to ease the notation.

vertices made only of vector currents. There were no mediators, either, and the fine structure constant was replaced by a **Fermi constant** G_F. This class of theories describes **contact interactions** because the final state is created by direct contact with the initial state, as in Fig. 10.10. The Fermi theory does not provide a faithful description of the process depicted in this figure because the matrix element is:

$$\mathcal{M} = G_F g_{\mu\nu} \left[\bar{u}_{\nu_e}(p_3)\gamma^\mu u_{e^-}(p_1) \right] \left[\bar{u}_{\mu^-}(p_4)\gamma^\nu u_{\nu_\mu}(p_2) \right] \equiv$$
$$G_F j_V^{e^- \to \nu_e} \cdot j_V^{\nu_\mu \to \mu^-} . \qquad (10.69)$$

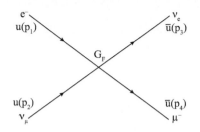

Fig. 10.10 The lowest order diagram for a $e^-\nu_\mu \to \mu^-\nu_e$ scattering in a contact interaction theory where the contact strength is the Fermi constant G_F. The (bi)spinors of the incoming and outgoing particles are indicated next to the particle name, together with the corresponding four-momenta.

$j_V^{e^- \to \nu_e}$ describes a vector current with a n electron in the initial state and a ν_μ in the final state. Equivalently, $j_V^{\nu_\mu \to \mu^-}$ describes a vector current with a muon–neutrino in the initial state and a muon in the final state. Equation 10.69 is just wrong because u_{ν_μ} has both a LH and a RH chirality component but the Goldhaber experiment testifies that the contribution of the RH component must be zero.

You can patch the Fermi theory with γ^5 by noting that:

$$P_L u = \frac{1}{2}(1 - \gamma^5)u. \qquad (10.70)$$

P_L is the chirality projector and selects the LH chirality component of a particle and the RH chirality component of an antiparticle. These components *do* interact by weak forces. That is great news because we know now that the chirality projector must enter the current to kill off the RH-chirality fermions and the LH-chirality anti-fermions. But,

$$\gamma^\mu P_L = \gamma^\mu \left[\frac{1}{2}(1 - \gamma^5) \right] = \frac{1}{2}(\gamma^\mu - \gamma^\mu \gamma^5). \qquad (10.71)$$

The first term of the right-side of eqn 10.71 is proportional to a vector current and the second to an axial current. And there is a minus sign between them! The "vector minus axial" (V-A) theory provides a faithful description of weak forces just replacing eqn.10.69 with:

$$\mathcal{M} = \frac{G_F}{\sqrt{2}} g_{\mu\nu} \left[\bar{u}_{\nu_e}(p_3)\gamma^\mu(1-\gamma^5)u_{e^-}(p_1) \right] \left[\bar{u}_{\mu^-}(p_4)\gamma^\nu(1-\gamma^5)u_{\nu_\mu}(p_2) \right] \equiv$$
$$G_F j_{V-A}^{e^- \to \nu_e} \cdot j_{V-A}^{\nu_\mu \to \mu^-} . \qquad (10.72)$$

The V-A theory is a current–current theory like QED but the currents are different and the photon propagator is missing. The $1/\sqrt{2}$ normalization factor is conventional . It is introduced to retain the same numerical value of the Fermi constant both in Fermi's and V-A theroy. For a long time, G_F had the status of a real fundamental constant: the intrinsic strength of weak forces. We will see in Chap. 12 that this is a hoax, even if G_F is still in use. The value of G_F can be extracted by the lifetime of the muon and is currently known with a precision of less than one part per million (ppm):

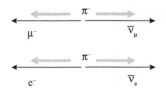

Fig. 10.11 The decay of a charged pion into a muon (top) or an electron (bottom) is forbidden if helicity is equivalent to chirality. The anti-neutrino is always produced with the spin parallel to the direction of motion (RH chirality that, for neutrinos and antineutrinos, is practically identical to helicity). Since J=0 in the initial state, angular momentum conservation implies that the muon and the electron must have the spin parallel to the direction of motion (RH helicity).

$$G_F = 1.1663787(6) \times 10^{-5} \text{ GeV}^{-2}. \tag{10.73}$$

Even more,

> the V-A theory is equivalent to the electroweak theory of the Standard Model at energies much smaller than the W boson masses. It can then explain the vast majority of scattering and decays of weakly interacting particles.

One of the earliest and most spectacular predictions of the V-A theory was the explanation of the decay of charged pions. We know already that charged pions belong to the lightest mesons octet. They are, therefore, light spin-0 mesons that decay only by weak interactions. A π^- can decay in a muon–antineutrino pair or an electron–antineutrino pair. Since $m_\mu \gg m_e$ the phase-space available for the muon decay mode is much less than the electron mode. As a consequence, one would expect $BR(\pi^- \to \mu^- \bar\nu_\mu) \ll BR(\pi^- \to e^- \bar\nu_e)$, which is in striking contradiction with experimental data. In one of the first experiments performed at CERN (1958), G. Fidecaro demonstrated that $BR(\pi^- \to e^- \bar\nu_e)$ is extremely small, but not zero. In 1963, Sudarshan and Marshak noted that if helicity and chirality were the same thing, the pion should be a stable particle. This is due to the fact that $\bar\nu_\mu$ and $\bar\nu_e$ are RH antiparticles, and angular momentum conservation requires the muon and the electron to be RH helicity states (see Fig. 10.11). Since RH chirality particles do not interact, the pion cannot decay into a charged lepton and a neutrino as long as the chirality is the same as the helicity. In this case, however, all particles are massive and there is a finite probability that a RH chirality muon has a component with LH helicity and fulfils angular momentum conservation. The size of this component is suppressed by the muon velocity as in eqn 10.35. Since $\beta_\mu \ll \beta_e$ or, equivalently, $m_\mu \gg m_e$, the helicity–chirality mismatch is much more likely for a muon than for an electron and is practically zero for a neutrino. Therefore, $BR(\pi^- \to \mu^- \bar\nu_\mu)$ must be larger than $BR(\pi^- \to e^- \bar\nu_e)$. The V-A theory predicts:

$$\frac{BR(\pi^- \to e^- \bar\nu_e)}{BR(\pi^- \to \mu^- \bar\nu_\mu)} = \frac{\Gamma(\pi^- \to e^- \bar\nu_e)}{\Gamma(\pi^- \to \mu^- \bar\nu_\mu^2)}$$

$$= \frac{m_e^2}{m_\mu^2} \left(\frac{m_\pi^2 - m_e^2}{m_\pi^2 - m_\mu^2} \right) \simeq 1.26 \times 10^{-4} \tag{10.74}$$

in excellent agreement with the findings of Fidecaro.

Exercises

(10.1) None of the mesons belonging to the lightest octet ($S = 0$, $l = 0$) are stable particles. Does it imply that all these mesons decay by weak interactions?

(10.2) The lifetime of the K^0 meson is much longer than the lifetime of the ρ particle. Explain the reason using the data from the Particle Data Group (https://pdg.lbl.gov).

(10.3) What is the probability to observe an RH-helicity neutrino in the Goldhaber experiment?

(10.4) The photon detectors in Wu's experiment were located outside the cryostat, while the electrons were observed inside. What is the rationale behind this choice?

(10.5) At what temperature should we perform the Goldhaber experiment to remove the correlation between the helicity of the most energetic photon and the helicity of the neutrino?

(10.6) Draw the Feynman diagram of a muon decay in the V-A theory.

(10.7) ** Using the crossing theorem of Exercise 7.9, write the corresponding matrix element in the V-A theory for $\mu^- \rightarrow e^- \bar{\nu}_e \nu_\mu$.

(10.8) ** Draw the Feynman diagram that produces the $\mu^- \rightarrow e^- \gamma$ decay.

(10.9) Assuming the conservation of the lepton family number, can you experimentally demonstrate the existence of the ν_τ from the decays of τ particles without observing the neutrinos?

(10.10) If we waive the assumption of the previous exercise, what can you prove about the ν_τ properties by the study of τ^- decays?

(10.11) Show that any photon emission probability $N(\theta)$ that depends on an odd power of $\cos\theta$ violates the conservation of parity. Here, N is the number of photons per second and θ is the polar emission angle.

(10.12) * Demonstrate that the dimension of the Fermi constant is eV^{-2}.

(10.13) * Show that the LH-chirality component of a massless anti-fermions is the RH helicity state.

(10.14) * Show that:

$$\mathcal{P} j_V^0 = j_V^0$$
$$\mathcal{P} j_V^i = -j_V^i$$

where $j_V = \bar{u}\gamma^\mu u$ is a vector current and $i = 1,2,3$.

(10.15) * For the axial current, demonstrate that:

$$\mathcal{P} j_A^0 = -j_A^0$$
$$\mathcal{P} j_A^i = j_A^i .$$

(10.16) ** A theorem by Dirac shows that the ψ of eqn 5.82 transform as $S\psi$ for a Lorentz boost along the z-axis, where

$$S = \begin{pmatrix} a & 0 & -b & 0 \\ 0 & a & 0 & b \\ -b & 0 & a & 0 \\ 0 & b & 0 & a \end{pmatrix}$$

in the Pauli–Dirac representation of the gamma matrices. In the previous formula,

$$a = \sqrt{\frac{1}{2}(\gamma + 1)}$$

$$b = \sqrt{\frac{1}{2}(\gamma - 1)}.$$

and γ is the Lorentz gamma of the particle. Using this theorem, show that $\bar{\psi}\psi$ is a Lorentz scalar, i.e. is invariant for Lorentz transformations. [Hint: show that $S = a\mathbb{I} - b\gamma^0\gamma^3$]

(10.17) ** The term "vector current" originates from the observation that $j^\mu \equiv \bar{\psi}\gamma^\mu\psi$ transforms as a Lorentz vector. Demonstrate this statement using the results of the previous exercise, i.e. show that:

$$j'^\mu = \Lambda^\mu_{\ \nu} j^\nu$$

where $\Lambda^\mu_{\ \nu}$ is a Lorentz transformation and j'^μ is the transformed vector current.

Radioactivity and cosmic engines

11.1 Radioactive decays

Radioactivity has been the first window available to particle physicists to explore matter below the atomic scale (10^{-10} m). Its discovery by H. Bequerel in 1896 marks the inception of nuclear physics. Despite a long and rewarding history, radioactive decays are extremely complex processes, involving a subtle interplay among strong, electromagnetic, and weak forces. Practical applications of radioactivity are great tools in medicine, biology, archeometry, material and earth sciences, and play an essential role in science and engineering since more than a century. Most radioactive processes are:

> a decay where a parent nucleus (A, Z) spontaneously decays into a daughter nucleus (A', Z') emitting either nuclear fragments or particles.

A notable exception is the aforementioned **electron capture** (EC), where an atomic electron contributes to the initial state (see Sec. 10.6.1). The description of radioactivity can be framed using the non-relativistic Fermi rule for $a \to b + c$ processes derived in Sec. 2.6.1. The complexity of the process mainly resides in the evaluation of the decay amplitude $\mathcal{T}$ and the phase space $\rho(E)$ of eqn 2.74. This definition of radioactivity appeared quite late in the history of nuclear science because radioactivity was discovered before quantum mechanics (QM). Still, radioactive decays are genuine QM processes because they are random processes whose occurrence does not depend on the history of the parent nucleus. In modern language, decays are an example of **Markov processes**: a sequence of possible events in which the probability of each event (e.g. the decay from the parent to the daughter at time t) only depends on the state attained at the previous stage of the sequence: the presence of the parent bound state (A, Z) at $t - dt$.

As anticipated in Sec. 2.6, the decay processes are driven by the decay amplitude $\Gamma_{i \to f}$, which is called the **decay constant** (or disintegration constant) $\lambda_{i \to f}$ in nuclear physics. The constant that describes the decay probability irrespective to the final state is:

$$\lambda \equiv \Gamma_{tot}. \tag{11.1}$$

A Modern Primer in Particle and Nuclear Physics. Francesco Terranova, Oxford University Press.
© Francesco Terranova (2021). DOI: 10.1093/oso/9780192845245.003.0011

Equation 2.95 then becomes the **radioactive decay law** or **Rutherford–Soddy law**:

$$N(t) = N_0 e^{-\lambda t} \tag{11.2}$$

derived on empirical grounds without reference to QM by E. Rutherford and F. Soddy in 1902. The **lifetime** of a nuclide is, therefore:

$$\tau \equiv \lambda^{-1}. \tag{11.3}$$

Like particles, the **half-life** $\tau_{1/2}$ of a nuclide is defined as the time needed to halve a sample of N_0 parent nuclei:

$$t_{1/2} \equiv \frac{\log 2}{\lambda}. \tag{11.4}$$

Note that the lifetime is actually a **mean lifetime** because:

$$\langle t \rangle = \frac{\int_0^\infty t |dN/dt| dt}{\int_0^\infty |dN/dt| dt} = \frac{1}{\lambda}. \tag{11.5}$$

A straightforward way to estimate λ is to count the number of decays occurring in a given time interval Δt. In spite of its simplicity, this method may trick the novice. $|\Delta N|$ can be expressed as:

$$|\Delta N| = N(t) - N(t - \Delta t) = N_0 e^{-\lambda t}(1 - e^{-\lambda \Delta t}). \tag{11.6}$$

The $|\Delta N|$ measurement method works well as long as $\Delta t \ll \tau$ so that eqn 11.6 can be expanded in a Taylor series at first order to give:

$$|\Delta N| \simeq \lambda N_0 e^{-\lambda t} \Delta t. \tag{11.7}$$

The term $\lambda N_0 e^{-\lambda t}$ is called the **activity** $\mathcal{A}$ of the sample:

$$\mathcal{A}(t) = \lambda N(t) = \mathcal{A}(0) e^{-\lambda t}. \tag{11.8}$$

So, $|\Delta N| = \mathcal{A} \Delta t$ provides a reliable estimate of the activity only in the $\Delta t \ll \tau_{1/2}$ approximation. Otherwise, the function that links the activity with ΔN is more complicated:

$$\Delta N = \int_{t_1}^{t_2 = t_1 + \Delta t} \mathcal{A}(t) \; dt. \tag{11.9}$$

If we aim at measuring very long lifetimes, we must resort directly to the radioactive decay law written in differential form:

$$\lambda = -\frac{dN/dt}{N} \tag{11.10}$$

and mesure dN/dt. In this case dN/dt is just the activity because:

$$\frac{d}{dt}\Delta N = \frac{d}{dt}\int_{t_1}^{t_2 = t_1 + \Delta t} \mathcal{A}(t) \; dt = \mathcal{A}(t). \tag{11.11}$$

but the measurement based on eqn 11.10 is difficult because we need to estimate N. In principle, there are no intrinsic limitations to measuring extremely long lifetimes, provided that N is very large. For instance, we claim the proton to be a stable particle against baryon number-violating processes (see Sec. 10.4.1) because decays like $p \to \pi^0 e^+$ have not been observed, even accumulating tens of kilotonnes of water. The SuperKamiokande experiment in Japan set an impressive limit to the proton lifetime against this decay mode: $\tau_p > 1.6 \times 10^{34}$ years at 90% confidence level. This time interval is 24 orders of magnitude longer than the age of the universe (1.38×10^{10} years). On the other hand, measuring very short lifetimes can be difficult. To measure, for instance, the 142 ns lifetime of an isomer of ^{57}Fe (see Fig. 11.1), we need to identify the appearance of the parent and the daughter and record the time elapsed after the parent decay. For ^{57}Fe*, it is possible to time tag the 122 keV and 14.4 keV photons arising from the

$$^{57}\text{Co} \to {}^{57}\text{Fe}^* + \gamma(122 \text{ keV}) \to {}^{57}\text{Fe} + \gamma(14.4 \text{ keV}) \qquad (11.12)$$

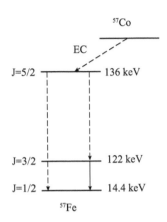

Fig. 11.1 The ^{57}Co isotope decays by electron capture to an intermediate state with total angular momentum equal to 5/2. This state can decay either to the ground state of ^{57}Fe or to a metastable state with 3/2 angular momentum and $BR \simeq 85\%$. The lifetime of such state is 142 ns and the isotope reaches the ground state by the emission of a 14.4 keV photon.

decay. The time difference between the detection of the 122 keV and the 14.4 keV photon is distributed according to an exponential probability density function (p.d.f.) $e^{-\lambda \Delta t}$, where λ is the decay constant of the ^{57}Fe*.

These methods, however, can only be applied to specific nuclides. Even smarter techniques have been delivered for $\tau \simeq 1$ ps down to 1 fs but measurements have an intrinsic limit set by the Heisenberg uncertainty principle. This limit is called the **intrinsic width** both in nuclear and atomic physics and corresponds to the decay width of the Breit–Wigner distribution of the parent state. The intrinsic width may swamp the time difference between the formation of the parent state and the appearance of the daughter state and makes the measurement of the lifetime impossible, whatever method the experimenter employs.

The SI unit of activity is the **Bequerel** (Bq) that corresponds to 1 decay per second. The corresponding natural unit (NU) is eV, although NU is rarely used in radioactivity. A deprecated unit of activity you may find in literature is the Curie, which corresponds to 3.7×10^{10} Bq.

11.2 Decay chains

Like ordinary particles, nuclides can decay in multiple modes and some of the daughters may be unstable, too. This phenomenon is quite common in radioactivity and gives rise to **radioactive chains**. Multiple decay modes of a single parent state are handled as the branching ratio (BR) in particle physics. If a parent decays in two modes, a and b, the total decay constant is $\lambda = \lambda_a + \lambda_b$ but the decay time we observe even selecting *only* the final state a or b will always be λ^{-1}. λ_a/λ is then equivalent to the BR of the particles and λ_a – the **partial decay constant** – plays the same role as the partial decay width in particle physics.

Chains of radioactive decays are more complicated. If the number of parent nuclei of the chain (N_1) is N_0 at $t = 0$ and no daughters are present at that time, we have:

$$dN_1 = -\lambda_1 N_1(t)dt$$
$$dN_2 = \lambda_1 N_1(t)dt - \lambda_2 N_2(t)dt. \qquad (11.13)$$

The size of the parent sample steadily decreases. The daughter sample is replenished by the decays of the parent and reduced by its own decays. This system of equations can be solved analytically by testing, for example, a solution of the form:

$$N_2(t) = Ae^{-\lambda_1 t} + Be^{-\lambda_2 t} \qquad (11.14)$$

together with the initial condition $N_2(0) = 0$. The results of this **two-step chain** with a single parent at $t = 0$ is:

$$N_1(t) = N_0 e^{-\lambda_1 t} \qquad (11.15)$$
$$N_2(t) = N_0 \frac{\lambda_1}{\lambda_2 - \lambda_1} \left(e^{-\lambda_1 t} - e^{-\lambda_2 t} \right). \qquad (11.16)$$

There are three cases of great practical interest:

$\boxed{\lambda_1 \ll \lambda_2}$ In this case the whole chain is driven by the parent. Its lifetime is so long that for every decay of the parent we find, shortly after, the decay of all the daughters. This is shown by eqn. 11.16 since $e^{-\lambda_1 t} \simeq 1$ and

$$N_2(t) \simeq N_0 \frac{\lambda_1}{\lambda_2} \left(1 - e^{-\lambda_2 t} \right). \qquad (11.17)$$

After a short transient, t becomes large and $\left(1 - e^{-\lambda_2 t} \right) \simeq 1$ so that:

$$N_2 \lambda_2 = N_1 \lambda_1 \simeq N_0 \lambda_1. \qquad (11.18)$$

In this condition, called **secular equilibrium**, the activity of the daughter is driven by the parent and the two activities are practically the same. Being in secular equilibrium is a great advantage, especially in long series (see Sec. 11.7), because a single measurement provides the activity of all the nuclides of the series.

$\boxed{\lambda_1 < \lambda_2}$ This case is less simple than secular equilibrium and the equality of the parent–daughter activities must be corrected by a time-dependent factor. Solving eqn. 11.16, we get:

$$\frac{\lambda_2 N_2}{\lambda_1 N_1} = \frac{\lambda_2}{\lambda_2 - \lambda_1} \left[1 - e^{-(\lambda_2 - \lambda_1)t} \right] \xrightarrow[t \to \infty]{} \frac{\lambda_2}{\lambda_2 - \lambda_1} . \qquad (11.19)$$

Unlike secular equilibrium, this condition – called **transient equilibrium** – results in time-dependent activities because $\mathcal{A}_1 \neq N_0 \lambda_1$ even at large t (see Fig. 11.2). On the other hand, transient equilibrium is valuable to measure the activity of the daughter because the daughter nuclei decay at nearly the same rate as the parent if t is sufficiently long.

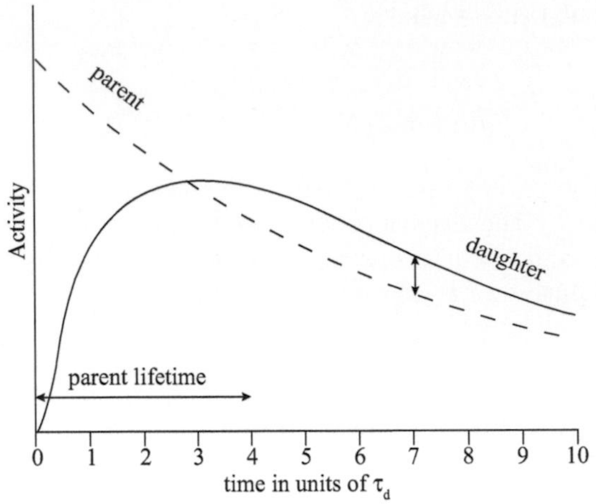

Fig. 11.2 Activity versus time (expressed in number of daughter's lifetimes) in the case of transient equilibrium. At large times, the daughter-to-parent ratio of activities (black arrow) is nearly constant and amounts to $\tau_p/(\tau_p - \tau_d) = \lambda_d/(\lambda_d - \lambda_p)$. τ_p and τ_d are the parent and daughter lifetimes, respectively.

$\boxed{\lambda_1 > \lambda_2}$ Here, the parent sample is quickly depleted and produces a fast accumulation of the daughter, which decays according to its own decay constant. We can compute the activity from eqn. 11.16 in the limit $t \to \infty$:

$$N_2(t) \to N_0 \frac{\lambda_1}{\lambda_1 - \lambda_2} e^{-\lambda_2 t} \qquad (11.20)$$

that is a simple exponential decay driven by λ_2.

11.2.1 Bateman equations*

Long decay chains are common in natural and artificial radioactivity. A long natural chain can be treated by the **Bateman equations** that hold for $1 \to 2 \to 3 \cdots \to n_f$ decays. If only one parent is present at $t = 0$ so that $N_1(0) = N_0$ and $N_i(0) = 0$ for $i \neq 1$, the activity of the n-th member is a function of the decay constants of the $1, 2 \ldots, n-1$ terms:

$$\mathcal{A}_n(t) = N_0 \sum_{i=1}^{n} C_i e^{-\lambda_i t}. \qquad (11.21)$$

The C_i Bateman's coefficients are given by:

$$C_i \equiv \frac{\prod_{j=1}^{n} \lambda_j}{\prod_{j=1}^{n} (\lambda_j - \lambda_i)}. \qquad (11.22)$$

The Bateman equation assumes a fixed number of parent nuclei at $t = 0$ but can be modified to account for sources that produce new parent nuclei. This is useful in **sample irradiation** where a target N_0 is hit by a flux of neutrons from a nuclear reactor or by particles produced at accelerators. In general, a research reactor can provide $\mathcal{O}(10^{14})$ neutrons/s/cm^2. However, even processes with cross-sections (σ) of the order of several barn are able to transform only a small fraction of the target used to generate the parents so that N_0 and the production rate can be considered as constants. The production rate R is given by:

$$R = N_0 \sigma I \qquad (11.23)$$

where I is the flux of the incident particles from the reactor or the accelerator. In this case, the first term of eqn. 11.2 must be corrected by a replenishment term Rdt:

$$dN_1 = Rdt - \lambda_1 N_1 dt \qquad (11.24)$$

and the the activity of an artificial chain with just one daughter is:

$$\mathcal{A}_1 = \lambda_1 N_1(t) \simeq R\left(1 - e^{-\lambda_1 t}\right) \qquad (11.25)$$

if $N_1(t=0)$ is small (see Exercise 11.20).

The asymptotic behaviour for small and large t is, respectively:

$$\mathcal{A}_1 \simeq R\lambda_1 t \quad \text{for } \lambda_1 t \ll 1 \qquad (11.26)$$
$$\mathcal{A}_1 \simeq R \quad \text{for } \lambda_1 t \to \infty. \qquad (11.27)$$

The first condition corresponds to the formation phase when $t \ll (\lambda_1)^{-1} = \tau_1$, the parent is accumulated by the formation reaction, and decay losses are negligible. The second condition corresponds to $t \gg \tau_1$ and the activity is driven by the production rate only.

Equation 11.27 shows that running the accelerator (or exposing the target to the reactor) for long periods generates a virtual source, whose activity is just R and the parent decay neutralizes any accumulation effect. This is the artificial analog of the secular equilibrium occurring in chains with long-living parents.

The Bateman equations can be generalized to account for multiple parent production effects – for example, when the sample is bombarded by a large neutron flux inside the core of a nuclear reactor and $N_2(t = 0)\dots N_{n_f}(t = 0)$ cannot be considered zero – (**Bateman–Rubinson equations**) and when a chain splits due to multiple decay modes of an isotope at a given step.

11.2.2 ^{14}C and the radiocarbon revolution

^{14}C is the isotope of choice for dating organic samples and is a spectacular application of the radioactive decay law. The technique was developed by W. Libby in 1946 and, since then, **radiocarbon dating** has been used extensively in geology, paleo-botanics, and climatology. Even more, the introduction of this method in archaeology represented a breakthrough in the field, the "radiocarbon revolution."

Libby was the first scientist to note that organic samples have a $^{14}C/^{12}C$ ratio that is a function of the age of sample. This is because organic samples are produced either by plants or animals. Cosmic rays (see Sec. 11.10) constantly produce ^{14}C in the earth's atmosphere by **neutron spallation**: $n + {}^{14}N \to {}^{14}C + p$. The ^{14}C atoms combines with oxygen to form carbon-dioxide molecules. A fraction of CO_2 is then absorbed by plants to carry on photosynthesis. The metabolism of chemoheterotrophs, that is, living organisms unable to perform photosynthesis like the vast majority of the animals, is driven by O_2 and the carbon that is eaten from the plants. As a consequence, both animals and plants have a $^{14}C/^{12}C$ ratio in equilibrium with the atmospheric ratio, $^{14}C/^{12}C \simeq \mathcal{O}(10^{-12})$. This number is the result of the production rate from neutron spallation and the decay of ^{14}C ($\tau_{1/2} \simeq 5730$ years), while both ^{12}C and ^{13}C are stable isotopes. The activity of a 1 g sample of carbon in the earth is $\simeq 0.25$ Bq (see eqn 11.25) and the decay is only due to $^{14}C \to {}^{14}N + e^- + \bar{\nu}_e$. When the organism dies, no carbon is exchanged with the atmosphere and the $^{14}C/^{12}C$ ratio decreases, halving every 5730 years. A precise measurement of the activity of the sample provides the time of the death of the organism and, hence, can be used to date the organic sample. What makes radiocarbon dating a revolution in archaeology is the fact that this technique can be applied to date samples from about 50,000 BC to ~1950 AD. This range covers most of human history and a good fraction of the Late Pleistocene. It encompasses, in particular, the Upper Paleolithic (50,000–10,000 BC) where anatomically modern humans developed language, agriculture, organized settlements, and created prehistoric art. Older ages are generally out of reach due to experimental limitations in the measurement of the activity. Dating recent artifacts is extremely complex because of the strong anthropogenic perturbations in the carbon cycle during the 20th century. The perturbations are due to burning large amounts of fossil fuels (oil, carbon, etc.) for industrial purposes since these fuels are very ancient and, hence, free of ^{14}C. Additional perturbations and biases arise from the fallout of ^{14}C during the nuclear weapon tests of the Cold War.

Within this time range, radiocarbon dating is quite reliable. The activities, however, must be corrected for changes in the ^{14}C production rate by cosmic rays in the last 50,000 years (solar cycles, local effects, etc.). These corrections can be done by calibrating the method with samples dated through independent techniques. Other corrections include isotopic fractionations due to the minuscule difference in the rate of uptake of ^{12}C, ^{13}C, and ^{14}C in photosynthesis, and "reservoir effects," that is, changes in the ^{14}C concentration due to the local environment (oceans, earth hemisphere, etc.). The most troublesome systematics may come from carbon contamination introduced at later ages.

This technique is very reliable for the study of well-preserved prehistoric samples and has reshaped our knowledge of humankind before the appearance of written documents. It is also very precise for historical dating, where it can be cross-checked by more traditional archaeological techniques. A breathtaking success of the radiocarbon method pioneered by Libby was the dating of the Dead Sea Scrolls, a set of Jewish manuscripts that influenced religious studies and the linguistics of the Old Testament.

Some readers probably know also the complex saga of the dating of the Shroud of Turin, a piece of linen cloth displaying an image of a man. It is one of the most popular relics of Christianity since it supposedly depicts Christ after his crucifixion. It is also one of the most loved relics in Europe and is shown to the public only on special occasions (e.g. during the COVID19 pandemic of 2020). In 1988, multiple radiocarbon measurements of fragments of the shroud indicated that the shroud was produced in the Middle Ages, confirming the dating from historical records: a 1390 AD document where Bishop Pierre d'Arcis reported the confession of an artist that the shroud was a forgery. Since then, radiocarbon data has been challenged by some scholars invoking unaccounted systematics due to contaminations (later repairs, bio-contamination, etc.) or other effects. Still, the dating performed in 1988 (1260–1390 AD at 95% confidence level) seems quite reliable and is incompatible with an ancient origin of the shroud (1st century AD).

11.3 Alpha decays

Unlike beta decays, whose understanding has progressed in parallel with the development of the Standard Model, the theory of alpha decay had a great start at the beginning of the century, marking a stunning achievement of QM, but progress has been much slower later on. Ab-initio calculations of α decays based on lattice quantum chromodynamics (QCD) are still in their infancy and we must rely on dedicated semi-empirical models.

Alpha decay is a spontaneous emission of an alpha particle (^{4}He) from a high-Z nuclide:

$$(A, Z) \rightarrow (A - 4, Z - 2) + {}^4\text{He} \equiv X \rightarrow Y + \alpha \tag{11.28}$$

The name is just a remnant of history.In 1900, E. Rutherford and P. Villard classified the "radiation" emitted by materials in three categories (α, β, and γ) and these forms of radiation were later identified as ^{4}He, $e^\pm$, and γ (high-energy photons), respectively. The kinematics of an alpha decay is simple because nuclei undergo a non-relativistic two-body decay. Most alphas are produced with kinetic energy $T_\alpha \simeq 5$ MeV even if a few isotopes can reach $T_\alpha \simeq 9$ MeV. The conservation of energy in the rest frame (RF) of the parent nucleus gives (in NU):

$$m_X = m_Y + T_Y + m_\alpha + T_\alpha \tag{11.29}$$

$$\mathbf{0} = \mathbf{p}_\alpha + \mathbf{p}_Y. \tag{11.30}$$

Recalling the definition of Q-value, we have:

$$m_X + m_Y - m_\alpha = Q = T_Y + T_\alpha = \tag{11.31}$$

$$\frac{|\mathbf{p}_Y|^2}{2m_Y} + \frac{|\mathbf{p}_\alpha|^2}{2m_\alpha} = \frac{p^2}{2m_\alpha}\left[1 + \frac{m_\alpha}{m_Y}\right]$$

where $p \equiv |\mathbf{p}_\alpha| = |\mathbf{p}_Y|$. Therefore, the measurement of the Q-value determines uniquely the kinetic energy of the final state:

$$T_\alpha = \frac{Q}{1 + m_\alpha/m_Y} \simeq Q\left(1 - \frac{4}{A}\right) \quad \text{if } m_\alpha \ll m_Y \tag{11.32}$$

$$T_Y = Q\frac{m_\alpha/m_Y}{1 + m_\alpha/m_Y} \simeq \frac{4Q}{A+4} \quad \text{if } m_\alpha \ll m_Y. \tag{11.33}$$

The approximation $m_\alpha \ll m_Y$ can be safely assumed for nearly all alpha decays. We will explain in the next section the reason for it.

11.3.1 Decay thresholds

^{4}He is a striking exception to the distribution of the binding energy per nucleon (see Fig. 9.7). The shell model explains this exception because ^{4}He is doubly magic. Therefore, the binding energy $B(Z, A) = B(2, 4)$ is very large (28.3 MeV) and corresponds to a binding energy per nucleon $\tilde{B}(Z, A) \equiv B(Z, A)/A$ of 7.1 MeV. An alpha decay is energetically allowed if:

$$Q = m_X - m_Y - m_\alpha > 0. \tag{11.34}$$

The **threshold condition** of eqn 11.34 can be expressed as a function of the binding energy per nucleon for a given nuclide. In this case:

$$Q = Zm_p + (A - Z)m_n - A \cdot \tilde{B}(Z, A)$$
$$- \left[(Z - 2)m_p + (A - Z - 2)m_n - (A - 4)\tilde{B}(Z - 2, A - 4)\right]$$
$$- \left[2m_p + 2m_n - 4\tilde{B}(2, 4)\right]$$
$$= A(\tilde{B}(Z - 2, A - 4) - \tilde{B}(Z, A)) - 4(\tilde{B}(Z - 2, A - 4) - \tilde{B}(2, 4))$$
$$= A(\tilde{B}_Y - \tilde{B}_X) - 4(\tilde{B}_Y - \tilde{B}_\alpha) > 0. \tag{11.35}$$

The difference $\tilde{B}_Y - \tilde{B}_X$ is small because $d\tilde{B}/dA$ in Fig. 9.7 is small for large values of A, too. To inflate the first term of eqn 11.35, we need A to be large. In practice, eqn 11.35 is fulfilled in nearly all nuclides for $A > 150$.

Using the semi-empirical mass formula (SEMF), we can give an analytical expression of the threshold for an α decay that reads:

$$Q = m_X - m_Y - m_\alpha = m(Z, A) - m(Z - 2, A - 4) +$$
$$-m(2, 4) \simeq \tilde{B}_\alpha - 4a_v + 4\left[\frac{2}{3}a_s + a_c Z\left(1 - \frac{Z}{3A}\right)\right]A^{-1/3} +$$
$$-4a_a\left(1 - \frac{2Z}{A}\right)^2. \tag{11.36}$$

The alpha decays are then common de-excitation modes for high A nuclides. This is why $m_\alpha \ll m_Y$ is such a good approximation.

11.3.2 The quantum mechanics of alpha decays

Alpha decays show a very clear correlation between the Q-value and the lifetime of the parent nuclide noted in 1911 by H. Geiger and J.M. Nuttal, a small increase of the Q-value reduces the lifetime by orders of magnitude. The **Geiger–Nuttal rule** is a logarithmic relation between the decay constant and the Q-value:

$$\log_{10} \tau_{1/2} \simeq a\frac{Z}{\sqrt{Q}} - b \tag{11.37}$$

where a and b are empirical parameters. The origin of alpha decays and the narrow range of kinetic energies available to the α inspired the first quantum mechanical explanation of radioactivity. This model was developed by G. Gamow and, independently, by R.W. Gurney and E. Condon in 1928. It is based on the hypothesis that ^{4}He is preformed inside a high-A nucleus because of its large binding energy. The mean effective potential is a 1-dimensional (1D) square well inside the volume of the nuclide and a Coulomb potential outside the nuclide. In modern terms, the Condon–Gurney potential is a rough approximation of the mean potential of the shell model complemented by an electrostatic correction because the alpha particle is positively charged and is pushed away by the $(A-4, Z-2)$ nuclide once it crosses the boundaries of the potential well. The potential is parameterized by the radius a of the nucleus, a minimum energy $-V_0$ (see Fig. 11.3), the Q-value of the reaction and the potential energy B of the barrier at $r = a$:

$$B = \frac{1}{4\pi\epsilon_0} \frac{zZ'e^2}{a}. \tag{11.38}$$

z and Z' are the charge of the alpha particle ($z = 2$) and the daughter ($Z' = Z - 2$), respectively.

In this model, the alpha particle is confined inside the barrier and can escape only by quantum tunneling. Classically, the particle cannot escape the potential well because of the missing kinetic energy $B - Q$. In QM, the transmission probability is non-zero but is exponentially suppressed as $B - Q$ increases. This is also hinted by the Geiger–Nuttal rule. The escape probability and then the lifetime can be computed solving the corresponding Schrödinger equation for a point-like particle of mass m_α. The equation can be solved analytically replacing the long-range part of the potential with a sequence of small rectangular barriers. The probability of crossing the barrier, expressed in SI, is:

$$P = e^{-2G} = \exp\left\{ -2\left(\frac{2m_\alpha}{\hbar^2}\right)^{1/2} \int_a^b [V(r) - Q]^{1/2} dr \right\}. \tag{11.39}$$

The G factor is called the **Gamow factor** and – for a Condon–Gurney potential energy like the one of Fig. 11.3 – is:

$$G = \sqrt{\frac{2m}{\hbar^2 Q}} \frac{zZ'e^2}{4\pi\epsilon_0} \left[\arccos(\sqrt{x}) - \sqrt{x(1-x)} \right] \tag{11.40}$$

where $x = a/b = Q/B$. For $x \ll 1$, the lifetime is (Krane, 1988):

$$\tau_{1/2} = 0.693 \frac{a}{c} \sqrt{\frac{mc^2}{2(V_0 + Q)}} \exp\left\{ 2\sqrt{\frac{2mc^2}{\hbar^2 c^2 Q}} \frac{zZ'e^2}{4\pi\epsilon_0} \left(\frac{\pi}{2} - 2\sqrt{\frac{Q}{B}} \right) \right\}. \tag{11.41}$$

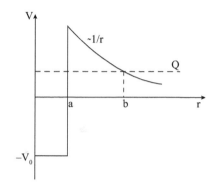

Fig. 11.3 1D potential in the Gamow–Condon–Gurney model of alpha decays. a is the radius of the parent nucleus, b is the distance of the α particle when the potential is equal to the Q-value of the reaction. The (Coulomb) potential beyond the well decreases as $1/r$. The minimum particle energy, corresponding to the ground state, is $-V_0$.

The logarithm of eqn 11.41 provides the correct functional form of the Geiger–Nuttal rule and quantitative insight on the dynamics of alpha decays. It also explains why alpha decays are so common in nature, while we barely see decays where other stable nuclei are emitted (e.g. ^{12}C): their Q-values are much larger than the alpha decay and, therefore, they are generally overwhelmed by the competing α emission. Still, **cluster decays** where the α is replaced by heavier nuclei have been observed since 1984 (for example, $^{223}Ra \to {}^{14}C + {}^{209}Pb$) and can be distinguished by spontaneous nuclear fission because they follow a pattern that resembles eqn 11.41. The spontaneous emission of nuclides lighter than the alpha, including the proton and the neutron, is also very rare because the nucleons are tightly bound in the nuclide compared with the alphas.

Despite these successes, the agreement of eqn 11.41 with experimental data is generally poor according to modern standards. It should not come as a surprise because this treatment is not based on the Fermi golden rule but on a rough parameterization of the effective potential. Besides, the angular momenta and deformations of nuclei are completely ignored.

11.3.3   Selection rules

The Gamow-Condon-Gurney model describes the process using the Q-value as the main driver for the transition probability. The Q-value encodes energy and momentum conservation but ignores the conservation of angular momentum and parity. ^4He is a spin-zero nuclide embedded in a system with a total angular momentum (spin of the parent nucleus) that can be different from zero. Furthermore, the alpha particle brings orbital momentum. The Gamow–Condon–Gurney approach can be supplemented to cope with angular momentum replacing the 1D finite well with a more realistic 3D spherical potential. Since the spatial wavefunction of the particle can be factorized as:

$$\psi(\mathbf{x}) = \psi(r, \theta, \phi) = R(r) Y_m^l(\theta, \phi) \tag{11.42}$$

for any central potential, and $Y_m^l(\theta, \phi)$ are universal functions (spherical harmonics) for any particle in a central potential, the lifetime can be corrected by orbital effects solving the corresponding Schrödinger equation for $R(r)$:

$$-\frac{\hbar^2}{2m_\alpha} \left(\frac{d^2 R}{dr^2} + \frac{2}{r} \frac{dR}{dr} \right) + \left[V(r) + \frac{l(l+1)\hbar^2}{2m_\alpha r^2} \right] R(r) = E R(r). \tag{11.43}$$

The inclusion of 3D wavefunctions and the approximation of spherical nuclei bring two additional pieces of information. First, the factorization of eqn 11.42 implies that the parity of the system is $(-1)^l$. Since alpha decays are due to the combined action of QED and QCD,

> parity must be conserved in any alpha decay. The parity of the parent and the daughter is provided by the shell model, while the parity of the alpha is $(-1)^l$. l is the orbital angular momentum of the particle since ^4He is spinless.

The

$$\text{CB} \equiv \frac{l(l+1)\hbar^2}{2m_\alpha r^2} \tag{11.44}$$

term of the potential produces an increase of the gap among the energy levels and acts as a **centrifugal barrier**. High l decays are suppressed compared with decays at low l. Alpha decays must fulfill angular momentum conservation:

$$|I_i - I_f| \leq l \leq I_i + I_f \tag{11.45}$$

where I_i and I_f are the nuclear spin of the parent and the daughter, and l is the orbital angular momentum of the alpha particle. Parity conservation implies:

$$\mathcal{P}_X = \mathcal{P}_Y(-1)^l \tag{11.46}$$

where $\mathcal{P}_X$ and $\mathcal{P}_Y$ are the intrinsic parity of the X and Y nuclides, respectively.

Example 11.1

The application of these laws is very effective in **even–even parent nuclei**, whose spin is always zero (see Sec. 9.4.4). In this case, $I_f = l$ and the only states that are allowed are "even spin and odd parity" or "odd spin and even parity": $0^+, 1^-, 2^+ \ldots$. Besides, most even–even nuclei have no low-lying states with negative parity. So even–even daughters generally have even spin-parity states: 0^+, 2^+, etc. If both the ground state and the closest excited states of the daughter fulfill energy conservation and the selection laws, the daughter can reach several excited states with different intensities. This phenomenon is called the **fine structure** of alpha decays. The excited states have a reduced intensity due to the lowering of the Q-value and the increase of the centrifugal barrier.

If a decay to the ground state is forbidden by these laws, it may occur through an excited state of the daughter. The spin of an **odd-A parent nucleus** is completely determined by the last uncoupled nucleon: the valence nucleon of the shell model. In most cases, the daughter in the ground state has a structure defined by a different valence nucleon. Such a mismatch in the structure of the parent and the daughter ground state tends to suppress the decays to excited states of the daughter that have a spin different from the spin of the parent.

11.4 Beta decays

Unlike alpha decays, the Standard Model provides a full-fledged theory to compute beta decay probabilities and lifetimes at the level of elementary fermions. A β^- **decay** is a

$$d \to u + e^- + \bar{\nu}_e \tag{11.47}$$

transition that changes the flavor of a quark and creates an antineutrino and an electron. It is, therefore, a pure semileptonic weak interaction process. Similarly

$$u \to d + e^+ + \nu_e \tag{11.48}$$

$$u + e^- \to d + \nu_e \tag{11.49}$$

correspond to a β^+ **decay** and an **electron capture**, respectively. Note that eqn 11.48 should be kinematically forbidden because $m_u < m_d$. On the other hand, quarks are always bound in color singlets (nucleons), which are in turn bound in stable nuclei. So, the β^+ decay *does* occur if we neglect neutrino masses and the binding energy is:

$$Q = m_N(Z, A) - m_N(Z - 1, A) - m_e > 0, \tag{11.50}$$

where $m_N(Z, A)$ is the mass of the nuclide. The binding of the quarks introduces nuclear effects, that is, non-perturbative QCD effects, that are difficult to handle. They make beta decays much more demanding than the decays studied in Chap. 10.

   The decay of eqn 11.47 can occur at hadron level, too. As already mentioned, the neutron decay occurs in vacuum ($n \to p\ e^-\ \bar{\nu}_e$) with a very long lifetime $\tau = 881.5$ s due to the feebleness of weak interactions and the small phase-space. $e^- + p \to \nu_e + n$ may occur, too, but the cross-section in vacuum is swamped by the $e^- + p$ electromagnetic interactions since both particles are charged. The vast majority of decays can be classified in the same way as for the quarks:

$$(A, Z) \to (A, Z + 1) + e^- + \bar{\nu}_e \qquad [\beta^- \text{ decay}] \tag{11.51}$$

$$(A, Z) \to (A, Z - 1) + e^+ + \nu_e \qquad [\beta^+ \text{ decay}] \tag{11.52}$$

$$e^- + (A, Z) \to (A, Z - 1) + \nu_e \qquad \text{EC.} \tag{11.53}$$

Note, in particular, that

> all beta decays produce nuclides with the same mass number A. The SEMF at fixed (odd) A is a parabola with a single global minimum, which represents the ground state of a pure beta decay chain. Even A chains run through two parabolas, corresponding to even-Z (at lower energy) and odd-Z (at higher energy) nuclides, respectively.

A pure even-A chain is shown in Fig. 11.4. In chains where alpha and beta decays occur, the atomic mass number A may change, too. We will discuss these mixed chains in Sec. 11.7.

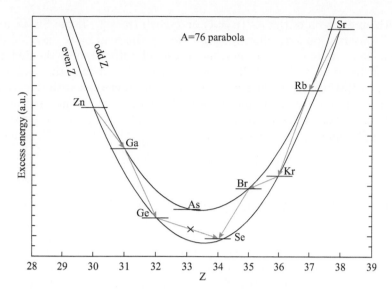

Fig. 11.4 Binding (excess) energy for $A = 76$ from the SEMF. Unlike odd-A, even-A chains show two parabolas at different energies and beta decays go from one to the other until the nucleus reaches the lowest state. The arrows show the allowed beta decays. Note, however, that the ^{76}Ge $\to$ ^{76}Se transition (dark gray arrow with the cross) is forbidden and can occur only by second order processes (see Sec. 11.5).

11.4.1   Kinematics of beta decays

A β^- decay is a three-body decay where the energy of the electron and the neutrino are not uniquely determined by kinematics (see Sec. 2.5). This holds also for the recoiling $(Z + 1, A)$ nucleus, which, in general, is not detected due to the large value of the mass ($\simeq A \times 1$ GeV) and the corresponding tiny velocity. The threshold for a β^- decay is given by eqn 11.50 as a function of the nuclide mass. If we want to employ the SEMF, eqn 11.50 must be recast in term of the atomic mass $m_A(Z, A)$ replacing $m_N(Z, A)$ with:

$$m_N(A, Z) = m_A(A, Z) - Zm_e + \sum_{i=1}^{Z} B_i^e \simeq m_A(A, Z) - Zm_e \quad (11.54)$$

so that

$$Q = (m_A(Z, A) - Zm_e) - (m_A(Z + 1, A) - (Z + 1)m_e) - m_e > 0. \quad (11.55)$$

The β^- **threshold** is then:

$$Q = m_A(Z, A) - m_A(Z + 1, A) > 0. \quad (11.56)$$

Following the same approach, the β^+ and **EC thresholds** are:

$$Q_{\beta+} = m_A(Z, A) - m_A(Z - 1, A) - 2m_e > 0 \quad (11.57)$$

$$Q_{EC} = m_A(Z, A) - m_A(Z - 1, A) - B_e(A, Z, n_{\text{shell}}) > 0. \quad (11.58)$$

$B_e(A, Z, n_{\text{shell}})$ is the electron-binding energy that depends on the atom (A, Z) and the energy of the shell n_{shell} where the initial-state electron is located. The overlap of the wavefunction is larger for electrons orbiting in a low principal quantum number n. So most electron captures occur with an electron of the K- ($n = 1$) or L-shell ($n = 2$). If the SEMF and the conservation laws allow for a β^- decay, this will most likely be the dominant mode due to the smaller threshold. Other modes dominate if the β^- decay is forbidden. If a pure beta chain is reaching the minimum of the SEMF parabola at fixed A (A, Z_{min}), an isotope with $Z < Z_{min}$ will undergo β^- decays to move toward higher Z values, while an isotope with $Z > Z_{min}$ will go through either EC or β^+ decays to move toward lower Z values. This trend is clearly visible in the Chart of Nuclides of Fig. 9.4.

11.4.2   The electron spectrum and the neutrino mass

The lifetime of a beta decay can be computed employing either the relativistic or – with some approximations – the non-relativistic golden rule. In the relativistic treatment of eqn 2.81, assuming the parent nuclide X at rest ($E_X = m_X$):

$$
\Gamma_{i \to f} = \frac{(2\pi)^4}{2E_a} \int \prod_{i=1}^{n} \frac{d^3\mathbf{p}_i}{(2\pi)^3 2E_i} |M|^2 \delta^4 \left(p_a^\mu - \sum_{j=1}^{n} p_j^\mu \right)
$$

$$
= \frac{1}{2m_X} \int \prod_{i=1}^{3} \frac{d^3\mathbf{p}_i}{(2\pi)^3 2E_i} |\langle p_1, p_2, p_3 |M| p \rangle|^2 (2\pi)^4 \delta^4 \left(p_a^\mu - \sum_{j=1}^{3} p_j^\mu \right)
$$

$$(11.59)$$

in NU. In this case $i = 1$ corresponds to the stable nuclide in the final state (Y) while $i = 2, 3$ are the electron (e) and the electron antineutrino (labeled $\bar{\nu}$ in the following), respectively. Since 1998, we have been aware that $\bar{\nu}_e$ has not a definite mass and $m_{\bar{\nu}}$ must be replaced by an effective neutrino mass described in Sec. 13.6.5. We can integrate out $d^3\mathbf{p}_Y \, \delta^3(\mathbf{p}_Y + \mathbf{p}_e + \mathbf{p}_{\bar{\nu}})$ to impose momentum conservation and get:

$$
\frac{1}{16(2\pi)^5 m_X E_Y} \int \left[\frac{d^3\mathbf{p}_e}{E_e} \frac{d^3\mathbf{p}_{\bar{\nu}}}{E_{\bar{\nu}}} \right] |\mathcal{M}|^2 \delta \left(E_0 - E_e - E_{\bar{\nu}} \right). \qquad (11.60)
$$

E_0 is the energy difference between E_X and E_Y, and $E_0 = m_X - E_Y$ in the laboratory frame (LAB). The kinetic energy of the recoiling nucleus:

$$
\sqrt{m_Y^2 + |\mathbf{p}_Y|^2} - m_Y \qquad (11.61)
$$

can be neglected because $m_Y \gg m_e$ and the initial-state energy E_0 is shared between the antineutrino and the electron: $E_Y \simeq m_Y$ and $E_0 \simeq m_X - m_Y$. We can also integrate part of the phase-space employing spherical coordinates: $d^3\mathbf{p}_e = \sin\theta \, d\theta \, d\phi \, |\mathbf{p}_e|^2 d|\mathbf{p}_e| =$

$4\pi|\mathbf{p}_e|^2 d|\mathbf{p}_e|$. In addition, $d^3\mathbf{p}_{\bar{\nu}} = 4\pi|\mathbf{p}_{\bar{\nu}}|^2 d|\mathbf{p}_{\bar{\nu}}| \simeq 4\pi|\mathbf{p}_{\bar{\nu}}|E_{\bar{\nu}}dE_{\bar{\nu}}$ because $|\mathbf{p}_{\bar{\nu}}|d|\mathbf{p}_{\bar{\nu}}| \simeq E_{\bar{\nu}}dE_{\bar{\nu}}$. If we integrate $\delta(E_0 - E_e - E_{\bar{\nu}})dE_{\bar{\nu}}$, we get $E_{\bar{\nu}} = \sqrt{|\mathbf{p}_{\bar{\nu}}|^2 + m_{\bar{\nu}}^2} = E_0 - E_e$ and $|\mathbf{p}_{\bar{\nu}}| = \sqrt{(E_0 - E_e)^2 - m_{\bar{\nu}}^2}$. The decay width is then:

$$\Gamma = \int \frac{\pi^2}{(2\pi)^5 m_X m_Y} \frac{|\mathbf{p}_e|^2}{E_e} \sqrt{(E_0 - E_e)^2 - m_{\bar{\nu}}^2} |\mathcal{M}|^2 d|\mathbf{p}_e|. \qquad (11.62)$$

Note that $\mathcal{M}$ is the Lorentz-invariant matrix element, which is related to the non-relativistic matrix element $\mathcal{T}$ by eqn 2.79 and:

$$|\mathcal{M}|^2 \sim E_e E_{\bar{\nu}} |\mathcal{T}|^2. \qquad (11.63)$$

The core of the original Fermi theory of Sec. 10.7.2 was that weak decays occur in a four-line vertex without off-shell particles and that the strength of the vertex is proportional to the Fermi constant G_F. We can then define:

$$|\mathcal{M}|^2 = E_e E_{\bar{\nu}} |\mathcal{T}|^2 \equiv E_e E_{\bar{\nu}} |\mathcal{M}_0|^2 \qquad (11.64)$$

where $|\mathcal{M}_0|^2$ does not depend on kinematical variables except for the Fermi constant and the rest mass of the particles. The lifetime can be recast as:

$$\Gamma = \int \frac{|\mathcal{M}_0|^2}{32\pi^3 m_X m_Y} |\mathbf{p}_e|^2 (E_0 - E_e) \sqrt{(E_0 - E_e)^2 - m_{\bar{\nu}}^2} \, d|\mathbf{p}_e| \qquad (11.65)$$

where $E_0 - E_e = E_{\bar{\nu}}$. This formula needs an electrostatic correction to account for the energy lost by the electron while escaping from the daughter nucleus, and is due to the attractive Coulomb force of $X(Z + 1, A)$. The correction $F(Z, E_e)$ can be computed analytically both in classical mechanics and in special relativity (SR).

The number of electrons emitted in a momentum interval between $|\mathbf{p}_e|$ and $|\mathbf{p}_e| + d|\mathbf{p}_e|$ is given by the probability of a decay where the electron momentum is in such interval times the number of parent nuclei. Using the (deprecated) notation $p \equiv |\mathbf{p}_e|$, we end up with the **Fermi–Kurie formula**:

$$dN = N(p)dp = d\Gamma \sim |\mathcal{M}_0|^2 p^2 (E_0 - E_e)^2$$

$$\times \ F(Z, E_e) \sqrt{1 - \frac{m_{\bar{\nu}}^2}{(E_0 - E_e)^2}} \, dp \qquad (11.66)$$

that provides the electron energy spectrum dN/dp for a β^- decay. The **Fermi–Kurie plot**,[1] proposed by F.N.D. Kurie in 1936 (see Fig. 11.5) is a plot of:

$$K(E_e) \equiv \sqrt{\frac{N(p)}{p^2 F(Z, E_e)}} \text{ versus } E_c \qquad (11.67)$$

and is a brilliant method used to test the Fermi hypothesis of a universal constant driving $|\mathcal{M}_0|^2$. In the Fermi theory, the plot is a straight line, whose slope depends on the nucleus. The **endpoint** of the Kurie plot is the value of the electron energy when $K(E_e) = 0$. If neutrinos were massless, the endpoint would be E_0. Otherwise, no electron can have an

[1]This plot is named "Kurie plot" in older textbooks and the name is still in use in literature.

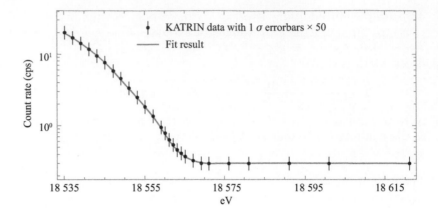

Fig. 11.5 The world most precise measurement of the electron energy in the proximity of the Fermi–Kurie endpoint, expressed as counts per second (cps $\sim K(E_e)$) versus energy in eV. The shape is consistent with massless neutrinos and provides an upper limit to the neutrino masses that enter the β^- decay of tritium: $^3\text{H} \rightarrow {}^3\text{He} + e^- + \bar{\nu}_e$ This decay has been chosen by KATRIN due to the low Q-value (18.57 keV) and relatively short half-life ($t_{1/2} = 12.32$ y). The error bars are inflated $\times 50$ to ease the reading. Reproduced under CC-BY-4.0 from Aker *et al.* 2019.

energy greater than $E_0 - m_{\bar{\nu}}$ and the shape of the plot is distorted in the proximity of the endpoint. Using this technique, early neutrino physicists were able to show that the mass of the electron antineutrino is "much smaller than the electron mass" since no deviation was observed. This is still true. To date, the most precise observation from the KATRIN experiment shows no deviation and puts a limit on m_ν of < 1.1 eV at 90% confidence level (Aker *et al.*, 2019). The sensitivity of this method is too coarse to reach the values suggested by neutrino oscillation experiments and cosmology (tens of meV) and discussed in Sec. 13.6. On the other hand, the Fermi–Kurie plot played a fundamental role in the construction of the $V - A$ theory and, later on, of the Standard Model. The systematic studies of beta decay and the linear slope of the Fermi–Kurie plot supported a constant value of $|\mathcal{M}_0|^2$, which only depends on the Fermi constant and the change of nuclear-spin in the initial and final state. This observation brought to the first interpretation of beta decays.

11.4.3 Classification of beta decays

The classification of beta decays is plagued by terms that date back to the original Fermi theory but are still in use and require a modern interpretation to avoid any misunderstanding. The main results of the Kurie analysis was the confirmation of Fermi's hypothesis on $|\mathcal{M}_0|$. Experimental data clearly support this statement:.

> the Kurie plot is linear in all beta decays where the leptons have no orbital momentum (**allowed decays**) and endorses the assumption of a constant $|\mathcal{M}_0|^2$ in most beta decays.

This outcome corroborated the Fermi theory using decays where the leptons had both antiparallel (**Fermi transitions**) and parallel (**Gamow–Teller transitions**) spin alignment.

11.4.4 Allowed transitions

Allowed transitions are beta decays where the final-state lepton pair has no orbital angular momentum and the spin of the two leptons can be either parallel to the direction of motion or antiparallel. For antiparallel spins, the total spin of the leptons is $S = 0$ and the nuclear spin[2] of the parent must be identical to the spin of the daughter: the eigenstates I_i and I_f must be such that $|I_i - I_f| = 0$ ($\Delta I = 0$). These decays are called allowed **Fermi decays**. If the spins are parallel $S = 1$. In this case, ΔI can be 0 or 1 but $0 \rightarrow 0$ occurs only in Fermi decays. The parallel spin condition corresponds to allowed **Gamow–Teller decays**. Without orbital momentum from the leptons, the parity of the initial and final state is $(-1)^l$ and $l_i = l_f$ from angular momentum conservation. Therefore,

> all allowed decays, that is decays where the leptons have no orbital angular momentum, conserve the parity of the nuclides and fulfill the $\Delta I = 0, 1$ selection rule.

[2] As in Chap. 9, we follow the conventions of nuclear physicists and call the total nuclear angular momentum I and the isospin T.

The Fermi (F) and Gamow–Teller (GT) decays are not mutually exclusive. The neutron decay $n \rightarrow p + e^- + \bar{\nu}_e$ is a mixed F + GT transition because $\Delta I = 0$ but $I \neq 0$.

11.4.5 Forbidden and superallowed transitions

Once more, the term "forbidden" is a remnant of the past and, in modern terms, should be interpreted as "suppressed." The **forbidden** decays are either Fermi or Gamow–Teller transitions where the leptons carry angular momentum and contribute to the total angular momentum of the final state. We can define forbidden Fermi transitions if the spins of the leptons are antiparallel ($S = 0$) and forbidden Gamow–Teller transitions if the spins of the leptons are parallel ($S = 1$). Forbidden decays are typically suppressed at 10^{-4} level for $l = 1$ and at 10^{-8} for $l = 2$. They are, therefore, of relevance only if the corresponding allowed transition is proscribed by one or more selection rules.

Among allowed decays, a special mention is due to **superallowed decays**. We emphasize these decays because they set another straightforward link between nuclear physics and the flavor symmetry of QCD. The superallowed decays are allowed decays where the initial and final nuclides are members of the same isomultiplet (see Sec. 9.6.1). In superallowed beta transitions the initial and final matrix elements are practically the same, the wavefunction are strongly overlapping and the beta decay lifetimes are generally short. A classical example is the **mirror decay** $^{41}_{21}\text{Sc}_{20} \rightarrow ^{41}_{20}\text{Ca}_{21} + e^- + \bar{\nu}_e$ where the number of protons and neutrons are swapped by a $p \rightarrow n$ transformation and the amplitude is nearly equally shared between Fermi and Gamow–Teller transitions.

11.4.6   A fresh look at GT transitions

The reader may wonder what is the link between the relativistic theory of weak interactions discussed in Chap. 10 with this classification, which is based on the original Fermi theory. The Fermi theory ignores the spins and their orientation and can explain quantitatively only Fermi transitions at the first order of perturbation theory (allowed transitions) or at higher orders if l is greater than 1. In this case, $I_f = I_i + l$. If the decay is "allowed," then $l = 0$. Besides, the Fermi theory is a parity-conserving theory like QED and QCD. Then, the parity of the whole system in a Fermi transition is $(-1)^l$ and must be conserved in the final state. This is the reason why Lee and Yang were seeking a Gamow–Teller transition to falsify parity conservation in weak forces. Even more, they were seeking for an "allowed" transition to benefit from the shorter lifetime. Lee, Yang, and Wu resorted to the $^{60}\text{Co} \rightarrow {}^{60}\text{Ni}^{**}$ decay, which is an allowed $5^+ \rightarrow 4^+$ Gamow–Teller transition. As a consequence, Gamow–Teller decays can be fully explained only by the V-A theory. But not all GT decays violate parity. Why?

The $V - A$ theory requires a vertex of the form:

$$\bar{\psi}_b \gamma^\mu (1 - \gamma^5) \psi_a \tag{11.68}$$

which corresponds to the current $j_{V-A}^{a \rightarrow b}$. This current can be split into two parts:

$$V \equiv j_V^\mu = \bar{\psi}_b \gamma^\mu \psi_a$$
$$A \equiv j_A^\mu = \bar{\psi}_b \gamma^\mu \gamma^5 \psi_a.$$

Using the results of Sec. 10.7.1, we have that

$$\mathcal{P}V = +V \;\; ; \;\; \mathcal{P}A = -A \tag{11.69}$$

where $\mathcal{P}$ is the parity operator. A first-order beta decay has two vertices[3] of the form $j_\mu j^\mu$ and then:

$$j_\mu j^\mu = (V + A)(V + A). \tag{11.70}$$

If we expand the product, the corresponding terms are:

$$VV \implies \mathcal{P}(VV) = +VV$$
$$VA \;\; \text{and} \;\; AV \implies \mathcal{P}(VA) = -VA \;\; \text{and} \;\; \mathcal{P}(AV) = -AV$$
$$AA \implies \mathcal{P}(AA) = AA.$$

The Fermi transitions belong to the "VV" class and do not represent a useful tool to study parity violation in weak processes. Even the Gamow–Teller transitions of the form "AA" do not help[4] because they have a double change of sign when we apply $\mathcal{P}$, and preserve parity conservation. The right tool is a subset of GT of the "VA" form because $\mathcal{P}(VA) = -VA$. The ^{60}C decay chosen by Lee, Yang, and Wu belongs to this class.

[3] In contact interactions, the two vertices are located in the same point, while, in QED, they are at the beginning and the end of the photon propagator.

[4] If we were following the historical development of nuclear physics, we should call GT only the transitions of the AA class because the GT definition was introduced before the discovery of parity violation. In a modern view, GT transitions should be intended as in Sec. 11.4.4: beta decays whose lepton spins are parallel.

Decay	ΔI	Parity change
Allowed $(l = 0)$	$0, \pm 1$	No
First forbidden $(l = 1)$	$0, \pm 1, \pm 2$	Yes
Second forbidden $(l = 2)$	$\pm 2, \pm 3$	No
N-th forbidden $(l = N)$	$\pm N, \pm(N + 1)$	Yes for l odd
		No for l even

Table 11.1 Selection rules for beta decays.

11.4.7 Selection rules for beta decays

The selection rules of beta decays can be derived from two sources: a simple application of angular momentum conservation and subtler constraint due to isospin. Angular momentum conservation provides the rules of Tab. 11.1. These rules follow from the results of the previous sections and are demonstrated in Exercise 11.15.

Other rules are enforced by the approximate $SU(2)_f$ flavor symmetry of QCD or, equivalently, by the isospin symmetry. Fermi decays change the value of T_z by one unit because of the $p \to n$ transition but do not change T. This can be demonstrated taking the non-relativitic limit of:

$$\mathcal{M} \sim \left[\bar{\psi}_d \gamma^\mu \psi_u\right] \left[\bar{\psi}_{\nu_e} \gamma_\mu \psi_e\right] \tag{11.71}$$

after replacing ψ_d and ψ_u with ψ_n and ψ_p. The original amplitude written by Fermi in the 1930s ignores the spins of the electron, neutrino and nucleons, and reads (Bertulani, 2007):

$$\mathcal{M} \sim \int \psi_p^* \hat{T}_\pm \psi_n \, d^3\mathbf{x}. \tag{11.72}$$

Here,

$$\hat{T}_+ = \hat{T}_x + i\hat{T}_y \tag{11.73}$$

$$\hat{T}_- = \hat{T}_x - i\hat{T}_y \tag{11.74}$$

where $\hat{T}_x$ and $\hat{T}_y$ are defined in eqs 5.15 and 5.16. The $\hat{T}_\pm$ operators are the "raising" and "lowering" operator (**ladder operators**) of isospin. They transform a particle with $T_z = -1/2$ into a particle with $T_z = +1/2$ and vice-versa. The ladder operators describe β^- and β^+ decays. For a real nucleus, this formula must be summed over all nuclides or, employing the shell model, over the valence nucleons, only:

$$|\mathcal{M}|^2 \simeq \sum_{m_f} \left| \int \psi_f^* \left(\sum_k \hat{T}_\pm^k \right) \psi_i \, d^3\mathbf{x} \right|^2 \tag{11.75}$$

In eqn 11.75, the sum over k is taken over all nucleons of the initial state ψ_i and the sum over m_f is taken over all possible values of the

[5]In the language of group representation theory (see Sec. 8.2.1), it means that the initial and final states must belong to the same vector space of the $2T+1$ irreducible representation of $SU(2)$.

magnetic quantum number m of the final-state nucleus. Equation 11.75 shows that the Fermi weak decays change T_z by one unit but does not change T. If $T_i \neq T_f$, $|\mathcal{M}|^2=0$. This constraint[5] generates the isospin selection rules for Fermi transitions. There are many $0^+ \to 0^+$ transitions, which are pure Fermi decays since $0^+ \to 0^+$ cannot be GT. These transitions are heavily suppressed by the aforementioned isospin rule if $T_i \neq T_f$.

11.5   Majorana neutrinos

To date, we have no clues as to whether or not a neutrino corresponds to its own antiparticle. Among elementary fermions, this possibility arises only for neutral fermions and has been considered as a viable option in 1937 by E. Majorana. The Standard Model incorporates the Dirac equation of Sec. 5.9, where all elementary fermions are different from their antiparticles but this assumption can be challenged. We have already shown that this model predicts particles that exist in the Dirac formalism but cannot interact with matter: the RH chirality neutrinos. The possibility that these particles do not exist at all is far from exotic. Even more, there is a vast literature claiming that the existence of Majorana particles and Majorana mass terms in the Standard Model Lagrangian can explain the smallness of neutrino masses with respect to the rest of the fermions (Bilenky, 2018). In principle, judging whether a neutrino is its own antiparticle should be straightforward. First, we note that if the neutrino were massless the distinction between Dirac and Majorana neutrinos would be an exercise in futility. The helicity of a massless particle is a Lorentz-invariant quantity and we can construct a successful weak interaction theory identifying ν_e^{RH} with $\bar{\nu}_e^{RH}$ and removing the sterile Dirac counterpart ν_e^{RH}. We can do the same with $\bar{\nu}_e^{LH}$ that is now renamed ν_e^{LH}. In this way, there are no more sterile particles in the theory and we have just two Majorana states. Alternatively, we can keep all four Dirac states but since ν_e^{RH} and $\bar{\nu}_e^{LH}$ are both sterile (no weak interactions) and massless (no gravitational interactions), they are completely decoupled from the rest of the universe and – by Occam's razor – we can claim they do not exist because no observable depends on them. This statement does not hold for massive neutrinos because an observer that moves faster than $\bar{\nu}_e^{LH}$ sees the opposite helicity (RH) and can perform an experiment to understand whether $\bar{\nu}_e^{RH}$ is equal to ν_e^{RH}.

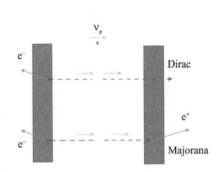

For instance (Strumia and Vissani, 2006), suppose she could put at rest a Dirac ν_e with its spin pointing toward right in LAB (see Fig. 11.6). If the experimentalist accelerates the neutrino on the left at ultra-relativistic energy, the neutrino can interact and produce an electron because helicity and chirality are practically the same at $\beta \simeq 1$ and the neutrino is a ν_e^{LH}. If she accelerates the neutrino on the right, no interactions are observed because ν_e^{RH} is a sterile particle. On the other hand, if the neutrino is a Majorana particle, the observer records

Fig. 11.6 A fictional experiment where the experimenter can slow down and accelerate neutrinos. The neutrino at rest shown on the top can be boosted on the right and on the left at an ultra-relativistic speed so that its LH helicity corresponds to a LH chirality. In the upper case (Dirac neutrinos), the observer records charged leptons only on the left. In the lower case, (Majorana neutrino) he sees charged leptons both on the left and on the right. The light gray lines show the spin orientation and the dark gray line the direction of the charged leptons. The neutrino trajectories are indicated by dashed arrows.

interactions in both cases: electrons after an acceleration on the left and positrons on the right.

The main problem is, once more, the small mass of these particles and the tiny cross-sections. We cannot really put at rest a neutrino because we cannot exert significant accelerations on it. A neutrino is nearly like a photon in SR: any inertial observer sees it at a speed very close to the speed of light.

At present, the only effective technique we are aware of to set this issue is by exploiting high-order processes in perturbation theory. The best approach was proposed by W.H. Furry in 1939 and is based on **double-beta decays**. These processes, whose Feynman diagrams are depicted in Fig. 11.7 (left), correspond to the emission of two neutrinos and two $\bar{\nu}_e$:

$$(A, Z) \to (A, Z + 2) + 2e^- + 2\bar{\nu}_e \qquad (11.76)$$

or, at the level of fundamental constituents:

$$2d \to 2u + 2e^- + 2\bar{\nu}_e \qquad (11.77)$$

so that two neutrons transform into two protons moving the daughter by two columns on the right in the Periodic Table. This phenomenon was predicted by Maria Goeppert-Mayer in 1935 but was observed only in 1950 in ^{130}Te by studying its isotopic abundance in ancient rocks. This decay has a well known half-life. The most precise measurement, performed by the CUORE experiment (Adams et al., 2021) in 2020, is:

$$\tau_{1/2} = 7.71^{+0.08}_{-0.06} \text{ (stat) } ^{+0.12}_{-0.15} \text{ (sys) } \times 10^{20} \text{ years} \qquad (11.78)$$

where "stat" and "sys" are the contributions from statistical and systematic errors. Contemporary physicists are not surprised by such a long lifetime because this is a second-order process compared with a single-beta decay and the decay constant is suppressed by G_F^2. Clearly, at the time of Goeppert-Mayer's proposal, this source of amplitude suppression was not known and – based mostly on phase-space considerations–people were optimistic about observing the process with a lifetime comparable to the usual beta decays described in Sec. 11.4.

We know now that these double-beta transitions are visible only if they are not overwhelmed by beta decay processes, which can be forbidden by energy conservation or selection rules. Double-beta decays have been observed in 14 isotopes although there are other 20 naturally occurring isotopes capable of decaying in this way, which has not been recorded to date (Tretyak and Zdesenko, 2002). The reason is the dominance of single-beta decay and the difficulty of discriminating a real double-beta decay from two single-beta decays occurring one after the other in a very short time. As a consequence, the observed decays are mostly in even–even nuclides, which are very stable in the shell model. Here, the loss of a neutron depletes a completed shell making a single decay to the ground state energetically forbidden and a single decay to the excited states highly suppressed by the large value of l.

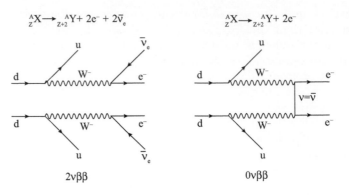

Fig. 11.7 Left: a double-beta decay process where two electrons and two antineutrinos are emitted in the final state ("$2\nu\beta\beta$"). Right: a neutrinoless double-beta decay were the Majorana neutrino is produced and reabsorbed in the off-shell line of the Feynman diagram ("$0\nu\beta\beta$"). In this case, the final state contains only two electrons.

Even if double-beta decay is a marginal process of interest mostly for nuclear structure studies, the observation of **neutrinoless double-beta decay** would be a smoking-gun proof that neutrinos are Majorana particles. Furry noted that this process can occur only if neutrinos are their own antiparticles. In modern terms (see Fig. 11.7 right), the two neutrinos flow through an off-shell line between two W bosons that mediate weak interactions in the Standard Model. Therefore, the final state is deprived of neutrinos:

$$(A, Z) \rightarrow (A, Z + 2) + 2e^- \tag{11.79}$$

if $\nu_e \equiv \bar{\nu}_e$. In a neutrinoless double-beta decay experiment, the signature is extremely clear: two electrons leave apart from the daughter and, since the recoil energy is negligible, the sum of the electron energy is just the Q-value of the process. For more than 70 years, a wealth of experiments have searched for this process, which is forbidden in the Standard Model due to violation of the lepton number. The most promising experiments are the ones using isotopes with a large isotopic abundance (e.g. ^{130}Te $\rightarrow$ ^{130}Xe), with outstanding energy resolution (germanium detectors that contain ^{73}Ge $\rightarrow$ ^{73}Se), with a very large mass (liquid scintillator detectors doped with ^{136}Xe $\rightarrow$ ^{136}Ba) or high Q-value (^{48}Ca $\rightarrow$ ^{48}Ti).[6] No ideal detector exists and we need to find the right trade-off for each candidate (Terranova, 2017). The enthusiasm toward these experiments has grown after 1998 because we know now that neutrinos are massive. A discovery of neutrinoless double-beta decay would probe the neutrino nature (Dirac versus Majorana), falsify lepton number conservation by two units, and provide information on the absolute mass of the ν_e. On the other hand, the ν_e is a superposition of different mass eigenstates and the effective mass entering this process might be zero due to cancellation effects (see Sec. 13.6.5). The link between the effective mass and the lifetime is further blurred by nuclear effects between the parent and the daughter transition amplitude, which cannot be treated by lattice QCD, yet, and rely on the semi-empirical models or the multi-body description mentioned in Sec. 9.6.3. The best limit on this process, forbidden in the Standard Model, is currently given by ^{136}Xe. Its half-time is $> 1.07 \times 10^{26}$ years (Gando *et al.*, 2016), 16

[6]The ^{48}Ca nuclide has the largest known Q-value among double-beta candidates but the natural abundance of this isotope is just 0.2%

orders of magnitude longer than the age of the universe. This should not discourage us because the neutrino masses inferred by oscillations point toward even longer lifetimes, which might become observable by neutrino experiments in the future.

11.6   Gamma decays

Gamma decays were discovered by P. Villard in 1900 and soon interpreted as the emission of high-energy photons (0.1–10 MeV) by nuclei. The identification of the "gamma radiation" (γ **rays**) with photons was established by E. Rutherford and E. Andrade in 1914. These experiments showed scattering phenomena similar to the diffraction of X-rays but with a photon wavelength much smaller than standard atomic transitions. The emission of photons eased the development of a nuclear theory for gamma decays, which was inferred by the semiclassical description of e.m. emission and, later on, by QED. Despite this simplification, strong interactions remain essential to understand the gamma transitions because they determine the initial and final state nuclides.

11.6.1   Classical and semiclassical theory

The classical theory of e.m. emission has been developed for time-dependent charge and current distributions. If the distributions depend on t, the moving charges produce e.m. fields that can be computed from the Jefimenko's equations of Sec. 5.7.1. The simplest applications are distributions that depend on $\sin \omega t$ and, therefore, produce waves with an angular frequency ω. The wave emission is computed using a **multipole expansion**. For instance, a system of two opposite sign charges $\pm q$ at distance d oscillating back and forth along a straight line (electric dipole) produces e.m. waves due to the acceleration of the charged particles. The e.m. waves depend on the electric dipole moment qd and the position of the charge at t: $z(t) = qd \cos \omega t$. From the solution of the Maxwell equations, the e.m. waves have a power $\sim \sin^2 \theta$, where θ is the angle between d and the wave-vector[7] $\mathbf{k}$. Similarly, a circular current loop of area A is a classical representation of a magnetic dipole whose e.m. radiation depends on $\mu = iA \cos \omega t$ and follows the same $\sin^2 \theta$ rule.

The **semiclassical theory** is built upon quantizing the source of the e.m. field, that is, the multipole. In this case, the observables are the probability of finding a photon at an angle θ and the average number of photons per unit time: the **average radiated power**. The decay probability per unit time of a nuclide is then given by:

$$\Gamma_{i \to f} = \int_V d^3 \mathbf{x} \, \psi_f^* m(\sigma L) \psi_i \tag{11.80}$$

where V is the volume of the nucleus, ψ_i and ψ_f are the initial and final state wavefunctions of the nucleus and $m(\sigma L)$ is the quantum operator representing the electric ($\sigma = E$) and magnetic ($\sigma = M$) multipole

[7]We use the notation $|\mathbf{k}| = 2\pi/\lambda$ where λ is the wavelength of the e.m. waves. Please note that in materials science and optics the wave-vector is defined as $|\mathbf{k}| = 1/\lambda$. In both definitions, the direction $\mathbf{k}/|\mathbf{k}|$ corresponds to the direction of propagation of the wave.

of order L. Unfortunately, this operator is quite complex to determine because it depends on the initial and final nuclear states. In the simplest case where only one nucleon is rearranged in the shells of the nuclide:

$$m(\sigma L) \sim er^L Y_{L,M}(\theta, \phi) \tag{11.81}$$

which reduces to ed for $L = 1$, as in the classical case. The quantum evaluation of the lifetime is a complex task but it formally resembles the classical result if we consider the decay amplitude $\Gamma_{\sigma L}$ (probability of photon emission per unit time) as the average power divided by the photon energy:

$$\Gamma_{\sigma L} = \frac{2(L+1)}{\epsilon_0 \hbar L[(2L+1)!!]^2} \left(\frac{\omega}{c}\right)^{2L+1} [m(\sigma L)]^2. \tag{11.82}$$

In this (SI) formula, the "double factorial" $n!!$ is a factorial where only the integers from 1 up to n that have the same parity (odd or even) as n are considered: $n!! = n(n-2)(n-4)\dots$. Once more, the complexity of the problem does not reside on the electromagnetic formulas, which have been known since the beginning of the 20th century, but on the initial- and final-state effects that determine the analytic form of the multipole operator, that is, non-perturbative QCD effects.

11.6.2 Selection rules

The photon is a spin 1 particle with just two polarizations and brings away an angular momentum L, whose minimum value is 1. Hence $0 \to 0$ transitions are forbidden. If the initial and final state nuclear spins are $\mathbf{I}_i$ and $\mathbf{I}_f$, $\mathbf{I}_f = \mathbf{I}_i + \mathbf{L}$, $|\mathbf{I}_{i,f}| = \sqrt{I_{i,f}(I_{i,f}+1)}\hbar$, and

$$I_i + I_f \geq L \geq |I_i - I_f|. \tag{11.83}$$

Parity conservation in the photon-nucleus final state generates additional constraints. In classical physics, the parity of an electric multipole is $(-1)^L$ and of a magnetic multipole is $(-1)^{L+1}$. We state without proof (Krane, 1988) that the same result holds in QM. We can now write the selection rules of gamma decay, for example for the first three multipoles ($L = 1$: dipole, $L = 2$: quadrupole, $L = 3$: sextupole):

Multipole	E1	M1	E2	M2	E3	M3
L	1	1	2	2	3	3
Parity change	yes	no	no	yes	yes	no

or, equivalently,

- $I_i + I_f \geq L \geq |I_i - I_f|$
- The parity of the initial nucleus is different from the parity of the final nucleus if the electric multipole is odd or the magnetic multipole is even.
- Otherwise the parity of the nucleus is unchanged.

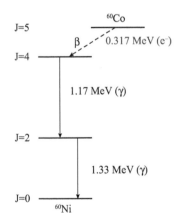

J=5 ^{60}Co

β --- 0.317 MeV (e$^-$)

J=4

1.17 MeV (γ)

J=2

1.33 MeV (γ)

J=0 ^{60}Ni

Fig. 11.8 Energy level of ^{60}Co $\to$ ^{60}Ni** $+ e^- + \bar{\nu}_e \to ^{60}Ni^* + \gamma \to$ ^{60}Ni$+\gamma$. The total angular momentum of the nuclide J (the nuclear spin) is indicated next to each energy level. The Q-value of the beta decay is 317 keV.

If the electric multipole L is odd, the parity is $(-1)^L = -1$ (parity change), but if the electric parity is even, $(-1)^L = +1$ and the parity is unchanged. It is the other way round for the magnetic multipoles. At the risk of being pedantic, I remind you that we should avoid confusing parity changes in nuclei with parity conservation in the whole system. The latter is a general property of QED and QCD and holds in all gamma decays.

In the experiment by Wu described in Sec. 10.5.3, the emission of the second photon at 1.33 MeV is quite easy to understand (see Fig. 11.8). Since it originates from a $2^+ \rightarrow 0^+$ decay, the photon changes the momentum by $\Delta L = 2$ without any change in the parity of the nuclide. Therefore, the only transition available is E2 and we can compute the angular distribution of E2 photons for perfectly polarized nuclei using QED. The 1.17 MeV photon from the $4^+ \rightarrow 2^+$ transition is more complicated because the transition can be due to $L = 2, 3, 4, 5, 6$. Since high multipoles are strongly suppressed, $L = 2$ is the main contribution and E2 is, again, the dominant term because parity is unchanged.

> As a rule of thumb, the lowest L usually dominates the gamma decay, $L+1$ being suppressed by $\sim 10^{-5}$ compared with L. Moreover, a magnetic multipole is suppressed by a factor ~ 100 with respect to the corresponding electric multipole. There are exceptions to these rules but they represent a good approximation if A is sufficiently large.

Finally, we may wonder if the angular distributions are the same for the classical and quantum cases. They may be quite different because of the quantization of the spin, which introduces the magnetic quantum number m. The closest analog is the $1 \rightarrow 0$ transition, where the initial state can have $m_i = -1, 0, 1$. Only the $1 \rightarrow 0$ gamma decay with $m_i = 0$ gives an angular distribution of the photon $\sim \sin^2 \theta$, where θ is measured along the direction of $\mathbf{I}_i$. This distribution is the same as the classical electric dipole transition. The two other distributions with $m_i = -1, 1$ are pretty different and look like $\frac{1}{2}(1 + \cos^2 \theta)$.

11.6.3   Gamma spectroscopy and the Mössbauer effect

Gamma spectroscopy is the nuclear counterpart of atomic spectroscopy and was essential to understand nuclear structure. At the same time, it is an astonishing tool for materials science, oncology, geology, and many other hard and life sciences. We will discuss materials science and geological applications in Sec. 11.7 but we focus here on an application of gamma spectroscopy that reached an exquisite precision in the measurement of nuclear energy levels, the **Mössbauer effect**.

As anticipated in Sec. 10.6.1, a photon emitted by a gamma decay cannot be re-absorbed by a nuclide of the same species because the photon lacks a tiny amount of energy taken by the atom recoil. We already demonstrated that this energy does not exceed 3 eV in the Goldhaber experiment but this number is still huge compared with the **intrinsic width** of most nuclear states with finite lifetime (τ). The intrinsic width is the broadening ΔE produced by the Heisenberg uncertainty principle: $\Delta E\, \tau \geq \hbar/2$. In Sec. 10.6.2, the photon source ($^{152}\text{Sm}^*$) is produced after an electron capture and is not at rest in LAB. The recoil energy is compensated, at least in the most energetic photon, by the Doppler shift due to the motion of the daughter nucleus. This is a very special case that is not suited for high-precision spectroscopy. For instance, Goldhaber was not able to scan the intrinsic energy spread of ^{152}Sm, which is given by the Breit–Wigner distribution of eqn 2.107:

$$\sigma(E_\gamma) = \frac{(\Gamma/2)^2}{\left[E_\gamma - (E_0 + E_R)\right]^2 + (\Gamma/2)^2} \tag{11.84}$$

where E_γ is the photon energy, Γ is the intrinsic width ($\Gamma = \tau^{-1}$), E_0 is the mean energy of the excited state, and E_R is the recoil energy. The scanning of the Breit–Wigner is equivalent to reaching the maximum attainable precision in the knowledge of the energy of a nuclear state and requires tuning of the photon energy by 10^{-6} eV for $\tau \simeq 1$ ns. Even worse, the linewidth of the Breit–Wigner is always blurred by the thermal Doppler shift of the target, which ranges from 0.1 eV at room temperature to 0.01 eV at 4 K.

We must find a way to increase or decrease the photon energy by a tiny amount using, for example, the Doppler shift and measuring the absorption probability. This probability gives eqn 11.84 if all sources of broadening are negligible except for the intrinsic broadening. A poor-man's solution is mounting the source in a spinning holder with an angular velocity of 10^5 rounds per minute to blue- or red-shift the photon by a tiny amount. Reaching this speed with high precision is extremely difficult in a macroscopic body.

In 1958, R. Mössbauer achieved a breakthrough in high-precision nuclear spectroscopy using a source embedded in a solid-state holder. The holder's lattice, made of thousands of atoms tied by chemical (e.m.) forces, works like a mass amplifier. The nucleus bounces inside the lattice, which has an enormous effective mass, and the recoil energy (eqn 10.48) turns out to be:

$$\frac{|\mathbf{p}_\gamma|^2}{2M_l} \simeq 0 \tag{11.85}$$

where M_l is the sum of all nuclear masses in the lattice and $\mathbf{p}_\gamma$ is the photon momentum. This mass amplification effect occurs only if the recoil energy is so small that the nucleus cannot be ejected outside the lattice. In Mössbauer's experiment, the ^{191}Ir source was embedded in a crystal lattice. ^{191}Ir decays emitting a photon at 129 keV with a recoil of $E_R = 0.047$ eV and the binding energy of the lattice is > 1 eV. The decay is, then, virtually recoil-less. The crystal also dumps the thermal Doppler shift and the experiment can be carried out at room temperature with a thermal broadening smaller than the intrinsic width of ^{191}Ir: 3×10^{-6} eV. The Breit–Wigner of ^{191}Ir is then scanned with a

Parent	$\tau_{1/2}(y)$ of the parent	last stable daugther	type
Thorium: ^{232}Th	1.41×10^{10}	^{208}Pb	$A = 4n$
Neptunium: ^{237}Np	2.14×10^{6}	^{209}Bi	$A = 4n + 1$
Uranium: ^{238}U	4.47×10^{9}	^{206}Pb	$A = 4n + 2$
Actinium: ^{235}U	7.04×10^{8}	^{207}Pb	$A = 4n + 3$

Table 11.2 The four natural radioactive series. ^{209}Bi decays in ^{205}Tl with a half-life of 1.9×10^{19} years and, therefore, can be considered a stable isotope.

rotating platform at the modest speed of 1.5 cm/s. **Mössbauer spectroscopy** has been used for decades in nuclear physics and, in particular, in the study of **nuclear hyperfine interactions** but is also employed in atomic physics and chemistry to investigate the variations induced by the chemical environment of the crystal. The *chef d'oevre* of Mössbauer spectroscopy is the demonstration of the red-shift of a photon in free-fall inside the gravitational field of the earth. This effect is a basic prediction of general relativity and a direct consequence of the equivalence principle (Cheng, 2010). R. Pound and G.A. Rebka observed for the first time this red-shift using an isomer of ^{57}Fe shown in Fig. 11.1. The isomer emits a 14.4 keV photon that falls to a 22 m high tower. Employing Mössbauer spectroscopy, Pound and Rebka were able to detect a gravitational red-shift of $\Delta E/E \simeq 5 \times 10^{-15}$ eV and opened the era of high precision tests of general relativity (Pound and Rebka, 1959).

11.7   Natural radioactivity

During the formation of the earth occurred about 4.5 billion years ago, many stable and unstable nuclides were formed. Most of unstable nuclides have either disappeared or regenerated at later stages (e.g. by cosmic rays – see Sec. 11.10) but a few of them have a lifetime much longer than the age of our planet and are still quite common in a large number of samples. They can produce decay chains but the number of independent chains does not exceed four. The reason is that neither beta nor gamma decays change the atomic mass number A of the atom, while alpha decays change A by four units. Hence, the independent **natural series** are the four series listed in Tab. 11.2 corresponding to $4n, 4n+1, 4n+2$, and $4n+3$ where n is an integer number. The neptunium series of Tab. 11.2 has a parent lifetime too short compared with the age of the earth, and does not contribute any more to natural radioactivity. The actinium series is also called the "^{235}U series" from the name of its parent nuclide. Fig. 11.9 depicts a simplified version of the

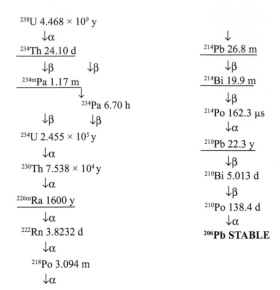

Fig. 11.9 The uranium (^{238}U) series. The underlined isotopes are the simplest to identify using gamma spectroscopy due to the large BR of the excited states. The chain is simplified removing minor branches of no relevance for gamma spectroscopy. The isotope half-life is indicated next to the name. Adapted from Gilmore, 2008.

uranium series where only alpha and beta decays are shown. Most of these nuclides emit photons from gamma decays too.

The underlined nuclides in Fig. 11.9 are the easiest to be identified by gamma spectroscopy. During the measurements, we record the spectrum of the photons by high-resolution detectors like germanium semiconductor detectors that can achieve an energy resolution of $\simeq 0.1\%$. In this way, we can determine the amount of material in the series even if it is present in traces. Ideally, a natural series is in secular equilibrium with its parent, which makes the measurement of the activity pretty simple. Unfortunately, samples have several sources of systematic errors. The most common source is the formation of radon, like the seventh nuclide of the chain in Fig. 11.9 (^{222}Rn). Radon is a gas at room temperature and is trapped in the source. If the source is damaged (e.g. a fracturing of the rock) the radon emanation breaks the secular equilibrium so that the photons produced after this chain step are no more visible. Similarly, groundwater can dissolve elements and remove them from the source. The easiest decays that can be measured are often the latest in the chain and, therefore, are particularly sensitive to the breaking of secular equilibrium. A common solution is to identify the loci of possible chain breaking, test the activity of the nuclides just before and after the candidate break, and correct the activities accordingly (Gilmore, 2008).

These applications of gamma spectroscopy are a powerful tool to characterize and date geological samples or perform material screening if systematics can be kept under control. The reader should be aware that radioactive materials can be found in the environment even outside the natural chains. There are about 15 additional isotopes with very long lifetimes that do not generate chains in secular equilibrium but can be used for dating and are visible in relatively large quantities.

A well-known example is ^{40}K that is found in wood, building materials, and in the metabolism of many animals (including humans). ^{40}K has $\tau_{1/2} = 1.28 \times 11^9$ years and decays both by β^- and electron capture (the BR of EC is $\simeq 11\%$). The EC produces an excited state that emits a 1460 keV photon and represents the standard signature of ^{40}K in gamma spectroscopy.

Finally, environmental radioactivity includes **cosmogenic nuclides** produced by cosmic rays like the aforementioned ^{14}C. Since 1950, we can also record **anthropogenic** radioactive isotopes, which are produced by nuclear reactor waste, fallout after reactor accidents, and tests of nuclear weapons.

11.8   Cosmic power sources

Energy production by nuclear reactions is not unique to humankind. The entire visible universe is fueled by nuclear reactions and produces sources whose power is unattainable on earth. These sources employ a combination of fusion reactions and radioactive decays, feed the energy cycle of the stars, the natural nucleosynthesis, and the production of cosmic rays.

11.8.1   Nuclear fusion

The SEMF shows very clearly that the derivative of the binding energy per nucleon is positive until ^{56}Fe. Therefore, the inverse process of the fission – the **nuclear fusion** – is energetically allowed by strong interactions at low A. Unlike fission, the process competes with the Coulomb repulsion of the two nuclei before strong interactions take place. The fusion of two nuclei occurs when the kinetic energy of the two parents overcomes the repulsive electrostatic force. The typical energy is $\simeq 5$ MeV, which is a small amount in the particle physics domain. However, if we want to use nuclear fusion as an energy source and exploit its enthalpy, we need a stable, macroscopic, thermodynamic system where fusion takes place in a finite volume. The simplest environments are hot gases that reach a temperature of $kT = 4.8$ MeV, that is, $T > 10^{10}$ K.

This impressive number can be reduced by two orders of magnitude by the interplay of quantum tunneling and the Maxwellian distribution of the particles. At high temperatures, the atoms form a plasma of nuclei and free electrons. The relative velocity distribution is given by the Maxwell-Boltzmann distribution:

$$P(v)dv = \sqrt{\frac{2m^3}{\pi(kT)^3}} \exp\left(-\frac{mv^2}{2kT}\right) v^2 \, dv \qquad (11.86)$$

where v is the relative velocity, k the Boltzmann constant, and m is the reduced mass of the two-nucleus pair: $m = m_1 m_2/(m_1 + m_2)$. Since the reaction occurs through QM tunneling, the cross-section has an exponential form:

$$\sigma(E) \sim \exp\left[-\sqrt{\frac{E_G}{E}}\right].$$ (11.87)

This formula is a generalization of the Gamow–Condon–Gourney theory for the alpha decay applied to two separate nuclei and the **Gamow energy** is defined (in SI) as:

$$E_G = 2mc^2(\pi\alpha Z_1 Z_2)^2$$ (11.88)

where Z_1 and Z_2 indicate the electric charge of the two nuclei in units of e. The nuclei have their own structure dictated by QCD and described by multi-body or semi-empirical models. These complex effects can be factorized in eqn 11.87 by a correction coefficient $S(E)/E$ so that the fusion cross-section reads:

$$\sigma(E) \simeq \frac{S(E)}{E} \exp\left[-\sqrt{\frac{E_G}{E}}\right].$$ (11.89)

The correction factor is expressed as $S(E)/E$ because the vast majority of nuclear cross-sections at low energy decreases with E.

We can now estimate the fusion rate (Bertulani, 2007; Martin, 2009):

$$R_{12} = n_1 n_2 \langle\sigma v\rangle$$ (11.90)

where the average velocity is computed weighting over the Maxwell-Boltzmann distribution of the relative velocities. Note that n_1 and n_2 are the number of nuclei that can undergo the fusion process. If the nuclei are the same ($n_1 = n_2 = n$) and cannot be distinguished, the formula must be corrected according to quantum statistics: $n_1 n_2 \to n(n-1)/2$. The final result for distinguishable nuclei is:

$$R_{12} = n_1 n_2 \sqrt{\frac{8}{\pi m (kT)^3}} \int_0^\infty S(E) \exp\left[-\frac{E}{kT} - \sqrt{\frac{E_G}{E}}\right] dE.$$ (11.91)

An increase of energy eases the crossing of the Coulomb barrier but the Maxwell-Boltzmann distribution penalizes the probability of finding particles with energy significantly larger than the mean energy at thermal equilibrium. The net result is that:

> if $S(E)$ is slowly varying so that the function inside the integral of eqn 11.91 is dominated by the exponential term, fusion occurs with a sizable probability only in a narrow range of energy called the **Gamow window**.

The range of the Gamow window is:

$$E_0 \simeq \left[\frac{1}{4}E_G(kT)^2\right]^{1/3} \pm \frac{4}{\sqrt{3}\sqrt[3]{2}} E_G^{1/6}(kT)^{5/6}.$$ (11.92)

In most practical cases (e.g. proton–proton fusion) the temperature is $\mathcal{O}(10^8)$ K, two orders of magnitude smaller than the naïve estimate made at the beginning of this section.

11.8.2   Fusion reactors and weapons

Fusion can be produced artificially on earth, too. As discussed in Sec. 9.8.3, the enthalpy can be extracted together with a large increase of entropy (uncontrolled reactions) to build fusion-based nuclear weapons called **thermonuclear weapons**. In these systems, fusion is achieved by a fission bomb that increases the temperature. The most common scheme is the **Teller–Ulam configuration** based on a staged implosion of a fission bomb. The implosion of the fission material compresses the fusion fuel and initiates the fusion reactions of deuterium nuclei:

$$d + d \to {}^3\text{He} + n + 3.27 \text{ MeV}. \tag{11.93}$$

The fission core is made of lithium deuteride because it is a stable solid compound and further increases the power of the bomb by the ancillary reaction:

$$n + {}^6\text{Li} \to {}^3\text{H} + \alpha. \tag{11.94}$$

A thermonuclear weapon is then a hybrid fission–fusion bomb that produces nuclear reactions by the fast increase of the temperature. These bombs are also called "hydrogen bombs" although $p - p$ fusion plays no role because its fusion cross-section is too small. Due to the large power, the reduced need for fissile material, and the compactness of the weapon, nearly all bombs available today are thermonuclear weapons based on the Teller–Ulam configuration.

The complexity of the task arises enormously if we want to control the fusion process, like in **fusion reactors** for the production of energy. These facilities are still under development and cannot replace conventional fission reactors yet. In a fusion reactor, the ideal fuel is deuterium–tritium (Morse, 2018):

$$d + {}^3\text{H} \to \alpha + n + 17.62 \text{ MeV} \tag{11.95}$$

which cannot be used in weapons due to the relatively short half-life of tritium (12.5 y). Unlike weapons, the large amount of enthalpy produced by eqn 11.95 must be extracted in a controlled way by confining the energy inside the reactor and, at the same time, reaching temperatures that no solid-state material can stand. The most promising option is magnetic confinement of hot plasma based on large superconducting magnets that are not in direct contact with the plasma. Other options include inertial confinement with high-power lasers.

In a self-sustaining reactor, the energy that is lost outside the reactor by radiation or neutron leakage (which are not magnetically confined) must be compensated by the energy produced by the fusion or deposited by the alpha particles. A common figure of merit to rate the quality of a reactor design is the **Lawson criterion**:

$$L \equiv \frac{\text{output power}}{\text{input power}} \simeq \frac{n_d \langle \sigma v \rangle t_c Q}{6kT}. \tag{11.96}$$

In the Lawson formula, Q is the energy released by the fusion process, n_d is the number of deuteron nuclei, σ is the cross-section of eqn 11.95

and t_c is the confinement time. All in all, the only parameters we can improve are the number of nuclei and the confinement time because the temperature is fixed by the Gamow window.

The most advanced fusion reactor prototype is ITER (Claessens, 2020), which is currently under construction in France. It is aimed at $L \simeq 10$ for $t_c \sim 20$ min. Even if it is not yet a device that can be used for industrial applications, it capitalizes decades of research and would represent a major leap toward fusion reactors. Compared with fission reactors, fusion reactors produce a much smaller amount of nuclear waste, require fuels that are available in large quantities, are safer, and environmentally more sustainable. On the other hand, they still have to overcome major scientific and engineering challenges even after ITER.

11.9   The lifecycle of a star

What cannot be (easily) done on earth has been done since 13.5 billion years ago – 200 million years after the Big Bang – in the rest of the universe. The majority of celestial bodies are not suitable to host life and most of them are a large collection of extreme environments at high temperatures and pressures. The life of a star like the sun (see Fig. 11.10) commences with the gravitational collapse of gases composed mainly of hydrogen, helium, and a tiny amount of heavier elements. The stars thus exploit a force (gravitation) that, despite its ultra-weak intensity, is always attractive and grows with the amount of mass. The faintness of gravitational interactions is compensated by the accretion of the star, which clusters masses larger than 10^{30} kg.

The gravitational force triggers an energy generation mechanism of outstanding power, which A. Eddington tentatively attributed to nuclear fusion in 1920. Some readers may be aware of Kelvin's attempt to explain this mechanism invoking chemical reactions. In this way, Lord Kelvin estimated the age of the sun that turned out to be 98 million years: a number in striking contradiction with geological data (and with Darwin's theory of evolution that Kelvin was fiercely challenging). The problem was solved in a satisfactory manner by Eddington, Gamow, and Teller in the 1920s following the claim of C. Payne-Gaposchkin that the stars are mostly composed of hydrogen. In 1939, Bethe showed that the "steam" of stellar engines is the fusion of two protons:

$$p + p \rightarrow d + e^+ + \nu_e + 0.42 \text{ MeV} \qquad (11.97)$$

that is, a weak scattering. This reaction dominates the energy process chain when the temperature of the star is $\sim 10^7$ K. Due to the large density and pressure, the probability of additional fusion chains is sizable. After producing enough deuteron, the star triggers the deuteron–proton fusion:

$$d + p \rightarrow {}^3\text{He} + \gamma + 5.49 \text{ MeV}. \qquad (11.98)$$

Note that this reaction is a pure strong interaction that occurs only when the star is enriched of deuteron. Later, another strong process fuses two helium isotopes producing a stable alpha particle:

$$^3\text{He} + {}^3\text{He} \to \alpha + 2p + 12.86 \text{ MeV}. \tag{11.99}$$

This process makes a substantial contribution to the energy production in the star due to the large binding energy of the alpha particle. Hence:

> the fusion chain in the stars is triggered by a weak process and fed by strong processes. The lifetime of a star is set by the weak rate of the parent process (proton–proton fusion) and depends on the mass of the star.

This result explains why Kelvin was off by orders of magnitude in estimating the age of the sun: he was not aware of weak forces. Since the initial process for a fusion chain is a weak process, the energy produced in a star is approximately constant in time and does not show strong variations or explosions, except at the end of its lifecycle. The thermal balance of the sun is obtained summing up all steps of the chains down to the stable isotopes. The energy balance is dominated by eqs 11.98 and 11.99, and amounts to:

$$4p \to \alpha + 2e^+ + 2\nu_e + 2\gamma + 24.68 \text{ MeV}. \tag{11.100}$$

The positrons further increase the energy production by annihilation (+1.022 MeV per positron). The two neutrinos escape from the star reducing the thermal budget by 0.52 MeV on average. The total energy produces by eqn 11.100 – the **p–p chain** – is 6.55 MeV per proton. We can describe this process using the terminology of artificial fusion reactors. The force that allows the fulfillment of the Lawson criterion is the self-gravity that compresses the fuel, increases the temperature, and triggers the fusion chain. While the main energy-loss sources in ITER are radiation losses and the escaping neutrons, these sources are trapped in the star due to the large mass of the system. The only particles that escape are the neutrinos, which are present only in the first stage of the p–p chain and in a few other chains that do not contribute significantly to the thermodynamics of the star. Electromagnetic radiation plays a role, too: the photons can reach the surface of the star in thousands of years and are emitted as black-body radiation at the temperature of the surface where the photon escapes, the **photosphere**. In the sun, the mean photosphere temperature is 5777 K. The escaping photons make the star a visible celestial object. Its **luminosity** is the average irradiated power and amounts to 3.8×10^{26} W in the sun. For most of its life, the star is a stable system where the outward pressure exerted by photons and high-temperature particles compensate for the inward gravitational pressure.

The sun is, therefore, a source of photons and neutrinos. Solar neutrinos were observed for the first time by R. Davis in the 1960s, together with J.N. Bahcall, who performed the calculation for the expected

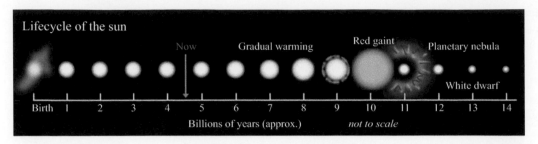

Fig. 11.10 The expected lifecycle of the sun. Note that the diameter of the red giant is not to scale. The actual diameter is ∼ 100 times longer than the current sun diameter. Reproduced from Wikimedia 2009, public domain.

number of neutrinos. The observed number was about 1/3 of Bachall's predictions and this anomaly was explained much later by neutrino oscillations (see Sec. 13.6.2). Real-time observation of neutrinos from the p-p and the subdominant carbon–nitrogen–oxygen (CNO) chain was reported in 2014 and 2020, respectively, by the Borexino experiment in Italy.

When a substantial part of the primary fuel of a star is exhausted, the energy production decreases and the gravitational pull increases the temperature of the star, triggering new fusion processes that create heavier nuclides, as shown by F. Hoyle in 1946. This process is called **stellar nucleosynthesis**. A sequence of gravitational collapses creates carbon, oxygen, and silicon. Finally, when the upper layer of the star collapses toward the core, even higher-A elements are produced up to nickel. Näively, one expects that no element should be produced beyond ^{56}Fe, which is the peak of the SEMF, but neutron and proton capture can create small quantities of heavy nuclides, too.

The death of a star critically depends on its mass. The sun is a middle-aged star (4.6 billion years) with a life expectancy of 10 billion years. In about 5×10^9 years, all the hydrogen of the core will be exhausted and the sun will be burning hydrogen in the surrounding halo, becoming a cooler but larger **red giant**. It will engulf several planets of the solar system, including the earth. In the end, half of its mass, rich in elements crucial for the development of planets and life (C, O, Si) will be dispersed and the remnants will form a cold (100,000 K) white dwarf where no fusion occurs but matter is compressed into a small volume similar to the earth. This remnant is mostly made of degenerate electrons since, without fusion, the Pauli exclusion principle is the only opponent to gravitational collapse.

11.10 Supernovas and cosmic rays

Stars slightly more massive than the sun (1.2–1.6 solar masses) transform into **neutron stars** at the end of their lifecycles. At these masses, the Pauli exclusion principle is not able to compensate for the gravitational pull any longer. The electrons are then captured by EC processes increasing the rate of $e^- + p \to n + \nu_e$ reactions. In this way, the ^{56}Fe limit can be overcome by

$$^{56}\text{Fe} + e^- \to {}^{56}\text{Mn} + \nu_e$$
$$^{56}\text{Mn} + e^- \to {}^{56}\text{Cr} + \nu_e. \qquad (11.101)$$

Later, however, high A nuclei start emitting neutrons to reach more stable states $^{A}\text{X} + e^- \to {}^{A-1}\text{Y} + \text{n} + \nu_e$ and the star becomes tiny (tens of km of diameter) and rich of neutrons.

Stars with a mass smaller than 1.4 solar masses (the **Chandrasekhar mass**) are the heaviest stable white dwarfs. Beyond this regime, the end-cycle of a star is much more complex and still an open area of research. The most spectacular event is a huge explosion caused either by prompt access to nuclear fusion fuel (for example, by accretion of a nearby red giant as in Type Ia supernovas) that produces a large amount of energy in a short time or, more frequently, by core collapse in stars much more massive than the sun (Type II supernovas). Here, the gravitational pressure is counterbalanced by the breaking of the nuclides and the transformation of the core into an incompressible Fermi gas of nucleons. In both cases, the corresponding shock wave creates a **supernova**: a transient stellar explosion sometimes visible in the sky even with the naked eye. In the aftermath of this breathtaking event further high-A nuclides are produced by rapid neutron capture (**r-process**) and a large number of particles are ejected in all directions, reaching the earth too.

The earth is hit by a steady flux of particles that were observed for the first time by V.F. Hess in 1912. These **primary cosmic rays** are composed mainly of protons (86%), alpha particles (11%), electrons (2%), and light nuclei. Traces of high-A nuclei up to uranium have also been recorded. The origin of these cosmic rays is still a work in progress but there is a consensus that the majority of these particles originates from supernova explosions in close and distant galaxies. Since cosmic rays are remnant of explosions far in space and time, and the original direction is blurred by galactic magnetic fields,

> the primary cosmic rays are an isotropic flux of protons and alpha particles that interact with the atmosphere creating a cascade of secondary particles. Light primary cosmic rays (e.g. positrons) are probably due to secondary interactions in the interstellar medium. Primary neutrinos and high-energy photons originate from point sources that may be traced back since neutral particles are not affected by magnetic fields.

The particles are generally of galactic origin but the spectrum extends up to 10^{20} eV. These ultra-high energy cosmic rays cannot originate from our galaxy because they are too energetic to be confined by the magnetic fields. The most astonishing feature of cosmic rays is the broad energy spectrum, whose flux decreases with energy. Below 1 GeV, the spectrum depends on local effects like the solar wind and the strength of the local geomagnetic field but above this energy it is very regular and follows a simple law:

- up to 10^{16} eV, the flux decreases as $N(E) \sim E^{-2.7}$. The region around 10^{16} eV is called the **knee**.
- above the knee (10^{16} eV) and before the **ankle** (10^{18} eV), the flux is steeper and decreases according to a $N(E) \sim E^{-3}$ law.
- above the ankle and below the **GZK cutoff** $E_{GZK} \simeq 4 \times 10^{19}$ eV, there is an additional change of slope: $N(E) \sim E^{-2.69}$. E_{GZK} is a very special value since at this energy cosmic rays interact with the photons of the cosmic microwave background (CMB) as they were the scattering centers of a fixed-target experiment, and the interstellar medium becomes opaque to cosmic rays.

The E_{GZK} cut-off was considered for the first time by K. Greisen, G. Zatsepin, and V. Kuzmin in 1966 and corresponds to a fixed target Δ^+ production similar to the one discussed in Sec. 8.4:

$$\gamma + p \to \Delta^+ \to p + \pi^0$$
$$\gamma + p \to \Delta^+ \to n + \pi^+. \tag{11.102}$$

Since the photon is massless, the four vectors in the initial state are $q^\mu = (q, 0, 0, q)$ and $p^\mu = \left(E = \sqrt{m_p^2 + |\mathbf{p}|^2}, \mathbf{p}\right)$ in LAB for the photon and proton, respectively. The invariant mass in NU is:

$$s = (E + q)^2 - (\mathbf{p} + \mathbf{q})^2 = m_p^2 + 2q(E - |\mathbf{p}| \cos\theta) \tag{11.103}$$

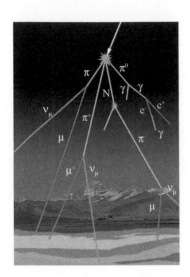

Fig. 11.11 A primary proton showering through the earth's atmosphere producing hadrons and leptons.

and depends on the scattering angle between the photon and the proton. The GZK threshold is given by eqn 11.103 setting $s = (m_p + m_{\pi^0})^2$ and $q = kT$ where $T = 2.7K$ is the temperature of the CMB. A more detailed calculation accounting for the θ distribution and the energy distribution of the photon in LAB provides the GZK threshold: $E_{GZK} \simeq 4 \times 10^{19}$ eV. The origin of the knee and ankle is not fully understood. It is likely due to the particle acceleration mechanisms during and after the shockwave of the supernova explosion, and – for the ankle – the onset of extragalactic cosmic rays.

The interaction of primary cosmic rays in the earth's atmosphere is equivalent to a fixed target scattering of protons with a column of air and it is depicted in Fig. 11.11. We then expect a large production yield of pions that mostly decay in flight. Due to time dilation, muons of energy larger than 3 GeV reach the ground and can be easily detected: they are the **hard components** of cosmic rays on earth. The remnants of the electromagnetic showers produced by $\pi^0 \to \gamma\gamma$ decays constitute the **soft component**. Despite the complexity of the process, for most applications, we can assume that the average energy of cosmic ray muons at the surface of the earth is $\simeq 4$ GeV and the intensity depends on the angle between the vertical and the muon direction (the zenith angle θ)

$$I = I_0 \cos^2\theta. \tag{11.104}$$

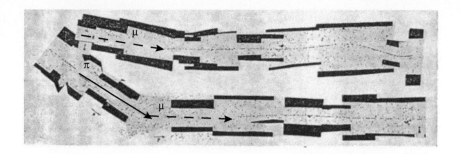

Fig. 11.12 (Lower track) A $\pi \to \mu \, \nu_\mu$ decay seen by a high-sensitivity nuclear emulsion developed by Kodak inc. The detector can measure the dE/dx and the trajectory but it can observe neither the neutrinos nor the particle sign (in this case, due to the lack of a strong magnetic field). The trajectories of the pion (black line) and the muon (dashed line) are rectilinear, except for multiple Coulomb scattering effects. The **decay kink**, that is, the angle between the π and μ trajectory, signals that part of the momentum is carried away by the neutrinos. (Upper track) the final part of the trajectory of the muon escaping the nuclear emulsion detector before decaying. Reproduced with permission from Lattes, *et.al.*, 1947. Copyright 1947 Nature Publishing Group.

The integral of the flux for $E > 1$ GeV is $\simeq 70$ m^{2}s^{-1}sr^{-1} corresponding to a well-known rule of thumb:

> the flux of cosmic muons on the earth surface is about one muon per minute per square centimeter for a horizontal muon detector.

The mean momentum of the muons is about 4 GeV, although this number may depend on the location of the observer on the earth's surface and solar cycles (Gaisser *et al.*, 2016). Cosmic rays played a fundamental role in the history of particle physics because they provided a wealth of new particles well before the invention of accelerators. The muon, the positron, and the pion were discovered this way.

11.11 Muons, pions, and the Yukawa meson*

Cosmic rays played quite a role in nuclear physics, too. In 1935, Y. Yukawa proposed a popular theory of strong forces where the mediator was a spin-0 massive meson. In modern terms, the Yukawa theory has not been completely superseded. First of all, QFT shows that the transition amplitudes for a massive boson M are suppressed by M^{-2} giving rise to short-range forces. The suppression due to massive mediators is nowadays exploited in the electroweak theory of the Standard Model, as we will see in Sec. 12.2. We know already that strong interactions in the non-perturbative regime of QCD look like a powerful short-range force.[8] Furthermore, the effective potential of Sec. 9.3 at

[8]Do not forget, however, that short-range forces in QCD are originated from confinement and not from the Yukawa meson.

intermediate distances can be studied assuming the exchange of a light color singlet and the lightest one is the pion: a spin-0 massive meson.

When C. Anderson and S. Neddermayer discovered the muon, the particle was interpreted as the Yukawa meson. The **discovery of the muon** was performed by simply measuring the dE/dx of the hard component of cosmic rays in a cloud chamber.

They noted that the new particle was depositing energy as if it had an "intermediate" mass, whose value was between m_e and m_p. This claim was confirmed a year later by another cloud chamber experiment done by J.C. Street and E.C. Stevenson. Since Yukawa predicted the pion to have a mass in between the electron and proton mass, the new particle was called "μ meson" and interpreted as the Yukawa meson. In 1947 a new would-be Yukawa meson was discovered by C. Powell, C. Lattes, and G. Occhialini using **nuclear emulsions**. These detectors are thick photographic plates with a large and uniform density of grains so that both the trajectory and the energy loss per unit length can be measured with a precision greater than cloud chambers, although the development and analysis of the photographs are much more cumbersome. This particle showed two clear kinks (see Fig. 11.12) that we can interpret today as a $\pi^- \rightarrow \mu^-\ (\bar{\nu}_\mu) \rightarrow e^-\ (\bar{\nu}_\mu\ \bar{\nu}_e\ \nu_\mu)$ decay chain with three undetected neutrinos.

The **discovery of the pion** challenged the interpretation of the "μ meson" as the Yukawa meson. The issue was solved in 1947 in a seminal experiment performed by M. Conversi, E. Pancini, and O. Piccioni. The experiment consisted of vertical blocks of magnetized iron selecting muons of a given sign. Before and after the blocks, they installed two Geiger–Müller counters put in coincidence. This set-up was mounted on top of an absorber and another Geiger counter was installed just below the absorber, as shown in Fig. 11.13. The scientists sought for coincidences in the first two counters to select positive or negative muons. The particle sign was selected changing the sign of the magnetic field. During the data taking, they detected events where the delay between the coincidence of the first two counters and the signal of the third counter was comparable to the lifetime of the muon at rest in vacuum (2.2 μs). This experiment is one of the earliest implementations of the **delayed coincidence** techniques, which is nowadays a standard tool in collider and fixed-target experiments. In this way, they could measure the half-life at rest of the μ^+ and μ^- separately. Muons are stopped in the absorbers by the energy loss of the Bethe formula and decay at rest. μ^-s do not experience any Coulomb barrier: they are stopped and can be absorbed inside the nucleus. If the muon were the Yukawa meson, the probability of a strong interaction should be very high and most of the mesons would be absorbed inside a nucleus before decaying. Such an absorption would be signaled by a "fake lifetime" much smaller than 2.2 μs. This fake lifetime results from muon decays in vacuum and the sudden disappearance of muons due to the absorption inside nuclei, and is much shorter than the lifetime in vacuum. Conversi noted that positive muons were stopped in the absorber after real decays at rest without

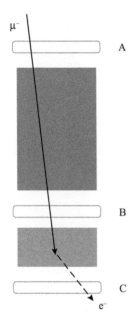

Fig. 11.13 Conceptual layout of the Conversi–Pancini–Piccioni experiment. The upper block (magnetized iron) selects a μ^- (black arrow), which stops in the absorber in a negligible time. The selection is done requiring the coincidence of counter A and B. This is equivalent to selecting a μ^- at rest in the absorber. When the muon decays, emits an electron (dashed arrow) that crosses C. The delay between the A+B coincidence and C is of the order of the muon lifetime. In this experiment, events were selected if the delay was between 1 μs and 4.5 μs. If the muon is not absorbed in a nucleus, its lifetime is the μ^- lifetime at rest (2.2 μm). Otherwise, the observed lifetime is seemingly shorter due to the muon disappearance inside the nucleus.

interacting with the nucleus because the μ^+ lifetime was exactly the same as for decays in vacuum. The negative muons, on the other hand, could be absorbed by the nuclei due to the lack of the Coulomb barrier and showed a reduced lifetime. Replacing the absorber with a low-Z material, no evidence of muon absorption in nuclei was observed and the lifetime turned out to be $\simeq 2.2~\mu$s both for positive and negative muons. Employing the Yukawa theory, they showed that if the muons were the mediators of strong interactions, the cross-section would be so strong that the scattering with the nuclei would be visible at any Z. This crack in the Yukawa theory was settled in 1949 when it became evident that the muon is just a decay product of the pion. The latest experiments employed high-sensitivity emulsions to detect minimum ionizing particles and reconstruct the whole pion decay chain.

At the end of the day, everyone was wrong. The mediators of the strong interactions are the gluons, not the pions. As noted in Sec. 9.3, the pion exchange is just an approximate model to describe interactions between nuclei at intermediate distances, and the muon is definitely not a meson. The π^- is a $\bar{u}d$ quark-bound state, while the μ^- is an elementary particle: a charged lepton that (mockingly) does not interact at all with matter by strong forces.

Further reading

Given the central role of radioactivity in applied physics and astrophysics, many textbooks on this subject are accessible to undergraduates. A general review and some historical background is given in Bertulani, 2007; Krane, 1998; and L'Annunziata, 2019. Gamma spectroscopy and its applications are presented in Gilmore, 2008, which emphasizes experimental techniques. Another classical textbook focused on radiation detectors is Knoll, 2010.

Nuclear medicine is a broad field of research not covered in this book: you can find a clear introduction in Chandra and Rahmim, 2017; and Prekeges, 2012. Nuclear astrophysics and cosmic rays are presented at Master's level in Perkins, 2009. A more general overview of nuclear fusion and its applications (fusion reactors, weapons) is available in Morse, 2018 and Murray and Holbert, 2019.

Exercises

(11.1) I bought a ^{22}Na source ($\tau_{1/2} = 2.6$ y), whose activity was 100 kBq in Jan 1, 2014. Compute the activity of this source on October 1, 2021.

(11.2) If we measure the activity of 1 g of natural samarium, we find 89 ± 5 alpha decays per second due to the decay of ^{147}Sm ($\tau_{1/2} = 1.06 \times 10^{11}$ y). Compute the isotopic abundance of ^{147}Sm and its uncertainty.

(11.3) Compute the total activity of your body, which contains 18% of ^{14}C and 0.2% of ^{40}K. The half-life of ^{40}K is 1.2×10^9 s.

(11.4) An apple a day keeps the doctor away, but what about bananas that are rich in ^{40}K? (isotopic abundance: 0.0117%, $\tau_{1/2} = 1.2 \times 10^9$ s) Compute the activity of 100 g of bananas assuming that each banana has half-gram of natural potassium.

In your opinion, do you take any risk if you eat a banana a day?

(11.5) To compute the age of a rock containing 1.8 g of ^{232}Th and 0.37 g of ^{208}Pb (stable isotope), the rock is pulverized and the α produced by the decays are collected as helium gas. Compute the age of the rock assuming that the helium extraction mechanism is fully efficient. [Hint: see Tab. 11.2]

(11.6) * A more appropriate unit to evaluate risks in radioactivity is the **Sievert** (Sv), corresponding to the absorbed dose expressed in J/kg – the **Gray** (Gy) – multiplied by the **equivalent dose** (H). H is a dimensionless coefficient that quantifies the biological potential damage for each type of radiation (α, β, γ, and neutrons at different kinetic energy) (Leo, 1994). What is the formula that gives the equivalent dose of an α and EC emitter? Is it the same for a β emitter?

(11.7) Compute the Q-value, the kinetic energy and the velocity of the daughter nucleus in a ^{230}Th $\rightarrow$ ^{226}Th $+ \alpha$ decay. [Note: the standard source of information for nuclear data is the US National National Institute of Standards and Technology (NIST). The **NIST** website provides the atomic weights and isotopic compositions for all elements (NIST, 2020).]

(11.8) * The probability of a radioactive decay is given by the Poisson distribution:

$$P(n) = e^{-\lambda} \frac{\lambda^n}{n!} \qquad (11.105)$$

where n is the number of observed decays in a given time interval and λ is the corresponding average probability. Explain why the process is Poissonian. If a sample of 10^{11} atoms has an activity of 2×10^{-11} Bq, what is the probability of observing five decays in 1 s?

(11.9) * A 1 g copper sample is irradiated for 20 minutes in a thermal neutron flux of 10^9 n/cm^{2}s. Natural copper is made of ^{63}Cu (isotopic abundance: 69%) and ^{65}Cu (31%). The irradiation produces:

$$^{63}\text{Cu} + n \rightarrow {}^{64}\text{Cu} \ (\tau_{1/2} = 12.7 \text{ hours})$$
$$^{65}\text{Cu} + n \rightarrow {}^{66}\text{Cu} \ (\tau_{1/2} = 5.1 \text{ min})$$

whose cross-sections are, respectively, 4.4 and 2.2 barn. What is the activity of each isotope at the end of the irradiation?

(11.10) * We want to measure the activity of a ^{22}Na source by counting the annihilation photons produced by the positron in the ^{22}Na $\rightarrow$ ^{22}Ne $+ e^+ + \nu_e$ decay. Demonstrate that taking 5 measurements of 5 minutes each gives the same precision as one measurement lasting 25 minutes. If the activity is 50 kBq, how many 511 keV photons are recorded in 5 minutes?

(11.11) ** A neutrinoless double-beta decay ($0\nu\beta\beta$) experiment records the decays of a 100 g sample of ^{76}Ge for 100 days without observing any $0\nu\beta\beta$ event. What is the (90% confidence level) upper limit on the half-life of this process that can be derived from the experiment? Why did they choose ^{76}Ge?

(11.12) * A pure α emitter produces an α at $t = 0$. Show that the probability of observing another α at time t is $P(t) = \mathcal{A}e^{-\mathcal{A}t}$, where $\mathcal{A}$ is the activity of the sample.

(11.13) Using the SEMF, compute the nuclear binding energy of ^{27}Mg, ^{27}Al, and ^{27}Si. Identify the most stable isotope, i.e. the isotope with the largest binding energy, and explain the reason for it.

(11.14) Many isotopes have multiple beta decay modes. Compute the Q-value of ^{196}Au for β^-, β^+ and EC.

(11.15) * Derive the selection rules of Tab. 11.1.

(11.16) * ^{187}Re is a beta emitter with a half-life of 41.2×10^9 years and a Q-value of 2.5 keV. Is it a good candidate for the neutrino mass measurement using the Kurie-plot technique? Is it better than tritium?

(11.17) Compute the GZK cutoff if, instead of producing a Δ^+, the photon–proton scattering produces a Roper resonance $N(1440)$ (see Sec. 8.6.1).

(11.18) * The μ^- produced by cosmic rays and reaching the earth surface have a non-zero polarization. Why?

(11.19) ** Employing the techniques of Chap. 3, design an experiment capable to identify the appearance of the GZK cutoff.

(11.20) * Proof eqn 11.25.

The electroweak theory

12.1 Weak interactions in the 1960s

During the 1960s, physicists realized that both weak and strong interaction data could be explained by the quark model and the V-A theory. On the contrary, they were reluctant to use quantum field theories (QFT) and the gauge principle for all fundamental interactions. The predictions of the V-A theory were clear and supported by the experiments:

- all beta and weak decays are explained by contact interactions described by four-line vertices like in Fig. 10.10. This theory does not need massless mediators as "weak photons" or "weak gluons."

- weak forces are the only forces that can change the flavor of leptons and quarks. A flavor-changing contact interaction occurs when, for instance, a μ^- transforms into a ν_μ but charge conservation is preserved. In modern jargon, these interactions are called **charged currents** (CC) and include decays like $\mu^- \to e^- \bar{\nu}_e \nu_\mu$ (see Fig. 12.1).

- weak interactions change the electric charge of particles but are decoupled from quantum electrodynamics (QED). Therefore, only fermions contribute to the contact interactions and there is no weak vertex that couples the photon to an elementary fermion.

- strong interactions[1] are decoupled both from QED and the V-A theory because their vertices do not depend on the color of quarks. Weak vertices can only change the flavor of quarks and, hence, their electric charge.

Still, nobody claimed that weak interactions were properly understood. In particular, people were astonished by the dimensional coupling constant of the V-A theory (G_F). In QED, the perturbative series is expressed in powers of the fine structure constant α, which is dimensionless. The corresponding weak constant – the Fermi constant – has a dimension of GeV^{-2}. The dimensional analysis of the perturbative series is not a technical detail. Dimensionless constants are a necessary (but not sufficient) condition for QFT to be mathematically consistent. In the language of QFT, the V-A theory is **non-renormalizable** and the cross-sections diverge at high energies. Given the value of G_F, the total cross-section of a $\nu_\mu + p \to \mu^- + X$ scattering diverges for $E_{\nu_\mu} \to \infty$, if X is a generic set of hadrons:

$$\sigma_{total} \to \infty \quad \text{for} \quad s \to \infty \tag{12.1}$$

[1] At that time, strong interactions were not dealt with QFT tools and QCD had yet to come.

A Modern Primer in Particle and Nuclear Physics. Francesco Terranova, Oxford University Press.
© Francesco Terranova (2021). DOI: 10.1093/oso/9780192845245.003.0012

This asymptotic behavior is unacceptable because the scattering probabilities must sum up to 1. In quantum mechanics (QM) and QFT, the scattering is described by a quantity called the **S-matrix**, which connects the initial-state particles (or quantized fields) with the final-state particles (fields). For instance, a particle in a 1-dimensional (1D) barrier has a finite probability to cross the barrier. Solving the Schrödinger equation, we get two solutions outside the barrier:

$$\psi_L(x) = Ae^{ikx} + Be^{-ikx} \tag{12.2}$$

$$\psi_R(x) = Ce^{ikx} + De^{-ikx}. \tag{12.3}$$

The first solution holds on the left side of the potential barrier, while the second solution describes the particle motion on the right side. In this simple case, the wave-vector is:

$$k = \sqrt{2mE} \tag{12.4}$$

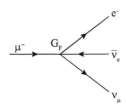

Fig. 12.1 Feynman diagram of a muon decay in the $V - A$ theory. In this theory, we only have four-line vertices with a dimensionful coupling constant G_F.

in natural units (NU). A represents the amplitude of the incoming wave and, as in standard optics, B and C are the amplitudes of the reflected and transmitted waves. If the incoming particle comes from the left at $t = 0$, then $D = 0$. The scattering amplitude is a linear superposition of all these waves and the linear coefficients are given by the S-matrix:

$$\begin{pmatrix} B \\ C \end{pmatrix} = \begin{pmatrix} S_{11} & S_{12} \\ S_{21} & S_{22} \end{pmatrix} \begin{pmatrix} A \\ D \end{pmatrix}. \tag{12.5}$$

S_{ij} are the elements of the S-matrix for the scattering of a particle in a 1D barrier. The incoming and outgoing states can be expressed as 2D vectors ψ_{in} and ψ_{out} like:

$$\psi_{out} = S\psi_{in} \tag{12.6}$$

where S is the 2×2 S-matrix and the vectors $(\psi_{in})^T$ and $(\psi_{out})^T$ are the row (A, D) and (B, C) vectors, respectively. The scattering to a finite wall is a special case but in general:

> the S-matrix must always be unitary for any value of the center of mass (CM) energy of the initial state.

In non-relativistic QM, it follows from the conservation of the probability currents. In our example, the current on the left of the barrier must be the same as the current on the right:

$$J = \frac{1}{2mi}\left(\psi^*\frac{\partial\psi}{\partial x} - \frac{\partial\psi^*}{\partial x}\psi\right) = \frac{k}{m}(|A|^2 - |B|^2) = \frac{k}{m}(|C|^2 - |D|^2) \tag{12.7}$$

which, in turn, gives:

$$S^\dagger S = \mathbb{I}. \tag{12.8}$$

Using either the scattering theory of QM (Exercise 12.15) or the same tools of QED, it is possible to show that the S-matrix violates this **unitarity bound** at $\sqrt{s} \simeq 300$ GeV. This discovery boosted the interest

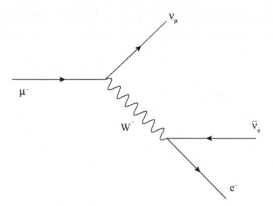

Fig. 12.2 Feynman diagram of a muon decay in the electroweak theory. Unlike V-A, the vertices couple fermions with spin-1 bosons in a way similar to QED and QCD. Note that the mass of the W (W^- in this diagram) is different from zero, while photons and gluons are massless.

toward models that are physically equivalent to the V-A theory below 300 GeV but mathematically consistent at higher energies. The winner of the race was a model now called the **electroweak theory**. This model incorporates both the V-A theory and QED.

In 1964, A. Salam and S. Weinberg put forward a baroque model that received very low attention until 1971. In the Salam–Weinberg model, the four-line vertex of the V-A was split into two vertices similar to the ones of QED (see Fig. 12.2). Weinberg speculated that the virtual particles exchanged between the vertices were **massive spin-1 bosons** similar to the photon but with a very large mass – of the order of $\mathcal{O}(100)$ GeV – and with an electric charge equal to ± 1: the **W$^+$** and **W$^-$ vector bosons**.[2]

The $W^\pm$ bosons are a kind of freak in QFT: they are massive and bring an electric charge. The electric charge of the mediator of weak interactions in the Salam–Weinberg model correlates weak and electromagnetic forces and sowed the seed of the electroweak theory.

[2]A **vector boson** is a boson with spin-1. A **scalar boson** is a boson with spin-zero. That's all. If you are intrigued by the name and want to get a deeper understanding of its origin, you should resort to Wigner's classification of particles, which is discussed, for instance, in Weinberg 2005a.

12.2 The electroweak theory

Salam and Weinberg were developing a far-fetched theory. In 1964, there was no QFT technique available to build a consistent theory based on massive mediators. We will discuss the solution of this problem (the **Higgs mechanism**) in Sec. 13.1. Despite this hindrance, the Salam–Weinberg model and a simplified model developed independently by S.L. Glashow in 1961 predicted the existence of a new particle: a neutral massive vector boson called the **Z^0 particle**. The discovery of the Z^0 in 1973 is discussed in Sec. 12.4 and marked the inception of the electroweak theory.

12.2.1 $SU(2) \times U(1)$ **gauge symmetry**

The electroweak theory is a gauge theory based on the $SU(2) \times U(1)$ group, that is, on rather simple groups, whose representations have already been discussed in Chaps 6 and 8. The number of generators is

$2^2 - 1 = 3$ – the generators of $SU(2)$ – plus the generator of $U(1)$. Therefore, we expect from the gauge principle four spin-1 massless mediators. They correspond to four fields called B^μ and W_j^μ (with $j = 1, 2, 3$) that are associated with the generators of $U(1)$ and $SU(2)$, respectively. To build the theory and write the corresponding Feynman diagrams, we need to compute how particles transform under the operators of this group. Since particles of different chiralities exhibit different weak interactions, the elementary fermions of the electroweak theory are 12 left-handed fermions ($e_L^-, \mu_L^-, \tau_L^-, \nu_{eL}, \nu_{\mu L}, \nu_{\tau L}, u_L, d_L, s_L, c_L, b_L, t_L$) and 12 right-handed fermions (e_R^-, μ_R^-, etc.).[3] The Dirac equation fixes the properties of the corresponding antiparticles, which are 12 right-handed ($e_R^+, \mu_R^+, \ldots, \bar{t}_R$) and 12 left-handed ($e_L^+, \mu_L^+, \ldots, \bar{t}_L$) antifermions. In 1964, physicists knew only a fraction of them but the theory can be straightforwardly extended to accommodate all elementary fermions.

The V-A theory shows that only LH fermions participate in weak interactions. This is embedded in the electroweak theory by dictating that a $SU(2)$ transformation only changes a e_L^- into a ν_{eL}. To prevent transformations from e_L^- into a $\nu_{\mu L}$, e_L^- and ν_{eL} transform as the fundamental representation of $SU(2)$, which is a 2D vector space. In this way, an electron cannot transform into a muon–neutrino or a muon and the lepton family number is conserved at the time of production ($t = 0$). In the original papers, Weinberg wrote that the $\mu^- \to \nu_e$ or $e^- \to \nu_\mu$ transitions cannot occur in an electroweak vertex because of conservation of the lepton family number. This claim has been updated, adding the time of production to comply with the findings of neutrino oscillations. The theory was originally designed assuming massless neutrinos, where oscillations do not occur. We present here the modern version of Weinberg's theory, where neutrinos are massive and oscillate. Sometimes, the updated theory is called the **minimally extended electroweak theory,** although most authors – including myself – retain the original name. We then write these particles in 2D vectors as:

$$\begin{pmatrix} \nu_{eL} \\ e_L^- \end{pmatrix} \quad ; \quad \begin{pmatrix} \nu_{\mu L} \\ \mu_L^- \end{pmatrix} \quad ; \quad \begin{pmatrix} \nu_{\tau L} \\ \tau_L^- \end{pmatrix} \quad ;$$

$$\begin{pmatrix} u_L \\ d_L \end{pmatrix} \quad ; \quad \begin{pmatrix} c_L \\ s_L \end{pmatrix} \quad ; \quad \begin{pmatrix} t_L \\ b_L \end{pmatrix} . \tag{12.9}$$

In the language of group representation theory, each vector generates a 2D vector space $\{2\}$. A $SU(2)$ transformation on the "electron-flavor" vector space is:

$$\begin{pmatrix} \nu_{eL} \\ e_L^- \end{pmatrix} \to e^{i\alpha_j T_j} \begin{pmatrix} \nu_{eL} \\ e_L^- \end{pmatrix} . \tag{12.10}$$

where T_j are the three generators of $SU(2)$ in the 2D representation, that is, the Pauli matrices.[4] Similarly, the V-A theory predicts that only RH antiparticles participate in weak interactions. In this case, we employ the conjugate transformations of eqn 8.35. The vector space is now $\{\bar{2}\}$ and:

$$\begin{pmatrix} e_R^+ \\ \bar{\nu}_{eR} \end{pmatrix} \to e^{-i\alpha_j T_j^*} \begin{pmatrix} e_R^+ \\ \bar{\nu}_{eR} \end{pmatrix} \tag{12.11}$$

where T_j^* is the complex-conjugate of T_j. In eqns 12.10 and 12.11, T_j and T_j^* are the generators of the $SU(2)$ group in the fundamental (2D) representation and its conjugate, respectively. The generators of the $SU(2)$ group in the fundamental representation are the three Pauli matrices (see eqn B.22) and those particles belong to **doublets** of $SU(2)$. Since $SU(2)$ is the same group used to described spin-1/2 particles or isospin, these particles have **weak isospin** $I^W = 1/2$. The particles in the upper part of the doublet (ν_{eL}, $\nu_{\mu L}$, and $\nu_{\tau L}$) are said to have the z component of weak isospin I_z^W equal to $+1/2$ and the particles in the lower part of the doublet (e_L^-, μ_L^-, and τ_L^-) have I_z^W equal to $-1/2$. The name has no special meaning in the electroweak theory and is related neither to the spin nor to the isospin. The weak isospin assignment of antiparticles is reversed and the upper antiparticle in the "electron-flavor" doublet is e_R^+ instead of $\bar{\nu}_R$, as shown in eqn 12.11. This rule holds for all antifermions (e^+, μ^+, ..., $\bar{t}$).

RH fermions and LH antifermions do not participate in weak interactions. We can account for it dictating that these particles do not transform under the $SU(2)$ group, except for a global phase. They can only be members of the 1D vector space $\{1\}$ or, equivalently, they are **singlets** of $SU(2)$:

$$U f_R = e^{ia} f_R \qquad (12.12)$$

for all RH fermions and LH antifermions if $U \in SU(2) \times U(1)$. Their weak isospin I^W is zero and, therefore, $I_z^W = 0$, too. These prescriptions give us the Feynman vertices and diagrams for the interactions that are mediated by the W_j^μ fields.

The real game starts when $U(1)$ comes into play. In the electroweak theory, the authors were forced to embed QED into the model because the charged currents change both the *flavor* and the *electric charge* of the particles. Then, the W_j^μ and B^μ fields must gang up to produce the correct QED interactions. If $U(1)$ were the QED group, it should not care about chirality because the strength of electromagnetic interactions does not depend on γ^5. This is not the case for the B^μ field. Here, all particles transform as the fundamental representation of $U(1)$, that is, they are singlets. We assign to each particle an **hypercharge**:

$$Y = 2(Q - I_z^W) \qquad (12.13)$$

where Q is the electric charge of the fermion. All fermions f transform as **singlets** of $U(1)$:

$$f_L \to e^{iqa} f_L, f_R \to e^{iqa} f_R. \qquad (12.14)$$

and a is an arbitrary real number. In this case, however, the coefficient q does not corresponds to the electric charge – as in QED – and is replaced by the hypercharge of the corresponding fermion:

$$f_L \to e^{iYa} f_L, f_R \to e^{iYa} f_R. \qquad (12.15)$$

This is why we customarily label the $U(1)$ group of the electroweak theory $U(1)_Y$. It is[5] the same group as $U(1)_{QED}$ from the mathematical

[5]We did the same in QCD when we defined $SU(2)_f$ and $SU(3)_f$ to describe the flavor symmetry, and $SU(3)_c$ for the color symmetry.

point of view but transforms particles according to their weak hypercharge. Similarly, the $SU(2)$ group of the electroweak theory is called $SU(2)_{IW}$ because the transformations depend on the weak isospin of the elementary fermion. The electroweak theory is therefore a gauge theory based on the **electroweak gauge group** $SU(2)_{IW} \times U(1)_Y$ or, in brief,

$$SU(2)_W \times U(1)_Y. \tag{12.16}$$

The reader may be astonished by eqn 12.13. Beginners often believe that the $U(1)$ group in the electroweak theory should be the group needed to include QED. Not at all. B^μ is an auxiliary field, which does not correspond to the QED vector field A^μ.

12.2.2 Field mixing

The next step toward the electroweak theory is to consider linear superpositions of the W_j^μ and B^μ fields based on an arbitrary parameter, the Weinberg angle. The group $SU(2)_W$ and $U(1)_Y$ have their own (unphysical) fine structure constants $\alpha_{SU(2)}$ and $\alpha_{U(1)}$ related with the (unphysical) charges g and g' by:

$$g = \sqrt{4\pi\alpha_{SU(2)}} \; ; \quad g' = \sqrt{4\pi\alpha_{U(1)}} \; . \tag{12.17}$$

The **Weinberg angle** or **weak mixing angle** is defined as:

$$\theta_W \equiv \arctan \frac{g'}{g} \; . \tag{12.18}$$

Its value can be determined experimentally and corresponds to:

$$\sin^2\theta_W = 0.23121 \pm 0.00004 \tag{12.19}$$

so that $\theta_W \simeq 29°$. The superposition of W_3^μ and B^μ through the Weinberg angle defines the **physical fields** that correspond to the actual particles. The superposition (omitting the μ superscipt) is:

$$\begin{pmatrix} Z^0 \\ A \end{pmatrix} = \begin{pmatrix} \cos\theta_W & -\sin\theta_W \\ \sin\theta_W & \cos\theta_W \end{pmatrix} \begin{pmatrix} W_3 \\ B \end{pmatrix} \tag{12.20}$$

For W_1 and W_2, the physical fields are W^+ and W^- defined as:

$$W^+ = \frac{1}{\sqrt{2}}(W_1 - iW_2) \tag{12.21}$$

$$W^- = \frac{1}{\sqrt{2}}(W_1 + iW_2). \tag{12.22}$$

If we mix up the unphysical fields according to eqn 12.20 and require the physical field A^μ to be the electromagnetic field of QED, we end up with:

$$g\sin\theta_W = g'\cos\theta_W = e = \sqrt{4\pi\alpha} \tag{12.23}$$

where α is the QED fine structure constant and e is the electron charge in NU. Equation 12.23 was derived by S. Weinberg and is called **electroweak unification** because is a relation between the fine structure constants g entering charged-current weak processes and α_{QED} or, equivalently, the electric charge of the electron.

> The field mixing accomplished by the Weinberg angle and the definition of hypercharge protects the electroweak theory from gross inconsistencies with QED: the vertices (see the next section) are independent of the chirality of the particles and their strength complies with QED.

The particles of the electroweak theory have then the same electric charge as in QED.

12.3 Feynman diagrams of the electroweak theory

Even if a rigorous derivation of the Feynman diagrams of the electroweak theory is outside the scope of this book, their definition is quite straightforward thanks to the experience gained with QED and QCD. Furthermore, an accurate derivation of the electroweak Lagrangian based on (classical) relativistic fields will be provided in Chap. 13.

12.3.1 W^+ and W^- fields

All weak interactions observed up to 1973 can be interpreted by replacing the contact interaction with an exchange of a virtual $W^\pm$ in the four line-vertex of the V-A theory. These vertices are shown in Fig. 12.3. The leptonic, semileptonic, and hadronic weak interaction processes, together with all radioactive beta decays, originate from them. The corresponding diagrams depend on the chirality of the particle because the vertex is proportional to P_L, the LH chirality projector of eqn 10.28. The operator selects the LH chiral component of the wavefunction (or the field in QFT). If the particle is RH, the projector zeroes the wavefunction, and the amplitude of the scattering vanishes. The projector also selects the RH component of antiparticles and zeroes the LH component. It embeds the finding of Lee and Yang into the electroweak theory, which is, then, a **chiral gauge theory**.

To be consistent with the V-A theory, there must be a link between the Fermi constant and the W propagator,[6] which has a form similar to eqn 6.79;

$$\sim -\frac{1}{q^2 - M_W^2} \xrightarrow{q^2 \ll M_W^2} \frac{1}{M_W^2} \tag{12.25}$$

where M_W is the mass of the W^- and of its antiparticle, the W^+. If the mass of the $W^\pm$ is large, the amplitude is suppressed by M_W^{-2} and the scattering looks like a contact interaction. The relationship between the Fermi constant and g is drawn by Fig. 12.3, eqs 12.25, and 10.72 (see Exercise 12.14), and reads:

$$G_F = \frac{\sqrt{2}g^2}{8M_W^2} . \tag{12.26}$$

[6] As noted in Sec. 7.6.3, the propagator also includes an imaginary part and depends on the gauge fixing. In the electroweak theory, the exact form of eqn 12.25 in **unitary gauge** is:

$$\frac{-i[g_{\alpha\beta} - \frac{q_\alpha q_\beta}{M_W^2}]}{q^2 - M_W^2 + i\epsilon} \tag{12.24}$$

Unitary gauge was introduced by S. Weinberg for this class of theories and is discussed in Sec. 13.1.3.

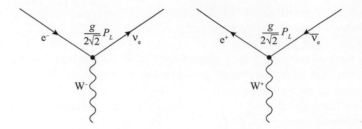

Fig. 12.3 Vertices between W^- and fermions (left) and between W^+ and antifermions (right). P_L projects a particle to its LH chirality component and zeroes the RH component. It also projects an antiparticle to its RH chirality component and zeroes LH antiparticles (see Sec. 10.7).

Since θ_W and G_F are known, we can predict the mass of the $W^\pm$ to be $\simeq 80$ GeV. This prediction makes sense because it is close to the scale where the V-A theory violates the unitarity bound. Besides, the electroweak theory brilliantly explains why weak interactions are weak:

> weak interactions are much fainter than electromagnetic interactions because the photon counterparts (the $W^\pm$) are very heavy and the corresponding Feynman diagrams are suppressed by $\sim M_W^{-2}$.

Contrary to expectations, the size of the electroweak charges g and g' is similar to e. The value of the $SU(2)$ fine structure constant results from eqs 12.19 and 12.23, and is:

$$\alpha_{SU(2)} = \frac{g^2}{4\pi} \simeq 0.034 > \alpha_{QED}. \tag{12.27}$$

If the $W^\pm$ particles were massless or, at least, very light, weak interactions would be even stronger than QED!

12.3.2   Neutral fields

The weak hypercharge assignment of eqn 12.13 is bizarre because it was tuned to clone the Feynman diagrams of QED. The vertices of the A^μ field of the electroweak theory are depicted in Fig. 12.4 (left). We can then identify A^μ with the **electromagnetic four-potential**. The corresponding particle is called **photon** and is drawn using the same symbol as the QED photon. Unlike the $W^\pm$, no chirality projector appear in the vertex because A^μ does not distinguish between LH and RH elementary fermions.

The electroweak theory predicts a new vector boson, which was not known at the time of the Glashow–Salam–Weinberg model. The vertex of this particle (Fig. 12.4 right) mixes up LH and RH fields in a way that depends on the weak isospin and hypercharge of the fermion. This is the price to pay for the fine-tuning of $\sin\theta_W$ and hypercharge to clone the QED diagrams. The coupling of the new neutral boson can be expressed as:

$$g_Z(\epsilon_L P_L + \epsilon_R P_R) \tag{12.28}$$

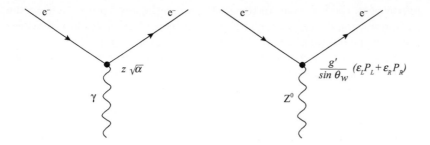

Fig. 12.4 Vertices between a photon and a fermion (left) and between a Z^0 and a fermion (right). We consider an electron for purpose of illustration. z is the fermion electric charge in units of e (-1 for the electron). P_L and P_R are the chirality projectors of eqs 10.28 and 10.29.

where

$$g_Z \equiv \frac{e}{\sin\theta_W \cos\theta_W} = \frac{g'}{\sin\theta_W} \qquad (12.29)$$

$$\epsilon_L = I_z^W - Q\sin^2\theta_W \;\; ; \quad \epsilon_R = -Q\sin^2\theta_W \qquad (12.30)$$

Hence, the formula corresponding to the vertex on the right of Fig. 12.4 is

$$g_Z(\epsilon_L \bar{u}_L \gamma^\mu u_L + \epsilon_R \bar{u}_R \gamma^\mu u_R). \qquad (12.31)$$

where, for instance, $u_L = P_L u$ is the LH bispinor of the initial-state fermion. Equation 12.31 is equivalent to,

$$\frac{g_Z}{2}(v\gamma^\mu - a\gamma^\mu\gamma^5) \; . \qquad (12.32)$$

γ^5 is the chirality operator and v, a are defined as:

$$v = I_z^W - 2Q\sin^2\theta_W$$
$$a = I_z^W. \qquad (12.33)$$

In eqn 12.33, I_z^W is the weak isospin of the fermion and Q the electric charge. P_L and P_R are the chirality projectors of eqns 10.28 and 10.29.

The Z^0 is a very unusual mediator: it employs chirality in a different way than charged currents and its coupling depends on the flavor of the fermions in the vertex through Q and I_z^W. As a consequence, even RH fermions can interact with this new boson. Note, however, that a particle must either have the "correct" chirality for weak interactions (LH for matter, RH for antimatter) or some electric charge to interact with the Z^0. A RH chirality neutrino or a LH chirality antineutrino cannot interact with any particle even in the electroweak theory. This particle has no QED interactions because its electric charge is zero and no QCD interactions since all color charges are zero (as any lepton). It cannot interact with the $W^\pm$ because has the wrong chirality, and does not couple with the Z^0 because $Q = 0$ and $I_z^W = 0$ and, therefore, both v and a in eqn 12.33 are zero. As a consequence,

> RH neutrinos and LH antineutrinos are **sterile** particles and – if they exist – only experience gravitational forces.

As noted in Sec. 11.15, we could give up the description of neutrinos as Dirac fermions introducing Majorana neutrinos to wipe these "useless" particles off the universe. In the 1960s, no sterile neutrino existed in the eletroweak theory because neutrinos were considered massless particles but now the existence of sterile neutrinos is an intriguing issue.

We have not yet discovered a mechanism by which sterile particles could be produced. Heavy RH neutrinos could be created at very high energy by the decays of unknown particles, for example, just after the Big Bang. If sterile neutrinos are produced, they can interact only by gravitational interactions or mix with standard neutrinos (see Sec.13.6). In any case, their existence remains speculative and they could even be a hoax if neutrinos are Majorana particles.

The Z^0 generates many new scattering processes. These processes are called **neutral current** (NC) reactions because they do not change the electric charge of the fermion in the vertex. One classical example is the **elastic scattering of muon neutrinos**:

$$\nu_\mu + e^- \to \nu_\mu + e^- \tag{12.34}$$

that proceeds only by a Z^0 exchange in the t-channel (Fig. 12.5). Most of the skepticism about the electroweak theory originated from the fact that, at that time, no neutral current event had ever been observed.

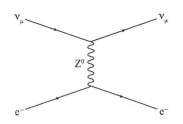

Fig. 12.5 Lowest-order Feynman diagram for the $\nu_\mu + e^- \to \nu_\mu + e^-$ elastic scattering. The scattering occurs by a Z^0 exchange in the t-channel.

12.4 The discovery of weak neutral currents

The electroweak theory joined the paradigms of particle physics in 1973. Since 1971, a CERN experimental group has been studying the elementary components of the protons using a pure weak probe (the ν_μ). The **neutrino beam** (Charitonidis *et al.*, 2021) was produced by a synchrotron accelerating 25 GeV protons that were sent toward a beryllium target (see Fig. 12.6). The scattering of high-energy protons produces a wealth of hadrons and, in particular, charged pions. Positive pions were partially focused toward the neutrino detector by a magnetic lens invented by S. Van der Meer and still called a **magnetic horn**. The horn was simply a conductor with cylindrical symmetry along the axis of the beam. As shown in Fig. 12.7, positive pions are focused by the magnetic field if they cross the conductor, while negative pions are defocused. Straight particles produced on the axis are undeflected. To change the sign of focused particles, the direction of the current flowing in the conductor must be reversed. Seventy meters after the horn, all charged particles were stopped by an iron shielding. In the meanwhile, a good fraction of pions had decayed into muons and neutrinos. Muons and undecayed pions are stopped in the shielding (**hadron dump**) by ionization losses or strong interactions, while nearly all neutrinos cross the shielding. More than 99% of the neutrinos are ν_μ that arise from the $\pi^+ \to \mu^+ \nu_\mu$ decay (branching ratio (BR)~100%). The ν_e mostly comes from kaon ($K^+ \to e^+ \pi^0 \nu_e$) of muon decays in flight ($\pi^+ \to \mu^+ \nu_\mu \to e^+ \nu_e \bar{\nu}_\mu \nu_\mu$) and, except for special cases (Longhin *et al.*, 2015), can be neglected.

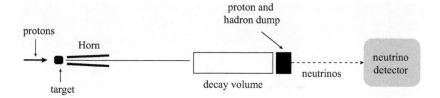

Fig. 12.6 Simplified layout of an **accelerator neutrino beam** like the beam serving the Gargamelle detector. The protons (black arrow) are extracted from the CERN Proton Synchrotron (PS) and impinge on a beryllium target ("target"). The pions are focused by a magnetic horn and the neutrinos are produced by the $\pi^+ \to \mu^+\nu_\mu$ decay in the decay volume. They cross the Gargamelle bubble chamber ("neutrino detector") after crossing the iron shielding ("proton and hadron dump").

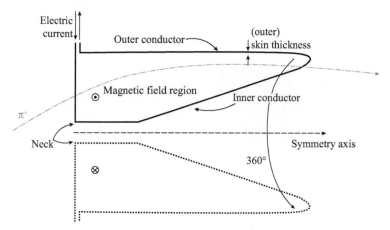

Fig. 12.7 Layout of a magnetic horn. The direction of the magnetic field inside the hollow volume of the conductor is shown in the figure for positive particle focusing. The axis of the horn corresponds to the axis of the beam (horizontal dashed line). The conductor is connected to a DC current generator and the current flows in the conductor only during the extraction of the proton bunches (10 μs up to a few ms). Image credit: M. Pari (Pari, 2021).

The neutrino detector – called Gargamelle, see Fig. 12.8 – consisted of a 15 tonne **bubble chamber**. The chamber was made up of a hollow steel cylinder filled with CF_3Br (freon). This type of detectors was invented by D.A. Glaser in 1952 and consisted of a volume filled with freon just below its boiling point. The working principle is nearly identical to the cloud chamber used by Anderson (see Sec. 3.3). When particles enter the chamber, a piston suddenly decreases its pressure. The freon reaches a superheated, metastable phase and the liquid vaporizes, forming tiny bubbles. The formation of the bubbles is accelerated by the presence of ions as and electrons, in cloud chambers. Therefore, the first visible bubbles are clustered along the trajectory of the charged particles that ionize freon. The momentum of the particles was measured using a magnetic field, which exceeded 2 T in Gargamelle. The photographs of the chamber thus provided information about the final-state particles produced in $\nu_\mu + p \to X$ scatterings and their momenta. The experimenters used a bubble chamber instead of a cloud chamber because of the higher density of liquids than gases. A high-density detector stops high-energy charged particles earlier and increases by three orders of magnitude the number of scattering centers. This is quite an asset given the tiny neutrino cross-section.

The reactions predicted by the V-A theory in Gargamelle are:

Fig. 12.8 The Gargamelle bubble chamber during the insertion inside the magnet yoke. The magnetic field is used to measure the momentum by particle bending. The photographs of the events are taken by the portholes visible on the walls of the cylinder. Reproduced under CC-BY-4.0 license (CERN archive).

- Charged current inelastic scatterings (see Fig. 12.9 left): $\nu_\mu + n \to \mu^- + p$. This well-known reaction is interpreted in the electroweak theory as the exchange of a W^+ that transforms a ν_μ into a μ^- and a d quark into a u quark. Therefore, a neutron is transformed into a proton.

- Charged current deep inelastic scatterings (see Fig. 12.9 right): if the four-momentum transferred by the W^+ is large, the proton or the neutron breaks apart. The quarks leave the broken nucleon and commence the hadronization process. The final state is thus $\nu_\mu + n$ (or p) $\to \mu^- + X$, where X is a generic set of hadrons.

If the electroweak theory is correct, the events observed by Gargamelle are enriched by a set of **neutral current** events:

- Neutral current inelastic scatterings (see Fig. 12.10 left): $\nu_\mu + n \to \nu_\mu + n$ and $\nu_\mu + p \to \nu_\mu + p$. Here, the Z^0 exchanged in the t-channel changes the trajectory of the ν_μ and kicks a neutron or a proton.

- Neutral current deep inelastic scatterings (see Fig. 12.10 right): if the four-momentum transferred by the Z^0 is large, the proton or the neutron is broken and hadronization takes place. In this case, the final state is $\nu_\mu + n$ (or p) $\to \nu_\mu + X$, where X is a generic set of hadrons.

Relativistic kinematics offer two ways to separate charged current (CC) from neutral current (NC) events. Gargamelle is not able to track neutrinos because these are neutral particles and do not deposit energy through

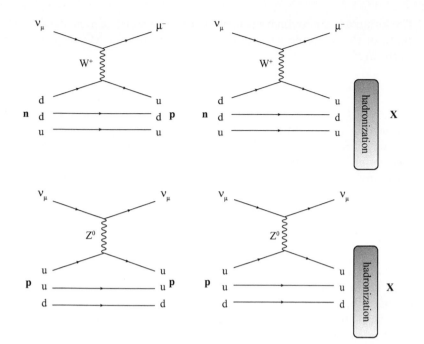

Fig. 12.9 Lowest order Feynman diagram of a CC scattering between a ν_μ and a neutron. In the left diagram, a d quark is transformed into a u quark by the W^+ and the initial-state neutron becomes a final state proton. The corresponding deep inelastic scattering is shown in the right diagram. In this case, the neutron breaks apart and hadronization takes place producing a set of hadrons X.

Fig. 12.10 Lowest-order Feynman diagram of an NC scattering between a ν_μ and a proton (left diagram). The corresponding deep inelastic scattering is shown in the right diagram. In this case, the proton is broken and hadronization takes place producing a set of hadrons X.

the Bethe formula (ionization losses). Besides, just a tiny fraction of the neutrinos crossing the detector interacts inside it due to the small cross-section. All CC events have muons in the final state that can be easily identified because they are minimum-ionizing particles (mip) for most of the range. In an elastic scattering, the W^+ is the only source of negative muons and the lack of the muon is a signature of NC events. Unfortunately, most events are deep inelastic scatterings and the jets produced by the hard scattering create pions or other hadrons that promptly decay into muons. These muons have a smaller energy than the ones produced by the incoming neutrino and, in general, are located close to the other hadrons of the jet. A particle selection based on energy and isolation with respect to other hadrons can mitigate this background, still keeping a good efficiency. However, the most effective technique to separate NC from CC deep inelastic events is the use of the **transverse momentum imbalance**. An incoming neutrino has an energy that is unknown to the experimentalist but its direction is approximately the direction of the axis of the horn (z-axis). Hence, $p^\mu_{\nu_\mu} \simeq (p, 0, 0, p)$ if we neglect the neutrino mass. The neutrino momentum in the plane perpendicular to z (**transverse momentum**) is zero in the LAB frame – and indeed in any frame, since the transverse momentum is Lorentz invariant. As a consequence, the sum of all transverse momenta $\mathbf{p}^j_T$ of the final particles ($j = 1, \ldots, N_f$) must be zero due to momentum conservation. In a CC event, where we observe all charged particles:

$$\sum_{j=1}^{N_f} \mathbf{p}^j_T \simeq 0 \tag{12.35}$$

The formula is approximate because neutral particles (e.g. neutrons) produced in the hadronic jet may escape detection. In a NC deep inelastic scattering:

$$\sum_{j=1}^{N_f} \mathbf{p}_T^j \neq 0 \qquad (12.36)$$

because the transverse momentum of the outgoing ν_μ is never observed. Between 1971 and 1973, Gargamelle recorded a significant excess of events unbalanced in transverse momentum[7] and fulfilling eqn 12.36. They provided the first (indirect) evidence of the existence of the Z^0.

[7]Even if the detail of the Gargamelle analysis is outside the scope of this book (Heidth and Pullia, 2013), it is worth stressing that the data were compelling and the results of Gargamelle were soon confirmed by other experiments. Glashow, Salam, and Weinberg were awarded the Nobel Prize in Physics for the electroweak theory in 1979.

12.5 The discovery of the Z^0

The Gargamelle team established the existence of the Z^0 in an indirect way, that is, by its contribution to the neutrino scattering. They studied the particle off-shell and made only a rough estimate of the mass.

In the late 1970s, C. Rubbia proposed the first collider able to attain the on-shell production of the electroweak vector bosons. He noted that the CERN Super-Proton-Synchrotron (SPS) could have been transformed in a proton–antiproton collider (S$p\bar{p}$S) with a $\sqrt{s} \simeq 540$ GeV. This value is much larger than $M_Z/2$ but the protons are composite particles and a quark transports only $\simeq 1/3$ of the overall momentum, as discussed in Sec. 7.7. Besides, the expected luminosity of the S$p\bar{p}$S was poor compared with modern standards and a higher energy benefits from a higher $p + \bar{p} \rightarrow Z^0 + X$ cross-section. Transforming the SPS into a collider has been far from trivial. The main issue is the production, steering, and cooling of the antiprotons before the injection into the SPS ring. Antiprotons are produced by fixed target experiments from high-energy protons together with many other hadrons; they must be selected with a mass spectrometer and have a large momentum spread. As a consequence, are not well-suited to be injected into the accelerator ring to reach a reasonable luminosity. As discussed in Sec. 4.5, colliders need **cooling** techniques to reduce the transverse momentum of the particles before injection. For the SPS, this issue was solved by Van der Meer employing **stochastic cooling**. While the antiprotons orbit inside the beampipe, they induce a current in the pick-up coils located along the pipe according to Faraday's law. This diagnostic tool can be used to get information on the average momentum and shape of the (anti)proton bunches. The pick-up coils drive a device that kicks the bunches by electric fields to reduce the transverse momentum. The technique works because the feedback signal that drives the kicker travels along the diameter of the ring and, therefore, arrives at the kicker before the arrival of the orbiting bunch. It then generates a **negative feedback** that penalizes particles with high transverse momentum. The kicks are applied hundreds of times for a few minutes until the machine achieves the desired emittance. Stochastic cooling was a major leap in accelerator science, improved the luminosity of the collider by orders of magnitude, and allowed for the discovery of the Z^0 particle.

A Z^0 is produced in a head-on collision if the invariant mass of the colliding quarks is close to 91 GeV. Since the protons are composite particles, we do not know the momentum of the quarks involved in the process. Furthermore, after the $p\bar{p}$ collision, there are many other soft particles produced by the hadronization of the remaining quarks, the underlying events of Sec. 4.6.4. For instance, if the Z^0 decays in a $\mu^+\mu^-$ pair, the observer in LAB sees the muons boosted with respect to the rest frame (RF) of the Z^0 and they are no more seen as two back-to-back particles (see Fig. 2.5). Since the size of the boost is unknown, how can we be sure that the muons originate from a Z^0? Relativistic kinematics helps. The scattering belongs to the class:

$$a + b \to M \to c + d \qquad (12.37)$$

where a and b are the incoming proton and antiproton. M is the (rest) mass of the Z^0 and c, d are the muon–antimuon pair. The invariant mass of the system is:

$$s = m_c^2 + m_d^2 + 2E_c E_d - 2\mathbf{p}_c \cdot \mathbf{p}_d . \qquad (12.38)$$

In this case, $m_c = m_d$ and $E_c \gg m_c \simeq 106$ MeV if the muon originates from a Z^0. The same consideration holds for E_d so that $E_c \simeq |\mathbf{p}_c|$ and $E_d \simeq |\mathbf{p}_d|$. This condition is very common in high-energy colliders where nearly all particles are ultra-relativistic. Then,

$$s_{fin} \simeq 2E_c E_d (1 - \cos\theta) \qquad (12.39)$$

where θ is the angle between the muons. Even if we do not know the RF of the Z^0, s_{fin} is equal to M^2 in any frame because is Lorentz invariant. Therefore, if c and d are the decay product of a Z^0,

$$2E_c E_d (1 - \cos\theta) = M^2 . \qquad (12.40)$$

The Z^0 was discovered by two detectors similar to the ones of Sec. 4.6: the UA1 and UA2 experiments at the CERN S$p\bar{p}$S. In these detectors, θ is measured by the trackers and the bending in the magnetic field provides $|\mathbf{p}_c| \simeq E_c$ and $|\mathbf{p}_d| \simeq E_d$. Using this information, the experimentalists can compute s and check if a peak appears at an energy of about 91 GeV. In a few years, the UA1 and UA2 Collaborations observed such a peak with remarkable statistical sensitivity. The peak testified for the on-shell production of the Z^0 and conclusively established its existence.[8]

The study of the Z^0 is much cleaner in a e^+e^- collider because of the lack of underlying events. Fig. 12.11 shows an elegant $Z^0 \to e^+e^-$ decay in the plane transverse to the beam direction. Here, the RF of the Z^0 is LAB and the two electrons are produced back-to-back and tracked from the interaction vertex up to the front-face of the electromagnetic calorimeters.

[8]This time, the Nobel Prize was deservedly awarded to the physicists that gave the most important contributions to the discovery, S. van Der Meer and C. Rubbia. Still, it is fair to say that the discovery of the Z^0 arose from a team of hundreds of physicists and engineers working in a collective undertaking. You will see in Chap. 13 that even the electroweak theory was a collective undertaking of many theoreticians that were inspired by the ideas of Glashow, Salam, and Weinberg. Collective undertakings, pioneered by the Manhattan Project, are now widespread in particle physics, genomics, neuroscience, astronomy, biology, and many other disciplines, given the complexity of contemporary science.

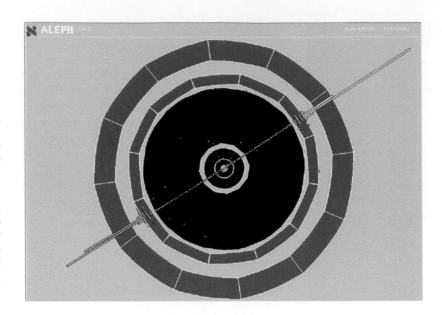

Fig. 12.11 A $e^+e^- \to Z^0 \to e^+e^-$ event seen in a e^+e^- collider (the ALEPH experiment at the LEP collider). The view is perpendicular to the beam direction (transverse view). The two electrons are produced back-to-back, cross the beampipe (not shown), are traced by a time projection chamber (in black) immersed in a solenoidal magnetic field, and impinge onto the electromagnetic calorimeters (red inner areas). The yellow bars are proportional to the deposited energy. Reproduced under CC-BY-SA license from CERN for the benefit of the ALEPH Collaboration (Barate *et al.*, 1999).

12.6 The discovery of the W bosons

A precise measurement of the mass of the W boson is of great interest because – as we will see in Sec. 13.2 – the Higgs mechanism applied to the electroweak theory predicts the ratio between the $W^\pm$ and Z^0 mass:

$$\frac{M_W}{M_Z} = \cos\theta_W \tag{12.41}$$

where θ_W is the Weinberg angle. Equation 12.41 is a very straightforward prediction of the Higgs mechanism and an effective way to test deviations that signal the existence of new particles. The first direct measurement of M_W was performed at the time of its discovery in 1983. Unfortunately, making precision measurements with the Ws is more difficult than Z^0. In proton–antiproton colliders, a quark–antiquark pair can produce Ws and Z^0s. All other quarks belonging to the same (anti)proton contribute to the underlying events and mess up the reconstruction of the kinematics. This is why in the 1980s, physicists mostly relied on the leptonic decay of the vector bosons.[9] Unlike the Z^0, the leptonic decays of the Ws are nearly invisible because one of the two outgoing particles is a neutrino: $W^+ \to e^+\nu_e$, $W^+ \to \mu^+\nu_\mu$, and $W^+ \to \tau^+\nu_\tau$. Even for such a difficult topology, relativistic kinematics provides quite a help.

Fig. 12.12 shows the lowest order Feynman diagram for $p\bar{p} \to W^+ + X \to \mu^+ + \nu_\mu + X$. X are the hadrons produced by the loosely interacting quarks and constitute the underlying events. In the RF of the W^+ (Fig. 12.13) the muon and the neutrino are produced back to back and the angle between the muon and the beam axis z is θ^*. In this frame, $\mathbf{p}_T = |\mathbf{p}|\sin\theta^*$. Since $\mathbf{p}_\mu \gg m_\mu$, $|\mathbf{p}_\mu| \simeq E_\mu$ and, assuming – without loss

[9]Given the large mass, a $W^\pm$ or a Z^0 has a large number of decay modes, whose BR is predicted by the electroweak theory. These modes are summarized in Table 12.1.

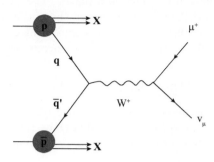

Fig. 12.12 Production of a W^+ in a $p\bar{p}$ collider. In the figure, which shows just a particular outcome, the W^+ decays in a $\mu^+ \nu_\mu$ pair. The quarks that do not participate in the hard scattering produce additional hadrons X. Note that the sum of the charges of the interacting quarks must be ± 1. Hence, a W^+ can be produced by a $u\bar{d}$ scattering and a W^- by a $d\bar{u}$ scattering. These particles can also be produced by the scattering of heavier quarks belonging to the "sea" (see Sec. 7.7).

of generality – that the transverse momentum is along the x-axis,

$$\mathbf{p}_T = p_x \simeq E_\mu \sin\theta^* = \frac{M_W}{2}\sin\theta^* \qquad (12.42)$$

so that

$$\sin\theta^* = \frac{2p_x}{M_W} . \qquad (12.43)$$

Note that $\mathbf{p}_T = p_x$ is Lorentz invariant if the W^+ is produced along z. This is an excellent approximation because the momentum of the incoming quarks is nearly parallel to z due to the large (anti)proton boost in that direction. Hence,

$$p_x = p_x^{LAB} = \frac{M_W}{2}\sin\theta^* . \qquad (12.44)$$

We can now compute the statistical distribution of the events when a W is produced:

$$\frac{dn}{dp_x} = \frac{dn}{d\theta^*}\frac{d\theta^*}{dp_x} = \frac{dn}{d\theta^*}\frac{2}{M_W\cos\theta^*} =$$

$$= \left[\left(\frac{M_W}{2}\right)^2 - (p_x^{LAB})^2\right]^{-1/2}\frac{dn}{d\theta^*} . \qquad (12.45)$$

If the accelerator is really producing Ws, the distribution has a singularity for $p_x^{LAB} = M_W/2$ so that:

$$\frac{dn}{dp_x} \to \infty \quad \text{for} \quad p_x^{LAB} \to \frac{M_W}{2} \qquad (12.46)$$

In a real experiment, the singularity (a Dirac delta) looks like a sharp peak because of the detector finite precision and the approximation $\mathbf{p}_W \simeq |p_W|\mathbf{k}$, in other words, the W is not exactly produced along z since the interacting quarks have a small transverse momentum. The observation of the peak provides evidence of the existence of the W boson and a measurement of its mass, since the maximum of the peak is located at $M_W/2$. This technique does not employ the neutrino momentum, replacing this quantity with a statistical distribution. The maximum of the

RF

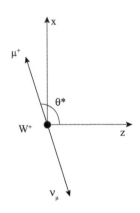

distribution is called the **Jacobian peak** because the singularity arises from the determinant of the Jacobian matrix when we change variables from p_x to θ^* in eqn 12.45. At the time of the discovery (Arnison *et al.*, 1983; Banner *et al.*, 1983), the best measurement of M_W from the Jacobian peak was $M_W = 81 \pm 5$ GeV. The current value from LEP and LHC is (Zyla *et al.*, 2020):

$$M_W = 80.379 \pm 0.012 \text{ GeV} \qquad (12.47)$$

This number should be compared with the outstanding precision obtained for the M_Z at the LEP e^+e^- collider by measuring the shape of the Breit–Wigner distribution with millions of events:

$$M_Z = 91.1876 \pm 0.0021 \text{ GeV.} \qquad (12.48)$$

12.7   Precision electroweak physics and N_ν

In 1971, M. Veltman and G. t'Hooft demonstrated that the electroweak theory is mathematically consistent at any order in perturbation theory and we can compute quantum corrections using the Feynman diagrams at any energy currently available at accelerators. This result simplifies enormously the search of new particles because their contributions may appear as modifications to the quantum corrections of the standard theory even if the experiment cannot produce them on-shell. The study of these **radiative corrections** was the greatest achievement of e^+e^- colliders in the 1990s and has been substantially improved after the discovery of the Higgs bosons at the Large Hadron Collider (LHC).

The electroweak theory fixes all couplings between elementary fermions and mediators. They can be checked by comparing cross-section predictions with experimental data. The physicists working at LEP obtained one of their greatest results by scanning the Breit–Wigner of the Z^0 and demonstrated that:

W^- (or W^+)		Z^0	
Decay	BR (%)	Decay	BR (%)
$W^- \to e^- \nu_e$	10.71 ± 0.16	$Z^0 \to e^- e^+$	3.363 ± 0.004
$W^- \to \mu^- \nu_\mu$	10.63 ± 0.15	$Z^0 \to \mu^- \mu^+$	3.366 ± 0.007
$W^- \to \tau^- \nu_\tau$	11.38 ± 0.21	$Z^0 \to \tau^- \tau^+$	3.370 ± 0.008
$W^- \to$ hadrons	67.41 ± 0.27	$Z^0 \to$ hadrons	69.91 ± 0.06
		$Z^0 \to$ invisible	20.00 ± 0.06

Table 12.1 Decay modes and BR for the Ws and the Z^0. $Z^0 \to$ invisible is the sum of the branching ratios of Z^0 decays into a neutrino pair ($\nu_e \bar{\nu}_e$, $\nu_\mu \bar{\nu}_\mu$, and $\nu_\tau \bar{\nu}_\tau$). $W^\pm \to$ invisible is forbidden by electric charge conservation.

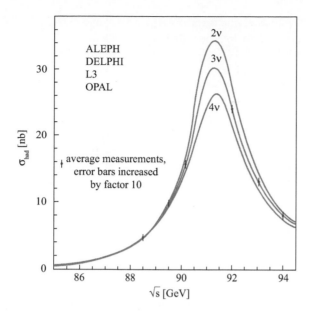

Fig. 12.14 Scanning of the Z^0 line-shape by the four LEP experiments: ALEPH, DELPHI, OPAL, and L3. The three lines indicate the expected line-shape for $N_\nu = 2, 3$, and 4. The errors have been inflated by a factor 10 to ease the reading. Reproduced with permission from Schael *et al.* 2006. Copyright 2006 Elsevier.

> there are only three neutrino flavors, if all possible neutrinos that couple with the Z^0 have a mass smaller than $M_Z/2 \simeq 45$ GeV.

This statement holds only in the framework of the electroweak theory but is extremely powerful and has profound implications in cosmology and particle physics. If we had an additional neutrino ν_x that couples with the Z^0 with a mass comparable to m_1, m_2, or m_3, it would contribute to the $Z \to$ invisible branching ratio (see Table 12.1). As a consequence, we would have an increase of the total decay width of the Z^0 and a deformation of the Breit–Wigner. The LEP data shows very clearly that the number of **light neutrino species** N_ν is compatible with 3 (see Fig. 12.14).[10]

$$N_\nu = 2.9840 \pm 0.0082 \qquad (12.49)$$

Since neutrinos are extremely light, it is very unlikely that the electroweak theory has more than three families: the lightest particle of the fourth family should have a mass $> M_Z/2 \simeq 45$ GeV, while the heaviest known neutrino mass eigenstate is < 1 eV! This also does not support the existence of a fourth type of "electron" or a quark heavier than the top because the Veltman–t'Hooft theorem holds only if each family is complete.[11] We can evade this constraint if the electroweak theory is wrong, the new particles have a very high mass or do not couple with the Z^0. For instance, a Z^0 cannot decay into sterile neutrinos and, therefore, sterile neutrinos do not contribute to the invisible width but might contribute to the total mass of the universe. Similarly, we cannot exclude a fourth generation where *all* particles are much heavier than $M_Z/2$ and have not been discovered by the LHC or other accelerators to date.

[10] N_ν is not necessarily an integer number because, if the neutrino mass is $\neq 0$, the available phase-space changes among flavors. This effect is completely negligible if the neutrino masses are small, as current data suggest.

[11] In Sec. 1.4.2 we mentioned that a complete family is made by a $SU(2)_W$ lepton doublet (e.g. ν_e and e^-) and a $SU(2)_W$ quark doublet (d and u). The heaviest particle (top quark) belongs to the third family made of ν_τ, τ^-, b and t quarks. Some authors speculated about "top-less" theories and theories with incomplete families before the discovery of the top quark in 1995, but no theory turned out to be either mathematically consistent or corroborated by experiments.

Fig. 12.15 Constraints on the Higgs mass (M_H) versus the top mass (m_t) in GeV. The constraints from radiative corrections at LEP and SLC are summarized by the solid gray line. Other indirect constraints from the W mass and the forward–backward asymmetries in the particle productions are shown in dashed and dotted lines. The direct measurement of M_H from the LHC (light long-dashed line) and of m_t from Tevatron and the LHC (dotted-dashed line) are also shown. The gray area is the allowed area from all indirect measurements, which is in agreement with the on-shell results. Reproduced with permission from Zyla *et al.* 2020.

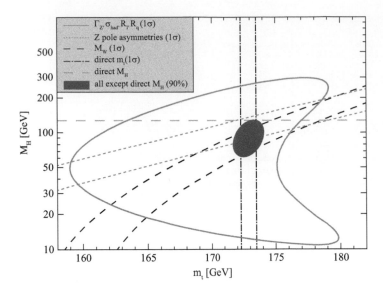

The study of radiative corrections and couplings at LEP and, nowadays, at the LHC established another fundamental property of the electroweak theory, **lepton universality**. This property follows from the electroweak vertices (see Figs 12.3 and 12.4) and reads:

> the couplings of leptons to gauge bosons ($\gamma, W^{\pm}, Z^0$) are flavour-independent.

This is the reason why the BR of the $W^{\pm}$ and Z^0 leptonic decays are the same in Table 12.1, except for small phase-space and helicity corrections. Precision electroweak physics allows measuring tiny variations of these couplings. They would signal a breakdown of the electroweak theory due, for example, to new flavor-dependent particles that contribute to the self-energy of the Feynman diagrams but cannot be produced on-shell. Needless to say, neither LEP nor the LHC have gained evidence of lepton-universality violations to date.

Another spectacular result of LEP was the determination of the masses of particles that had not been observed yet at that time: the top quark and the Higgs boson. These particles contribute as m_t^2 and $\log m_H$ to the radiative corrections of Z^0's Breit–Wigner (Leader and Predazzi, 1996). In particular, the LEP results showed that the Higgs boson should lay between 100 and 300 GeV. This finding boosted the race toward the LHC, which was designed to produce on-shell Higgs bosons with large statistics. The power of radiative correction physics is depicted in Fig. 12.15. The plot shows the precision in the Higgs and top quark mass obtained by combining all indirect measurements (dark area) and the corresponding direct (on-shell) discovery measurements from the Tevatron (the top quark) and the LHC (the Higgs). The LEP and SLC data are shown in the solid gray line and are in excellent agreement with the direct measurements performed from 1995 (top) and 2012 (Higgs).

Exercises

(12.1) Solve the potential well of eqn 12.3 for $D \neq 0$ in the initial state.

(12.2) Compute the ratio Γ_e/Γ_μ, where Γ_e and Γ_μ are the partial width of the $Z^0 \to e^+e^-$ and $Z^0 \to \mu^+\mu^-$ decays, respectively. Neglect the masses of the fermions.

(12.3) Compute the ratio $\Gamma_e/\Gamma_{\text{invisible}}$ neglecting the electron mass.

(12.4) Rewrite the Feynman diagram of the electron capture (EC) according to the electroweak theory.

(12.5) If a Z^0 is produced in a e^+e^- collider, the two final-state particles (leptons or jets) are back to back. Demonstrate that this statement does not hold in the case of a **radiative return**, i.e. when one of the initial electrons emits a radiation photon with energy E_γ. Compute the angle between the leptons if the photon momentum is along the positron axis (z-axis) and $|\mathbf{p}_\gamma| = 10$ GeV.

(12.6) What is the fermion that has the largest coupling to the Z^0? Is the Z^0 coupling of a LH top quark bigger than a LH up quark?

(12.7) * In the Gargamelle experiment, a neutron produced by a neutrino interaction in the hadron dump can enter the detector and mimic a NC event. How does the neutron background distribute in Gargamelle as a function of z, where the z axis is the propagation axis of the neutrinos.

(12.8) * Can you reduce the rate of neutron background mentioned in the previous exercise increasing the thickness of the hadron dump?

(12.9) * The operators of the Gargamelle bubble chamber must lift the piston when neutrinos cross the detector. If the proton accelerator is a synchrotron, how can they compute the time of arrival?

(12.10) In the analysis of the Gargamelle data, the isolation selection is employed to reject muons produced by the jet. Is this selection more effective if we increase the incoming neutrino energy?

(12.11) * Compute the width of the observed Z^0 peak in UA1 for a $Z^0 \to e^+e^-$ sample if the resolution of the calorimeter is:

$$\frac{\sigma}{E} = \frac{15\%}{\sqrt{E(\text{GeV})}} \oplus 2\% \,.$$

Does it correspond to the intrinsic width of the Z^0? Is the width different for a $Z^0 \to$ hadrons sample?

(12.12) A W^+ cannot decay in a $\mu^+\nu_e$ pair due to the conservation of lepton family number at the production time. Can a Z^0 decay into a $e^+\mu^-$ pair?

(12.13) * Using the PDG (Zyla *et al.*, 2020), inspect the BR of the τ lepton. Is it possible to identify $Z^0 \to \tau^+\tau^-$ decays in a e^+e^- collider? Is this task easier at the LHC?

(12.14) * Demonstrate eqn 12.26.

(12.15) * Consider a $\nu_e e^- \to \nu_e e^-$ scattering in the V-A theory and neglect the electron mass. Using the dimensional analysis, shows that cross-section increases as $G_F k^2$, where k is the sum of the initial state four-momenta. What happens if $\sqrt{s} \to +\infty$?

(12.16) * The electroweak theory shares with QCD the **self-boson couplings**, which are a common feature of all gauge theories based on non-abelian groups. In the electroweak theory, the self-boson couplings are $W^+W^-\gamma$, $W^+W^-Z^0$, $W^+W^-\gamma\gamma$, $W^+W^-Z^0Z^0$, $W^+W^-\gamma\gamma$, $W^+W^-\gamma Z^0$, and $W^+W^-W^+W^-$. Write some Feynman diagrams that can demonstrate the existence of these **triple** and **quartic** couplings in e^+e^- colliders and at the LHC.

(12.17) ** What is the BR of the $Z^0 \to \gamma\gamma$ decay?

(12.18) ** How can you measure the $W^\pm$ mass in a e^+e^- collider? Is this measurement more precise at the LHC?

13 At the forefront of the Standard Model

13.1 The Higgs mechanism

The electroweak theory presented in Sec. 12.2 is a striking exception among gauge theories. We have seen in Sec. 6.6.2 that there is a deep connection between the gauge principle and the number of degrees of freedom in the spin-1 mediators. The gauge invariance of quantum electrodynamics (QED) reduces the number of the possible spin orientations of the photon from three ($S_z = -1, 0, +1$) to two ($S_z = \pm 1$) and this feature is a unique signature of massless bosons. A gauge theory with massive bosons cannot get rid of the unphysical $S_z = 0$ polarization and the price to pay is too high. The amplitude of reactions like $e^+ e^- \to W^+ W^-$ increases with s up to violating the unitarity bound. A. Komar and A. Salam showed in 1960 that these theories are mathematically inconsistent even if the corresponding fine structure constants $g/4\pi$ and $g'/4\pi$ are dimensionless.

Many theorists attempted to construct a gauge theory with massive bosons, and some of these models – like the technicolor models (Hill and Simmons, 2003) – survived until the late 1990s and are still a source of inspiration for theories beyond the Standard Model.

The right track was inspired by the findings of P.W. Anderson on superconductivity and ported to particle physics in 1964 by R. Brout, F. Englert, P. Higgs, G. Guralnik, C.R. Hagen, and T. Kibble. Their findings can be summarized as follows:

> the massive mediators of gauge theories (i.e. the massive vector bosons) can be generated by a scalar field, provided that the Lagrangian of the theory is gauge invariant but the ground state of the theory is not.

This statement originates a new class of gauge theories that are improperly called **spontaneously broken gauge theories** (SBGT). Once more, the choice of the name is unfortunate: an SBGT like the electroweak theory is a *true* gauge theory because the Lagrangian is invariant under the gauge group, but the ground state of the theory, called the **vacuum state**, is not, because:

$$U|\mathbf{0}\rangle \neq e^{i\alpha}|\mathbf{0}\rangle \tag{13.1}$$

if U belongs to the gauge group and α is a real constant.

A Modern Primer in Particle and Nuclear Physics. Francesco Terranova, Oxford University Press.
© Francesco Terranova (2021). DOI: 10.1093/oso/9780192845245.003.0013

Even if this result is deeply rooted in quantum field theory (QFT) – which has no place in a Primer – it can be clearly illustrated by the classical theory of relativistic fields.

13.1.1   Relativistic fields

We will focus here on the simplest gauge theory based on relativistic fermions (QED), putting aside the issue of quantization. A complex-valued scalar potential is enough to enable the Higgs mechanism and generate a massive photon plus a remnant spin-0 field.

We already know that a relativistic spin-1/2 fermion is described by the Dirac equation. We interpret here the solutions of eqn 5.82 as classical fields instead of QM wavefunctions.[1] The Dirac equation is the equation of motion of a Lagrangian but, unfortunately, analytical mechanics gives us tools to derive the Lagrangian only for a finite number of variables. Nonetheless, we want a Lagrangian that works for a continuum system because the fields are functions that assign a quantity to every point in space-time. That is a harmless complication. The Lagrangian formalism can be extended to a continuum set of variables defining a **Lagrangian density**. For a finite-variable system:

$$L = L(q_i, \dot{q}_i) \;\; \text{with} \;\; i = 1, 2, \ldots, N \tag{13.2}$$

and the Euler–Lagrange equation:

$$\frac{d}{dt}\left(\frac{\partial L}{\partial \dot{q}_i}\right) - \frac{\partial L}{\partial q_i} = 0 \tag{13.3}$$

generates the equations of motion (Taylor, 2005). Similarly, a Lagrangian density $\mathcal{L}$ such that:

$$\mathcal{L} = \mathcal{L}(\psi_i, \partial_\mu \psi_i) \;\; ; \;\; L \equiv \int_{\mathbb{R}^3} d\mathbf{x}^3 \mathcal{L}(\psi_i, \partial_\mu \psi_i) \tag{13.4}$$

gives the relativistic equations of motion by the corresponding Euler–Lagrange equation:

$$\partial_\mu \left(\frac{\partial \mathcal{L}}{\partial(\partial_\mu \psi_i)}\right) - \frac{\partial \mathcal{L}}{\partial \psi_i} = 0. \tag{13.5}$$

$\mathcal{L}$ is called a Lagrangian density because of eqn 13.4 (right) and has a dimension of eV^4 in natural units (NU). The Lagrangian L plays a minor role in classical field theories or QFT because depends on space only, and does not ease the writing of the theory in covariant form. To get acquainted to this formalism, I encourage the readers to show that the **Dirac Lagrangian density**:

$$i\bar{\psi}\gamma^\mu \partial_\mu \psi - m\bar{\psi}\psi = \bar{\psi}(i\gamma^\mu \partial_\mu - m)\psi \tag{13.6}$$

gives the Dirac equation through eqn 13.5, and the **Maxwell Lagrangian density**:

$$\mathcal{L} = -\frac{1}{4}F^{\mu\nu}F_{\mu\nu} \tag{13.7}$$

[1] This can be done without loss of generality because quantization issues arise only when we compute high-order radiative corrections.

gives Maxwell's equations if $F^{\mu\nu}$ is the electromagnetic tensor of eqn 6.7 (see Exercises 13.13 and 13.14).

The Lagrangian density[2] of a free electron and a non-interacting photon is:

$$\bar{\psi}(i\gamma^\mu \partial_\mu - m)\psi - \frac{1}{4}F^{\mu\nu}F_{\mu\nu} \tag{13.8}$$

but, if we modify the Lagrangian to satisfy the invariance of $\mathcal{L}$ for local $U(1)$ transformations, we get the QED Lagrangian:

$$\mathcal{L}_{QED} = \bar{\psi}(i\gamma^\mu \partial_\mu - m)\psi - \frac{1}{4}F^{\mu\nu}F_{\mu\nu} + e\bar{\psi}\gamma^\mu\psi A_\mu \tag{13.9}$$

where the last term accounts for the fermion–photon vertices of Fig. 6.2.

There is no easy way to put a massive photon inside the QED Lagrangian without spoiling the gauge symmetry. The photon mass term:

$$\frac{1}{2}m_\gamma^2 A_\mu A^\mu \tag{13.10}$$

is the Maxwell counterpart of the electron mass term in Dirac's equation:

$$- m\bar{\psi}\psi \tag{13.11}$$

and the photon term is incompatible with the $U(1)$ gauge symmetry.

Example 13.1

Equation 13.10 spoils the gauge symmetry of QED because the electromagnetic field changes by the classical gauge transformations of eqn 6.29. The gauge-transformed Lagrangian is then:

$$\mathcal{L}'_{QED} + \frac{1}{2}m_\gamma^2 A'_\mu A'^\mu = \mathcal{L}'_{QED} + \frac{1}{2}m_\gamma^2(A_\mu - \partial_\mu\alpha(x^\mu))(A^\mu - \partial^\mu\alpha(x^\mu))$$

$$= \mathcal{L}_{QED} + \frac{1}{2}m_\gamma^2(A_\mu - \partial_\mu\alpha(x^\mu))(A^\mu - \partial^\mu\alpha(x^\mu))$$

where all primed quantities are gauge transformed and QED is gauge invariant: $\mathcal{L}'_{QED} = \mathcal{L}_{QED}$. Therefore,

$$\mathcal{L}'_{QED} + \frac{1}{2}m_\gamma^2 A'_\mu A'^\mu - \left(\mathcal{L}_{QED} + \frac{1}{2}m_\gamma^2 A_\mu A^\mu\right) =$$

$$\frac{1}{2}m_\gamma^2(A_\mu - \partial_\mu\alpha(x^\mu))(A^\mu - \partial^\mu\alpha(x^\mu)) - \frac{1}{2}m_\gamma^2 A_\mu A^\mu \neq 0$$

13.1.2   Spontaneous symmetry breaking

It is now time to introduce the scalar field that produces the vacuum state. The field is a complex-valued function like:

$$\phi = \frac{1}{\sqrt{2}}(\phi_1 + i\phi_2) \tag{13.12}$$

where we split the real and complex part of the field into ϕ_1 and ϕ_2, respectively. This field contributes to the Lagrangian with a peculiar potential, chosen to produce a gauge-asymmetric vacuum:

$$V(\phi) \equiv \mu^2 \phi^* \phi + \lambda(\phi^* \phi)^2 . \tag{13.13}$$

We can interpret the scalar field as a spinless particle, whose equation of motion is given by eqn 5.74, that is, the Klein–Gordon equation. The corresponding Lagrangian is thus:

$$\mathcal{L} = \frac{1}{2}(\partial_\mu \phi)^*(\partial^\mu \phi) - V(\phi) \tag{13.14}$$

$$= \frac{1}{2}(\partial_\mu \phi_1)(\partial^\mu \phi_1) + \frac{1}{2}(\partial_\mu \phi_2)(\partial^\mu \phi_2) - \frac{1}{2}\mu^2(\phi_1^2 + \phi_2^2) - \frac{1}{4}\lambda(\phi_1^2 + \phi_2^2)^2 .$$

Note that eqn 13.14 does not change if we redefine ϕ by a unit phase: $\phi \to e^{i\alpha}\phi$. Why does eqn 13.13 suit our needs? The minimum value of the potential is reached when the squared sum of ϕ_1 and ϕ_2 is equal to $-\mu^2/\lambda$. Fig. 13.1 shows that the number of minima collapses to 1 if $\mu^2 > 0$ but is an infinite set (a circle in the ϕ_1, ϕ_2 plane) if $\mu^2 < 0$. Here,

$$V(\phi_1, \phi_2) = V_{min} = -\frac{\mu^4}{2\lambda} + \frac{\mu^4}{4\lambda} = -\frac{\mu^4}{4\lambda} \tag{13.15}$$

when

$$\phi_1^2 + \phi_2^2 = -\frac{\mu^2}{\lambda} . \tag{13.16}$$

We jokingly call eqn 13.13 the **Mexican hat** potential[3] if $\mu^2 < 0$ and $\lambda > 0$. The sign of λ determines whether the potential is concave down ($\lambda < 0$) or upper convex ($\lambda > 0$) and λ is positive in the Higgs mechanism because V must have finite minima.

Equations 13.14 and 13.16 encode the core of the **Higgs mechanism**:

[3]To be fair, the Mexican hat potential was used for the first time in particle physics by Y. Nambu and J. Goldstone in their pioneering studies of spontaneous symmetry breaking in strong interactions.

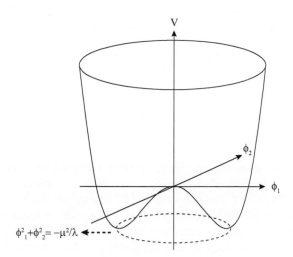

$\phi_1^2 + \phi_2^2 = -\mu^2/\lambda$

Fig. 13.1 The complex scalar potential employed in the Higgs mechanism. The potential has an infinite set of minima for $\mu^2 < 0$ located along the $\phi_1^2 + \phi_2^2 = -\mu^2/\lambda$ circle. Redrawn from Thomson, 2013.

> the Lagrangian of the scalar complex field is invariant for a global phase change – the $U(1)$ global symmetry. The physical ground state is one of the points belonging to the circle in eqn 13.16. Once the system reaches its minimum energy, only one point laying in the circle is selected. The selection breaks the $U(1)$ symmetry (just) for the ground state because a global change of phase moves the ground state from one minimum to another.

The symmetries belonging to this class are called **spontaneously broken**. This is a misnomer. The symmetry of the theory is still there because the Lagrangian is invariant for a global phase change but the ground state depends on the phase. You may be puzzled by an asymmetric choice arising from a symmetric Lagrangian but this is a common occurrence in classical physics too. If you slide a marble around the rim of a Mexican hat, there is no way to predict where the marble stops, unless we have a perfect knowledge of the initial conditions and the system is fully deterministic. Particle physics does not fulfill any of these provisions.

13.1.3   The appearance of the remnant field

We now want to perform the same spontaneous symmetry breaking with the local $U(1)$ symmetry: the gauge symmetry of QED.

First, we need to modify the Lagrangian of the scalar field to match the $U(1)$ *local* invariance. This is not granted a priori because the complex scalar Lagrangian includes a derivative ∂_μ that produces additional terms when:

$$\phi \rightarrow e^{iq\alpha(x^\mu)}\phi. \tag{13.17}$$

In relativistic gauge theories, we sidestep this issue introducing a **covariant derivative**:

$$D_\mu \equiv \partial_\mu + iqA_\mu. \tag{13.18}$$

This method is equivalent to what we did in non-relativistic gauge theories (see Sec. 6.4) when we computed the semi-classical Hamiltonian. The result is astonishing because the new Lagrangian

$$\mathcal{L} = \frac{1}{2}(D_\mu\phi)^*(D^\mu\phi) - V(\phi) \tag{13.19}$$

is invariant for the transformation of eqn 13.17 (see Exercise 13.20) if the A^μ field transforms like the classical electromagnetic four-potential:

$$A'^\mu = A^\mu - \partial^\mu\alpha(x^\mu). \tag{13.20}$$

The way I presented the covariant derivative looks like a spell but there is no magic here. We already know that a local symmetry is a tighter condition than the corresponding global symmetry and the change of the Lagrangian (or Hamiltonian) brings a correlation term between the electromagnetic (e.m.) field and the particle field. In SR, this correlation can be neatly embedded in the definition of the covariant derivative, while – in the semi-classical theory – the identification of the correct change is more cumbersome. This is one of the reasons why the gauge principle has been discovered first in relativistic theories and then ported to non-relativistic QM.

The Lagrangian that describes the free motion of an electromagnetic field A^μ and a scalar field ϕ retaining the $U(1)$ gauge symmetry is:

$$-\frac{1}{4}F^{\mu\nu}F_{\mu\nu} + (D_\mu\phi)^*(D^\mu\phi) - V(\phi) \tag{13.21}$$

where D^μ is given by eqn 13.18 and $V(\phi)$ by eqn 13.13.

We now have the two bishops for a grand endgame. Expand the scalar field around its vacuum state:

$$\phi(x^\mu) = \frac{1}{\sqrt{2}}\left(\sqrt{\frac{-\mu^2}{\lambda}} + \epsilon(x^\mu) + i\tilde{\epsilon}(x^\mu)\right) \equiv \frac{1}{\sqrt{2}}\left(v + \epsilon(x^\mu) + i\tilde{\epsilon}(x^\mu)\right). \tag{13.22}$$

We call

$$v \equiv \sqrt{\frac{-\mu^2}{\lambda}} \tag{13.23}$$

the **vacuum expectation value** (VEV) of the scalar field ϕ. The VEV corresponds to the value of $|\phi| = (\phi^*\phi)^{1/2} = (\phi_1^2 + \phi_2^2)^{1/2}$ when $V(\phi)$ reaches the minimum. $\epsilon(x^\mu)$ and $\tilde{\epsilon}(x^\mu)$ are small displacements of the field along the real and complex axes of $\phi(x^\mu)$, and are real-valued scalar fields. Then, we insert eqn 13.22 in eqn 13.19 and we go through the algebra (Exercise 13.21) to express $\mathcal{L}$ as a function of the new fields and the VEV. Dropping x^μ to ease the reading, the final result is:

$$\mathcal{L} = \frac{1}{2}(\partial_\mu\epsilon)(\partial^\mu\epsilon) - \lambda v^2\epsilon^2 + \frac{1}{2}(\partial_\mu\tilde{\epsilon})(\partial^\mu\tilde{\epsilon}) \tag{13.24}$$
$$-\frac{1}{4}F^{\mu\nu}F_{\mu\nu} + \frac{1}{2}q^2v^2A_\mu A^\mu + qvA_\mu(\partial^\mu\tilde{\epsilon}) - V(\epsilon,\tilde{\epsilon},A^\mu)$$

where $V(\epsilon,\tilde{\epsilon},A^\mu)$ accounts for triple and quartic couplings of the $\epsilon,\tilde{\epsilon}$, and A^μ fields. Equation 13.24 is a specific realization of the **Nambu–Goldstone theorem:**[4]

> the spontaneous breaking of a local symmetry produces a massive scalar field, a massless boson (the **Goldstone boson**), and a massive gauge field.

Note that we have just rewritten eqn 13.21 in the form of eqn 13.24. Therefore, eqn 13.24 still fulfils the local $U(1)$ symmetry but its "photon" (A^μ) is massive thanks to the $\frac{1}{2}q^2v^2A_\mu A^\mu$ term. Novices should watch out for any source of misunderstanding: choosing a VEV has nothing to do with choosing a gauge. We still can simplify eqn 13.24 by fixing the gauge of the A^μ field in a smart way. Indeed, that is a good idea if we

[4]Nambu was awarded the Nobel Prize in Physics in 2008 for this theorem. The Nambu–Goldstone theorem has many applications ranging from phonons in fluids and solids to QCD. For instance, it provides an elegant explanation of the smallness of pion masses in the two quark model and grounds the chiral perturbation theory mentioned in Sec 9.6.3 (Donnelly *et al.*, 2017).

want to get rid of unphysical particles like the Fadeev–Popov ghosts of QCD (see Sec. 7.6.3).

What are the ghosts of this dummy theory (let us call it spontaneously broken QED)? A simple "ghost counting" method that originates from Nambu and Goldstone's papers is the counting of the degrees of freedom. Equation 13.21 has four degrees of freedoms: two from ϕ_1 and ϕ_2, and two from the $S_z = \pm 1$ polarization of the massless photon A^μ. Equation 13.24 has three degrees of freedom coming from the massive photon ($S_z = -1, 0, +1$), one from ϵ – which is massless but also spinless – and one from $\tilde{\epsilon}$, five degrees of freedom – one more than eqn 13.21. The additional degree of freedom must be unphysical. By inspection of eqn 13.24, we spot the ghost as the sixth term of the Lagrangian, a direct coupling between A^μ and $\tilde{\epsilon}$ proportional to qv. This term is a weird spontaneous transformation of a spin-1 particle into a spin-0 particle. Clearly, we can live with this ghost adding the corresponding (unphysical) Feynman diagrams because the final result is ghost-independent, like the Fadeev–Popov ghosts in QCD. But we can also fix a gauge that throw out ghosts by taking:

$$\alpha = -\frac{\tilde{\epsilon}}{gv} . \tag{13.25}$$

This gauge fixing is called **unitary gauge** and leads to a Lagrangian deprived of $\tilde{\epsilon}$. We can rewrite the field perturbation around the VEV as:

$$\phi = \frac{1}{\sqrt{2}}\left(v + \epsilon + i\tilde{\epsilon}\right) \simeq \frac{1}{\sqrt{2}}(v+\epsilon)\left(1 + \frac{i\tilde{\epsilon}}{v+\epsilon} + \mathcal{O}(\epsilon^2, \tilde{\epsilon}^2)\right) \tag{13.26}$$

$$= \frac{1}{\sqrt{2}}(v+\epsilon)e^{\frac{i\tilde{\epsilon}}{v+\epsilon}} \simeq \frac{1}{\sqrt{2}}(v+\epsilon)\,e^{i\tilde{\epsilon}/v}$$

where the two $\simeq$ arise from adding or neglecting higher-order terms in the Taylor expansion of the exponential function. In unitary gauge, a gauge transformation happily cut out the ghost, canceling $\tilde{\epsilon}$:

$$\phi \to \phi' = \exp\{-iq\tilde{\epsilon}/qv\}\phi = \exp\{-i\tilde{\epsilon}/v\}\phi \tag{13.27}$$

and, using eqn 13.26,

$$\phi' \simeq e^{-i\tilde{\epsilon}/v}\frac{1}{\sqrt{2}}\,(v+\epsilon)\,e^{i\tilde{\epsilon}/v} = \frac{1}{\sqrt{2}}\,(v+\epsilon). \tag{13.28}$$

The name of the game is that all fields are real and there are no commutation issues. We can then change the order of the products at will. Further,

fixing the gauge is equivalent to choosing a real-valued scalar field

$$\phi(x^\mu) = \frac{1}{\sqrt{2}}(v + \epsilon(x^\mu)) \equiv \frac{1}{\sqrt{2}}(v + h(x^\mu)) . \tag{13.29}$$

Unitary gauge is a great choice in spontaneously broken gauge theories because all fields that appear in the Lagrangian are physical, including the "remnant field" $h(x^\mu)$. We changed the name of $\epsilon(x)$ into $h(x^\mu)$ to signify that the remnant field is a physical field, a spin-0 massive particle called the **Higgs boson**.

In our dummy theory, the Lagrangian that contains only physical fields is:

$$\mathcal{L} = \frac{1}{2}(\partial_\mu h)(\partial^\mu h) - \lambda v^2 h^2 - \frac{1}{4}F^{\mu\nu}F_{\mu\nu} + \frac{1}{2}q^2 v^2 A_\mu A^\mu$$

$$+ q^2 v A_\mu A^\mu h + \frac{1}{2}q^2 A_\mu A^\mu h^2 - \lambda v h^3 - \frac{1}{4}\lambda h^4 + \frac{\lambda^4 v}{4}. \tag{13.30}$$

If you like, you can ignore the constant term $\lambda^4 v/4$ because the equations of motion come from the derivatives of the Lagrangian and the derivative of a constant is zero. In eqn 13.30 we find a massive photon with $m_\gamma = qv$. There is also a massive Higgs boson with $m_H = \sqrt{2\lambda}v$ that comes from the $-\lambda v^2 h^2$ term of eqn 13.30 (see Exercise 13.22). The Feynman diagram catalog is richer than QED because of triple and quartic couplings between the h and A^μ fields (see Exercise 13.15).

13.2 Higgs in the electroweak theory

The **Higgs** or **Brout–Englert–Higgs mechanism** has been devised in the 1960s to transform the electroweak theory into an SBGT and, therefore, solve the problem of the vector boson masses, which otherwise compromise the $SU(2) \times U(1)$ gauge symmetry.

As any SBGT, it is based on complex scalar fields. In this case, the field is a $SU(2)$ doublet of complex fields:

$$\phi \equiv \begin{pmatrix} \phi^+ \\ \phi^0 \end{pmatrix} = \frac{1}{\sqrt{2}} \begin{pmatrix} \phi_1 + i\phi_2 \\ \phi_3 + i\phi_4 \end{pmatrix}. \tag{13.31}$$

The Lagrangian of this field, corresponding to four independent real fields, is the same of eqn 13.14 and reads:

$$\mathcal{L} = \frac{1}{2}(\partial_\mu \phi)^\dagger(\partial^\mu \phi) - V(\phi) = \frac{1}{2}(\partial_\mu \phi)^\dagger(\partial^\mu \phi) - \mu^2 \phi^\dagger \phi - \lambda(\phi^\dagger \phi)^2. \tag{13.32}$$

We replaced the complex conjugate ϕ^* with the conjugate (or hermitian) transpose $\phi^\dagger \equiv (\phi^T)^*$ because ϕ is now a two-component column vector. $V(\phi)$ is called the **Higgs potential** and has multiple degenerate minima reached when:

$$\phi^\dagger \phi = \frac{1}{2}(\phi_1^2 + \phi_2^2 + \phi_3^2 + \phi_4^2) = -\frac{\mu^2}{2\lambda} \equiv \frac{v^2}{2}. \tag{13.33}$$

The parameter v is called the **VEV of the Standard Model**. The vacuum is chosen randomly among the points fulfilling eqn 13.33 and, without loss of generality, can be written as:

$$V_{min} = \frac{1}{\sqrt{2}} \begin{pmatrix} 0 \\ v \end{pmatrix} \tag{13.34}$$

so that the scalar fields can be expanded in a neighborhood of V_{min} as

$$\phi = \frac{1}{\sqrt{2}} \begin{pmatrix} \phi_1 + i\phi_2 \\ v + \epsilon + i\tilde{\epsilon} \end{pmatrix} . \tag{13.35}$$

In the electroweak theory, Weinberg exploited the unitary gauge to rewrite the ϕ field using only real-valued physical fields:

$$\phi = \frac{1}{\sqrt{2}} \begin{pmatrix} 0 \\ v + h(x^\mu) \end{pmatrix} . \tag{13.36}$$

Once more, we can restore the $SU(2) \times U(1)$ symmetry for the electroweak Lagrangian with the scalar field by defining a covariant derivative. In the case of the electroweak theory, the covariant derivative reads (Peskin and Schroeder, 1995):

$$D^\mu \equiv \partial^\mu + ig\mathbf{T} \cdot \mathbf{W}^\mu + ig'\frac{Y}{2}B_\mu . \tag{13.37}$$

Please note that this definition applies to the unphysical fields W_j^μ and B^μ defined in Sec. 12.2.1, and not to the electroweak fields describing the interactions $(A^\mu, Z^\mu, W^{\pm\mu})$. Hence, g and g' are the unphysical charges of eqn 12.17. $\mathbf{T} = \boldsymbol{\sigma}/2$ are the three generators of $SU(2)$ and $\boldsymbol{\sigma}$ is the vector of the three Pauli matrices.

We reach the final result with a lengthy calculation by replacing the derivatives in the electroweak Lagrangian with the covariant derivatives and substituting ϕ with eqn 13.36. The result is worth the effort and lies at the core of modern particle physics. We can summarize it as follows:

- The physical vector bosons have a mass that comes from the $(D_\mu\phi)^\dagger(D^\mu\phi)$ term of the Lagrangian. The masses of the physical fields $W^{\pm\mu}, A^\mu, and\ Z^\mu$ are:

$$M_W = \frac{1}{2}gv \ ; \ m_\gamma = 0 \ ; \ M_Z = \frac{v}{2}\sqrt{g^2 + g'^2} \tag{13.38}$$

where we identified the physical A^μ field with the photon field. A^μ is, therefore, the quantum counterparts of the classical A^μ electromagnetic potential defined in eqn 6.16.

- Using the electroweak unification formula (eqn 12.23) and eqn 13.38, we can compute the ratio between the Z and W mass:

$$\frac{M_W}{M_Z} = \cos\theta_W \tag{13.39}$$

where θ_W is the Weinberg angle. Equation 13.39 is a fundamental prediction of the Higgs mechanism that has been tested with high precision at LEP and SLC well before the discovery of the Higgs boson.

- M_W can be experimentally measured and g can be derived by the Fermi constant through eqn 12.26. Since $M_W = gv/2$, the VEV of the Standard Model turns out to be (Zyla *et al.*, 2020):

$$v = 246.22 \text{ GeV} \qquad (13.40)$$

- the remnant field of the spontaneous breaking of $SU(2) \times U(1)$ corresponds to a spin-0 particle called the **Higgs boson**, whose mass is:

$$m_H = \sqrt{2\lambda}v. \qquad (13.41)$$

Since λ is unknown the Higgs mass must be measured by the experiments. Before the discovery of the Higgs, the radiative corrections of the electroweak theory discussed in Sec. 12.7 supported a value of m_H in the 100-300 GeV range. The actual value of m_H measured at the LHC is in agreement with this expectation and testifies for the mathematical consistency of the electroweak theory. The Higgs boson couples with all massive particles and with the vector bosons, and the Higgs couplings have been tested at the LHC too.

The parts of the Lagrangian that give charged and neutral currents are what we expect from the vertices of Figs. 12.3 and 12.4. For charged currents (CC),

$$\mathcal{L}_{CC} = \frac{g}{2\sqrt{2}} \left[\left(\bar{q}^U \gamma^\mu (1 - \gamma^5) q^D + \bar{l}^U \gamma^\mu (1 - \gamma^5) l^D \right) W_\mu^+ + \text{h.c.} \right] \quad (13.42)$$

where q^U are the (bi)spinors of the "up-type quarks" u, c, t, that is, the fields describing the particles located in the upper part of the $SU(2)$ doublet ($I_z^W = +1/2$). These quarks have $+2/3$ electric charge in units of e. q^D are the (bi)spinors of the "down-type" quarks d, s, b, located in the lower part of the doublet ($I_z^W = -1/2$) with $Q = -e/3$. Similarly l^U and l^D correspond to the "up-type" (ν_e, ν_μ, ν_τ) and "down-type" (e^-, μ^-, τ^-) leptons. "h.c." stands for "hermitian conjugate": there is no way to keep $\mathcal{L}_{CC}$ invariant for $SU(2) \times U(1)$ transformations without adding the conjugate fields. If we write explicitly the h.c. of eqn 13.42, we have:

$$\mathcal{L}_{CC} = \frac{g}{2\sqrt{2}} \left[\left(\bar{q}^U \gamma^\mu (1 - \gamma^5) q^D + \bar{l}^U \gamma^\mu (1 - \gamma^5) l^D \right) W_\mu^+ \right.$$
$$\left. + \left(\bar{q}^D \gamma^\mu (1 - \gamma^5) q^U + \bar{l}^D \gamma^\mu (1 - \gamma^5) l^U \right) W_\mu^- \right]. \qquad (13.43)$$

Hence, the first term of eqn 13.42 produces the $d \to u$ transitions and the hermitian conjugate the $u \to d$ transitions[5].

For NC, we have the QED interaction Lagrangian producing fermion–fermion–photon vertices:

$$\mathcal{L}_{NC}^\gamma = Q \sum_j \bar{f}_j \gamma^\mu f_j A_\mu \qquad (13.44)$$

where $Q = ze$ is the fermion charge and f_j the (bi)spinor of the fermion.

[5]The CC for antiparticles are also given by eqn 13.43 but do not forget that the "up-type antiquarks" are now $\bar{d}, \bar{s}$, and $\bar{b}$, and the "up-type antileptons" are e^+, μ^+, and τ^+. Further, we warn the readers that eqn 13.43 is not yet the Standard Model Lagrangian that describes CC because we are neglecting the fermion mixing discussed in Secs. 13.6 and 13.7.

The Z^0 Lagrangian is

$$\mathcal{L}_{NC}^Z = \frac{e}{2\sin\theta_W \cos\theta_W} \sum_f \left[\bar{f}^U \gamma^\mu (v_f - a_f \gamma^5) f^U \right.$$
$$\left. + \bar{f}^D \gamma^\mu (v_f - a_f \gamma^5) f^D \right] Z_\mu. \qquad (13.45)$$

In eqn 13.45, $f^{U,D}$ are all fermions of type "up" (ν_e, u, c, etc.) and "down" (e^-, d, s, etc.). $v_f^{U,D}$ and $a_f^{U,D}$ are the corresponding couplings defined in eqn 12.33, and there is no h.c. because the initial and final state fermion in the vertex is the same. The fields f are bispinors that depend on x^μ because describe relativistic fermions. The NC Lagrangian represents vertices where the Z^0 interacts with LH and RH fermions, provided that the initial- and final-state fermion is the same. It also generates the neutrino interactions observed in Gargamelle (see Sec. 12.4).

13.3   Fermion masses

An astonishing result obtained while transforming the original Glashow–Salam–Weinberg model in an SBGT is the appearance of mass terms in the Lagrangian for elementary fermions, too. These terms originate from a part of $\mathcal{L}$ that can be harmlessly introduced because fulfills the $SU(2) \times U(1)$ symmetry. It is called the **Yukawa Lagrangian** and brings with it several free parameters. We have already shown that a fermion mass term has no room in the Standard Model (SM) because is not gauge invariant. We can check it by splitting the Dirac term into its LH and RH chirality components:

$$- m\bar{\psi}\psi = -m(\bar{\psi}_R \psi_L + \bar{\psi}_L \psi_R). \qquad (13.46)$$

Equation 13.46 in inconsistent with the $SU(2)$ symmetry because mixes LH and RH components. On the other hand, the **simplified (lepton) Yukawa Lagrangian**:

$$\mathcal{L}_f = -\lambda_f \left[(\bar{\nu}_{fL} \quad \bar{f}_L) \begin{pmatrix} \phi^+ \\ \phi^0 \end{pmatrix} f_R + \bar{f}_R \begin{pmatrix} \phi^{+*} & \phi^{0*} \end{pmatrix} \begin{pmatrix} \nu_{fL} \\ f_L \end{pmatrix} \right] \qquad (13.47)$$

satisfies the gauge symmetry and generates the mass of the neutrinos and the "down-type" leptons (f is the lepton flavor, $f = e, \mu, \tau$). λ_f is a free parameter called the **Yukawa coupling** of the Higgs field to the lepton of flavor f. Since the second term of eqn 13.47 is just the hermitian conjugate of the first term, $\mathcal{L}_f$ can be shortened as:

$$\mathcal{L}_f = -\lambda_f \left[(\bar{\nu}_{fL} \quad \bar{f}_L) \begin{pmatrix} \phi^+ \\ \phi^0 \end{pmatrix} f_R + \text{h.c.} \right]. \qquad (13.48)$$

We can write a similar "simplified" Lagrangian for quarks but we postpone this subject when we write the real Yukawa Lagrangian, which includes the phenomenon of fermion mixing.

If we apply the spontaneous symmetry breaking to eqn 13.48, we end up with:

$$\mathcal{L}_f = -\frac{\lambda_f}{\sqrt{2}}v(\bar{f}_L f_R + \bar{f}_R f_L) - \frac{\lambda_f}{\sqrt{2}}h(x^\mu)(\bar{f}_L f_R + \bar{f}_R f_L). \qquad (13.49)$$

That is great news because eqn 13.49 is gauge invariant and the terms proportional to $\bar{f}_L f_R$ or $\bar{f}_R f_L$ are mass terms of the lepton with flavor f. Since the unitary gauge puts a zero in the upper fields of the $SU(2)$ doublet (eqn 13.36), eqn 13.49 cannot be used to generate the mass of the ν_e and all "up-type" fermions of eqn 12.9. The way out is similar to what we did for the two-quark model in Example 8.2. We introduce a conjugate field doublet:

$$\phi_c = -i\sigma_2\phi^* = \begin{pmatrix} -\phi^{0*} \\ \phi^- \end{pmatrix} = \frac{1}{\sqrt{2}}\begin{pmatrix} -\phi_3 + i\phi_4 \\ \phi_1 - i\phi_2 \end{pmatrix}. \qquad (13.50)$$

We write ϕ_c in a more compact notation labeling the $SU(2)$ doublet of LH leptons as L and the RH singlet as R. The simplified Yukawa Lagrangian is then:[6]

$$\mathcal{L} = -\lambda_f\left[\bar{L}\boldsymbol{\phi}R + (\bar{L}\boldsymbol{\phi}R)^\dagger\right] + \lambda_f\left[\bar{L}\boldsymbol{\phi}_c R + (\bar{L}\boldsymbol{\phi}_c R)^\dagger\right]$$

and spontaneous symmetry breaking does the rest of the job, generating the mass term through eqn 13.49 with $f = \nu_e, \nu_\mu, \nu_\tau$. Clearly, there is no physics in the choice of the vacuum. The zero in eqn 13.34 is a matter of taste and we can make a different choice getting the same Yukawa Lagrangian.

If we consider the up-type and down-type particles, the simplified Yukawa Lagrangian for leptons ($\mathcal{L}_{SYL}$) after symmetry breaking reads:

$$\begin{aligned}
\mathcal{L}_{SYL} = & -\frac{\lambda_e v}{\sqrt{2}}\bar{e}_L e_R - \frac{\lambda_\mu v}{\sqrt{2}}\bar{\mu}_L \mu_R - \frac{\lambda_\tau v}{\sqrt{2}}\bar{\tau}_L \tau_R \\
& -\frac{\lambda_{\nu_e} v}{\sqrt{2}}\bar{\nu}_{eL}\nu_{eR} - \frac{\lambda_{\nu_\mu} v}{\sqrt{2}}\bar{\nu}_{\mu L}\nu_{\mu R} - \frac{\lambda_{\nu_\tau} v}{\sqrt{2}}\bar{\nu}_{\tau L}\nu_{\tau R} \\
& -\frac{\lambda_e}{\sqrt{2}}h(x^\mu)\bar{e}_L e_R - \frac{\lambda_\mu}{\sqrt{2}}h(x^\mu)\bar{\mu}_L\mu_R - \frac{\lambda_\tau}{\sqrt{2}}h(x^\mu)\bar{\tau}_L\tau_R \\
& -\frac{\lambda_{\nu_e}}{\sqrt{2}}h(x^\mu)\bar{\nu}_{eL}\nu_{eR} - \frac{\lambda_{\nu_\mu}}{\sqrt{2}}h(x^\mu)\bar{\nu}_{\mu L}\nu_{\mu R} - \frac{\lambda_{\nu_\tau}}{\sqrt{2}}h(x^\mu)\bar{\nu}_{\tau L}\nu_{\tau R}. \quad (13.51)
\end{aligned}$$

The fermion masses are then:

$$m_f = \frac{\lambda_f v}{\sqrt{2}} . \qquad (13.52)$$

Like the Higgs mass, the "λ_f" free parameters douse one's enthusiasm.

> The Yukawa sector allows for massive fermions but does not predict the mass of any fermion because a new free parameter appears in the Lagrangian each time we put a new fermion.

[6]Please note that σ_2 is a 2×2 matrix. Therefore, $\bar{L}$ is a two-component row vector and ϕ_c is the two-component column vector of eqn 13.50.

Even worse, these free parameters span an enormous range because the fermion masses go from tens of meV (neutrinos) up to hundreds of GeV (the top quark)! On the other hand, the coupling of the Higgs to the fermions are uniquely defined because they are drawn by the:

$$-\frac{\lambda_f}{\sqrt{2}}h(x^\mu)(\bar{f}_L f_R + \bar{f}_R f_L) \tag{13.53}$$

term, and the Yukawa coupling λ_f now is fixed by measuring the fermion mass ($\lambda_f = \sqrt{2}m_f/v$). At the end of the day,

> the Higgs boson is the only particle whose fermion couplings depend on the size of the fermion mass,

which is not so strange since the Higgs boson is the remnant field of a mass generation mechanism.

13.4   The Standard Model of particle physics

At the end of this (admittedly tough) race through relativistic fields, we have all the ingredients to define what is the **Standard Model** of particle physics rigorously.

> The Standard Model is a set of two theories that explain strong and electroweak interactions, respectively. The first theory – quantum chromodynamics – is a non-abelian gauge theory based on the $SU(3)_c$ group. The second theory – the electroweak theory – is a spontaneously broken gauge theory based on the $SU(2)_W \times U(1)_Y$ group.

The two theories share the same elementary fermions but are physically decoupled. A QCD scattering does not depend on the electric charge or chirality of the fermion. Similarly, color charges play no role in the electroweak theory. Since both theories are based on the same methods (QFT and the gauge principle), many particle physicists have been tempted to find a theory that unifies strong and electroweak interactions. This fascination is fed by quantitative findings, too. If we extend the Landau formula (eqn 6.93) to the electroweak theory and we employ the Gross–Politzer–Wilczek formula of eqn 7.25, we can compute the running of all fine-structure constants at high energy. This famous result, shown in Fig. 13.2 and first drawn by U. Amaldi, W. de Boer, and H. Furstenau in 1991, demonstrates that the three coupling constants of the Standard Model get closer and closer and reach the minimum distance at about 10^{16} GeV. This scale, much bigger than any scale accessible at colliders, is called the **grand unification scale**. In the same paper (Amaldi *et al.*, 1991), the authors also considered a specific class of theories – the **supersymmetric theories** (Weinberg, 2005*b*) – that were predicting the unification of the couplings *exactly* at the grand

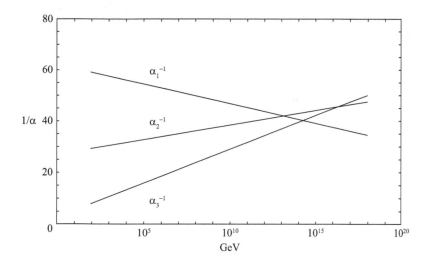

Fig. 13.2 Running of the three SM couplings as a function of $\sqrt{s}$ (in GeV). In this plot, $\alpha_1 = (g'^2/4\pi)(5/3)$, $\alpha_2 = g^2/4\pi$ and $\alpha_3 = \alpha_s$. The 5/3 factor accounts for the renormalization of g' at high-energy due to the running of the fine structure constants (see Sec. 6.9). Figure adapted from Amaldi, de Boer and Furstenau, 1991.

unification scale for all alphas. It is worth stressing that the most promising theories developed along this line were falsified by the experiments at the Large Hadron Collider (LHC), which observed none of the predicted particles (Giudice, 2019). The Standard Model thus remains the only game in town in terms of experimental verification and mathematical consistency.

As usual, we overlooked the historical developments and you may be confused by the different names of the theory (Glashow–Weinberg–Salam model, electroweak theory, SBGS, Salam–Weinberg model, etc.). The source of confusion comes from SBGS, which was developed in the late 1950s and found many applications in physics well before electroweak interactions were discovered. In 1967, A. Salam and S. Weinberg noted that SBGS could be applied to their $SU(2) \times U(1)$ weak interaction model, establishing the electroweak theory of the Standard Model (Weinberg, 2004).

Even if we presented the Standard Model in the framework of relativistic field theory, this theory is perfectly consistent with QM. M. Veltman and G. t'Hooft demonstrated it in 1971. They shared the Nobel Prize for Physics in 1999 for the proof that the Standard Model with the spontaneous breaking of the gauge symmetry is renormalizable and, therefore, predicts finite cross-sections at any energy. But this is definitely not the end of the story. We will discuss the limitations of the Standard Model and how we are moving beyond them in Sec. 13.8.

13.5 The discovery of the Higgs boson

The search for the Higgs boson, that is, its on-shell observation, has been a large cooperative effort that lasted more than 30 years and culminated in 2012 with the ATLAS and CMS discovery: the most spectacular success and the greatest challenge of the LHC.

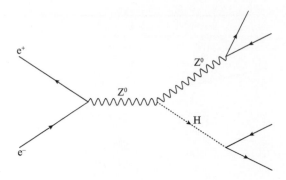

Fig. 13.3 Higgs production in a e^+e^- collider (**Higgs-strahlung**). The rightmost Z^0 and H decay immediately into a particle-antiparticle pair (or other channels for the Higgs). If $\sqrt{s}$ is sufficiently high, the Z^0 and H are produced on-shell and the invariant mass of the decay particles are m_H and M_Z, respectively. In this case, the only off-shell particle is the leftmost Z^0, where $p^2 \neq M_Z^2$.

The Standard Model uniquely predicts all couplings of the Higgs with elementary fermions and bosons. The couplings to vector bosons can be pinned down from the $(\partial_\mu \phi)^*(\partial^\mu \phi)$ term of eqn 13.32 because the Higgs field shows up in the form

$$VV(v + h)^2 = VVv^2 + VVh^2 + 2VVvh. \tag{13.54}$$

These terms correspond respectively to the Higgs mass, the quartic coupling between two vector bosons and two Higgs bosons, and the triple boson–boson–Higgs coupling. Therefore, early attempts at LEP focused on "Higgs-strahlung" (Fig. 13.3), which has a pretty high cross-section if m_H is small enough. Unfortunately, the maximum energy of LEP was limited by radiation losses to 206 GeV and the LEP physicists missed such a historical discovery for a fistful of GeVs. The Higgs production threshold at LEP was:

$$E_{thr} = E_{LEP} - M_Z = 115 \ \ \text{GeV} \tag{13.55}$$

and the LEP experiments did not observe the Higgs below this threshold putting a limit on m_H of 115 GeV.[7]

At hadron colliders, the Higgs boson can be produced in several ways but the dominant modes involve the heaviest particles because the heaviest the fermion the largest the coupling. From the Yukawa Lagrangian, the size of the coupling is:

$$\lambda_f = \sqrt{2}\frac{m_f}{v} \tag{13.56}$$

and from eqn 13.32 the Higgs–W–W coupling is gM_W. The big players in the production of the Higgs in pp and $p\bar{p}$ colliders are the top quark and the massive vector bosons. The "gluon–gluon" diagram is the leading contributor to the Higgs production at the LHC even if other diagrams (e.g. the fusion of two Ws) are also relevant. The gluon–gluon diagram is shown in Fig. 13.4. This figure gives an intuitive explanation of why the LHC is a pp instead of a $p\bar{p}$ collider: at very large energies particle production is no more dominated by the three valence quarks but by the gluons and quarks from the "sea" that produce the bulk of the proton mass. Therefore, we do not need to resort to stochastic cooling or other

[7]They observed an unexplained excess of four jet events that sparked a fierce debate on whether LEP should continue or stop to leave room for the LHC. The debate was cooled by the companies in charge of the construction of the new machine, which intimidated CERN with delay penalties. And the CERN Director said: "let there be the LHC."

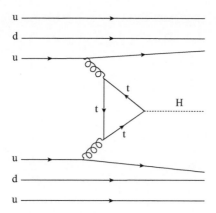

Fig. 13.4 The leading diagram producing a Higgs at the LHC. The Higgs originates from the second-order loop of the heaviest known fermion (t) after the emission of two gluons. The emission occurs from any of the proton quark.

sophisticated techniques to produce and inject the anti-protons, and we can exploit proton–proton collisions.[8] Similarly, the decays are dominated by $H \to W^+W^-$, $H \to Z^0Z^0$, $H \to t\bar{t}$ (if they were kinematically allowed) and $H \to b\bar{b}$.

Table 13.1 summarizes the main Higgs decay modes and the branching ratio (BR). By the time of writing, all the decays in heavy particles have been observed by the LHC experiments (ATLAS and CMS) but the discovery of the Higgs in 2012 was driven by a minor channel, whose BR is just 0.2% for $m_H \simeq 125$ GeV: $H \to \gamma\gamma$ (see Fig. 13.5). This evidence was strengthened by a more classical channel, $H \to ZZ^* \to l^+l^-l'^+l'^-$. Z^* stands for an off shell Z^0 that promptly decays in two fermions. This mode is allowed below the kinematic threshold $2M_Z$ by the Heisenberg uncertainty principle, as discussed in Sec. 6.6. Why was the two-photon decay so important?

The production rate of the Higgs at the LHC is given by:

$$N = \int dt \int_0^1 dx_1 dx_2 \mathcal{L}(t)\, f_g(x_1)f_g(x_2)\sigma(gg \to H)BR \qquad (13.57)$$

if the final-state particles are produced on-shell. $g(x_1)$ and $g(x_2)$ are the parton distribution functions of Sec. 7.7: the probability that a parton – in this case a gluon – brings a fraction of the total proton momentum between x and $x + dx$. These functions are generally known at the 10% level. $\mathcal{L}(t)$ is the instantaneous luminosity, which is integrated over the running time of the machine. BR is the branching ratio of the observed decay mode. Unlike LEP, the number of particles produced by the underlying events is extremely high, so the best observation strategy is a trade-off between clean final states and a high cross-section.

The $H \to \gamma\gamma$ decay mode was considered with great care at the time of the construction of the experiments. The signature is extremely clear, two energetic photons at high $\mathbf{p}_T$ isolated from low-energy jets. The detector designers decided to exploit at most the constraint on the invariant mass of the two photons seeking for a bump in the photon distribution of hard radiation or other background events. This is the reason why the ATLAS and CMS electromagnetic calorimeters studied in Sec. 3.8.2 are

[8]This is a special case of the **Pomeranchuk theorem**. It states – under very general hypotheses – that the difference between the scattering cross-section of a particle–antiparticle pair and a two-particle pair goes to zero if $s \to \infty$ (Bogoliubov *et al.*, 1975).

$H \to \gamma\gamma$	2.27×10^{-3}
$H \to ZZ^*$	2.62×10^{-2}
$H \to W^+W^{-*}$	2.14×10^{-1}
$H \to \tau^+\tau^-$	6.27×10^{-2}
$H \to b\bar{b}$	5.82×10^{-1}
$H \to c\bar{c}$	2.89×10^{-2}
$H \to Z\gamma$	1.53×10^{-3}
$H \to \mu^+\mu^-$	2.18×10^{-4}

Table 13.1 The BR of the Higgs boson computed from the SM for $m_H = 125$ GeV. The theoretical uncertainties range between 1 and 5%. Data from Zyla *et al.*, 2020.

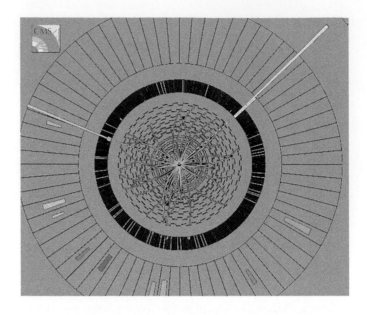

Fig. 13.5 A $H \to \gamma\gamma$ decay seen from the transverse plane of the CMS experiment. The soft tracks are visible in the inner part (tracker). The two largest energy deposits visible in the upper half of the image are not associated with any track and stop in the e.m. calorimeter (photons). The height of the white stripes are proportional to the γ energy. Reproduced with permission from (CMS, 2012). Copyright 2012 CERN.

[9]The Run II of the LHC was carried out at $\sqrt{s} = 14$ TeV, the design center-of-mass energy.

[10]An off-shell particle has $p^2 \neq m^2$ and, in this case, the invariant mass of the lepton pair coming from the Z^* can be significantly smaller than M_Z.

[11]In particle physics, we are used to calling "evidence" a signal whose significance exceeds 3σ, i.e. the probability of being a statistical fluctuation is less than 0.3%. A "discovery" (5σ significance) requires a probability of 0.00006%. We should not take this terminology too seriously. Systematic errors can be strongly underestimated because a spurious effect has been overlooked. In practice, evidence becomes a discovery when more than one experiment, possibly using different techniques affected by different systematics, confirms the original finding. This is also the rationale of having multiple experiments working at the same collider.

among the most precise devices ever built for high-energy γ or $e^{\pm}$. The segmentation of the calorimeter plays a central role in the matching of the energy deposit with the reconstructed tracks in the inner part of the detector, and vetoes a large number of charged particles produced close to the candidate photons. This method also offers a good resolution on the invariant mass, which should be compatible with an on-shell particle. Still, the expected width of the Higgs boson is so small ($\Gamma_H \simeq 4$ MeV) that, even today, the measurement is limited by the finite resolution of the detectors and there is no hope to get Γ_H in hadron colliders.

ATLAS and CMS published their first observations after accumulating 20 fb^{-1} of integrated luminosity at $\sqrt{s} = 8$ TeV. At that time, LHC had produced nearly 4×10^5 Higgs bosons. Besides the two-photon channels, the $H \to ZZ^*$ channel offered another important piece of evidence. $H \to ZZ^*$ is shorthand of $H \to Zl^+l^-$ and the star above the Z means that one of the two Z^0 is produced off-shell. During the LHC Run I at $\sqrt{s} = 8$ TeV, the Higgs was below the kinematic treshold for $H \to ZZ$ and only $H \to ZZ^*$ is allowed.[9] This was quite an hassle because they could not use the invariant mass of the leptons as a constraint for background reduction.[10] On the other hand, the observation of four high-$\mathbf{p}_T$ leptons or two high-energy photons are by far the easiest signatures in a hadron collider. Combining several decay channels, both ATLAS and CMS published evidence for a new particle at $m = 125$ GeV with a statistical significance exceeding five sigmas.[11] They were not able to claim that the particle was the SM Higgs boson because the statistics were not enough to measure the spin and parity of the new particle, which has been measured later on. At the time of writing, most of the decay channels of the Higgs have been observed. The Higgs mass is:

$$m_H = 125.25 \pm 0.147 \text{ GeV} \qquad (13.58)$$

and the measured branching ratios are consistent with the expectations from the Standard Model. The LHC is now moving toward precision physics, testing triple and quartic couplings, and evaluating radiative corrections to glimpse deviations from the Standard Model that might become evident at higher energies, as happened in the past for the V-A theory.

13.6   The Yukawa sector of the Standard Model: leptons

We already anticipated that neutrinos are a superposition of particles with different masses. This is a recent finding that had to be incorporated in the Standard Model but the theory is already equipped with the appropriate formalism. The formalism of lepton mixing was inherited from hadrons, where mixing was discovered by M. Gell-Mann, M. Levy, and N. Cabibbo in the early 1960s. We present the Cabibbo mechanism in its modern form starting from leptons because they are not plagued by QCD effects. Quarks will be included in Sec. 13.7.

For the sake of simplicity, let us consider only two flavors of neutrinos (for instance, ν_e and ν_μ) and two mass eigenstates, m_1 and m_2. If a ν_e is produced at $t = 0$, the wavefunction is:

$$\psi = a\psi_1 + b\psi_2 \tag{13.59}$$

where ψ_1 and ψ_2 are the wavefunctions of the mass eigenstates and a, b are real constants. Neutrinos fulfill both the Dirac and the Klein–Gordon equations. If a neutrino freely streams in vacuum, the solutions of eqn 5.74 are:

$$\psi(t) = a\psi_1 \exp\{-ip_1 \cdot x\} + b\psi_2 \exp\{-ip_2 \cdot x\}. \tag{13.60}$$

a and b cannot be generic numbers if the unitarity of the wavefunction must be preserved:

$$\int_{\mathbb{R}^3} d^3\mathbf{x} |\psi(x^\mu)|^2 = 1 \tag{13.61}$$

at any time. Hence $a^2 + b^2 = 1$ and we can parameterize eqn 13.60 as:

$$\psi(t) = \cos\theta \ \psi_1 \exp\{-ip_1 \cdot x\} + \sin\theta \ \psi_2 \exp\{-ip_2 \cdot x\} \tag{13.62}$$

where θ is called the **mixing angle**. At the time t, the wavefunction is quite different from the original one because the m_1 and m_2 components move at different speeds. For instance, if $m_1 < m_2$, ψ_1 lags behind ψ_2 and:

$$\psi(t) = \exp\{-ip_1 \cdot x\} \left[\cos\theta \ \psi_1 + \sin\theta \ \psi_2 \exp\{i(p_1 - p_2) \cdot x\}\right]. \tag{13.63}$$

If the two neutrino mass eigenstates propagate in the same direction, $\mathbf{p}_1 = p_{z1}\mathbf{k}$ and $\mathbf{p}_2 = p_{z2}\mathbf{k}$, where the propagation axis is the z axis and $\mathbf{p}_1, \mathbf{p}_2$ are the neutrino three-momenta in the laboratory frame (LAB).

Example 13.2

Equation 13.63 can be computed analytically if the neutrino produced at energy E is ultra-relativistic and the detector is unable to spot the minuscule time difference between the arrival of the mass eigenstates. In this case, the neutrino interaction (e.g. a CC) will be recorded at the time T and $x = (T, 0, 0, L)$ in NU, where L is the source-to-detector distance or, in the jargon, the **baseline** of the experiment. Then,

$$(p_1 - p_2) \cdot x = (E_1 - E_2, 0, 0, p_{z1} - p_{z2}) \cdot (t, 0, 0, L)$$
$$= (E_1 - E_2)t - (p_{z1} - p_{z2})L . \tag{13.64}$$

Since the neutrinos are ultra-relativistic and reach the detector at $t \simeq L$:

$$(p_1 - p_2) \cdot x = L(E_1 - E_2) - \frac{p_{z1}^2 - p_{z2}^2}{p_{z1} + p_{z2}} L$$
$$= -\frac{m_2^2 - m_1^2}{p_{z1} + p_{z2}} L \simeq -\frac{\Delta m^2 L}{2p}$$

where we used $E_1 = E_2$, $p_{z1} + p_{z2} \simeq 2p$ with $p \equiv (p_{z1} + p_{z2})/2$ (the mean value of the two momenta). Δm^2 is defined as $m_2^2 - m_1^2$. The phase mismatch is now:

$$\psi_1 \cos\theta + \psi_2 \sin\theta \exp\left\{-i\frac{\Delta m^2}{2p}L\right\} \tag{13.65}$$

and we can compute the probability of seeing a ν_μ in the detector instead of a ν_e by computing (Exercise 13.16):

$$P(\nu_e \to \nu_\mu) = \sin^2 2\theta \sin^2 \frac{\Delta m^2 L}{4p}$$
$$\simeq \sin^2 2\theta \sin^2 \frac{\Delta m^2 L}{4E}. \tag{13.66}$$

Equation 13.66 is called the **two-family oscillation formula** and was derived by several authors in 1976. This derivation is disputable because neutrinos are not plane waves of equal energy but real wavepackets with energy and position spread. The full wavepacket treatment is more cumbersome but provides the same result (Giunti *et al.*, 1991). The derivation in the SM is even more troublesome by, again, all differences are unobservable in any realistic experiment (Beuthe, 2002; Blasone *et al.*, 2005). The mismatch between the wavepackets of the mass eigenstate originates the phenomenon of **neutrino oscillations**, which consists of observing a flavor at the detector different from the original flavor at source (**appearance**) or, equivalently, an apparent loss of neutrinos of the same flavor of the source because in the two flavor approximation:

$$P(\nu_\mu \to \nu_\mu) = 1 - P(\nu_\mu \to \nu_e) \tag{13.67}$$

(**disappearance**). The discovery of neutrino oscillations has a decade-long history. In the following, we will discuss the experiment that provided for the first time a compelling evidence of oscillations: the SuperKamiokande (SK) experiment in Japan.

13.6.1 Cherenkov light

SuperKamiokande is based on a detection principle that I did not mention in Chap. 3 because of its small signal yield. It plays, however, an important role in particle identification and energy measurements. **Cherenkov radiation** was discovered by P. Cherenkov in 1958 and is commonly observed, for example, as a faint blue glow in water-based nuclear reactor cores.

> Cherenkov radiation is an electromagnetic radiation emitted when a charged particle crosses a dielectric medium at a speed greater than the phase velocity of light in that medium.

It can be explained using classical electrodynamics and is the optical analog of a supersonic boom (Jackson, 1998). Fig. 13.6 shows the formation of waves using the Huygens–Fresnel principle.[12] The figure also demonstrates that the emission angle θ of the wavefront is:

$$\cos \theta = \frac{1}{n\beta} \tag{13.68}$$

where n is the refraction index of the medium and β the speed of the particle. For ultra-relativistic particles, $\cos \theta$ is just n^{-1}. This light is faint compared with light emission in a scintillator and peaks in the extreme visible or ultraviolet. For condensed materials, the typical loss is 10^{-3} MeV cm^2 g^{-1} (Leo, 1994). A particle, whose speed is slower than the speed of light in the medium (c/n), cannot produce Cherenkov light. Nevertheless, this faint light is a wonderful tool to identify particles with the same momenta, if n is properly chosen or can be tuned: the lightest particles are above the minimum speed and produce light; the heaviest ones are below this speed and do not produce any photon. The tunability is available only for gases where n depends on the pressure. In water, $n = 1.333$ and the minimum energy to emit Cherenkov photons – the **Cherenkov threshold** – is 0.8 MeV for electrons, 160 MeV for muons and 1.4 GeV for protons.

13.6.2 The discovery of neutrino oscillations

Neutrino detectors require an enormous amount of mass and, hence, the detector material must be cost-effective. The cheapest is water, which can exploit Cherenkov radiation to observe charged particles. This idea led M. Koshiba to commence the construction of a large neutrino observatory (Kamioka), the first experiment that recorded neutrinos emitted by a supernova explosion (see Sec. 11.10). The explosion was seen in

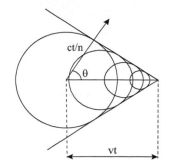

Fig. 13.6 A charged particle crosses a medium with refraction index n and the velocity v of the particle is faster than the Cherenkov threshold: $v > n/c$. The e.m. waves emitted create a coherent wavefront at the angle θ corresponding to the emission of Cherenkov light.

[12]We remind that the Huygens-Fresnel principle describes diffraction by considering every point on a wavefront as a source of wavelets. These wavelets spread out at the same speed as the source wave. The diffracted wavefront is a line tangent to all wavelets (Brooker, 2003; Sharma, 2006).

1987 from the Large Magellanic Cloud (SN1987a) and Kamioka collected 12 antineutrinos in about 13 s. Koshiba was awarded the Nobel Prize for Physics in 2002 for this observation and the success of Kamioka boosted the construction of an even larger water Cherenkov detector: SuperKamiokande.

This enormous detector was built to study other supernovas explosions (have yet to observed), neutrinos produced from the sun, from cosmic rays, and to search for violation of the baryon number in the $p \rightarrow e^+\pi^0$ decay (**proton decay**). SK achieved evidence of neutrino oscillations by observing the rate of ν_μ and ν_e CC events produced by the cosmic rays in the atmosphere. If oscillations exist and Δm^2 lies between 10^{-2} and 10^{-3} eV^2, a neutrino traveling through the earth[13] experiences a significant mismatch between the mass-eigenstate wavefunctions. On the other hand, the flavor of the neutrinos produced on top of the detector ($\langle L \rangle \simeq 15$ km) stays unchanged. In 1998, SK reported a very strong reduction of the ν_μ flux coming from below with respect to the expectations and to the ν_μ coming from above (Fukuda *et al.*, 1998). Instead, the ν_μ flux from above was in excellent agreement with the calculations. This difference was one order of magnitude larger than any deviation that might be due to local (geomagnetic, overburden) effects or other systematics. A few years later, the statistics of ν_μ became so large that SK was able to see the actual sinusoidal **oscillation pattern**: the disappearance of ν_μ was in excellent agreement with the two-family formula of eqn 13.67:

the number of ν_μ CC events with energies around 1–10 GeV showed a $\sin^2 aL/E$ pattern, which is peculiar to neutrino oscillations. The SK data indicated that $4a = |\Delta m^2|$ was about 2.5×10^{-3} eV^2. No significant deficit was observed for ν_e CC.

The ν_μ CC events are observed every time the muon is above the Cherenkov threshold and produces a ring of photons with an angle $\theta \simeq n^{-1}$ for $\beta \simeq 1$. The particles are identified from these rings, which provide the direction, too. If the particle is fully contained, SK can measure the energy from the range or the total number of photons. In a ν_μ CC, the ring is isolated from the rest of the light produced by particles after the breaking of the nucleon in deep inelastic CC events. Both the isolation and $\mathbf{p}_T$ technique can be employed in Superkamiokande to select a pure sample of charged currents although the resolution is much lower than Gargamelle. SK employs 22,500 tons of high-purity water and is located deep underground to avoid background induced by cosmic rays. Most of the protons are below the Cherenkov threshold, and cannot be observed. The electrons can be identified by the corresponding e.m. showers, which produces a fuzzier ring than muons. The Cherenkov photons are recorded by photomultipliers surrounding the water container (see Fig. 13.7)

A few people were skeptical about the SK results, given the limited imaging capability of the detector. However, a detailed experimental

[13] $L \simeq 100 - 10000$ km since the earth radius is 6371 km.

Fig. 13.7 The SuperKamiokande detector during maintenance. The tank is half-filled with water and the operators – sitting in a rubber boat – check the 13,000 photomultipliers (visible on the walls) surrounding the tank. The photomultipliers record the amount of Cherenkov photons and their time of arrival to reconstruct the neutrino interaction event. Image credit: Kamioka Observatory, ICRR (Institute for Cosmic Ray Research), University of Tokyo, Japan.

campaign carried out at accelerators with a smaller prototype validated the particle identification and many control samples substantiated the existence of the deficit. All doubts faded when SK observed the disappearance of ν_μ produced by an accelerator, that is, an artificial neutrino beam mostly made of ν_μ. The beam was generated at KEK and pointed toward SuperKamiokande ($L \simeq 250$ km). These findings were also confirmed by MACRO in Italy and MINOS in the United States. At that time, SK was not able to measure the appearance of new flavors, although they excluded $\nu_\mu \to \nu_e$ oscillations by the isotropic flux of ν_e leaving $\nu_\mu \to \nu_\tau$ as the most likely option (Migliozzi and Terranova, 2011). In 2015, SK with atmospheric neutrinos and OPERA, located in Italy and receiving neutrinos from CERN (Agafonova *et al.*, 2015), demonstrated that the leading oscillations observed by SK were – as expected – $\nu_\mu \to \nu_\tau$.

In the same year, T. Kajita, the leader of the atmospheric neutrino analysis in SK, received the Nobel Prize for the discovery of neutrino oscillations.

13.6.3 Leptonic Yukawa Lagrangian and mixing matrices

Many years have passed since the SuperkamioKande discovery and we have now nearly all the information to fix the Standard Model Lagrangian in the Yukawa sector for the neutrinos. Equation 13.51 does not account for mixing and oscillation because the "historical" Standard Model was designed for massless neutrinos. Nonetheless, it is already equipped with the right formalism because fermion mixing was observed in quarks in the 1960s.

Equation 13.51 expresses the fermions in their **mass basis**, which is not appropriate to describe CC and NC. These interactions produce

[14]The kinetic terms are the parts of the Lagrangian where only fields and their derivatives appear. For the photon and the Dirac fermions, they are:

$$-\frac{1}{4}F_{\mu\nu}F^{\mu\nu} \quad \text{and} \quad i\bar{\psi}\gamma^{\mu}\partial_{\mu}\psi,$$

respectively.

leptons of a given flavor, not mass. We, then, turn to the piece of the Standard Model Lagrangian that determines the W–fermion couplings. This is given by the **kinetic term**[14] of the Dirac Lagrangian by replacing the derivative with the covariant derivative defined in eqn 13.37. The result is just eqn 13.42.

Going back to the Yukawa Lagrangian that generates the lepton mass, we can rewrite eqn 13.51 in a compact matrix form and label the flavor states with a prime to distinguish flavor eigenstates from mass eigenstates:

$$\mathcal{L}_{YL} = -\frac{v + h(x^{\mu})}{\sqrt{2}}\bar{l}'_L Y'^l l'_R - \frac{v + h(x^{\mu})}{\sqrt{2}}\bar{\nu}'_L Y'^{\nu}\nu'_R + \text{h.c.} \tag{13.69}$$

where l' and ν' are now vectors of flavor fields:

$$l' \equiv \begin{pmatrix} e' \\ \mu' \\ \tau' \end{pmatrix} \quad , \quad \nu' \equiv \begin{pmatrix} \nu'_e \\ \nu'_\mu \\ \nu'_\tau \end{pmatrix}. \tag{13.70}$$

Y'^l, Y'^{ν} are generic 3×3 complex matrices. If the flavor and mass eigenstates were the same, as people believed before the discovery of neutrino oscillations, Y'^l and Y'^{ν} would be diagonal. Fortunately, linear algebra is full of resources and we can resort to a theorem stating that a general complex Y' matrix can be diagonalized by a **biunitary transformation**:

$$V_L^{\dagger}Y'V_R = Y \tag{13.71}$$

where V_L and V_R are appropriate unitary matrices and Y is diagonal with real and positive coefficients. We can define a new set of mass states l and ν that incorporate the unitary matrices:

$$l_{L,R} \equiv V_{L,R}^{\dagger}l'_{L,R} \;\; ; \;\; \nu_{L,R} \equiv V_{L,R}^{\dagger}\nu'_{L,R} \tag{13.72}$$

$$l'_{L,R} \equiv V_{L,R}l_{L,R} \;\; ; \;\; \nu'_{L,R} \equiv V_{L,R}\nu_{L,R} \tag{13.73}$$

so that the full **lepton Yukawa Lagrangian** ($\mathcal{L}_{YL}$) is again in diagonal form like the simplified Lagrangian of eqn 13.47:

$$\mathcal{L}_{YL} = -\frac{v + h(x^{\mu})}{\sqrt{2}}\bar{l}_L Y^l l_R - \frac{v + h(x^{\mu})}{\sqrt{2}}\bar{\nu}_L Y^{\nu}\nu_R + \text{h.c.} \tag{13.74}$$

and is described by two diagonal matrices Y^l and Y^{ν}.

There are forces in nature where this redefinition plays no role. For instance, nobody cares about the mismatch between flavor and mass eigenstates in QED, where the vertices always couple particles of the same flavor. In a QED electron scattering:

$$\bar{e}'_L\gamma^{\mu}e'_L = \bar{e}_L V_L^{\dagger}\gamma^{\mu}V_L e_L = \bar{e}_L\gamma^{\mu}e_L \tag{13.75}$$

because $V_L^{\dagger}V_L = \mathbb{I}$ and the Vs commute with γ^{μ} (the Dirac matrices act on the bispinors and are flavor-independent). The same statement holds in QCD, too. But the result is different in flavor-changing interactions.

A charged-current vertex between an "up-type" lepton (l^U, i.e. a neutrino), a "down-type" lepton (l^D, i.e. a charged lepton) and a W^- reads:

$$\bar{l'}^U_L \gamma^\mu l'^D_L W^-_\mu = \bar{l}^U_L \gamma^\mu V^{U\dagger}_L V^D_L l^D_L W^-_\mu \qquad (13.76)$$

and the amplitude depends on the product:

$$V = V^{U\dagger}_L V^D_L = V^{\nu\dagger}_L V^l_L \qquad (13.77)$$

or, equivalently, on its inverse:

$$V_{PMNS} = V^{-1} = V^{D\dagger}_L V^U_L = V^{l\dagger}_L V^\nu_L. \qquad (13.78)$$

> The V_{PMNS} matrix is called the **Pontecorvo–Maki–Nakagawa–Sakata** matrix. The amplitude of CC events and, in general, weak processes that include the exchange of a $W^\pm$ depend on the corresponding matrix element of V_{PMNS}.

Since the neutrino mass is small, we can record the oscillations in neutrinos much more easily than charged leptons. The charged lepton oscillations are unobservable due to the enormous size of their squared-mass difference and it is natural to interpret V_{PMNS} as the mixing matrix among neutrino mass eigenstates, considering the charged lepton flavor states equal to their mass eigenstate. Note, anyway, that this interpretation is done without loss of generality or additional hypotheses.

The same consideration holds for quarks, too. Unfortunately, the designer of the electroweak theory put the d, s, and b quarks on the lower side of the $SU(2)_W$ doublet, that is, they assigned $I^w_z = -1/2$ to these particles, while the neutrinos have $I^w_z = +1/2$ ("up-type" particles). The quark-mixing hypothesis was developed under the assumption that the mixing quarks were the "down-type quarks" because, at that time, the c-quark had not been discovered. This convention still holds and we define the quark mixing matrix as V instead of V^{-1} to remove this inconsistency in notation:

$$V_{CKM} = V = V^{U\dagger}_L V^D_L . \qquad (13.79)$$

> Flavor-changing charged currents in quarks depends on V_{CKM}: the **Cabibbo–Kobayashi–Maskawa** matrix. V_{CKM} and V_{PMNS} describe the same phenomenon: a flavor eigenstate is a superposition of mass eigenstates.

These matrices have a long history. The technique we used to derive them is inherited from the 1973 work of M. Kobayashi and T. Maskawa but the concept of fermion mixing dates back to M. Gell-Mann and M. Levy and was successfully employed by N. Cabibbo in 1963. At that time, neutrinos were considered massless and, therefore, mixing was introduced for quarks only. In 1957, B. Pontecorvo introduced the concept of neutrino oscillations that, in turn, requires fermion mixing. Z. Maki, M. Nakagawa, and S. Sakata wrote down the V_{PMSN} five years later.

13.6.4 The PMNS mixing matrix*

V_{PMNS} is the product of two 3×3 matrices but there is no hope of measuring 18 complex parameters because most of them are unphysical. First, in all processes known to date, the amplitudes depend on the product $V_L^{l\dagger} V_L^{\nu}$ and not on $V_L^{l\dagger}$ and V_L^{ν}, separately. Furthermore, several complex phases can be canceled by a redefinition of the global phase of the fields, which plays no role in the scattering amplitudes. As an example, we show that, if V_{PMSN} were a 2×2 matrix, it would depend only on one parameter: the θ angle of eqn 13.66.

Example 13.3

A 2×2 complex matrix has eight real parameters, which are reduced imposing the unitarity conditions $U^\dagger U = U U^\dagger = \mathbb{I}$. Solving the corresponding equations, we can write the matrix as:

$$U = \begin{pmatrix} a & b \\ -e^{i\alpha} b^* & e^{i\alpha} a^* \end{pmatrix} \tag{13.80}$$

[15] Incidentally, this is not an $SU(2)$ matrix because $\det U = e^{i\alpha} \neq 1$

with $|a|^2 + |b|^2 = 1$.[15] Changing variables, eqn 13.80 can be recast as:

$$U = e^{i\alpha/2} \begin{pmatrix} e^{i\alpha_1} \cos\theta & e^{i\alpha_2} \sin\theta \\ -e^{-i\alpha_2} \sin\theta & e^{-i\alpha_1} \cos\theta \end{pmatrix}. \tag{13.81}$$

We introduce two differences of angles: $\alpha_1 = \psi + \Delta$ and $\alpha_2 = \psi - \Delta$ so that

$$U = \begin{pmatrix} e^{i(\alpha/2+\psi)} & 0 \\ 0 & e^{i(\alpha/2-\psi)} \end{pmatrix} \begin{pmatrix} \cos\theta & \sin\theta \\ -\sin\theta & \cos\theta \end{pmatrix} \begin{pmatrix} e^{i\Delta} & 0 \\ 0 & e^{-i\Delta} \end{pmatrix}. \tag{13.82}$$

In charged currents, where the vertices hold a quadratic form like $\bar{\psi} U \psi$, we can redefine the fields on the left by a global phase as:

$$\begin{pmatrix} \tilde{\psi}_1(x^\mu) \\ \tilde{\psi}_2(x^\mu) \end{pmatrix} \to \begin{pmatrix} \tilde{\psi}_1(x^\mu) e^{i(\alpha/2+\psi)} \\ \tilde{\psi}_2(x^\mu) e^{i(\alpha/2-\psi)} \end{pmatrix} \tag{13.83}$$

and, on the right, as:

$$\begin{pmatrix} \psi_1(x^\mu) \\ \psi_2(x^\mu) \end{pmatrix} \to \begin{pmatrix} \psi_1(x^\mu) e^{i\Delta} \\ \psi_2(x^\mu) e^{-i\Delta} \end{pmatrix}. \tag{13.84}$$

In this way, the mixing matrix U only depends on θ and has the form:

$$U = \begin{pmatrix} \cos\theta & \sin\theta \\ -\sin\theta & \cos\theta \end{pmatrix}. \tag{13.85}$$

---

[16] This demonstration holds if fermions are described by the Dirac equation, as in the Standard Model. If neutrinos are Majorana fermions, the independent phases are $N(N-1)/2 = 3$.

In general, an $N \times N$ complex unitary matrix can always be parameterized with $N(N-1)/2$ mixing angles and $N(N+1)/2$ complex phases (Giunti and Kim, 2007). Using the same procedure described in Example 13.3, the physical phases are only $(N-2)(N-1)/2$. Since the Standard Model has three families, that is, three "up-type" fermions and three "down-type" fermions, $N = 3$ and V_{PMNS} can be parameterized with three angles and one phase.[16] This is generally written as:

$$V = \begin{pmatrix} 1 & 0 & 0 \\ 0 & c_{23} & s_{23} \\ 0 & -s_{23} & c_{23} \end{pmatrix} \begin{pmatrix} c_{13} & 0 & s_{13}e^{-i\delta_{CP}} \\ 0 & 1 & 0 \\ -s_{13}e^{i\delta_{CP}} & 0 & c_{13} \end{pmatrix} \begin{pmatrix} c_{12} & s_{12} & 0 \\ -s_{12} & c_{12} & 0 \\ 0 & 0 & 1 \end{pmatrix}$$

$$= \begin{pmatrix} c_{12}c_{13} & s_{12}c_{13} & s_{13}e^{-i\delta_{CP}} \\ -s_{12}c_{23} - c_{12}s_{13}s_{23}e^{i\delta_{CP}} & c_{12}c_{23} - s_{12}s_{13}s_{23}e^{i\delta_{CP}} & c_{13}s_{23} \\ s_{12}s_{23} - c_{12}s_{13}c_{23}e^{i\delta_{CP}} & -c_{12}s_{23} - s_{12}s_{13}c_{23}e^{i\delta_{CP}} & c_{13}c_{23} \end{pmatrix}.$$

$$(13.86)$$

In eqn 13.86, the three angles are labeled θ_{12}, θ_{23}, and θ_{13} and $c_{ij} \equiv \cos\theta_{ij}$, $s_{ij} \equiv \sin\theta_{ij}$. In this parameterization, which has been in use for more than 20 years, $\theta_{ij} \in \left[0, \frac{\pi}{2}\right]$ and $\delta_{CP} \equiv \delta \in [0, 2\pi]$. Physical observables are parameterization-independent and the range of the angles is fixed without loss of generality.

The SM does not provide any guidance on the values of the three angles and δ_{CP}, and their measurement is a tremendous challenge. In the last 20 years, many neutrino experiments have constrained these parameters but, for instance, we do not know yet δ_{CP}, whether θ_{23} is slightly larger or slightly lower than $45°$ (the **octant ambiguity**) and whether $\Delta m_{23}^2 \equiv m_3^2 - m_2^2$ is negative or positive. As we will see in the next section, the situation of the V_{CKM} is much brighter because we do not need to use neutrino oscillations, which come with the tiny neutrino cross-sections. The current best values of the PMNS are listed in Table 13.2. All these measurements come from neutrino oscillation data (see Example 13.4) and are nearly free from QCD systematic uncertainties. Even the most challenging measurement – the δ_{CP} phase that signals a possible violation of the CP symmetry in the leptonic sector of the SM – could be done by just measuring:

$$R \equiv \frac{P(\nu_\mu \to \nu_e)}{P(\bar{\nu}_\mu - \bar{\nu}_e)} \qquad (13.87)$$

using accelerator neutrino beams (Mezzetto and Terranova, 2020). $R \neq 1$ would demonstrate that the CP symmetry is violated by leptons as it is in quarks. We will discuss this important item and its cosmological implications in Sec. 13.7.

θ_{13}	$8.57^{+0.13}_{-0.12}$
θ_{23}	$49.0^{+1.1}_{-1.4}$
θ_{12}	$33.44^{+0.78}_{-0.75}$
δ_{CP}	195^{+51}_{-25}
Δm_{21}^2	$7.42^{+0.21}_{-0.20} \times 10^{-5}$
$|\Delta m_{23}^2|$	$2.514^{+0.028}_{-0.027} \times 10^{-3}$

Table 13.2 Current values of the parameters of V_{PMNS} and the quadratic mass difference between two neutrino mass eigenstates (Esteban *et al.*, 2020). The units are degrees for the angles and eV^2 for the quadratic mass differences.

Example 13.4

Equation 13.66 can be extended to three neutrino flavors using, for example, QM in the Dirac formalism (Giunti and Kim, 2007). In this case, a neutrino wavefunction is written as:

$$|\nu_\alpha\rangle = \sum_k U_{\alpha k}^* |\nu_k\rangle \qquad (13.88)$$

where $U_{\alpha k}^*$ is the α, k element of the PMNS. $\alpha = e, \mu, \tau$ is the flavor index, and $k = 1, 2, 3$ is the mass eigenstate index.

In plane wave approximation, $|\nu_k\rangle$ evolves as:

$$|\nu_k(t)\rangle = e^{-iE_k t}|\nu_k\rangle. \tag{13.89}$$

Note that in this case, we made the assumption the the neutrino is produced at fixed momentum, not energy. This assumption is as disputable as the equal-energy assumption of Example 13.2 but gives the same final result, as shown in this Example. The flavor state at t is then:

$$|\nu_\alpha(t)\rangle = \sum_k U_{\alpha k}^* e^{-iE_k t}|\nu_k\rangle. \tag{13.90}$$

The conservation of lepton family number requires that the neutrino is produced in a well-defined flavor state at $t = 0$. Hence, the neutrino wavefunction at $t = 0$ is

$$|\nu_\alpha(0)\rangle \equiv |\nu_\alpha\rangle. \tag{13.91}$$

Since U is unitary,

$$|\nu_k\rangle = \sum_\alpha U_{\alpha k}|\nu_\alpha\rangle \tag{13.92}$$

and we can replace this sum in eqn 13.90 to get

$$|\nu_\alpha(t)\rangle = \sum_{\beta=e,\mu,\tau} \left(\sum_k U_{\alpha k}^* e^{-iE_k t} U_{\beta k}\right)|\nu_k\rangle. \tag{13.93}$$

The transition amplitude is then:

$$\mathcal{A}_{\nu_\alpha \to \nu_\beta}(t) = \langle\nu_\beta|\nu_\alpha(t)\rangle = \sum_k U_{\alpha k}^* U_{\beta k} e^{-iE_k t} \tag{13.94}$$

and the oscillation probability in vacuum is ($L = t$ in NU):

$$P_{\nu_\alpha \to \nu_\beta}(t) = |\mathcal{A}_{\nu_\alpha \to \nu_\beta}(t)|^2 = \sum_{k,j} U_{\alpha k}^* U_{\beta k} U_{\alpha j} U_{\beta j}^* e^{-i(E_k - E_j)L}. \tag{13.95}$$

In the equal momentum approximation, we recover the same result of the equal energy approximation for ultra-relativistic neutrinos:

$$E_k \simeq E + \frac{m_k^2}{2E} \implies E_k - E_j \simeq \frac{\Delta m_{kj}^2}{2E} \tag{13.96}$$

because $E \simeq |\mathbf{p}|$.

It is instructive to recast eqn 13.95 detaching the real and imaginary part:

$$P_{\nu_\alpha \to \nu_\beta}(t) = \delta_{\alpha\beta} - 4\sum_{k>j} \mathrm{Re}[U_{\alpha k}^* U_{\beta k} U_{\alpha j} U_{\beta j}^*] \sin^2\left(\frac{\Delta m_{kj}^2 L}{4E}\right)$$

$$+ 2\,\mathrm{Im}[U_{\alpha k}^* U_{\beta k} U_{\alpha j} U_{\beta j}^*] \sin\left(\frac{\Delta m_{kj}^2 L}{2E}\right). \tag{13.97}$$

Since an antineutrino has the same mass eigenstates of the neutrino and the only change is:[17]

---

[17] Since the mass of an antiparticle is the same of a particle, novices are tempted to identify $|\nu_k\rangle$ with $|\bar\nu_k\rangle$. Please, don't do it. These wavefunctions are solutions of the Dirac equation: $|\nu_k\rangle$ is the particle solution and $|\bar\nu_k\rangle$ is the anti-particle solution. They represent different particles (a Dirac fermion and a Dirac anti-fermion) with the same mass m_k.

$$|\bar{\nu}_\alpha\rangle = \sum_k U_{\alpha k}|\bar{\nu}_k\rangle \tag{13.98}$$

the oscillation probability is the same except for the imaginary part, which changes sign:

$$P\bar{\nu}_\alpha \to \bar{\nu}_\beta(t) = \delta_{\alpha\beta} - 4\sum_{k>j} \text{Re}[U^*_{\alpha k}U_{\beta k}U_{\alpha j}U^*_{\beta j}]\sin^2\left(\frac{\Delta m^2_{kj}L}{4E}\right)$$

$$-2\,\text{Im}[U^*_{\alpha k}U_{\beta k}U_{\alpha j}U^*_{\beta j}]\sin\left(\frac{\Delta m^2_{kj}L}{2E}\right). \tag{13.99}$$

$\nu_\alpha \to \nu_\beta$ is the CP-conjugate states of $\bar{\nu}_\alpha \to \bar{\nu}_\beta$ (see Exercise 13.11). Equation 13.99 is the transition probability we obtain if we apply the partity and C-parity operators to $\nu_\alpha \to \nu_\beta$. The transition probability should be the same if CP were a symmetry of the electroweak interactions. In conclusion,

---

CP is violated in neutrinos, that is in the lepton sector of the electroweak theory, if $P(\nu_\alpha \to \nu_\beta)(t) \neq P(\bar{\nu}_\alpha \to \bar{\nu}_\beta)(t)$ or, equivalently, if $\delta_{CP} \neq 0$ or π.

This is the reason why we are used to label the δ phase as δ_{CP}.

13.6.5 Effective neutrino masses

The measurement of absolute neutrino masses is performed using flavor eigenstates because we cannot select a wavefunction that contains only one mass eigenstate. For instance, the drop in the Fermi–Kurie endpoint of Sec. 11.4.2 is searched in a $^3\text{H} \to {}^3\text{He} + e^- + \bar{\nu}_e$ decay, where $\bar{\nu}_e$ is a flavor eigenstate and has no definite mass. If we replace the flavor eigenstates with the mass eigenstates, we can show (Giunti and Kim, 2007) that the origin of the drop in the endpoint is an **effective neutrino mass** m_β depending on the mixing angles of the PMNS. In particular:

$$m_\beta \equiv \left[c^2_{13}c^2_{12}m^2_1 + c^2_{13}s^2_{12}m^2_2 + s^2_{13}m^2_3\right]^{1/2} \tag{13.100}$$

or, equivalently,

$$m^2_\beta = \sum_{k=1}^3 |U_{ek}|^2 m^2_k \tag{13.101}$$

We do not know yet what is the heaviest eigenstate (the **neutrino mass hierarchy** problem). If it is the same as quarks, measuring the absolute mass of neutrinos might be even more challenging because the leading term in eqn 13.100 is m_1, the lightest mass eigenstate.

If neutrinos are Majorana particles, the PMNS matrix gains two additional phases. We cannot measure the Majorana phases using neutrino oscillations because they cancel out in any $\nu_\alpha \to \nu_\beta$ transition. Unlike neutrino-less double beta decay (Fig. 11.7 right), oscillations do not help to understand whether neutrinos are their own antiparticles. However, the off-shell neutrinos running in Fig. 11.7 (right) are flavor eigenstates, too. We then need an **effective Majorana mass** $m_{\beta\beta}$ that accounts for the mixing and the (three!) phases of the PMNS. This mass is:

$$m_{\beta\beta} \equiv |m_1 c_{13}^2 c_{12}^2 + m_2 c_{13}^2 s_{12}^2 e^{i\phi_2} + m_3 s_{13}^2 e^{i\phi_3}| \tag{13.102}$$

where ϕ_2 and ϕ_3 are functions of δ_{CP} and the additional **Majorana phases** η_1 and η_2 of the PMNS[18]. Eqn 13.102 is equivalent to

$$m_{\beta\beta} = \left| \sum_{i=1}^{3} U_{ek}^2 m_k \right|. \tag{13.103}$$

Again, having m_1 as the lightest eigenstate is an issue but other suppression effects may arise if the phases conspire to reduce the size of $m_{\beta\beta}$.

The lack of SM predictions in the Yukawa sector has, therefore, important experimental consequences.

13.7 The Yukawa sector of the Standard Model: quarks and CP violation

The mixing between mass and flavor eigenstates was discovered in quarks 30 years before neutrino oscillations. Quark mixing was an attempt to save **quark universality**: the counterpart of lepton universality in the quark sector. It originates from the 1963 works of N. Cabibbo to understand the decay of "strange" particles,[19] whose amplitude was very different from any process explained by the V-A theory. Reading Cabibbo's papers in modern language, we can conjecture that the s-quark wavefunction is a superposition of mass eigenstates whose eigenvalues are m_d and m_s. The superposition is due to a mixing angle called the **Cabibbo angle** θ_c. After the discovery of the charm quark, Cabibbo's ideas further evolved. If four quarks exist, their flavor states are mixed with their mass eigenstates:

$$d' = \cos\theta_c d + \sin\theta_c s$$
$$s' = -\sin\theta_c d + \cos\theta_c s$$

where the primed quantities are the flavor eigenstates and the unprimed quantities are the mass eigenstates. Unlike the PMNS matrix, the CKM matrix is nearly diagonal. We do not know the reason for it because the SM is not a predictive theory in the Yukawa sector. Still, if we rely on experimental data, we can write an approximate two-family form of the CKM neglecting the contributions of the heaviest quarks (b and t) as:

$$\begin{pmatrix} d' \\ s' \end{pmatrix} \simeq \begin{pmatrix} V_{ud} & V_{us} \\ V_{cd} & V_{cs} \end{pmatrix} \begin{pmatrix} d \\ s \end{pmatrix} \tag{13.104}$$

Without knowing anything about the c-quark, we can measure the first four elements of the Cabibbo–Kobayashi–Maskawa matrix V_{CKM}. Example 13.3 shows that these elements only depend on the Cabibbo angle

[18] The two additional phases can be parameterized multiplying eqn 13.86 by the diagonal matrix:

$$\begin{pmatrix} e^{i\eta_1} & 0 & 0 \\ 0 & e^{i\eta_2} & 0 \\ 0 & 0 & 1 \end{pmatrix}$$

If m_1 is the lightest mass eigenvalue (**normal ordering**), $\phi_2 = 2(\eta_2 - \eta_1)$ and $\phi_3 = -2(\delta_{CP} + \eta_1)$. Other possibilities are discussed in Zyla *et al.*, 2020.

[19] Strange particles are hadrons containing at least one s quark. They were called this way because their decay amplitude is much smaller than ordinary mesons. There is nothing strange about it, indeed: light strange mesons must change the flavor of s to decay into lighter mesons ($s \to u$ in CC vertices) and the only flavor-changing interactions are the weak forces, which are much fainter than strong and e.m. forces due to the large mass of the Ws.

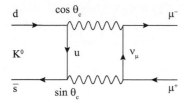 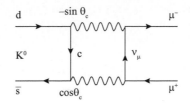

Fig. 13.8 The two diagrams that contribute to the $\mu^+\mu^-$ decay of a neutral kaon. This process is strongly suppressed because the amplitudes of the diagrams have opposite signs ($\sin\theta_c \to -\sin\theta_c$). This cancellation is due to the GIM mechanism.

that can be measured with exquisite precision using, for example, a superallowed beta decay like those in Sec. 11.4.5:

$$|V_{ud}| \simeq \cos\theta_c = 0.97370 \pm 0.00014. \tag{13.105}$$

It corresponds to $\theta_c \simeq 13.1°$ (Zyla *et al.*, 2020). In 1971, other physicists predicted the existence of the c quark using the ideas behind V_{CKM}. In 1970, S. Glashow, J. Iliopoulos, and L. Maiani showed that:

> unlike charged currents, flavor-changing neutral current are strongly suppressed in weak interactions due to the **Glashow–Iliopoulos–Maiani (GIM) mechanism.**

The GIM mechanism is at work in Fig. 13.8 where the "box diagram" with off-shell u quarks is nearly canceled by the corresponding diagram with the c quark.[20]

Again, that makes perfect sense to us. The amplitude of the first diagram must be proportional to:

$$\mathcal{M} \sim g^4 \sin\theta_c \cos\theta_c \tag{13.106}$$

but the amplitude of the second diagram is:

$$\mathcal{M} \sim g^4(-\sin\theta_c)\cos\theta_c \tag{13.107}$$

and the corresponding amplitude is nearly zero ($BR \simeq 6.8 \times 10^{-9}$). The GIM mechanism was proposed in 1970, three years before the discovery of the c quark!

Leaving history aside, we can write down the full **Yukawa Lagrangian** for the quark sector. From eqn 13.74, we have:

$$\mathcal{L}_{YQ} = -\frac{v + h(x^\mu)}{\sqrt{2}}\bar{q}_L^{\prime U} Y^{\prime U} q_R^{\prime U} - \frac{v + h(x^\mu)}{\sqrt{2}}\bar{q}_L^{\prime D} Y^{\prime D} q_R^{\prime D} + \text{h.c.} \tag{13.108}$$

where U are the "up-type" quark, that is, the quarks that are in the upper position in the $SU(2)$ doublet: u, c, and t. D are the "down-type" quark, which lie in the lower position in the $SU(2)$ doublet: d, s, and b. Both types have weak isospin $I^W = 1/2$ but d, s, and b have $I_z^W = -1/2$ and u, c, and t have $I_z^W = +1/2$. As in the lepton sector, we redefine $q \to q'$ to diagonalize the Yukawa matrices $Y^{\prime U}$ and $Y^{\prime D}$. The Yukawa matrices do not affect the e.m. and strong interactions of quarks but impacts on the W vertices. Using the definition of the Cabibbo–Kobayashi–Maskawa matrix in eqn 13.79, we can write the CC vertex as:

[20]The cancellation would be perfect if only four degenerate quarks existed.

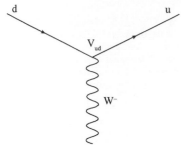

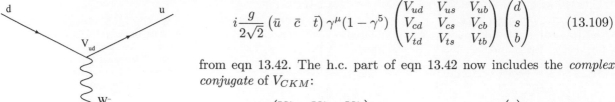

$$i\frac{g}{2\sqrt{2}}\begin{pmatrix}\bar{u} & \bar{c} & \bar{t}\end{pmatrix}\gamma^{\mu}(1-\gamma^5)\begin{pmatrix}V_{ud} & V_{us} & V_{ub} \\ V_{cd} & V_{cs} & V_{cb} \\ V_{td} & V_{ts} & V_{tb}\end{pmatrix}\begin{pmatrix}d \\ s \\ b\end{pmatrix} \qquad (13.109)$$

from eqn 13.42. The h.c. part of eqn 13.42 now includes the *complex conjugate* of V_{CKM}:

$$i\frac{g}{2\sqrt{2}}\begin{pmatrix}V_{ud}^* & V_{us}^* & V_{ub}^* \\ V_{cd}^* & V_{cs}^* & V_{cb}^* \\ V_{td}^* & V_{ts}^* & V_{tb}^*\end{pmatrix}\begin{pmatrix}\bar{d} & \bar{s} & \bar{b}\end{pmatrix}\gamma^{\mu}(1-\gamma^5)\begin{pmatrix}u \\ c \\ t\end{pmatrix}. \qquad (13.110)$$

Fig. 13.9 The vertex among a d, u, and W^- accounting for quark mixing. The strength of the vertex is the same as in Fig. 12.3 ($g/2\sqrt{2}$ times the LH projector P_L) but is weighted by the V_{ud} term of the CKM matrix.

To ease the comparison with Fig. 12.3, the modified vertex is shown in Fig. 13.9 for a $u - d - W^-$ vertex. These formulas are the modern realization of the Cabibbo mixing because the weak coupling constant g must be corrected with a term V_{ij} or V_{ij}^*. In 1973, Kobayashi and Maskawa pointed out that, $V_{ij} \neq V_{ij}^*$ if the matrix is 3×3 and δ_{CP} is different from 0 or π.

> In the Standard Model, both V_{CKM} and V_{PMNS} are unitary complex matrices described by three mixing angles and one complex phase (δ_{CP}) that explains CP-violating amplitudes.

Measuring the terms of the CKM is relatively simple from the point of view of the statistics, which is large thanks to the large hadronic cross-sections, but nightmarish for the treatment of systematics. Systematics arise from non-perturbative QCD corrections and are handled semi-empirically, by lattice QCD, or chiral perturbation theory, like in nuclear physics. Even if the quark sector is plagued by QCD effects, the results obtained since 1964 on the CKM are astonishing.

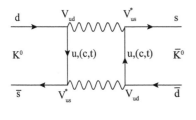

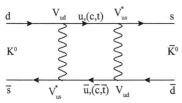

Fig. 13.10 Feynman diagrams producing $K^0 \leftrightarrow \bar{K}^0$ oscillations. Both of them are second-order diagrams where the exchange of the virtual $W^{\pm}$ occurs either as an s-channel (upper diagram) or t-channel (lower diagram). The off-shell fermions can be u, c, and t quarks. In the figure we also indicated the corresponding CKM matrix elements for off-shell u and $\bar{u}$ quarks.

13.7.1   CP violation

The existence of CP violation in the PMNS is not yet established and is on the agenda of the biggest neutrino facilities currently under construction (DUNE in the United States and HyperKamiokande in Japan). On the contrary, CP violation in the CKM was discovered nine years before the CKM matrix!

The main tool used for the discovery is **meson oscillation**, which – despite the name – has quite a different origin than neutrino oscillations because

> the phase mismatch $\Delta m^2 L/4E$ among quarks is negligible in any realistic experiment due to the large quark mass difference.

Quarks undergo thousands of oscillation cycles in the hadronization phase and, then, the squared-sinusoidal pattern averages to $1/2$:

$$P(q_\alpha \to q_\beta) \sim \sin^2\frac{\Delta m^2 L}{4E} \simeq \frac{1}{2}. \qquad (13.111)$$

The oscillation pattern disappears for any realistic L and the quark with flavor d produced after a decay is already a time-independent mixture of d and s mass eigenstates:

$$d_f = \cos\theta_c \, d + \sin\theta_c \, s \qquad (13.112)$$

if only four quarks exist (**two-family approximation**). And,

$$d_f = V_{ud} \, d + V_{us} \, s + V_{ub} \, b \qquad (13.113)$$

if we account for all quarks.

13.7.2 Meson oscillations

In a sense, CP violation (CPV) in quarks was discovered "too early." In 1964, the SM was in its infancy and people tried to pin down all potential sources of CPV in meson oscillation and decay. Unlike neutrinos,

> the meson oscillations $K^0 \leftrightarrow \bar{K}^0$, $D^0 \leftrightarrow \bar{D}^0$, $B^0 \leftrightarrow \bar{B}^0$, etc. result from second-order Feynman diagrams that transform a particle into its antiparticle.

The wavefunction of K^0 is given by the three-quark model in Sec. 8.7. D^0 and B^0 are made by heavy quarks, whose mass is $> \Lambda_{\mathrm{QCD}}$. B^0 can be considered as a $d\bar{b}$ bound state and D^0 as a $c\bar{u}$ bound state, although the wavefunction cannot be extracted by the three-quark model.

The second-order diagrams that produce $K^0 \leftrightarrow \bar{K}^0$ oscillations are depicted in Fig. 13.10. The off-shell particles can be any of the "up-type" quarks (u, c, t) and any of the "down-type" antiquarks $(\bar{u}, \bar{c}, \bar{t})$ even if we depicted the diagrams only for the u and $\bar{u}$ quarks. The vertices depend on the CKM entries and their complex conjugate. Therefore, the amplitude of a $K^0 \leftrightarrow \bar{K}^0$ driven, for example, by off-shell c and t quarks is proportional to:[21]

$$\mathcal{M}_{K^0 \leftrightarrow \bar{K}^0} \sim V_{cd} V_{cs}^* V_{td} V_{ts}^* \qquad (13.114)$$

while

$$\mathcal{M}_{\bar{K}^0 \leftrightarrow K^0} \sim V_{cd}^* V_{cs} V_{td}^* V_{ts} \qquad (13.115)$$

and the two amplitudes are different if

$$V_{cd} V_{cs}^* V_{td} V_{ts}^* \neq V_{cd}^* V_{cs} V_{td}^* V_{ts}. \qquad (13.116)$$

This phenomenon was observed for the first time by J. Cronin and V. Fitch in 1964 by studying the propagation and decay of neutral kaons. If K^0 were a pure mass eigenstate with no quantum corrections, it would propagate as a standard unstable particle:

$$|K^0(t)\rangle = |K^0(t=0)\rangle \exp\left\{-\frac{t}{2\tau}\right\} \exp\{-iEt\} \qquad (13.117)$$

[21] To identify this diagram, replace the leftmost u quark in Fig. 13.10 top with a c quark, replace the rightmost u quark with a t quark, and update the corresponding CKM entries. Similarly, the diagram in Fig. 13.10 bottom must be changed replacing $u \to c$, $\bar{u} \to \bar{t}$ and updating the CKM entries.

where E is the particle energy and $\tau = \Gamma^{-1}$ is the lifetime ($E = m$ in the particle rest frame). The system is a superposition of K^0 and $\bar{K}^0$, whose wavefunction cannot be factorized in $|K^0\rangle$ and $|\bar{K}^0\rangle$. We can easily describe this system because $|K^0(t = 0)\rangle \equiv |K^0\rangle$ and $|\bar{K}^0(t = 0)\rangle \equiv |\bar{K}^0\rangle$ are eigenstates of the $\mathcal{CP}$ operator and we can construct the physical states as a linear combination of them.

Example 13.5

K^0 and $\bar{K}^0$ have $J = 0$ and intrinsic parity $P = -1$ from the three-quark model. Then,

$$\mathcal{P}|K^0\rangle = -|K^0\rangle \;\; ; \;\; \mathcal{P}|\bar{K}^0\rangle = -|\bar{K}^0\rangle \qquad (13.118)$$

where $\mathcal{P}$ is the parity operator. Unfortunately, they are not eigenstate of the C-parity operator but we can compute the C-parity transformed state from the wavefunctions of Fig. 8.7:

$$\mathcal{C}|K^0\rangle = \mathcal{C}|\{d\bar{s}\}\rangle = e^{i\xi}|\{\bar{d}s\}\rangle = e^{i\xi}|\bar{K}^0\rangle \qquad (13.119)$$

where ξ is an arbitrary phase. We choose to fix ξ as π ($e^{i\xi} = -1$), even if other conventions are still in use.[22] Then,

$$\mathcal{CP}|K^0\rangle = |\bar{K}^0\rangle \;\; \text{and} \;\; \mathcal{CP}|\bar{K}^0\rangle = |K^0\rangle \qquad (13.120)$$

and the linear combinations:

$$K_1 = \frac{1}{\sqrt{2}}(K^0 + \bar{K}^0) \;\; ; \;\; K_2 = \frac{1}{\sqrt{2}}(K^0 - \bar{K}^0) \qquad (13.121)$$

are CP-eigenstates, whose eigenvalues are $+1$ and -1, respectively.

---

[22] Some authors employ $\xi = 0$ without loss of generality.

Cronin and Fitch showed that K_1 and K_2 are not exactly the physical states because – using their language – the CP symmetry is mildly violated in weak interaction by an unknown force. To establish this major result,[23] they tested the flavor composition of the physical states as a function of time. To measure the flavor, they exploited the hadronic decays of the kaons.

[23] Cronin and Fitch were awarded the Nobel Prize in 1980 for their seminal experiment on CP violation in neutral kaons.

Example 13.6

A neutral kaon can decay in two and three neutral pions. Both particles are spinless and have an odd intrinsic parity from the three-quark model. Conservation of angular momentum requires the orbital momentum l of the $K^0 \to \pi^0\pi^0$ final state to be zero. Then,

$$\mathcal{P}(\pi^0\pi^0) = (-1)^l P_{\pi^0} P_{\pi^0} = +1 \qquad (13.122)$$

where $P_{\pi^0} = -1$ is the intrinsic parity of the π^0. The three-quark model provides the intrinsic C-parity of π^0 ($C_{\pi^0} = +1$), too. Hence,

$$\mathcal{C}(\pi^0\pi^0) = \mathcal{C}(\pi^0)\mathcal{C}(\pi^0) = +1. \tag{13.123}$$

Note that the factorization of the $\pi^0\pi^0$ system works well here because the spin and angular momentum are both zero and $|\pi^0\pi^0\rangle = |\pi^0\rangle|\pi^0\rangle$. In conclusion,

$$CP(\pi^0\pi^0) = +1 \tag{13.124}$$

and the same consideration holds for a pair of charged pions where:

$$CP(\pi^+\pi^-) = +1 \;. \tag{13.125}$$

The CP eigenstate of a $\pi^0\pi^0\pi^0$ requires the composition of a pion pair with an angular momentum l_{12} and the study of the motion of its center-of-mass around the third π^0 as in Sec. 8.6.1. Since the kaon is spinless, $l_{12} = l_3$ to conserve angular momentum[24] and

$$\mathcal{P}|\pi^0\pi^0\pi^0\rangle = [P_{\pi^0}]^3(-1)^{l_{12}}|\pi^0\pi^0\rangle(-1)^{l_3}|\pi^0\rangle = -|\pi^0\pi^0\pi^0\rangle. \tag{13.126}$$

For the C-parity, the angular momenta play no role and

$$\mathcal{C}|\pi^0\pi^0\pi^0\rangle = [C_{\pi^0}]^3|\pi^0\pi^0\pi^0\rangle = |\pi^0\pi^0\pi^0\rangle. \tag{13.127}$$

A $\pi^+\pi^-\pi^0$ system shows the same behavior. Therefore,

$$CP(\pi^0\pi^0\pi^0) = -1 \ \text{ and } \ CP(\pi^0\pi^+\pi^-) = -1. \tag{13.128}$$

[24]To be precise (pedantic), we should say that the first and second angular momenta are the vectors $\mathbf{L}_{12}$ and $\mathbf{L}_3$. If we measure $|\mathbf{L}^2|$ for a vector $\mathbf{L}$, we get $\sqrt{l(l+1)}$ in NU. If $\mathbf{L} = \mathbf{L}_{12} + \mathbf{L}_3$ and we need $\mathbf{L} = 0$, then $l = 0$ and $l_z = 0$. When we combine $l_{12} \otimes l_3$, the $l = 0$ and $l_z = 0$ conditions are fulfilled only if $l_{12} = l_3$.

Cronin and Fitch measured the lifetime of the K_1 and K_2 states, which are expected to decay in two and three pions, respectively, if CP is conserved. The lifetime is very different due to phase-space: $\tau_{K_1} \simeq 0.9 \times 10^{-10}$ s and $\tau_{K_2} \simeq 0.5 \times 10^{-7}$ s, respectively. This is why they dubbed K_1 as K_S (K-short) and K_2 as K_L (K-long). Thanks to the large difference in the lifetime, the experimentalists were able to select a pure K_L beam just waiting for the decay of all K_S.

Cronin and Fitch observed that

> even in a pure K_L beam, we can observe a small number of $\pi\pi$ decays: $K_L \to \pi^0\pi^0$ or $K_L \to \pi^+\pi^-$. This number amounts to
>
> $$\frac{K_L \to \pi\pi}{K_L \to \pi\pi\pi} \simeq 2 \times 10^{-3} \tag{13.129}$$
>
> where π can be charged or neutral (e.g. $\pi\pi$ is the sum of the $\pi^0\pi^0$ and $\pi^+\pi^-$ decays).

The result of Cronin and Fitch implies that $K_L \neq K_2$ and K_L is slightly polluted by K_1:

$$|K_L\rangle = \frac{1}{1+|\epsilon|^2}|K_2 + \epsilon K_1\rangle \qquad (13.130)$$

with $|\epsilon| \simeq 10^{-3}$. A further experiment performed in the 1980s at CERN (CPLEAR) confirmed that $|\epsilon| = (2.264 \pm 0.035) \times 10^{-3}$ and its complex phase – not to be confused with δ_{CP} – is $43.19° \pm 0.73°$ (Thomson, 2013). Therefore, there must be a source of CP violation that shows up in the oscillation of the neutral kaon.

This source of CPV is clear to us thanks to eqn 13.116. The outcome of the Cronin and Fitch experiment can be interpreted as evidence for a non-zero value of δ_{CP}. In 1964, however, a tiny but non-zero violation of CP observed *only* in kaon oscillations was a deep mystery.

13.7.3 B mesons

If CP violation arises in the SM by the complex phase of the CKM, it must show up in a large number of processes, and people were puzzled that the only piece of evidence was kaon oscillation. Indeed, the first evidence of CP violation originating from the decays of the mesons (**direct CP violation**) and not from oscillation (**indirect CP violation**) was established in the 1990s by NA31, NA48, and KTeV experiments at CERN and Fermilab. Still, there was no convincing proof that the CKM could explain all of CP effects in quark physics. This proof came in the following decade, starting from the BABAR results in 2001. What does a "convincing proof" mean? In the CKM mechanism, we expect CPV in oscillation, decay, and even interference between the oscillation and decay diagrams. We expect CPV in systems different from the kaon meson, too. Finally, we can test that the CKM is really a complex unitary matrix. This proof was provided a few years ago by the **B factories**: high-luminosity colliders[25] running around the resonances of $b\bar{b}$ bound states that decay in a $B^0\bar{B}^0$ pair. The closest resonance above the $B^0\bar{B}^0$ threshold is called $\Upsilon(4s)$ because is an exited state with $n = 4$, $l = 0$ (*s*-state). These machines were **asymmetric e^+e^- colliders** running at $\sqrt{s} = 10.58$ GeV, which corresponds to $m_{\Upsilon(4s)}$. The B factories are asymmetric because the energy of the electron is larger than the positron. The rest frame of the $\Upsilon(4s)$ is thus boosted with respect to LAB, and the B mesons are boosted, too. In this way, the tracker of the collider detector can measure a B in flight to study $B^0 \leftrightarrow \bar{B}^0$ oscillations and decays.

The PEP-III B-factory was complemented by a detector called BABAR.[26] The inner part of the detector was instrumented with a high-precision silicon tracker. Since the B mesons are produced in-flight, the tracker can identify the decay vertices of each meson, which have a mean distance from the e^+e^- interaction point of about 200 μm. The $B^0\bar{B}^0$ pair is an entangled state and a measurement produces the collapse of the wavefunction, as discussed in Secs. 3.9 and B.4. Therefore, if a B^0 meson decays emitting a lepton, the sign of the lepton uniquely identifies the parent particle.

[25]To get the sense of "high luminosity," we can look at the record achievements of PEP-II, the accelerator serving BABAR. PEP-II started running in 1999, reached the incredible value of $\mathcal{L}(t) = 1 \times 10^{34}$ cm^{2}s^{-1} in 2005 accelerating an electron current > 2 A. BABAR performed the physics analyses using data from an integrated luminosity of 557.4 fb^{-1} corresponding to hundreds of millions of $B\bar{B}$ pairs. This machine was eventually decommissioned in 2009.

[26]A pun from *B-B*-bar.

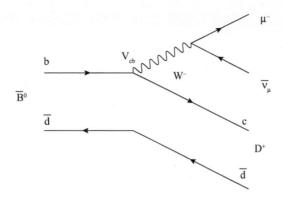

Fig. 13.11 Leading Feynman diagram of the $\bar{B}^0 \to D^+ \mu^- \bar{\nu}_\mu$ semileptonic decay.

Example 13.7

Consider a semileptonic weak decay whose final state has a μ^-. If the parent is a B meson, it *must* be a $\bar{B}^0$. This statement holds because $\bar{B}^0$ is the lightest $b\bar{d}$ state and cannot decay without changing the flavor of the b. The decay will be either toward a $\bar{c}$ or a $\bar{u}$. The vertex of the $b \to \bar{c}$ (see Fig. 13.11) is weighted by V_{cb} and $b \to \bar{u}$ by V_{ub}. Since experimental data indicates $|V_{cb}| \gg |V_{ub}|$, the leading semileptonic mode is:

$$\bar{B}^0 \to D^+ \mu^- \bar{\nu}_\mu \tag{13.131}$$

because D^+ is a pure $\bar{d}c$ state. Due to the collapse of the wavefunction, in the very moment we observe the $\bar{B}^0 \to D^+ \mu^- \bar{\nu}_\mu$, we are sure that the other particle *must* be a B^0. Similarly, if we see a μ^+, the corresponding parent must be a B^0. This technique is called **flavor tagging**.

The masterpiece of BABAR and its Japanese competitor BELLE was the study of $B \to \psi K_S$ decays, where the ψ particle (also called J/ψ) is the $c\bar{c}$ ground state. The ψ was simultaneously discovered in 1974 by two experiments led by S. Ting, who called it J, and B. Richter, who called it ψ. In postgraduate courses, you will hear a lot about the "November revolution", when the first particle containing the charm quark was discovered. The discovery confirmed the predictions of the GIM mechanism of Sec. 13.7 and opened the post-$SU(3)$ era of the quark model. Richter and Ting saw for the first time a "heavy quark," a quark whose mass is bigger than Λ_{QCD}.

Here, we start noticing that the ψ is a **charmonium**. It belongs to the spin-1 multiplet of the four-quark model and has odd parity and even C-parity. What is, then, the CP parity of the ψK_S state?

Example 13.8

The charmonium is a particle-antiparticle pair and the C parity is given by $(-1)^{l+s} = -1$, as discussed in Sec. 5.7.3. Since the intrinsic parity is odd, $\mathcal{CP}\psi = (-1)(-1)\psi$. Therefore, the ψ is a CP-even state. Both B^0 and K_S are spinless particles and the spin of the ψ is 1. Hence, the conservation of angular momentum in the $B^0 \to \psi K_S$ decay implies that the orbital momentum of the final state must be $l = 1$. As a consequence,

$$\mathcal{CP}|\psi K_S\rangle = (-1)^l \mathcal{CP}|\psi\rangle \mathcal{CP}|K_S\rangle = (-1)^1(+1)|\psi\rangle(+1)|K_S\rangle = -|\psi K_S\rangle \tag{13.132}$$

if we identify K_S with K_1 neglecting tiny CPV effects in kaons. The CP-parity of the whole system is opposite to the intrinsic CP-parity of each particle.

Suppose now that an observer in LAB sees a semileptonic decay with the emission of a μ^-. The parent is then a $\bar{B}^0$ and he can be sure that the other B meson is a B^0.

He can also compute the corresponding t_{tag}: the decay time of the $\bar{B}^0$ measured in its rest frame (proper decay time). After t_{tag}, the other particle – the B^0 – is not yet decayed and can oscillate into a $\bar{B}^0$. We do not know if, at the time of the decay, this particle is a B^0 or a $\bar{B}^0$. In any case, the particle is unstable and can decay at t_{rec} in ψK_S. The observer can thus measure t_{rec}: the proper time when this decay occurs. This phenomenon involves CP violation both in oscillations and in decay. An observable sensitive to CP violation occuring in the interference of the oscillation and decay diagrams is:

$$\mathcal{A}(\Delta t) = \frac{\Gamma(\bar{B}^0(\Delta t) \to \psi K_S) - \Gamma(B^0(\Delta t) \to \psi K_S)}{\Gamma(\bar{B}^0(\Delta t) \to \psi K_S) + \Gamma(B^0(\Delta t) \to \psi K_S)} \tag{13.133}$$

where Γ is the decay amplitude and Δt is $t_{rec} - t_{tag}$. If $\mathcal{A} \neq 0$, CP is violated not only by kaons but also by B mesons. Even more, the SM predicts:

$$\mathcal{A}(\Delta t) \simeq \sin(\Delta M \Delta t) \sin(2\beta). \tag{13.134}$$

Equation 13.134 shows a sinusoidal pattern that depends on the difference ΔM between the mass eigenstates of the B mesons (usually called B_H and B_L, with $m_{B_H} > m_{B_L}$) and Δt. Even if the full derivation of eqn 13.134 is beyond the scope of this book, there is no doubt that $\sin 2\beta$ must be a parameter linked to CP violation. If the CKM mechanism is the only source of CP violation in quarks, $\sin 2\beta$ must be a function of the CKM elements and the complex phase δ_{CP}. This is, indeed, the case:

$$\frac{V_{td}}{|V_{td}|} = e^{-i\beta} \tag{13.135}$$

$$V_{CKM} \pm 1 \text{ sigma}$$

$$
\begin{pmatrix}
0.97401 \pm 0.00011 & 0.22650 \pm 0.00048 & 0.00361^{+0.00011}_{-0.00009} \\
0.22636 \pm 0.00048 & 0.97320 \pm 0.00011 & 0.04053^{+0.00083}_{-0.00061} \\
0.00854^{+0.00023}_{-0.00016} & 0.03978^{+0.00082}_{-0.00060} & 0.999172^{+0.000024}_{-0.000035}
\end{pmatrix}
$$

$$V_{PMNS} \ 3 \text{ sigma range}$$

$$
\begin{pmatrix}
0.801 \rightarrow 0.845 & 0.513 \rightarrow 0.579 & 0.143 \rightarrow 0.155 \\
0.234 \rightarrow 0.500 & 0.471 \rightarrow 0.689 & 0.637 \rightarrow 0.776 \\
0.271 \rightarrow 0.525 & 0.477 \rightarrow 0.694 & 0.613 \rightarrow 0.756
\end{pmatrix}
$$

Fig. 13.12 Comparison between the CKM (top) and PMNS (bottom) matrices. The numbers are the absolute values of the matrix elements. The PMNS is still poorly known and plagued by a few inconsistencies among data. Therefore, we indicated the values of the CKM entries with the standard notation (mean $\pm$ one σ range), while the PMNS entries are given as "3σ range": the probability that the true value of the PMNS entry is inside this range is 99.73% ("3σ"). The CP phase is given by eqn 13.136 for CKM and is nearly unknown (3σ range: $107° - 403°$) for PMNS. Data from Zyla *et al.*, 2020 and Esteban *et al.* 2020.

and $\beta = 0$ or π if V_{td} is a real number. The combined data from the K and B mesons, together with the measurement of the Cabibbo angle (or, equivalently, $|V_{ud}|$) give:

$$\delta_{CP} = (68.5^{+2.6}_{-2.5})° \tag{13.136}$$

assuming the parameterization of eqn 13.86. The BABAR and BELLE results are amazing because they demonstrate in one shot that:

- CP is violated in systems different from kaons and, in particular, in the B system;
- CP is violated in the oscillation, decay, and interference diagrams between oscillation and decay;
- the CP asymmetry follows eqn 13.134 as predicted by the Standard Model.

This is what we mean by "convincing proof"! The size of the CKM and its pattern is all but a curiosity. Unlike the PMNS, this matrix is nearly diagonal and all off-diagonal terms are small. The further the term from the diagonal, the larger the suppression, up to $\sin^3 \theta_c$ for $|V_{td}|$ (see Fig. 13.12). Conversely, nearly all terms of the PMNS are large and the suppression of off-diagonal terms does not exceed 10^{-1}. Once more, we have no explanation of this feature because the Standard Model does not make predictions on the parameters of the Yukawa sector.

On the other hand, sources of CP violation may have profound cosmological implications. CP violation is a necessary (but not sufficient) condition to create an asymmetry between matter and antimatter in the early universe. It might explain why the universe we observe now is matter-dominated despite the (nearly) perfect symmetry between particles and antiparticles. The results of the B-factories and their outstanding precision demonstrated that

> δ_{CP} is a source of CP violation in the early universe but the size of CP violation in the quark sector is too small to explain why the universe is matter-dominated.

This field is still vibrant. In the last ten years, the LHCb experiment at the Large Hadron Collider provided a wealth of information on the CKM mechanism. The Standard Model predictions were probed by observing CPV in new systems as the oscillations of the B_s and D mesons, by direct observation of T-parity violation (see Sec. 5.8.1), and through stringent tests of the unitarity of the CKM. A novel B-factory is running in Japan (SuperKEKB serving Belle II) and new results come at a steady pace.

Conversely, we do not know yet if a PMNS source of CP violation exists because the uncertainty on its δ_{CP} phase is too large. Neither do we know if this source produces our matter-dominated universe. This is the first interplay between particle physics and cosmology we encounter and others will follow.

13.8   Light and shadow of the Standard Model*

The Standard Model cannot be the end of the story. It is a very successful theory and is a "theory" even if is called a "model" for historical reasons. Still, I like to think that this lapse has a meaning. If you are acquainted with general relativity, you are probably annoyed by the inelegance of the Standard Model, even if relying on beauty or elegance led many physicists astray (Hossenfelder, 2018). We know for sure that the Standard Model is not a "Theory of Everything" because it does not include gravity. It is packed with free parameters that must be experimentally measured (the Higgs mass, the Weinberg angle, the whole Yukawa sector, etc.) and does not answer many basic questions like those in Table 1.1. On the contrary, it is mathematically consistent at least up to 10^{16} GeV. Maybe it is consistent even above but the λ of eqn 13.32 – measured through m_H – is rather small, and there is a finite probability that at 10^{16} GeV or, equivalently, at a temperature of 10^{29} K, the SM vacuum moves to another vacuum jumping through the top of the Higgs potential. In jargon, this is called a **false vacuum** because the SM vacuum is higher than other possible vacua and a transition may occur. Particle physicists realized this possibility in the 1980s but in 2012 the discovery of the Higgs boson demonstrated that the Standard Model vacuum is **metastable** and that in the early universe – if the Standard Model still holds – these transitions would be unavoidable. This is a clear example of how fragile is Standard Model's consistency.

Another example, partially linked to the previous one, is the statement that the Standard Model must be an **effective theory**. An effective theory is a theory that holds up to an energy scale Λ. The Standard Model is an effective theory for sure if Λ is the Planck mass of eqn 1.22 because it ignores gravity. More likely, it is an effective theory for $\Lambda \simeq 10^{16}$ GeV for the aforementioned issue of the metastable vacuum. This is not striking at all: many theories are approximations of more fundamental theories. Thermodynamics is an approximation of statistical mechanics and classical physics is an approximation of QM. And classical physics breaks down when a few-keV photon exhibits Compton scattering. Still, classical physics and thermodynamics are consistent because all the complexity of the microscopic world is hidden by a few free parameters that we need to measure experimentally. These examples may have exceptions. Nobody can ensure a priori that thermodynamics is the effective theory of statistical mechanics. It worked in that instance but, in general, the parameters that encode the information of physics at high energy may be correlated with the dynamics of the effective theory occurring at low energy (Giudice, 2008). In this case, we will not understand what is going on unless we become aware of the high-energy, more fundamental theory. Unfortunately, this seems the case of the Standard Model. In the language of QFT, when a model is an effective theory, we can compute any physical process involving particles with momenta smaller than Λ – the energy scale where the theory breaks apart – by replacing the full theory with a truncated version of it. In this way, we do not have to master gravity at the Planck scale to compute the $e^+e^- \to \mu^+\mu^-$ cross-section, even if the Planck scale may play (and probably plays) a role to determine the mass of the electron or the muon. For us, m_e and m_μ are free parameters of the Standard Model that we experimentally measure without understanding it. This works even for the Higgs boson: at a given energy scale μ, quantum corrections increase as μ^2 for $\Lambda \to \infty$ but, in the computation of high-order diagrams, there are counterterms of the same size and opposite sign. A light Higgs (125 GeV) is a fortunate coincidence arising from the difference of two huge terms that have been carefully fine-tuned by the mastermind of the fundamental theory. In this sense, the dynamics of Λ affects the low-energy scale of the electroweak theory. Quantifying rigorously such an idea has been extremely difficult and was tried by A. Eddington, P. Dirac, K.G. Wilson, S. Weinberg, and many others. The most successful definition has the form of a heuristic principle called **naturalness**, introduced by G. 't Hooft in 1979:

> a theory valid up to a scale Λ can be expressed by N dimensionless parameters $p_1, \ldots p_N$ if we measure them in units of Λ. p_j $(j = 1, \ldots, N)$ can be much smaller than 1 only if setting $p_j = 0$ adds a new symmetry to the theory.

Despite the arbitrariness in the definition of "much smaller than 1," this heuristic principle holds in the vast majority of physical theories but not in the Standard Model. Since a massless Higgs does not create a new symmetry in the Standard Model Lagrangian, this theory is **unnatural** according to t'Hooft's definition. Note that this holds only for the Higgs. If we put to zero the masses of the fermions, the SM has a new symmetry called **chiral symmetry**.[27] In a natural theory, the Higgs mass should be of the order of Λ, while $m_H \ll M_{Planck}$ or even $\ll 10^{16}$ GeV because the counterterms are fine-tuned (by tens of orders of magnitude!) to cancel the quantum corrections that would bring $m_H \to \Lambda$. This is why the vast majority of theories beyond the Standard Model predicts new

[27]We employ the chiral limit when, for example, we approximate to zero the mass of the quarks in quantum chromodynamics. More rigorously, a **chiral symmetry** is the freedom to change the global phase of a LH fermion independently of its RH counterpart.

particles at the TeV scale: they want the Standard Model to be natural, as nearly all theories in particle physics are.

A heuristic principle is not a theory and may have exceptions. If no new particles are found at the LHC, the Standard Model is that exception, and the exception is raised by the Higgs mechanism. All these issues were considered domestic affairs of the Standard Model in the 20th century. The situation has changed dramatically in the last 20 years.

13.9   Particle physics and cosmology

The Supernova Cosmology Project and the High-z Supernova Search Team reshaped our understanding of the universe in 1998. Their findings were then confirmed using different techniques based on the Cosmic Microwave Background (CMB) anisotropies and in the simulations of galaxy clustering. This impressive amount of data point toward a striking statement:

> the speed at which distant galaxies move away from us increases with time and, therefore, the expansion of the universe is accelerating.

Furthermore, we can trace this expansion by looking at light produced by Type Ia supernovas (see Sec. 11.10) and, in general, by visible matter. Looking at the gravitational motion of the galaxies or other subtler effects,[28] the early claim of F. Zwicky in 1933 is now indisputable (Sanders, 2014):

> in the universe, the mass made up of Standard Model particles, which is generally composed of baryons, neutrinos, and (massless) photons, is a small fraction of the whole mass of the universe. The origin of this **dark matter** is presently unknown.

[28] General relativity predicts bending of the light trajectory caused by intense gravitational fields (**gravitational lensing**). The measurement of the bending identifies massive celestial object even if they do not produce any kind of radiation.

[29] Inflation is not an undisputed statement: several authors considered alternative explanations within or outside ΛCDM. Inflation is, however, embedded in the most accredited versions of ΛCDM because of the excellent agreement with observational cosmology.

These two statements can be combined in a single paradigm, called Λ Cold Dark Matter (ΛCDM), which represents the counterpart of the Standard Model in cosmology (Weinberg, 2008; Dodelson and Schmidt, 2020). This paradigm is based on general relativity (GR) as the Standard model is based on QFT but the two paradigms are disconnected and, partially, in contradiction. Time in the ΛCDM universe starts with the Big Bang, where the volume of the universe $V \simeq 0$. Just after the Big Bang, the volume increases exponentially (**inflation**) until $t \simeq 10^{-32}$ seconds.[29] Later on, the universe expands due to post-inflationary acceleration but is slowed down by the gravitational pull. Gravitation is always attractive and, if not counteracted by other forces, shrinks matter toward a single point. The ΛCDM paradigm claims that, after 10^9 years, the gravitational pull is overwhelmed by an unknown type of energy (**dark energy**) that is proportional to the volume of the universe and, today, has a density of 7×10^{-27} kg/m^3. Many cosmological observables (CMB, galaxy distributions and formation, etc.) can be explained by noting that dark energy is formally equivalent to a small **cosmological constant** Λ and GR predicts that the late time acceleration of the universe is due to Λ (Weinberg, 2008; Ruiz-Lapuente *et al.*, 2010). In ΛCDM, the matter of the universe is mostly composed of non-relativistic dark matter (Bertone, 2013). CDM stands for **cold dark matter**, matter that was not relativistic when it started streaming through the universe independently of other forms of matter or forces except gravity.

> In the ΛCDM paradigm, the energy content of the universe at our time (1.38×10^{10} years) is dominated by dark energy (68.3%), dark matter (26%), and baryons (5%). Other SM particles (photons and neutrinos) are practically negligible.

In the past, this paradigm was called the "concordance model" because it was able to fit a large number of observables. More recently, it predicted the statistics of gravitational lensing and the polarization features of the CMB, which are corroborated by present data. The ΛCDM supporters claim that

- the SM plays an immaterial role in the universe at large scales;
- the origin of dark energy is beyond the capabilities of the SM or its "natural" extensions – in the sense of Sec. 13.8; and
- dark matter is not a SM form of matter.

ΛCDM has a poor reputation among particle physicists because we do not appreciate playing an insignificant role in the universe. More serious criticisms are that ΛCDM assumes GR to be true at any scale and that, all in all, the ΛCDM description of the universe is a set of free parameters whose origin is unknown. The obscure origin of the ΛCDM parameters is (deservedly) called the **dark matter–dark energy** problem. Still, defeating ΛCDM is far from trivial: any extended version of GR conceived to explain the accelerating universe is either inconsistent or poorly matches the astronomical data. Dark matter follows all prescriptions of GR and a cosmological constant is naturally allowed in GR.[30]

Solving the dark matter–dark energy problem will likely be the core business of the next generation of physicists, and the subject is inappropriate for an introductory book. Nonetheless, we can use this tremendous challenge to glimpse at the future of particle and nuclear physics. Even if predicting what is going to happen in science is often an exercise in futility, the path that scientists are following seems wise to me. We do not believe anymore that a Theory of Everything (if any) will emerge from a single source of information like, for instance, high-energy colliders.

[30]The cosmological constant was introduced by A. Einstein in 1917 to build a model of the universe that reconciles GR with the "obvious" fact that the universe is static in space and time. Well before the Big Bang theory, he understood that adding Λ was "his worst mistake." Otherwise, he could have added the prediction of the expanding universe to his terrific portfolio of discoveries. This achievement is attributed to A. Friedmann, who was more open-minded about GR than its founding father.

It will result from a systematic exploration of particle physics, nuclear physics, astrophysics, astronomy, and cosmology data, accompanied by a deeper understanding of QFT, GR, or even QM. We will give concrete examples here, at the end of this book.

13.10   Natural dark matter

The problem of the origin of dark matter might be linked to the naturalness issues of the Standard Model. The most popular "natural" extensions of the Standard Model are **supersymmetric theories** (SUSY). In all these theories, the Lagrangian has a new symmetry because we have now a boson counterpart for each elementary fermion and vice-versa. The number of elementary particles is doubled but the new particles solve the problem of the Higgs mass naturalness (Weinberg, 2005b). Quantum corrections are now much fainter and there is no fine-tuning of two large numbers to get a small and finite difference. Naturalness in the sense of t'Hooft explains this property saying that the new symmetry forbids an ultra-heavy Higgs boson and keeps $m_H \simeq \mathcal{O}(M_Z)$. On the other hand, this symmetry must be (spontaneously?) broken. Otherwise, we would have discovered the super-partners of SM particles already in the previous century. There are many realizations of SUSY but all of them predict a light stable particle – typically one of the **neutralinos**: the fermion counterparts of the vector bosons W^+, W^-, Z^0 and the Higgs boson. The neutralino is a light massive fermion that interacts weakly, an ideal candidate for dark matter (Bauer and Plehn, 2019). This coincidence is known as the "WIMP miracle", where WIMP stands for "weakly interacting massive particles." The neutralinos, or, more generally, the WIMP are the most popular dark matter candidates since the 1980s (Bertone, 2013).

The WIMP cross-section with matter depends on the actual SUSY model we consider and the spin of the target. Still, WIMP are food for particle physicists because the WIMP *do* interact with matter by weak forces and not only by gravity. In 1985, M.W. Goodman and E. Witten realized that WIMP could be detected by low-energy elastic scattering with nuclei, as in fixed-target experiments. In this case, the process is $\chi + N \to \chi + N$ where χ is a generic WIMP. The rate of scattering off a target nucleus is:

$$\frac{dR}{dE_r} = \frac{\rho_0 M}{m_N m_\chi} \int_{v_{min}}^{v_{esc}} dv \; v f(v) \frac{d\sigma}{dE_r} \; . \qquad (13.137)$$

In eqn 13.137, R is the interaction rate per nucleus, E_r is the nuclear recoil energy and M is the mass of the detector. m_N and m_χ are the masses of the nucleus and the WIMP, respectively. v is the WIMP velocity in LAB, where the nucleus is at rest, and $f(v)$ parametrizes the velocity distribution of the WIMP in our galaxy measured in LAB. The WIMP density in the galaxy must be of the order of $\rho_0 \simeq 0.3$ GeV cm^{-3} if WIMP are the missing mass that explains the gravitational motion of stars inside the Milky Way. The velocity distributions must be integrated between the minimum velocity that produces a recoil and the escape velocity. If a WIMP is too fast, the gravitational field of our galaxy cannot bound it anymore, and the particle free-streams in the intergalactic space.

We can (maliciously) note that the WIMP is a sort of "ether" that permeates the galaxy and flows crossing the earth. Since the earth revolves around the sun once per year, we should see an annual modulation of R in eqn 13.137, which so far has been observed only by the DAMA/LIBRA experiment in Italy, but – at the time of writing – has not been confirmed by any other experiment. If the cross-section is too small to observe annual modulations, we can still detect WIMP with large mass detectors. The minimum velocity producing a recoil is given by (Schumann, 2019):

$$v_{min} = \sqrt{\frac{E_r m_N (m_N + m_\chi)^2}{2(m_N m_\chi)^2}} = \sqrt{\frac{E_r m_N}{2\mu^2}} \tag{13.138}$$

where μ is the reduced mass of the nucleus–WIMP pair. The escape velocity is estimated by the strength of the potential well due to the gravitational pull of the Milky Way and is $v_{esc} \simeq 544$ km/s. Therefore, the number of nuclear recoils observed in a detector is:

$$N = T \int_{E_l}^{E_h} dE_r \epsilon(E_r) \frac{dR}{dE_r} \tag{13.139}$$

where T is the running time. E_l and E_h are the threshold energy and the maximum recoil energy $2\mu^2 v_e^2 / m_N$, respectively. For a non-relativistic WIMP, E_l is just $m_\chi v_{min}^2 / 2$. Finally, $\epsilon(E_r)$ is the detector efficiency, which is a function of E_r and is zero below a given threshold: the minimum energy that produces a signal recorded by the experimenters.

At present, the most precise experiments of this type are **xenon dark matter experiments**. Experiments like XENON1T, located deep underground at the Gran Sasso laboratories in Italy, record both the charge and scintillation light produced by particles inside the detector fiducial volume. XENON1T is a 2-tonnes double-phase Time Projection Chamber, whose volume is filled with liquid xenon (LXe) at 177 K inside a two-bar pressure vessel. When a particle interacts with Xe, releases energy by producing ionization electrons and scintillation light. The electrons drift toward the anode located on top of the cryostat. In this area, the Xe is in gaseous form and the electrons undergo avalanche multiplication around wires, like in a multi-wire proportional chamber. This signal is rather slow because the electrons drift at $v \simeq 1$ mm/μs. The photons produced during the avalanche formation reach a set of photomultipliers located above the anode (see Fig. 13.13) and are recorded as a light signal proportional to the number of electrons created after the avalanche multiplication. The corresponding signal is called "S2" (secondary scintillation). Well before S2, the scintillation light produced directly inside the volume reaches the photomultipliers located above the anode or below the cathode with a n/c speed, where n is the refraction index of LXe. This smaller but faster signal is called "S1" (primary

Fig. 13.13 Left: working principle of the XENON1T experiment. A WIMP-nucleus recoil produced inside the TPC creates S1 photons (dashed lines) that bounce up to the top or bottom photomultipliers (PMT) thanks to a Teflon™(PTFE) reflector. The S2 signal is produced on top of the TPC, inside the gas-xenon volume, and the photons are mostly seen by the top PMTs. Right: signal observed by the PMT for a candidate WIMP, a background electron or photon, and a background neutron. Reproduced under CC-BY-4.0 from Aprile *et al.* 2017*a*.

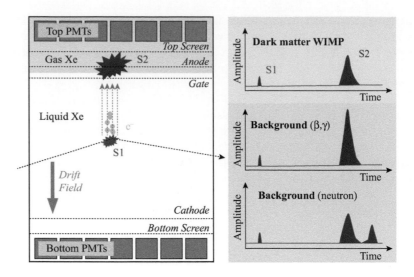

[31] If you are curious (and brave) enough to glance at original literature, you will find the energy deposited by recoils and electrons indicated as keV$_{nr}$ and keV$_{ee}$, respectively. This notation signals that the energy release mechanism is different between nuclear recoils and electrons. The different mechanisms impact on S1 through quenching effects that must be kept well under control.

scintillation). XENON1T exploits the time difference and the amplitude of S1 and S2 to tag potential elastic scattering between a WIMP and a nucleus. The scattering produces a recoiling nucleus that releases energy in the 1–100 keV range.[31] The ratio S2/S1 depends on the type of interactions: if ionization dominates, the ratio is high, while S2/S1 is lower if scintillation is the leading energy-loss mechanism. Also, the electrons producing S2 can recombine with Xe during their drift toward the anode. The recombination rate is different between nuclear recoils and electrons (or photons). In this way, the experimentalists can identify a pure sample of nuclear recoils. Since we have no evidence of WIMP yet, the detector must be calibrated with Standard Model nuclear recoils. They are produced by neutron sources, where the neutron bounces on the nucleus, and the recoil energy is measured together with the neutron reinteraction. Xenon dark matter experiments are extremely challenging because the expected cross-section from SUSY models ranges from 10^{-42} to 10^{-47} cm^2 and strongly depends on the WIMP mass (see Fig. 13.14). At such a low rate, there is a plethora of potential background due to natural radioactivity, including radon, neutrons produced by cosmic rays or surrounding materials, radioactive contaminations in the TPC components, etc. The same consideration holds for liquid argon (LAr) experiments: LAr is much cheaper than LXe and can better exploit the signal shape for background rejection than LXe but is plagued by a naturally occurring isotope (^{39}Ar) producing a beta activity of 1 Bq/kg. This is a huge rate for these kinds of searches. Therefore, ^{39}Ar must either be tagged as background or removed before the insertion inside the TPC.

We can handle these difficulties by building large volume detectors deep underground so that the outer shell of Xe shields the ultra-pure inner core. All materials must be extremely radiopure, including the TPC components and the photosensors. Furthermore, LXe produces UV scintillation light at $\lambda = 178$ nm, which is difficult to be observed by

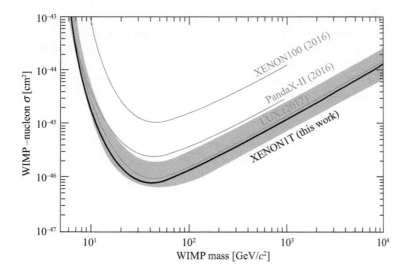

Fig. 13.14 Limits on WIMPs from XENON1T. The limit is expressed as WIPM interaction cross-section per nucleon (in cm^2/nucleon) versus WIMP mass (in GeV). The interactions can also depend on nuclei and, in particular, on the nuclear spin. Here, they are considered spin-independent. The black line is the 90% C.L. limit from XENON1T ("this work"). The bands are the 68 and 95% CL limits. Limits from other experiments are also shown. Reproduced under CC-BY-04 from Aprile *et al.* 2017*b*. Copyright 2017 American Physical Society.

standard photosensors. That is even worse for LAr where $\lambda = 128$ nm. The physicists working with LXe must use low-radioactivity photomultipliers sensitive to ultraviolet (UV). For LAr, they resort to wavelength-shifter materials that absorb the photons and re-emit them uniformly in space at a more amenable λ. This smart solution comes at the expense of efficiency because photons can be absorbed or reflected by those materials. Despite these difficulties, experiments like XENON1T were able to exclude many WIMP models. Fig. 13.14 shows the 2017 results from XENON1T, where no evidence of WIMP has been reported. The plot is then a limit on the WIMP cross-section. The assumptions of the model (v, $f(v)$, etc.) are those mentioned at the beginning of the section. Furthermore, the WIMP cross-section may depend on the spin of the nucleus. In this plot the interactions were assumed to be spin-independent. We remind that a 90% limit of 10^{-45} cm^2/nucleon for a 100 GeV WIMP means that the probability that a 100 GeV WIMP exists and interacts with a nucleon with a cross-section $> 10^{-45}$ cm^2/nucleon is less than 10%.

Most of the models tested in XENON1T are also excluded by the LHC because the WIMP and their parents should be produced on-shell and observed at the LHC experiments. Still, there are several WIMP models where dark matter experiments are more effective than high-energy colliders.

13.11 The discovery of gravitational waves

Some paths to solve the dark matter–dark energy problem are disconnected from the issue of naturalness. First, we do not know the premises where general relativity holds. Especially if we explore the largest scales or the most extreme conditions. Besides, we are too confident about our knowledge of celestial bodies, their number, type, and dynamics.

A marvelous example is the direct observation of gravitational waves in 2016 and the birth of multimessenger astronomy in 2017. Even if this word is a bit overused, multimessenger physics is a great tool. It allows the study of a celestial body or a transient effect, like the explosion of a supernova, using a broad range of observables: visible and UV light, X- and gamma-rays, neutrinos, and, since 2016, gravitational waves. The combination of this information provides a more realistic description of these bodies and an estimate of their number inside the galaxies. We will see in a minute, that the universe looks to us much richer after 2016, and new data shed new light on its matter and energy content.

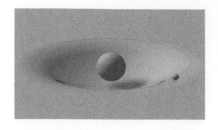

Fig. 13.15 Geometrical interpretation of a Keplerian orbit in GR. The large mass M in the center warps the space, and the straight motion of the test mass is bent around M. The observer in M records a test mass with an elliptical orbit. Image credit: G. Ceruti.

13.11.1   Waves in GR

General relativity has been conceived as a classical theory of fields and offers an elegant geometrical explanation of the attractive force due to gravitation. In GR, mass and energy are sources of perturbation of the Minkowski space-time described in Sec. 2.2.1. Fig. 13.15 shows an over-simplified description of the motion of a test mass m in space when space is curved by the presence of a large gravitational mass M. The body follows the shortest path from the start to the endpoint (**geodesic**) but the reference frame – the Minkowsky space-time - is now curved. The same phenomenon can be described in more classical terms: a body of mass m pulled by a force does not run along a straight trajectory at a constant speed because its reference frame is not inertial. The force, here, is gravitation and the two descriptions of the motion of m are equivalent except for the concept of "inertial frame" that can be thrown away in the geometrical interpretation.

The situation changes if the size of the test mass $m = M_2$ is comparable with $M = M_1$ because now also M_2 warps the space-time and the motion of M_1 and M_2 depends on the final space-time curvature. For instance, if M_1 and M_2 are two binary stars their motion is slightly different than what predicted by Newtonian mechanics because there is a new energy loss mechanism that enters the game: the motion of the stars create waves in space-time that propagate far from the stars bringing part of their energy. This happens normally in e.m. because the e.m. field has an energy that can leave the system through the boundaries by electromagnetic waves.[32] This new energy loss mechanism was observed in 1974 looking at the trajectories of a neutron star and a pulsar, the **Hulse–Taylor binary**, while they get closer and closer. R.A Hulse and J.H. Taylor showed that the energy loss rate was compatible with GR predictions, provided that the energy leaves the system by the production of **gravitational waves**. In this sense, gravitational waves (GW) were indirectly proved 50 years ago.

[32]The rate of energy loss by radiation can be computed from Maxwell's equations or, equivalently, the Poynting theorem (Griffiths, 2017). In GR, computing this rate is quite a job. The Einstein equations – the analog of Maxwell's equations – describe how matter changes the curvature of space-time but they are non-linear equations. If the gravitational field is weak, they can be solved by linearization, otherwise, we must resort to numerical simulations.

13.11.2   Gravitational interferometers

Making a direct observation of a gravitational wave is a different story. There are many sources of GW in the universe and can be observed by

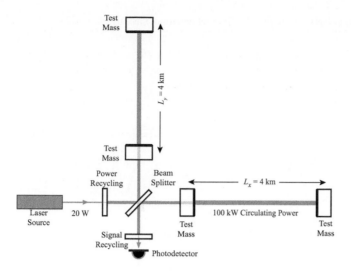

measuring the oscillation amplitude of a rigid body in LAB. But the size of the oscillation is laughably small: a 4 km rod is shortened by 10^{-18} m, which is orders of magnitude smaller than the size of a nucleus! Several methods were developed to reach such a level of precision. The only one that achieved the goal – 50 years after K. Thorne's proposal – was **gravitational interferometry**.

> A gravitational interferometer is a Michelson–Morley interferometer made of two perpendicular arms. Each arm is equipped with a Fabry–Perot cavity that reflects the light hundreds of times, increasing the effective length of the arm.

When the interferometer (see Fig. 13.16) is balanced, the electric field at the photosensor is zero because of destructive interference between the two laser beams. A gravitational wave creates a displacement in one bar, only, which appears as a non-zero electric field $\mathbf{E}(t)$ at the photosensor. This technique works because the difference in the path of the two rays crossing the arms increases with the size of the gravitational displacement, while nearly all other sources of noise do not depend on L.

Behind this simple concept, there are, at least, two major technology breakthroughs. The first one is the **Pound–Drever–Hall** laser stabilization mechanism. This is essential in ultra-sensitive interferometers because the frequency of a laser generally wanders around a mean value. The spread is due to temperature variations, local variations of the laser gain, and other imperfections. R.V. Pound, R. Drever, and J.L Hall developed a technique where the frequency of the laser light is stabilized using negative feedback coming from an optical cavity more stable than the laser. Note that this technique is conceptually similar to van der Meer's stochastic cooling in accelerators (Sec. 12.5) because we do not attempt to reduce the source of the spread but we act on the light (or particles) to reduce the spread by negative feedback. While stochastic

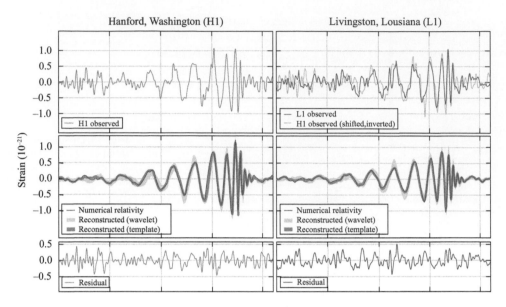

Fig. 13.17 The signal observed during GW150914 by the two LIGO interferometers expressed as strain versus time. The signal is chirped – the frequency increases with time as a chirp – because the two black holes get closer and the change in angular mass (moment of inertia) affects the orbital frequency. The dimensionless strain $\Delta L_{arm}/L_{arm}$ is the space change within an arm of the interferometer in comparison to the total length of the arm. This length is defined by the light path between the two test masses ($L_{arm} \gg L$ thanks to the Fabry–Perot cavity). The uppermost dark line is the signal observed at Hanford (left) and Livingstone (right). The middle plot is the expectation from numerical integration of the Einstein equations ("numerical relativity"). The difference between data and expectation ("residual") is shown in the bottom plot. Reproduced from Abbott *et al.* 2016*b* under CC-BY-3.0 license.

cooling is a niche market, the Pound–Drever–Hall laser stabilization is the cornerstone of modern metrology. The second breakthrough is the development of a **vibration isolation** that decouples the interferometer from any mechanical wave produced in the earth (tides, earthquakes, wind, etc.). The system is based on an active plus a passive isolator. Active isolators are commonly used in industry and produce a feedback signal that cancels the seismic noise. This is not enough for gravitational interferometers, where the active system is coupled to a passive isolator (**optics suspensions**) made of four correlated pendula that hang the mirrors where the light is reflected or split. Finally, a gravitational interferometer is never operated in standalone mode but works in coincidence with other interferometers to further reduce spurious signals and point to the source of the gravitational wave.

The discovery of gravitational waves occurred in 2016 and the wave GW150914 was detected by the two LIGO interferometers[33] located at Hanford, WA, and Livingston, LA, in the United States (Abbott *et al.*, 2016*b*). They employ a stabilized 1064 nm laser, whose power is amplified inside the arms up to 100 kW. The signal at the photosensor appears as a chirped oscillation, which is the standard signature for gravitational waves. The chirped signal for GW150914 is shown in Fig. 13.17. This chirp was recorded for 0.2 ms first at Livingston and then at Hanford. The lapse was compatible with a wave traveling at $v = c$. Simulations

[33]GW150914 stands for "gravitational wave of September 14, 2015".

suggest that the signal originated from the merger of two black holes of about 30 solar masses each. The total energy released by the gravitational wave loss mechanism was about three solar masses. A photon of about 50 keV was observed by FERMI, a telescope equipped with particle detectors for the observation of high-energy photons, but was not uniquely attributed to GW150914. No neutrinos were observed by the largest detectors in water (ANTARES) and ice (IceCube). Even the first observation was a bit of a surprise: the rate of black hole binaries estimated by astronomers was so small that the signal would have been extremely unlikely.

13.11.3 Multimessenger astronomy

In 2020, we already have a catalog (GWTC-2) of 50 gravitational wave signals that originated from a pair of black holes, neutron stars, or black hole–neutron star binaries. This catalog is reshaping our knowledge of massive objects in the universe and shows, for instance, that the black hole population is larger than expected. This also matches the observations of the late 1990s demonstrating that nearly every massive galaxy has a black hole inside. For our galaxy, a supermassive black hole of about 4×10^6 solar masses is located in the galactic center. In general, massive black holes are much more common than our 20th-century estimates. Nonetheless, they cannot be considered a prime candidate for dark matter, which requires an even larger population or a crack in GR that we have not observed to date.

Even the precision on the location of the source has improved significantly: since 2017, LIGO and VIRGO – another gravitational interferometer located in Italy – have been recording events together. The large level arm between the United States and Europe revealed the location of the source of GW170814 – the first event observed by all three interferometers. LIGO and VIRGO reconstructed the position of the source with a precision twenty times better than GW150914. GW170814 marks the inception of **multimessenger astronomy** because the source was observed by gravitational waves, gamma rays, X-rays, standard telescopes, and radio-telescopes (De Angelis, 2018). Only neutrinos were missing. On the other hand, in 2018 IceCube detected an extremely high-energy (290 TeV) neutrino of extragalactic origin associated with photons recorded by FERMI and other experiments on earth. The interpretation of these events requires an exquisite knowledge of the Standard Model to ascertain the acceleration mechanisms. Conversely, high-energy extragalactic neutrinos test the fundamentals of the Standard Model, like the CPT theorem, the unitarity of cross-sections, neutrino oscillations, etc.

It is unlikely that the dark matter–dark energy problem will be solved without gravitational waves and multimessenger physics because they explore the universe at scales that cannot be reached on earth. But it is difficult to prove anything sound in the universe if we do not firmly ground hypotheses in our back yard.

Further reading

You have now learned the fundamentals of particle and nuclear physics and, if you have gone through all chapters, you can even access some of the research literature. You still have to deepen your understanding of the quantum origin of forces and become acquainted with the computational tools needed to evaluate the amplitudes. Master's and graduate-level textbooks will give you this opportunity. The common language of our discipline is QFT, and you appreciated only a flavor of it (covariant formalism, relativistic fields, Feynman diagrams, etc.). A clear introduction to this subject at the postgraduate level is given by Maggiore, 2005. Schwartz, 2013 and Peskin and Schroeder, 1995 provide a graduate-level overview covering both the foundation of QFT and the SM. The "classical" QFT textbooks of my generation are the first two volumes of Weinberg, 2005a. They explore the meaning of the theories up to the Standard Model with no discount on rigorous proofs, including the CPT theorem and the t'Hooft–Veltman renormalization of the Standard Model. Weinberg's books require a strong mathematical background and sometimes, the author uses non-standard notations: not for the faint of heart.

Among the books that require only a basic knowledge of QFT, I recommend Thomson, 2013; Peskin, 2019; Barr *et al.*, 2019; and Larkosky, 2019. The first chapters, of these books recap the fundamentals – the content of this book – but add topics that are appropriate for postgraduate or advanced undergraduate students in particle physics. You can delve into nuclear physics using the aforementioned Bertulani, 2007. A classical and clear textbook is Krane, 1988, albeit is a bit obsolete and weak on particle–nuclear physics connections. Hans, 2011 and D'Auria, 2018 are well-thought out compromises between completeness and modernity. Bryan, 2021 emphasizes the many applications of nuclear physics. My favorite graduate-level book is Donnelly *et al.*, 2017 where contemporary nuclear physics is introduced starting from QCD. Astrophysics is well-presented in Perkins, 2009. You can get a flavor of multimessenger astrophysics in Spurio, 2018 and De Angelis, 2018. The literature on GR is vast and I used here Rindler, 2006 and Cheng, 2010. The classic textbooks are Wald, 1984; Weinberg, 1972; and the monumental Misner *et al.*, 1973. They are great fun but all at the graduate level and obsolete in several parts. Carroll, 2019 is clear and up-to-date but requires a good background in maths. ΛCMD is presented in detail in Dodelson and Schmidt, 2020. Modern cosmology is masterfully exposed in Weinberg, 2008. In my opinion, the most updated book on gravitational waves at the graduate level is Andersson, 2020. There are, of course, many popular science books describing the central discoveries in our field, including breakthroughs in the 21st century: the Higgs boson (Flores Castillo, 2016), neutrino oscillations (Close, 2012), and gravitational waves (Binetruy, 2018).

Exercises

(13.1) Write a scalar potential with a false vacuum.

(13.2) What are the color charges of the Higgs boson?

(13.3) If a transition occurs from the false to the true vacuum what happens to the SM?

(13.4) Estimate the energy and angular resolution needed to measure the Higgs width Γ_H.

(13.5) * Why can the Higgs have an intrinsic parity ($+1$) even if parity is violated by weak interactions? Has the Z^0 an intrinsic parity?

(13.6) Assume that only $\nu_\mu \rightarrow \nu_\tau$ oscillations cause the ν_μ CC rate reduction observed by SuperKamiokande. Compute the rate reduction for a 2 GeV ν_μ beam produced on the other side of the Earth, reaching SuperKamiokande from below (zenith angle $\theta_z = \pi$). What happens if $\theta_z = 3\pi/4$ or $\theta_z = 0$?

(13.7) Both the D^+ and the K^+ can decay into $e^+\pi^0\nu_e$ but $\mathrm{BR}(D^+ \rightarrow e^+\pi^0\nu_e) \ll \mathrm{BR}(K^+ \rightarrow e^+\pi^0\nu_e)$. Why?

(13.8) * **Quantum regeneration** is the re-appearance of particles that disappeared by decays. This feature is due to superposition effects. For a pure beam of K^0 – corresponding to a 50%–50% superposition of K_1 and K_2 – compute the percentage of regenerated K^0 by a target that absorbs 10% of the K^0 and 30% of the $\bar{K}^0$. The target is very thin and located at $L = 30c\tau_{K_1}$, where τ_{K_1} is the K_1 lifetime. [Hint: consider non-relativistic particles and note that $\tau_{K_2} \simeq 600\tau_{K_1}$]

(13.9) What is the interaction (strong, e.m., weak) that produces the regeneration effect mentioned in Exercise 13.8?

(13.10) **Cabibbo suppression** occurs when the size of the CKM entries appearing in the decay vertices are small. Draw the Feynman diagram of a Cabibbo suppressed decay. [Hint: look at the numerical values of Fig. 13.12.]

(13.11) What is the CP conjugate process of $\nu_\mu \to \nu_e$?

(13.12) * $K \leftrightarrow \bar{K}^0$ oscillations occur in SM but $n \leftrightarrow \bar{n}$ are forbidden. Why?

(13.13) * Derive the Dirac equation from eqs 13.5 and 13.6.

(13.14) * Derive Maxwell's equations from eqs 13.5 and 13.7 using the results of Chap. 6.

(13.15) * Identify the triple and quartic vertices of the Higgs boson by inspecting eqn 13.30.

(13.16) * Compute the probability to see a ν_μ instead of a ν_e in a neutrino detector deriving eqn 13.66. Perform the derivation assuming the existence of two neutrino flavors, only (ν_e and ν_μ). [Hints: use the trigonometric product-to-sum identities to gather the terms with the same angles.]

(13.17) Show that
$$\frac{\Delta m_{23}^2 L}{4E} \simeq \frac{1.27\Delta m_{23}^2 [\text{eV}^2]}{E[GeV]}. \tag{13.140}$$

Equation 13.140 gives the **oscillation phase** of the neutrino mass eigenstates m_2 and m_3.

(13.18) * Using eqn 13.97, demonstrate that:
$$P(\nu_\mu \to \nu_e) = \sin^2 2\theta_{13} \sin^2 \theta_{23} \sin^2 \frac{\Delta m_{23}^2 L}{4E}$$

if $\Delta m_{12}^2 L/4E \to 0$ and $\delta_{CP} = 0$. Show that this condition may occur, for instance, if neutrinos are produced by cosmic rays at a few GeV. Does this result hold if $\delta_{CP} \neq 0$?

(13.19) * Demonstrate that eqn 13.66 holds also for three neutrino families if $\theta_{13} = 0$ and $\theta_{23} = 0$. Do we need to assume $\Delta m_{23}^2 \equiv m_3^2 - m_2^2 = 0$ or $\delta_{CP} = 0$ to get this result?

(13.20) ** Show that eqn 13.19 is invariant for the transformation of eqn 13.17 if the A^μ field transforms like the classical electromagnetic field:
$$A'^\mu = A^\mu - \partial^\mu \alpha(x^\mu).$$

(13.21) ** Demonstrate eqn 13.24. [Hint: insert eqn 13.22 in eqn 13.19 and write $\mathcal{L}$ as a function of the new fields and the VEV]

(13.22) * The Lagrangian (density) of a real scalar field is
$$\frac{1}{2}(\partial_\mu \phi)(\partial^\mu \phi) - \frac{m^2}{2}\phi^2. \tag{13.141}$$

Show that this particle satisfies the Klein-Gordon equation. As a consequence, the **mass of a scalar particle** is located in front of the term of the Lagrangian proportional to ϕ^2. Using eqn 13.30, demonstrate that the mass of the Higgs boson in the theory of Sec. 3.1.3 is $m_H = \sqrt{2\lambda}v$.

A Special relativity

Special relativity (SR) arose from a set of glaring inconsistencies noted by many authors in Maxwell's equations and some experiments. Physicists responded to this challenge with several models that originated most of the equations of SR even if they were formulated in a classical framework. In 1905, A. Einstein pinned down the assumption that creates these inconsistencies and published a theory capable to re-interpret all previous results in a consistent and elegant way. The faulty assumption is hidden in Newton's definition of **inertial frame**.

Both in classical mechanics and SR:

> an inertial frame is a frame where each point-like particle pulled by a null force moves in a straight line at constant velocity.

The null force may also be the vector sum of all forces pulling the particle. To go from one inertial frame to another, Newton applied a rotation and a translation that brings the axes of the first frame on top of the axes of the second frame and corrected for the relative velocity of the two frames. The transformations among frames in classical theory date back to G. Galilei. For the frame of Fig. A.1, there is no need of rotations or translation because the first frame moves along the z axis of the second frame at velocity $\mathbf{v} = v\mathbf{k}$. In this case, the **Galilean transformations** are:

$$
\begin{aligned}
x' &= x \\
y' &= y \\
z' &= z - vt
\end{aligned}
\tag{A.1}
$$

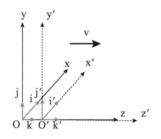

Fig. A.1 The simplest Lorentz transformation. The reference system O' moves with respect to O with a velocity v along the z-axis. All axes are parallel and the origin of O moves along the z- (or z'-) axis. The unit vectors are labeled $\mathbf{i},\mathbf{j},\mathbf{k}$ and $\mathbf{i'},\mathbf{j'},\mathbf{k'}$ for O and O', respectively. The corresponding Lorentz transformation is a Lorentz boost along z.

where (x', y', z') are the particle coordinates in the new reference frame O' and (x, y, z) are the ones in the old frame O. These are only a special case of Galilean transformation, the **Galilean boost** along z. We can have systems where $\mathbf{v}$ is not along z and we need three boosts because $\mathbf{v} = v_x\mathbf{i} + v_y\mathbf{j} + v_z\mathbf{k}$ so that:

$$
\begin{aligned}
x' &= x - v_x t \\
y' &= y - v_y t \\
z' &= z - v_z t
\end{aligned}
$$

or, equivalently, $\mathbf{x'} = \mathbf{x} - \mathbf{v}t$. Finally, the origin of the two systems could be different or the axes could be rotated. In this general case, we need to apply a rotation and a rigid translation together with the boosts. In conclusion,

> Galilean transformations can be uniquely written as the composition of a rotation, a translation, and a boost. These transformations do not change the relative distance between two points.

The last statement ($|\mathbf{x}'_B - \mathbf{x}'_A| = |\mathbf{x}_B - \mathbf{x}_A|$) follows from the definition of distance in Euclidean geometry:

$$|\mathbf{x}_B - \mathbf{x}_A| \equiv \sqrt{(x_B - x_A)^2 + (y_B - y_A)^2 + (z_B - z_A)^2} \qquad \text{(A.2)}$$

and the definition of Galilean transformations. For the special case of a boost along z we get $|\mathbf{x}'_B - \mathbf{x}'_A| = |\mathbf{x}_B - \mathbf{x}_A|$ just rewriting $\mathbf{x}'$ as a function of $\mathbf{x}$ using eqn A.1 but this result holds in general. That is not particularly amazing: Newton and Galilei knew that Euclidean geometry is a nice tool to describe physical systems and there is no reason why Pythagoras's theorem should not work. This is the rationale of the definition of Euclidean distance by eqn A.2.

The set of all Galilean transformations forms a group[1] and is called the **Galilean group**. The number of parameters that define the group is nine: three boosts (v_x, v_y, v_z), three rotation angles, and three rigid translations in space. The position of the particle is uniquely defined by three space coordinates plus the time variable t. "Plus the time variable t" switched on a light in Einstein's mind.

[1] I mean a mathematical group defined, e.g. in Example 7.1.

A.1 What is relative in relativity?

Classical physics is plagued by an apparently harmless assumption: time passes at the same pace in any inertial frame. This means that the correct form of eqn A.1 is:

$$t' = t$$
$$x' = x$$
$$y' = y$$
$$z' = z - vt \qquad \text{(A.3)}$$

and t looks like a **universal time**. To make sense of Maxwell's equation, Einstein claimed that:

> time is relative because $t' = t$ is a wrong assumption. Time intervals depend on the motion of the observer or, equivalently, on the inertial frames. Conversely, the speed of light in vacuum (c) is universal because each inertial observer measures the same light-speed independently of the frame.

The theory of SR is named this way because of the relativity of time. Indeed, we should call it "the theory of universal velocity of light in vacuum."[2]

We then need to find the correct form of eqn A.3 that preserves the numerical value of the speed of light in vacuum. For the inertial frame considered in eqn A.3, the new transformations read (Rindler, 2006):

[2] We usually say that Einstein made this assumption because of the result of the Michelson–Morley experiment and the inconsistencies arising from the concept of "ether": a fluid permeating the universe where the light travels at speed c. This makes sense because such a strong statement requires overwhelming experimental evidence. But Einstein was not working in the science establishment in 1905. Many authors doubt that Einstein was fully aware of these experiments, which are never cited in his seminal paper. He probably made this assumption just because a universal speed $c = (\epsilon_0\mu_0)^{-1/2}$ naturally arises in Maxwell's equations. This is one of the instances supporting the myth that pure thought suffices to understand nature, a myth that has been torn off by quantum mechanics.

$$t' = \gamma \left(t - \frac{vx}{c^2} \right)$$
$$x' = x$$
$$y' = y$$
$$z' = \gamma(z - vt) \tag{A.4}$$

where $\gamma = (1 - \beta^2)^{-1/2}$ and $\beta = v/c$ is the ratio between the speed of O' with respect to O and the speed of light in vacuum. We can invert eqn A.4 to describe a point in O as a function of its coordinates in O' in two ways. Either by brute force, solving eqn A.4 as a function of x, y, z, t, or noting that an observer in O' sees O moving with a velocity $-v$. In this case, we can just reverse the sign of v and replace the primed with the unprimed quantities of eqn A.4. Equation A.4 is the SR counterpart of the Galilean boost along z and is called a **Lorentz boost** along z, not Einstein's boost because the formula was already derived by H. Lorentz in an ether-based model of electromagnetism. We can produce all Lorentz boosts moving O' by an arbitrary velocity **v**. We then have to add the rotations and the rigid translations. The set of transformations produced by a Lorentz boost and a rotation of the frame axes constitutes a group called the **restricted Lorentz group**. More generally:

- the transformations that include boosts and rotations make up the **proper orthochronous Lorentz group**, that is the **restricted Lorentz group**. This group is connected: we can reach every inertial frame belonging to the group making a continuous change of **v** and the rotation angles starting from our inertial frame.

- if we apply a parity transformation to the previous group ($\mathbf{x} \to -\mathbf{x}$), we find another connected group: the **improper orthochronous Lorentz group**. In essence, this is equivalent to choosing a left-handed axis orientation instead of right-handed axes.

- if we apply a T-parity transformation to the restricted Lorentz group ($t \to -t$), we find another connected group: the **proper non-orthochronous** (or antichronous) **Lorentz group**.

- the last connected group comes from the simultaneous application of $\mathbf{x} \to -\mathbf{x}$ and $t \to -t$ to the restricted Lorentz group. This is the **improper non-orthochronous Lorentz group**.

- the union of the four groups is called the **Lorentz group**, which is a disconnected group. We need six continuous parameters to define the inertial frame (three boosts and three rotations) plus the parity and T-parity that have been applied.

- the Lorentz group extended to include rigid translations of the axis's origin along x, y, and z is called the **Poincaré group**. This group needs three additional continuous parameters to fix the inertial frame than the Lorentz group.

What are the features shared by these groups? First,

$$\gamma \to +\infty \quad \text{if} \quad v \to c . \tag{A.5}$$

Therefore,

> the speed of a particle cannot exceed $c = 2.99792459 \times 10^8$ m/s. A more general statement resulting from SR is that we cannot transport information at a speed greater than the speed of light in vacuum.

The generalization that brings the concept of **information** into the game is subtle.

Example A.1

Consider a group of soccer fans aligned in a queue at a distance of 1 m each, just before the start of an important match. The first supporter raises his hand at $t = 0$. The second at $t = 1$ ns, the third at $t = 2$ ns, and so on. An observer in the same inertial frame as the soccer fans will measure a stadium wave with a phase velocity of 1 m/1 ns $\simeq 3c$. Much faster than the maximum velocity of SR!

This example does not harm SR. It teaches us that we can construct observables that move faster than light in vacuum but these observables bring no information around. In the previous example, the fans raise their hand looking at their clock without caring about the hands of their fellows. To get the wave, they must arrange the performance well before the start of the match. On the other hand, if a particle moves faster than light, it can scatter and a new physical event is *caused* by the particle. These superluminal particles have never been observed and probably do not exist at all. They have been considered in other theories than relativity and are very popular in science fiction movies. They are called **tachyons**.

The second feature shared by the Lorentz and Poincaré groups is the following. In all inertial frames, an observer that measures the straight trajectory of a light-ray leaving the origin of the reference frame finds:

$$(ct)^2 - x^2 - y^2 - z^2 = 0. \tag{A.6}$$

This is equivalent to saying that the speed of light in vacuum is the same for any inertial observer. We can prove that:

> the Poincaré group is the most general group that fulfills the founding principle of SR: eqn A.6.

In SR, $(ct)^2 - x^2 - y^2 - z^2$ plays a role similar to the Pythagorean theorem to define the distance of two events in space-time.[3]

[3]For the sake of completeness, we add that the Poncaré group is the most general group preserving Minkowsky space-time isometries: the distance between two events in Minkowsky space does not change after a transformation of the Poincaré group. We discuss this important topic in Sec. 2.2.1.

A.2 Consequences of the universal speed c

Dropping the universal time in favor of a universal speed implies many changes in physics. In 1905, they were considered small corrections to classical physics, while they play a central role in science and technology in the 21st century. The most popular consequences are time dilation, space contraction, and the relativistic Doppler effect.

A.2.1 Time dilation

Let us consider (without loss of generality) an inertial frame belonging to the restricted Lorentz group like the system O' boosted along the z-axis of eqn A.4. Time dilation is the different elapsed time measured by two clocks located in O and O', respectively. If the clocks are at rest in their frame, the time dilation is due to the relative velocity between the frames and the universal nature of c. We prove it through eqn A.4. The size of time dilation arises from the time component of eqn A.4. A time interval $T_B - T_A$ in O is recorded in O' as:

$$T'_A = \gamma \left(T_A - \frac{vx(T_A)}{c^2} \right)$$
$$T'_B = \gamma \left(T_B - \frac{vx(T_B)}{c^2} \right)$$

where $x(t)$ is the position of the clock at time t in the frame O. Since the clock remains at rest in their inertial frame, it follows $x(T_A) = x(T_B)$. Then,

$$\Delta T' = T'_B - T'_A = \gamma(T_B - T_A) = \gamma \Delta T. \qquad (A.7)$$

This feature of SR is called **time dilation** because $\gamma > 1$ and can be generalized as follows:

> the time interval measured by any observer is larger than the interval measured in the comoving frame of the clock. The time measured in this frame is slowed down compared to any other observer and is called the proper time τ.

The lifetime difference of a muon at rest and in motion in the laboratory frame (LAB) is a striking confirmation of this SR prediction. More recently, the two-clock experiment proposed by Einstein in 1905 was carried out with atomic clocks at moderate velocity and the result is in perfect agreement with the prediction of relativity (Chou *et al.*, 2010). Time dilation also affects the **simultaneity** of two events: events occurring at the same time in O may be recorded at different times in O'. Despite Newton's opinion, simultaneity is not universal, either.

A.2.2 Length contraction

The contraction of lengths is another striking feature of the Lorentz transformations and was derived for a rigid body (a rod) in 1892 – well before SR – by H. Lorentz and G.F. FitzGerald to explain the puzzling results of the Michelson–Morley experiment. The theorem can be proven using the space part of eqn A.4. The measurement is done *at the same time* by an observer that sees the object is in motion. This observer records a length $L = z_2 - z_1$, where z_1 and z_2 are the first and last point of the rod. To compute the **proper length** of the rod, which is the length measured in the rest frame of the rod, we introduce a new frame O', the rest-frame comoving with the rod. Equation A.4 gives:

$$z_1' = \gamma(z_1 - vt_1)$$
$$z_2' = \gamma(z_2 - vt_2)$$

Since the measurement was done at the same time in O, $t_1 = t_2$ and $L' = \gamma L$. This effect is called length contraction because all observers measure shorter lengths than the observer located in the rest frame of the rigid body:

$$L = \frac{L'}{\gamma} \qquad\qquad (A.8)$$

and $L < L'$ because $\gamma > 1$.

The length of an object is not a universal quantity and the maximum length is measured by an observer comoving with the object.

Lorentz and FitzGerald used the concept of "rigid body" in a way that is incompatible with relativity. In classical mechanics, if we apply a shrinking force to a rigid body, the body *instantaneously* exerts an equivalent force that cancels the shrinking force and zeroes the strain. Rigid bodies do not exist in SR because information cannot be instantaneously transferred but runs at most at a speed c. The body is shrunk but the strain disappears in a time $\simeq L/c$, which is generally very short and, of course, consistent with SR. Classical rigid bodies exist as long as we neglect the L/c lapse.

A.3 Relativistic dynamics*

Even if the concept of force is marginal in quantum mechanics and particle physics, it can be smoothly used in SR. This is particularly simple if we employ the covariant formalism of Sec. 2.1.1 Covariance can be used to define all quantities in relativistic mechanics following the same procedure employed in classical mechanics.

To define velocity in SR, we first need a definition of an infinitesimal change in the position of the particle. This is:

$$\text{(Newton)} \quad d\mathbf{x} = dx\,\mathbf{i} + dy\,\mathbf{j} + dz\,\mathbf{k}$$
$$\text{(Einstein)} \quad dx^\mu = (dx^0, dx^1, dx^2, dx^3)$$

where we labeled "Newton" the classical definition of $d\mathbf{x}$ and "Einstein" the corresponding quantity in SR.

Both in SR and classical physics, the velocity is the time derivative of the trajectory. Since the trajectory is parameterized by an universal quantity, we must replace the time with an invariant quantity, whose value is the same for each inertial observer. The **velocity** or, in SR, **four-velocity** is:

$$\text{(Newton)} \ \ \mathbf{v} = \frac{d\mathbf{x}}{dt} \ ; \ \ \text{(Einstein)} \ \ v^\mu = \frac{dx^\mu}{d\tau} \tag{A.9}$$

where τ is the proper time, that is the time computed in the comoving frame of the particle. Why is τ universal or, equivalently, invariant for Lorentz transformations? The intuitive reason is that we are defining τ in a specific reference frame – the rest frame – and all observers must compute the proper time in two steps: Lorentz-transform their system to the frame comoving with the clock and, then, switch on the clock. If they all follow this procedure, they reach the same inertial frame and get the same time interval.

We can formally prove this statement using the covariant formalism. The position of the particle in space in the comoving frame is $\mathbf{x} = \mathbf{0}$ and $x^\mu = (c\tau, 0, 0, 0)$. The squared "norm" (dx^2) of dx^μ in the Minkowsky space is given by eqn 2.38, and you can check that dx^2 is equal to $d\tau^2$ by computing:

$$d\tau^2 = dx^2 = g_{\mu\nu}dx^\mu dx^\nu \tag{A.10}$$

where $g_{\mu\nu}$ is the metric tensor of eqn 2.31. $d\tau^2$ is a Lorentz-invariant quantity in the covariant formalism because each norm is invariant, too. In this way, we have shown that $d\tau$ is Lorentz invariant without resorting to the Lorentz transformations. As noted in Chap. 2, the covariant formalism establishes how a quantity changes after a Lorentz transformation without applying the transformation.

v^μ is a (contravariant) vector because dx^μ is a vector and $d\tau$ is a scalar (i.e. a Lorentz-invariant quantity). For a general Lorentz transformation $\Lambda^\mu{}_\nu$:

$$v'^\mu = \frac{dx'^\mu}{d\tau'} = \frac{\Lambda^\mu{}_\nu dx^\nu}{d\tau} = \Lambda^\mu{}_\nu v^\nu \tag{A.11}$$

which is exactly the definition of a (contravariant) vector. What is the link between velocity in SR and classical velocity? All in all, the experiments measure the velocity of the particles from a ratio between space and time in their inertial frame. This system is the laboratory frame and, in many cases, is *not* the rest frame of the particle. In the rest frame, $\mathbf{x} = \mathbf{x}_\parallel = \mathbf{x}_\perp = \mathbf{0}$ and from eqn A.4:

$$\frac{d\tau^2}{dt^2} = \gamma^{-2} = 1 - \frac{v^2}{c^2}. \tag{A.12}$$

The same result holds for a general transformation of the Poincaré group. Then,

$$v^1 = \frac{dx}{d\tau} = \frac{dx}{dt}\frac{dt}{d\tau} = v_x\gamma$$
$$v^2 = \frac{dy}{d\tau} = \frac{dy}{dt}\frac{dt}{d\tau} = v_y\gamma$$
$$v^3 = \frac{dz}{d\tau} = \frac{dz}{dt}\frac{dt}{d\tau} = v_z\gamma$$

and, therefore,

$$v^\mu = \gamma(c, v_x, v_y, v_z) = (\gamma c, \gamma\mathbf{v}). \tag{A.13}$$

$\mathbf{v}$ is the classical velocity measured by the experiment in the laboratory frame and eqn A.13 provides the link with the (four-)velocity of SR.

We define the **acceleration** as the time (or proper time in SR) derivative of the velocity:

$$(\text{Newton})\ \ \mathbf{a} = \frac{d\mathbf{v}}{dt} \ \ ; \ \ (\text{Einstein})\ \ a^\mu = \frac{dv^\mu}{d\tau} \tag{A.14}$$

and

$$a^\mu = \frac{dv^\mu}{d\tau} = \frac{dv^\mu}{dt}\frac{dt}{d\tau} = \gamma\frac{dv^\mu}{dt}. \tag{A.15}$$

Here, the link with classical acceleration is subtler because γ can change over time:

$$a^\mu = \gamma\left(\dot{\gamma}c, \dot{\gamma}\mathbf{v} + \gamma\mathbf{a}\right) \tag{A.16}$$

where $\mathbf{a}$ and $\mathbf{v}$ are the classical acceleration and velocity measured in LAB and $\dot{\gamma} \equiv d\gamma/dt$. Note that, if $a^\mu \neq 0$, the proper frame is not inertial. Conversely, the laboratory frame can be considered inertial in all cases of interest in this book.[4]

The **rest mass** of a particle is defined as the mass measured in the proper frame. In classical physics, there is no difference between mass and rest mass but this distinction is mandatory in SR. Particle physicists do not employ the time-honored definition of "relativistic mass" and we label the rest mass simply as m. The "relativistic mass" is thus γm and the celebrated Einstein **mass–energy relation** "$E = mc^2$" is written as $E = \gamma mc^2$ or $E = \gamma m$ in natural units (NU). The definition of **momentum** is then:

$$(\text{Newton})\ \ \mathbf{p} = m\mathbf{v} \ \ ; \ \ (\text{Einstein})\ \ p^\mu = mv^\mu \tag{A.17}$$

and

$$p^\mu = mv^\mu = \gamma m(c, \mathbf{v}). \tag{A.18}$$

Equation A.18 is a very useful relation between the momentum $\mathbf{p}$ measured by the experiments using, for example, the bending of a charged particle in a magnetic field and the corresponding four-momentum in SR. The first component p^0 of the four-momentum is $p^0 = \gamma mc$ and $p^0 c$ is called the **energy** of the particle[5] because:

$$E \equiv p^0 c = \gamma mc^2 \tag{A.19}$$

and reproduces the Einstein energy–mass relation.

[4]Our laboratory is not inertial because the earth spins around its axis and the sun. These small gravitational corrections are negligible in the scattering and decay of particles, which are dominated by strong, electromagnetic, and weak forces.

[5]We need to multiply by c because in SI the energy is measured in Joule and the momentum in Kg m/s. In NU, the energy is just p^0.

A.4   Newton's laws of motion in SR

What about the three Newton's laws of mechanics in SR? Do they still hold? The first law reads "if the net force on an object is zero, then the velocity of the object is constant":

$$\sum_i \mathbf{F}_i = 0 \iff \dot{\mathbf{v}} = 0. \tag{A.20}$$

This law holds in SR because is the definition of inertial frame: an inertial frame is a frame of reference where an object interacting with no forces (or with forces that sum up to zero) is either at rest or moves in a straight line at a constant velocity.

The second law reads "the rate of change of momentum of a body over time is directly proportional to the applied force, and occurs in the same direction of the force":

$$\mathbf{F} = \frac{d\mathbf{p}}{dt} = m\mathbf{a}. \tag{A.21}$$

The second equality of eqn A.21 does not hold in SR because the inertial mass is different from the rest mass. In particular, if the particle velocity gets closer to the speed limit c, the inertial mass must grow toward infinity so that no force can increase the speed of the particle:

$$m_{inertial} \equiv \gamma m \tag{A.22}$$

and[6]

$$m_{inertial} \to \infty \quad \text{for} \quad \gamma \to \infty. \tag{A.23}$$

As expected, the inertial mass diverges for $v \to c$. The **force** is better defined as:

$$\text{(Newton)} \quad \mathbf{F} = \frac{d\mathbf{p}}{dt} \quad ; \quad \text{(Einstein)} \quad K^\mu = \frac{dp^\mu}{d\tau}. \tag{A.24}$$

The relation between the three-force $\mathbf{F}$ (eqn A.24, left) experienced by a particle in the laboratory frame and the three-acceleration $\mathbf{a}$ can be obtained carrying on the derivatives until:

$$\mathbf{F} = \gamma^3 m\mathbf{a}_\parallel + \gamma m\mathbf{a}_\perp \tag{A.25}$$

where $\mathbf{a}_\parallel$ and $\mathbf{a}_\perp$ are the components of the acceleration parallel and perpendicular to the direction of motion.

The third Newton's law reads "if one object exerts a force $\mathbf{F}_1$ on a second object, then the second object simultaneously exerts a force $\mathbf{F}_2$ on the first one, and the two forces are equal in magnitude and opposite in direction: $\mathbf{F}_1 = -\mathbf{F}_2$." This law contradicts SR if the position and size of the two bodies are different because the law implies the transmission of information (the deformation due to the force) at infinite velocity. Once more, the very concept of a rigid body is in contradiction with SR. In classical mechanics, this law is equivalent to the conservation of

[6]In old textbooks, this quantity was the aforementioned "relativistic mass" but this name has been dropped off. The meaning is much clearer if we call m the rest mass and γm the inertial mass.

momentum in an isolated system made of point-like particles. Conservation of momentum holds in SR, too, because is a consequence of the invariance of the system for a rigid translation of the reference frame, as explained in Sec. 5.4. The corresponding law in SR is:

$$\sum_{i=1}^{N_i} (p_i^{ini})^\mu = \sum_{j=1}^{N_f} (p_j^{fin})^\mu. \tag{A.26}$$

The first and second sums run over all particles in the initial and final state, respectively. This law is called the **conservation of four-momentum** in SR.

Finally, the four-momentum is a four-vector and transforms in a way similar to (ct, x, y, z). For the system of Fig. A.1:

$$\begin{aligned}
E' &= \gamma \left(E - vp_z \right) \\
p'_x &= p_x \\
p'_y &= p_y \\
p'_z &= \gamma \left(p_z - \frac{Ev}{c^2} \right).
\end{aligned} \tag{A.27}$$

We can derive eqn A.27 from the definition of energy and momentum combined with eqn A.4. Even better, we can use the covariant formalism of Sec. 2.1 and rewrite the Lorentz transformation for the system of Fig. A.1 in matrix form as:

$$\begin{pmatrix} ct' \\ x' \\ y' \\ z' \end{pmatrix} = \begin{pmatrix} \gamma & 0 & 0 & -\beta\gamma \\ 0 & 1 & 0 & 0 \\ 0 & 0 & 1 & 0 \\ -\beta\gamma & 0 & 0 & \gamma \end{pmatrix} \begin{pmatrix} ct \\ x \\ y \\ z \end{pmatrix} = \begin{pmatrix} \gamma ct - \beta\gamma z \\ x \\ y \\ \gamma z - \beta\gamma ct \end{pmatrix}. \tag{A.28}$$

The four-momentum transforms by the same matrix of eqn A.28 because both x^μ and p^μ are four-vectors:

$$\begin{pmatrix} E'/c \\ p'_x \\ p'_y \\ p'_z \end{pmatrix} = \begin{pmatrix} \gamma & 0 & 0 & -\beta\gamma \\ 0 & 1 & 0 & 0 \\ 0 & 0 & 1 & 0 \\ -\beta\gamma & 0 & 0 & \gamma \end{pmatrix} \begin{pmatrix} E/c \\ p_x \\ p_y \\ p_z \end{pmatrix} = \begin{pmatrix} \gamma E/c - \beta\gamma p_z \\ p_x \\ p_y \\ \gamma p_z - \beta\gamma E/c \end{pmatrix}. \tag{A.29}$$

These formulas are easier to remember than eqn A.4. They are even neater in NU where $c = 1$, time and space have the same dimension (eV^{-1}), and mass, momentum, and energy are all measured in eV. SR mixes time with space as well as energy and momentum if space and momentum are in the direction of the boost (z and p_z in our case). This is the reason why using the same units for time and space (or energy and momentum) is always a good deal.

A.5 Relativistic and non-relativistic particles

SR must reproduce classical mechanics if the speed of the particles is much slower than c. These particles are called **non-relativistic**. The energy of a non-relativistic particle is:

$$E = \gamma mc^2 = \frac{1}{\sqrt{1 - v^2/c^2}} mc^2 = \left(1 - \frac{v^2}{c^2}\right)^{-1/2} mc^2$$

$$\simeq \left(1 + \frac{v^2}{2c^2}\right) mc^2 = mc^2 + \frac{1}{2}mv^2$$

where we expanded γ in a Taylor series at first order assuming $v/c \ll 1$.

> The relativistic energy is just the **rest energy** of the particle mc^2 plus the classical kinetic energy
>
> $$E_{kin} = \frac{1}{2}mv^2 \qquad (A.30)$$
>
> if v is much smaller than c.

If $v \ll c$ does not hold, we can still define the **kinetic energy** in SR as

$$E_{kin} = E - mc^2. \qquad (A.31)$$

For a relativistic particle,

$$E = \sqrt{|\mathbf{p}|^2 c^2 + m^2 c^4} \qquad (A.32)$$

and

$$\beta = \frac{|\mathbf{p}|c}{E} \quad ; \quad \gamma = \frac{E}{mc^2} \qquad (A.33)$$

as shown in Sec. 2.3.2.

The **velocity addition formula** is a rule that gives the velocity of a particle seen by two different inertial frames. In classical mechanics, this rule has a trivial form. For two inertial frames O and O' moving with a relative velocity $\mathbf{v}$, the velocity of the particle $\mathbf{u}'$ seen by an observer O' is just:

$$\mathbf{u}' = \mathbf{u} - \mathbf{v} \qquad (A.34)$$

where[7] $\mathbf{u}$ is the velocity of the particle in O. We can prove eqn A.34 taking the derivative of eqn A.3 with respect to time for the two reference frames. For Lorentz transformations, the velocity-addition formula is more complicated and reads:

$$u' = \frac{u - v}{1 - \left(\frac{vu}{c^2}\right)} \qquad (A.35)$$

if all particles and LAB move along the z-axis. We can prove eqn A.35 by the proper time derivatives of eqn A.4. Once more, this formula

[7]Equation A.34 is discussed in the famous example of a ball (particle) falling inside a boat (O system) seen from the shore (O' system) in Galilei's "Dialogue Concerning the Two Chief World Systems" written in 1692.

was already available in literature before 1905. It was partially derived by Lorentz for an ether-based model that should have explained the puzzling results of Fizeau's experiment (Miller, 1998).

The most general form of the velocity-addition formula when $\mathbf{v}$ is along the z-axis is:

$$u'_x = \frac{\sqrt{1 - \frac{v^2}{c^2}} u_x}{1 - \frac{v}{c^2} u_z}$$

$$u'_y = \frac{\sqrt{1 - \frac{v^2}{c^2}} u_y}{1 - \frac{v}{c^2} u_z}$$

$$u'_z = \frac{u_z - v}{1 - \frac{v}{c^2} u_z}. \qquad (A.36)$$

We can express $\mathbf{u}$ as a function of $\mathbf{u}'$ just reversing the sign of v, as done for the Lorentz transformations.

A.6 Relativistic Doppler effect

When the speed of a particle saturates toward c, the energy increases because of the increase of γ and the inertial mass γm goes to infinity. But what happens if the particle has zero rest mass ($m = 0$) like a photon? In this case, the four-momentum of the photon as seen by its source (O') is:

$$p'^{\mu} = (p'c, 0, 0, p') \qquad (A.37)$$

if the photon propagates along the z-axis because $E = \sqrt{|\mathbf{p}|^2 c^2 + m^2 c^4} = pc$ when $m = 0$. The photon is light. Therefore, any observer sees photons moving at speed c. The velocity addition formula of eqn A.35 ensures that if the photon velocity at source (O') is c, its velocity in LAB (O) is the same. This result would have puzzled Galilei: if we shoot a cannonball from a moving ship (O') at velocity $\mathbf{u}'$ against a buccaneer (O), the ball in O moves *faster*. Its velocity is $\mathbf{u} = \mathbf{u}' + \mathbf{v}$ because of eqn A.34, the Galilean velocity-addition formula. And the buccaneer will note it at his own expense. But if we shoot a photon (in vacuum), the speed of the photon is c in any reference frame. How can the buccaneer record the difference? Performing a Lorentz boost of eqn A.37 along z for a boat traveling at speed $v\mathbf{k}$ and approaching the buccaneer from the left, we get

$$p^{\mu} \equiv (E/c, p_x, p_y, p_z) = (\gamma E'/c + \beta\gamma p', 0, 0, \gamma p' + \beta\gamma E'/c). \qquad (A.38)$$

There is a plus sign because the buccaneer sees the ship approaching him and the sign of the speed is the same as the photon. The physics meaning of this sign is that the photon increases its *energy*, not the speed! The faster the ship, the bigger the energy. The photon energy in O is $E = \gamma p'c(1 + \beta)$, where βc is the speed of the ship and we used $E' = p'c$.

In QM,[8] the energy of a photon is the Planck constant times the light frequency: $E = h\nu$. Then,

[8]The relativistic Doppler formula can be derived without reference to photons and QM by using electromagnetic waves. We employ photons because this feature of SR is important in particle and nuclear physics as an energy loss/gain mechanism for massless particles.

$$E = h\nu = \gamma p'c(1 + \beta) \quad ; \quad E' = h\nu' = p'c \qquad (A.39)$$

and the ratio of the frequencies is:

$$\frac{\nu}{\nu'} = \gamma(1 + \beta) = \frac{1 + \beta}{\left(1 - \frac{v^2}{c^2}\right)^{1/2}} = \frac{1 + \beta}{\sqrt{(1 - \beta)(1 + \beta)}} = \sqrt{\frac{1 + \beta}{1 - \beta}} > 1.$$
$$(A.40)$$

In summary,

> a photon emitted by a source moving toward the observer
> is recorded with a higher frequency with respect to a pho-
> ton emitted by a source at rest. The photon is **blue-shifted**.
> Conversely, if the source moves away from the observer, the fre-
> quency recorded by the observer is smaller and the photon is
> **red-shifted**.

In SR, the motion of the source produces an energy loss if the source
moves away and an energy gain if it moves closer to the observer. If the
frequency of the photon emitted by the source at rest is known (e.g. the
light emission of a star), we can determine the direction of the source by
the observation of its light. The **relativistic Doppler shift** formula is
then:

$$\nu_{obs} = \nu_s \sqrt{\frac{1 + \beta}{1 - \beta}} \qquad (A.41)$$

where ν_{obs} is the observed frequency and ν_s is the frequency when the
source is at rest for the observer. If β is positive, the light is blue-shifted
and the source is approaching the observer. If β is negative, the light
is red-shifted and the source is moving away from the observer. It goes
without saying that eqn A.41 is a central tool in cosmology and astronomy because traces the motion of any celestial object that emits
electromagnetic waves (radio waves, infrared, visible, and ultraviolet
light, X-rays, and γ rays).

A.7   Special and general relativity*

SR is called "special" because is a particular case of the general relativity
(GR) theory: SR can describe the motion of particles as long as gravitational effects are neglected. This is perfectly meaningful in particle and
nuclear physics, where gravity is much fainter than the faintest "weak
force". The approach followed by SR is identical to classical physics: we
define a set of reference frames where null forces give no accelerations
to a point-like particle and we call each member of this set an inertial
frame. Working on an inertial frame is always a good choice because we
can have complete control of the forces exerted on a particle. However,
we cannot forget that inertial frames are approximations. The earth is
an inertial frame as long as we neglect gravitational effects and my house
is as good as the earth until an earthquake occurs. Both classical mechanics and SR can handle accelerated particles: the trajectory is derived

by solving the equation of motion in an inertial frame. Sometimes, we are forced to work in non-inertial frames, for instance when we measure the proper time of accelerating particles. But every time we record the time, we boost our frame in a "momentarily inertial frame comoving with the particle." We measure there the proper time, velocity, and acceleration and, then, we go back to our frame. In this way, we can detect, for instance, that a rocket is moving at a constant acceleration in its (non-inertial) rest frame just using SR.

Some problems cannot be solved by SR, even in particle physics, but they are extremely rare and can often be handled with fictitious forces. Again, fictitious forces are common in classical physics, too. The Coriolis force and the centrifugal force are well known examples. SR is a special case of GR because, in the general theory, both inertial and accelerated frames are treated on the same footing and we could even give up the definition of inertial frame. GR has strong consequences: even the navigation systems of smartphones would not work without a GR correction and the entire dynamics of the universe at large scale would be in contradiction with experimental data. As discussed in Sec. 13.11, GR is already in the scope of particle and nuclear physics but plays a minor role in this Primer.

B

The principles of quantum mechanics

Quantum mechanics (QM) is the paradigm describing microscopic systems for more than a century. QM effects are clearly visible in the macroscopic world too, but the strangest features of the theory are particularly evident at the scale of atoms and particles. We summarize here the founding principles of QM employing its most popular description: the **Dirac–von Neumann** mathematical formulation. The most impressive feature of QM is that particle trajectories and, in general, the outcome of any measurement cannot be uniquely predicted: if we perform two different measurements of *exactly* the same system, the outcome is – in general – not the same. However, we can access the statistical distribution of the outcomes, which can be computed from the theory. An experiment designed with perfect knowledge of the Hamiltonian and initial conditions, and performed with instruments much more precise than the width of the observed distributions gets non-deterministic results. As a consequence, uncertainties on the outcome of measurements are inbred to QM, which is a genuine **non-deterministic theory**.

Even today, some models wish to explain QM from deterministic theories. However, the most convincing models[1] were falsified thanks to the seminal works of J.S. Bell, who demonstrated in 1964 that

> any local hidden-variable theory is incompatible with the predictions of QM.

Since 1972, many discrepancies between deterministic theories and QM, called violations of **Bell inequalities**, have been experimentally observed (Bell and Gao (eds.), 2016).

[1]We often group these models in the class of "hidden variable theories" because uncertainty in QM is attributed to additional degrees of freedom we are not aware of.

B.1 States

In classical physics, the state of a particle is given by its position in space-time and by the set of observables (velocity, acceleration, momentum, charge, etc.) associated with that particle at the time when the measurement is done. In QM, the state of a particle (or a system) is a function $\psi(\mathbf{x}, t)$ that contains the maximum amount of information about the particle. In the Dirac–von Neumann description:

a quantum state ψ is called a **wavefunction** and is a **ray** in the $L^2(\mathbb{R}^3)$ **Hilbert space**.

Scary. Well, that is less weird than it sounds. ψ is a complex-valued function of $\mathbf{x}$ that changes in time and may depend on other variables that describe the particle like spin, angular momentum, electric charge, the color charges covered in Chap. 7, etc. Hence, $\psi \in \mathbb{C}$. A "ray" is the set of all functions that differ by a complex phase, only. ψ, $i\psi$ and, in general, $e^{i\alpha}\psi$ with $\alpha \in \mathbb{R}$ are all members of the same ray.

A Hilbert space is a vector space where instead of vectors we use functions. We can play with Hilbert spaces as we play with standard vector spaces and enjoy all the theorems of linear algebra (Larson and Falvo, 2009). For instance, in linear algebra, the **sum of two vectors** is $\mathbf{a} + \mathbf{b} = (a_1, a_2 \ldots, a_N) + (b_1, b_2 \ldots, b_N) = (a_1 + b_1, a_2 + b_2, \ldots, a_N + b_N)$ if the dimension of the vector space is N. Similarly, we can define the **scalar product** of two vectors in a 3D real vector space as $\mathbf{a} \cdot \mathbf{b} = a_x b_x + a_y b_y + a_z b_z$. The generalization to a complex vector space with N dimensions (ND) is:

$$\mathbf{a} \cdot \mathbf{b} \equiv a_1^* b_1 + a_2^* b_2 + \ldots + a_N^* b_N. \tag{B.1}$$

We can also define the squared **norm** of a vector as:

$$|\mathbf{a}|^2 = a_1^* a_1 + a_2^* a_2 + \ldots + a_N^* a_N = |a_1|^2 + |a_2|^2 + \ldots |a_N|^2 \tag{B.2}$$

where a_1^* is the complex conjugate of a_1 and $|a_1| \equiv \sqrt{a_1^* a_1}$ is the modulus (or norm) of the complex number a_1. Finally, all linear operators (linear transformations) of a vector are described by matrices.

$$\mathbf{b} = M\mathbf{a} = \begin{pmatrix} m_{11} & m_{12} & \ldots & m_{1N} \\ m_{21} & m_{22} & \ldots & m_{2N} \\ & & \ldots & \\ m_{N1} & m_{N2} & \ldots & m_{NN} \end{pmatrix} \begin{pmatrix} a_1 \\ a_2 \\ \ldots \\ a_N \end{pmatrix} = M^i{}_j a^j = b^i \tag{B.3}$$

where we wrote the product of M with $\mathbf{a}$ in matrix notation and (rightmost term) using the Einstein summation convention of Sec. 2.1.1.

What is the difference between a vector space and a Hilbert space? The "vectors" of a Hilbert space are functions! So, the number of components of a function of x are $\psi(x_1), \psi(x_2) \ldots$ for every $x \in \mathbb{R}$.

The number of components of a "vector" in a Hilbert space is infinite. Even worse, the number of components has the cardinality of the continuum: the size of the set of all real numbers.

Nothing changes for functions that depend on $\mathbf{x}$. In this case, the components are again the set of all real numbers but the set[2] is labeled $\mathbb{R}^3$ because each element is described by a standard vector with three real components: $\mathbf{x} = (x, y, z)$.

Since the number of components of an element in a Hilbert space is infinite, sums must be replaced by integrals. For instance, the scalar

---

[2]The readers that are into maths probably know that the product of infinite sets has the same size (**cardinality**) as the largest set. For instance, the set of integer numbers $\mathbb{N}$ has the same size of the set of signed integer numbers $\mathbb{Z}$ or the set of fractions m/n, called $\mathbb{Q}$. Since we can define a unique algorithm that counts all these numbers in an infinite time, mathematicians and computer scientists call these sets **countable** and label their size as $\aleph_0$ (aleph-with-zero). Sets with the cardinality of the continuum are **uncountable** and the size is labeled $\aleph_1$ or $\mathfrak{c}$. Hilbert spaces have been a great discovery in mathematics because they deal with vector spaces of uncountable dimension. D. Hilbert directly contributed to the Dirac–von Newmann axioms of QM: he claimed QM was "too complicated to be left to physicists."

product (also called **inner product**) of two vectors is replaced by:

$$\langle\psi_1|\psi_2\rangle \equiv \int_{\mathbb{R}^3} d^3\mathbf{x}\ \psi_1(\mathbf{x})^*\psi_2(\mathbf{x}) \tag{B.4}$$

and the squared **norm** of a function is then:

$$||\psi||^2 \equiv \langle\psi|\psi\rangle = \int_{\mathbb{R}^3} d^3\mathbf{x}\ \psi(\mathbf{x})^*\psi(\mathbf{x}) = \int_{\mathbb{R}^3} d^3\mathbf{x}\ |\psi(\mathbf{x})|^2\ . \tag{B.5}$$

There is a lot of physics hidden inside these formulas and it is time to dig it out. We will see in a minute that:

$$\int_V d^3\mathbf{x}\ |\psi(\mathbf{x})|^2 \tag{B.6}$$

is the probability of detecting the particle inside the volume V. If the particle is somewhere in the universe, then:

$$\int_{\mathbb{R}^3} d^3\mathbf{x}\ |\psi(\mathbf{x})|^2 = 1. \tag{B.7}$$

Therefore, we are interested only in Hilbert spaces where functions are normalized to 1. The "norm" is defined by the integral of $|f|^2$ as in eqn B.5. Mathematicians call this Hilbert space $L^2(\mathbb{R}^3)$ because the integral[3] is performed for the squared modulus of the complex-valued function over $\mathbb{R}^3$. Finally, since the norm of ψ is the same if we only change the function by a phase, like $e^{i\alpha}\psi$, all functions belonging to the same ray describe exactly the same system. This is the reason why we used the arcane word "ray" instead of "function" in the definition of the state.

[3]The L in $L^2(\mathbb{R}^3)$ stands for "Lebesgue integral" because we may have convergence issues using the standard Riemann integral. We leave these subtleties to advanced QM courses.

B.1.1   Bra-ket notation

QM uses a very handy notation introduced by P. Dirac in 1939 and called the **bra-ket** or **Dirac notation**. Here, an element of the vector space ψ is written as $|\psi\rangle$ and it is called a **ket**. A **bra** is a linear operator that maps an element of the Hilbert space to a complex number. The corresponding symbol is $\langle\psi|$. If we were in a standard vector space, a "ket" would be a column vector and a "bra" a row vector so that

$$\langle\mathbf{a}|\mathbf{b}\rangle = \sum_i a_i^* b_i. \tag{B.8}$$

This is why we wrote the scalar product in a Hilber space as $\langle\psi_1|\psi_2\rangle$ in eqn B.4.

B.2   Observables

A physics quantity that can be measured by an instrument or a detector is called an **observable**. In QM,

> observables are described by **hermitian operators** $\hat{O}$. In particular, if a state (wavefunction) ψ is an eigenstate (eigenfunction) of $\hat{O}$, a measurement of the observable $\hat{O}$ on that system provides only one outcome: the corresponding eigenvalue.

Once more, we need to translate this formal statement in something sensible. We already noted that in linear algebra, linear operators are described by matrices. In a Hilbert space, a linear operator is an infinite-dimensional matrix written as $\hat{O}$. The hat signifies that the operator is on a Hilbert space and not on a finite-dimensional vector space. These operators are linear as their standard counterparts:

$$\hat{O}(af(x) + bg(x)) = a\,\hat{O}f(x) + b\,\hat{O}g(x) \tag{B.9}$$

where a and b are complex numbers. The Dirac notation can be used also with linear operators on the Hilbert space. In this case:

$$\langle f|\hat{O}g\rangle = \langle \hat{O}^\dagger f|g\rangle \tag{B.10}$$

$O^\dagger$ (read: O-dagger) is the transpose conjugate of an infinite matrix: $(O^T)^*$. The superscript T indicates that the matrix is transposed: rows are exchanged with columns. We can define the **expectation value** of an observable in the same way as:

$$\langle \hat{O}\rangle \equiv \int_{\mathbb{R}^3} d^3\mathbf{x}\,\psi^*(\mathbf{x})\hat{O}\psi(\mathbf{x}) \tag{B.11}$$

where ψ is the wavefunction of the system.

In standard vector spaces, a hermitian operator is a matrix such that:

$$(H^T)^* = H. \tag{B.12}$$

Equation B.12 is equivalent to:

$$M^i{}_j = (M^j{}_i)^*. \tag{B.13}$$

In a Hilbert space, where $\hat{H}$ has an infinite number of components, it is better to define hermitian operators using the definition of scalar product. A hermitian operator is an operator such that:

$$\langle f|\hat{H}g\rangle \equiv \int_{\mathbb{R}^3} f(\mathbf{x})^*\hat{H}g(\mathbf{x}) = \int_{\mathbb{R}^3} \hat{H}^\dagger f(\mathbf{x})^*g(\mathbf{x})$$
$$= \langle \hat{H}^\dagger f|g\rangle = \langle \hat{H}f|g\rangle. \tag{B.14}$$

Hence, for a hermitian operator:

$$H = H^\dagger. \tag{B.15}$$

$\hat{H}^\dagger$ is called the **hermitian conjugate** or **adjoint** of the operator $\hat{H}$:

$$\langle f|Hg\rangle = \langle H^\dagger f|g\rangle \tag{B.16}$$

Observable	Operator
x	$\hat{x} \equiv x$
y	$\hat{y} \equiv y$
z	$\hat{z} \equiv z$
p_x	$\hat{p}_x \equiv -i\hbar\frac{\partial}{\partial x}$
p_y	$\hat{p}_y \equiv -i\hbar\frac{\partial}{\partial y}$
p_z	$\hat{p}_z \equiv -i\hbar\frac{\partial}{\partial z}$
$\mathbf{L}$	$\hat{\mathbf{L}} = -i\hbar(\mathbf{r} \times \nabla)$
$\mathbf{S}$	$\hat{\mathbf{S}}$
$\mathbf{J}$	$\hat{\mathbf{L}} + \hat{\mathbf{S}}$

Table B.1 List of the most important observables in QM: position in x, y, z (first three rows), momentum component in x, y, z (next three row), angular momentum $\mathbf{L}$, spin $\mathbf{S}$, total angular momentum $\mathbf{J}$. Note that for the angular momenta and spin we used a compact notation instead of writing explicitly the three components. $\mathbf{r}$ is the position of the particle with respect to the origin, as in classical physics, and is commonly expressed in spherical coordinates.

and, therefore, if the operator is hermitian (**self-adjoint**):

$$\langle f|Hg\rangle = \langle Hf|g\rangle. \tag{B.17}$$

In practice, a hermitian operator hops from the ket to the bra without getting the dagger. The reason why Dirac and von Neumann chose hermitian operators to describe observables is that:

> the eigenvalues of a hermitian operator are always real numbers. The hermitian operators chosen to describe observables in QM are the subset of hermitian operators, whose eigenstates form a complete basis of the Hilbert space.

This statement means that if we select an infinite set of functions such that $\hat{H}f = af$, we can be sure that a is a real number and that any function of the Hilbert space – including functions that are not eigenstates of $\hat{H}$ – can be written as:

$$\psi = \int da \ \alpha(a)f(a) \tag{B.18}$$

where a is a (real) eigenvalue of $\hat{H}$, $f(a)$ is the corresponding eigenstate, and $\alpha(a)$ is a complex number that depends on a. This is equivalent to saying that the eigenstates $f(a)$ form a **complete basis** for the Hilbert space like in linear algebra, where any vector can be expressed as a linear combination of the vectors belonging to the basis. Table B.1 lists the most important observables in QM and the corresponding hermitian operators. For instance, the operator corresponding to the x position of the particle is just the operator that multiplies ψ by x. The eigenfunctions are labeled $|x\rangle$ because $\hat{x}|x\rangle = x|x\rangle$.

B.2.1   Angular momentum and spin

You probably noticed that in Tab. B.1 I did not write explicitly the spin operator as a function of operators that have a classical counterpart. The spin is an intrinsic angular momentum of the particle, which is not associated to its mechanical state. However, spin is a genuine source of angular momentum because the spin operators have the same properties of the angular momentum operators. These properties are:

- the components of $\hat{\mathbf{L}} \equiv (\hat{L}_x, \hat{L}_y, \hat{L}_z)$ do not commute. While $\hat{x}\hat{y} = \hat{y}\hat{x}$, you can check that the **commutator** $[\hat{L}_a, \hat{L}_b] \equiv \hat{L}_a\hat{L}_b - \hat{L}_b\hat{L}_a$ is not zero for the angular momentum:

$$[\hat{L}_x, \hat{L}_y] = i\hbar\hat{L}_z \ \ ; \ \ [\hat{L}_y, \hat{L}_z] = i\hbar\hat{L}_x \ \ ; \ \ [\hat{L}_z, \hat{L}_x] = i\hbar\hat{L}_y. \tag{B.19}$$

- using just the algebraic properties of $\hat{\mathbf{L}}$ we can show that the eigenvalues of $\hat{L}^2 \equiv |\hat{\mathbf{L}}|^2 = \hat{L}_x^2 + \hat{L}_y^2 + \hat{L}_z^2$ are $l(l+1)\hbar^2$, where l is an integer number ≥ 0. Similarly, the eigenvalues of any component of $\hat{\mathbf{L}}$ (e.g. $\hat{L}_z$) are $m\hbar$ were m is an integer number running from $-l$ to $+l$. l is called the **orbital quantum number** and m is the **magnetic quantum number**.[4] Note also that $\hat{L}^2$ commutes with $\hat{L}_x, \hat{L}_y$, and $\hat{L}_z$.

[4]As discussed in Sec. 8.2, states with equal m are degenerate if there are no special directions in space. This is a necessary condition for the rotational invariance of the system. If you create a special direction using a magnetic or an electric field, this degeneracy is broken. This is why m is called for historical reasons the magnetic quantum number. Anyway, it is worth stressing that m is a property of the angular momentum and has nothing to do with magnetic fields.

- the spin operator has exactly the same properties, even if spin does not originate from the mechanical motion of the particle. In particular, for a spin-1/2 particle, the eigenvalues of $\hat{S}^2$ are $s(s+1)\hbar^2 = (3/4)\hbar^2$ and the eigenvalues of the $\hat{S}_z$ component are $\pm(1/2)\hbar$.

For the special case of spin-1/2 particles, we can express the spin operator as 2×2 matrices applied to a (non-relativistic) spinor. It means that ψ can be written as a two 2D column vector, whose components correspond to two different orientation of S_z (or any other axis):

$$\psi = \begin{pmatrix} \psi_\Uparrow \\ \psi_\Downarrow \end{pmatrix} \tag{B.20}$$

Once the axis is chosen, the **spin-1/2** operators are

$$\hat{\mathbf{S}} = \frac{1}{2}\hbar\boldsymbol{\sigma} \tag{B.21}$$

where $\boldsymbol{\sigma} = (\sigma_x, \sigma_y, \sigma_z)$ are the **Pauli matrices**:

$$\sigma_x = \begin{pmatrix} 0 & 1 \\ 1 & 0 \end{pmatrix} \;\; ; \;\; \sigma_y = \begin{pmatrix} 0 & -i \\ i & 0 \end{pmatrix} \;\; ; \;\; \sigma_z = \begin{pmatrix} 1 & 0 \\ 0 & -1 \end{pmatrix} . \tag{B.22}$$

This approach can be generalized for any value of the spin, although QM shows that only integer (boson) or half-integer (fermions) are possible. For a particle with spin s, the Pauli matrices have $(2s+1)(2s+1)$ entries and $\hat{S}_z$ is:

$$\hat{S}_z = \hbar \begin{pmatrix} s & 0 & 0 & \dots 0 \\ 0 & s-1 & 0 & \dots 0 \\ 0 & 0 & & \dots \\ 0 & 0 & & \dots -s \end{pmatrix} . \tag{B.23}$$

You can find $\hat{S}_y$ and $\hat{S}_z$ in most QM textbooks like, for instance, Griffiths, 2018. We can sum the angular momentum and the spin of a particle to get $\hat{\mathbf{J}}$. In this case the eigenvalues are,

$$J^2 = j(j+1)\hbar^2 \;\; ; \;\; J_z = m_j\hbar \tag{B.24}$$

where j and m_j are the **total angular momentum** and **total angular momentum projection quantum number**, respectively.

Similarly, we can combine the wavefunctions of several particles to get the overall spin or angular momentum. We used this technique and its generalization in many contexts, especially in Chap. 8. The master formula for the angular momenta and, in general, for all quantities related to rotational symmetries was derived by two mathematicians - A. Clebsch and P. Gordan - and is called the **Clebsch–Gordan formula**. For the combination of two particles with angular momentum j_1 and j_2, the combined wavefunction with angular momentum j is

$$|j, j_z, j_1, j_2\rangle = \sum_{j_{z1}, j_{z2}} |j_1, j_{z1}, j_2, j_{z2}\rangle \langle j_1, j_{z1}, j_2, j_{z2}|j_1, j_2, j, j_z\rangle \tag{B.25}$$

where $\langle j_1, j_{z1}, j_2, j_{z2} | j_1, j_2, j, j_z \rangle$ are real numbers. They are called **Clebsch–Gordan coefficients** and can be derived by the study of the rotation group.[5] These coefficients are tabulated, for example, in Cohen-Tannoudji, *et al.*, 1991. In Dirac's notation, $|j, j_z, j_1, j_2\rangle$ is a wavefunction made of two particles with angular momentum j_1 and j_2, respectively. This wavefunction has two quantum numbers: j and j_z.

B.2.2   The Dirac delta function

Two functions are orthogonal in a Hilbert space if:

$$\langle f(x) | g(x') \rangle = \delta(x - x'). \tag{B.26}$$

In a standard vector space, the scalar product of two orthogonal vectors is zero. In Hilbert spaces and, therefore, in QM, we need a function that gives zero except when the positions x and x' are the same. This is the famous **delta function** $\delta(x - x')$, which is 0 if $x \neq x'$ and infinite if $x = x'$. As anticipated in Sec. 1.2, the delta function is not an ordinary function because we want:

$$\langle x' | x \rangle = \delta(x - x') \tag{B.27}$$

but if the delta function inside the $\langle x' | x \rangle$ integral were an ordinary function that is always zero (but in a point), the integral would be zero, too. Even if it is possible to define rigorously the delta function (Hoskins, 2009), we stick on the original heuristic definition of Dirac, imposing:

$$\int_{\mathbb{R}} dx \, \delta(x) = 1 \tag{B.28}$$

and defining the function as the limit of Gaussian functions of smaller and smaller width. This poorman's definition is enough to demonstrate the basic properties of the delta function and, in particular, the features used in this book:

$$\delta(x) = \delta(-x) \tag{B.29}$$

$$\delta(\alpha x) = \frac{\delta(x)}{|\alpha|} \tag{B.30}$$

$$\delta(f(x)) = \sum_i \frac{\delta(x - x_i)}{|f'(x_i)|} \quad \forall i \text{ where } f(x_i) = 0 \tag{B.31}$$

$$\int_{\mathbb{R}} dx \, f(x) \delta(g(x)) = \sum_i \frac{f(x_i)}{|g'(x_i)|} \quad \forall i \text{ where } g(x_i) = 0. \tag{B.32}$$

B.3   Time evolution of quantum states

If we know ψ at the time $t = 0$, how does the function change in time? The answer is an axiom of QM:

the time evolution of a quantum system is given by the solution of the Schrödinger equation:

$$i\hbar\frac{\partial\psi}{\partial t} = \hat{H}\psi \tag{B.33}$$

where $\hat{H}$ is the hermitian operator associated with the total energy of the system, the **Hamiltonian operator**.

The corresponding solution is:

$$\psi(t) = \exp\left[-\frac{i}{\hbar}\hat{H}t\right]\psi(0) \equiv \hat{U}(t)\psi(0). \tag{B.34}$$

$\hat{U}(t)$ is the **time-evolution operator** and is an example of unitary operator. **Unitary operators** are defined by:

$$U^\dagger U = \mathbb{I} \implies U^\dagger = U^{-1}. \tag{B.35}$$

We discuss unitary operators in Chap. 5 because they are instrumental in describing symmetries in QM.

For a system that has a classical counterpart, we can compute $\hat{H}$ using the **canonical quantization**.

A system with a classical counterpart can be quantized replacing the observables in the Hamiltonian with the corresponding quantum operators of Tab. B.1.

For instance, a single particle in a scalar potential is described by the following Schrödinger equation:

$$i\hbar\frac{\partial}{\partial t}\psi(\mathbf{x}, t) = \left[-\frac{\hbar^2}{2m}\nabla^2 + V(\mathbf{x})\right]\psi(\mathbf{x}, t) \tag{B.36}$$

because the total energy is the sum of the kinetic energy ($|\mathbf{p}|^2/2m$) and the potential energy[6]. Unfortunately, this method does not work for genuinely quantum systems, where the classical counterpart is missing. For instance, we used the gauge principle to write the electroweak Hamiltonian (or, equivalently, the Lagrangian) before quantization. This method is heuristic because we cannot ensure a priori that the gauge principle is the right way to describe fundamental interactions (although, at present, is the only one that works). Sometimes, we can resort to other methods than canonical quantization, which are outside the scope of this book (Ali and Englis, 2005).

[6]In QM, we call $V(\mathbf{x})$ the "potential" even if the correct name should be the "potential energy function". You should not confuse it with the potential of an electric field, which is the potential energy per unit charge.

B.4 Measurements and the uncertainty principle

Since ψ is the solution of a differential equation, its time evolution is *deterministic*. Where does the intrinsic uncertainty of QM originate?

It comes from the last – and most problematic – axiom of Dirac and von Neumann, the definition of measurement.

Consider a wavefunction ψ describing a particle in space and time. If we want to perform a measurement of the particle position in x at time t, we can express ψ as a linear combination of position eigenfunctions $|x\rangle$:

$$\psi = \int_{\mathbb{R}} a(x)|x\rangle = \int_{-\infty}^{+\infty} a(x)|x\rangle. \tag{B.37}$$

Then,

> a position measurement on x is a **projector** that causes the collapse of the wavefunction into $|\tilde{x}\rangle$, where $\tilde{x}$ is the measured position. The probability of such occurrence is $|\langle\tilde{x}|\psi\rangle|^2 = |a(\tilde{x})|^2$.

This axiom is nowadays called the **generalized statistical interpretation** of QM even if it dates back to the pre-history of QM. A projector is an operator $\hat{P}$ that selects only one component of the series that describes the function:

$$\hat{P}\psi = \hat{P}\left(\int_{-\infty}^{+\infty} a(x)|x\rangle\right) = a(\tilde{x})|\tilde{x}\rangle \tag{B.38}$$

and, of course, $P^2 = PP = P$. This is the most shocking feature of QM: a measure perturbs the system in such a way that its time evolution will be no more $\hat{U}\psi$, where $\hat{U}(t)$ is the time-evolution operator. It will be $\hat{U}|\tilde{x}\rangle$! If the set of eigenvalues – the **spectrum** of the hermitian operator – is discrete, we record a single number. If it is continuous, we record an interval centered in $\tilde{x}$, whose width depends on the precision of the instrument.

Well, if you are not shocked, you probably did not get the point. There are two major issues here. The first one is **entanglement**. If we have a two-particle system like a $\pi^0 \to e^+e^-$ decay,[7] a possible final state ψ is

$$\psi = \frac{1}{\sqrt{2}}\left(|\Uparrow\rangle_{e^-}|\Downarrow\rangle_{e^+} - |\Downarrow\rangle_{e^-}|\Uparrow\rangle_{e^+}\right) \tag{B.39}$$

where the factor $1/\sqrt{2}$ normalizes the function as in eqn B.7. $|\Uparrow\rangle$ is a function describing a particle with $S_x = (+1/2)\hbar$ (spin up) and $|\Downarrow\rangle$ has spin down. Suppose that one particle goes on the left and the other on the right moving at a speed close to c. After a few minutes, an observer makes a measurement (see Fig. B.1) and discovers that the electron has spin up. At that very moment, the function collapses to $|\Uparrow\rangle_{e^-}|\Downarrow\rangle_{e^+}$ and the observer measuring the positron will *surely* see the particle with spin down. The perturbation of the system due to the first measurement instantaneously "causes" a change of the possible outcomes of the positron measurements. This "paradox" was raised in 1935 and is called the **Einstein–Podolsky–Rosen** (EPR) paradox. The EPR seems a paradox because suggests a violation of special relativity (SR). But this is perfectly consistent with QM in the generalized

[7]This story is D. Bohm's simplified version of the Einstein–Podolsky–Rosen paradox. I do not know, however, if he realized that $BR(\pi^0 \to e^+e^-)$, that is, the probabilty of the occurrence of this decay mode compared with other decay modes, is 6.5×10^{-8}!

statistical interpretation. QM is a causal theory consistent with SR but is a non-local theory, too. The electron observer cannot influence the measurement of the electron just because she does not know a priori if the spin is up or down. The correlation is created by the measurement but this information cannot be sent instantaneously to the other observer to meddle in the outcome of his measurement. For instance, she cannot say (instantaneously) to him "please, do not perform the measurement: I know the result already!". The observers will discover the anticorrelation when they will meet again because the anticorrelation is implied by the form of eqn B.39. This interpretation was tested experimentally using the aforementioned Bell's theorem that falsifies the existence of hidden properties we are not aware of when we produce the e^+e^- pair. We used this "spooky action-at-a-distance" – in Einstein's words – many times in this book and in particular, in Sec. 13.7.3.

The second issue is due to the actual measurement. The experiment that records the measurement is a physical system made of a person, a detector, and a particle. All of them fulfill the Schrödinger equation and, therefore, the ψ of the observer + detector + particle evolves in a deterministic way. How can we get such a discontinuous collapse? To be honest, we do not have a smoking gun proof similar to Bell's theorem. In the last 20 years, a consensus has grown that this effect might be explained by **decoherence** among the parts of the observer + detector + particle wavefunction. But the equivalent of Bell's theorem is still to come (Weinberg, 2012).

The most important corollary of the measurement axiom is the **Heisenberg uncertainty principle** that puts an intrinsic limit to the possibility of measuring two observables with infinite precision. The general version of the principle reads:

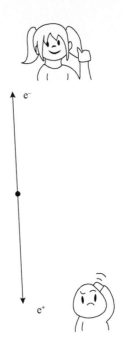

Fig. B.1 A simplified version of the EPR paradox. A neutral pion decays in an electron-positron pair. The spins along x of the electron and positron in the entangled state of eqn B.39 are measured by two observers that cannot be causally connected in SR. Image credit: V. Terranova.

---

if $\hat{A}$ and $\hat{B}$ are two hermitian operators such that $[\hat{A}, \hat{B}] \neq 0$ (**conjugate observables**), then

$$\sigma_A^2 \sigma_B^2 \geq \left(\frac{1}{2i} \langle [\hat{A}, \hat{B}] \rangle \right)^2 \qquad \text{(B.40)}$$

where σ_A and σ_B is the uncertainty on the measurement of the observable associated to $\hat{A}$ and $\hat{B}$, respectively.

For the special case of $\hat{A} = \hat{x}$ and $\hat{B} = \hat{p}_x$, we have:

$$\sigma_x \sigma_{p_x} \geq \frac{\hbar}{2}. \qquad \text{(B.41)}$$

Another well-known statement is the energy-time uncertainty principle:

$$\sigma_E \sigma_t \geq \frac{\hbar}{2} \qquad \text{(B.42)}$$

but a note of caution is in order. In QM, we do not have a hermitian operator whose observable is time. Equation B.42 holds if we precisely define what we mean by σ_t. Loosely speaking, σ_t is the time it takes the system to change substantially (Griffiths, 2018). We implicitly assumed this definition when we discussed the Breit–Wigner distribution in Sec. 2.6.4.

B.5   The Dyson series[*]

The Schrödinger equation can be solved only for very simple systems. For instance, a free particle with minimum uncertainty in space x and momentum p_x is:

$$\psi = A \exp\left[-a\frac{(x - \langle x \rangle)^2}{2\hbar}\right] \exp\left[i\frac{\langle p_x \rangle(x - \langle x \rangle)}{\hbar}\right] \tag{B.43}$$

where a and A are real parameters fixed by eqn B.7. This wavefunction is a **Gaussian wavepacket**. In Secs. 5.6.1 and 12.1, we show the solution of other simple cases like the hydrogen atom or the nuclear well. However, analytical solutions are rare and we need to resort to approximations or numerical methods. The most important technique in particle and nuclear physics is the **perturbative method**, where the Hamiltonian is split in a non-perturbed part H_0 and a small perturbation H_1. The breakthrough that led to the Feynman diagrams is based on the same principle but uses the so-called interaction picture.

In most of the book, we described QM using the **Schrödinger picture**, where all operators are time-independent and all wavefunctions are time-dependent. QM can be described consistently also in the **Heisenberg picture**, where the time dependence is moved to the operators. In this picture, the Schrödinger equation is replaced by an equation of motion for the operators that reads:

$$i\hbar\frac{\partial}{\partial t}\hat{O} = [\hat{H}, \hat{O}]. \tag{B.44}$$

In the **interaction picture** both the operators and the wavefunction depend on time. In this case, we can write the Schrödinger equation in the usual way:

$$i\hbar\frac{\partial}{\partial t'}\psi = \hat{H}\psi \tag{B.45}$$

but the solution is now an integral over time of the (time-dependent) Hamiltonian:

$$\psi(\mathbf{x}, t) = \exp\left[-\frac{i}{\hbar}\int_{t_0}^{t} dt'\ \hat{H}(t')\right]\psi(\mathbf{x}, t_0). \tag{B.46}$$

The perturbative series arises from the Taylor expansion of eqn B.46:

$$\exp\left[-\frac{i}{\hbar}\int_{t_0}^{t} dt'\ \hat{H}(t')\right] = 1 - \frac{i}{\hbar}\int_{t_0}^{t} dt_1\ \hat{H}(t_1)$$

$$+ \left(-\frac{i}{\hbar}\right)^2 \int_{t_0}^{t} dt_1 \int_{t_0}^{t_1} dt_2\ \hat{H}(t_1)\hat{H}(t_2) + \dots$$

$$= 1 + \sum_{j=1}^{+\infty}\left(-\frac{i}{\hbar}\right)^j \int_{t_0}^{t} dt_1 \int_{t_0}^{t_1} dt_2 \cdots \int_{t_0}^{t_{j-1}} dt_j \hat{H}(t_1)\hat{H}(t_2)\cdots\hat{H}(t_j) \tag{B.47}$$

and is called the **Dyson series**. Since the Hamiltonians at different time do not commute, the series can be rewritten time-ordering the integrals so that $t_1 < t_2 \ldots < t_j$. Each term of the series corresponds to the application of j Hamiltonians to the wavefunction and the whole series is a unitary operator. These multiple applications of $\hat{H}(t)$ were interpreted by Feynman as diagrams with an increasing number of vertices, as discussed in Sec. 6.6.

References

Abada, A., Abbrescia, M., AbdusSalam, S.S., Abdyukhanov, I., Abelleira Fernandez, J. et al. (2019). FCC physics opportunities: Future circular collider conceptual design report, volume 1. *Eur. Phys. J.C*, **79**(6), 474.

Abbott, B.P., Abbott, R., Abbott, T.D., Acernese, M.R. Abernathy F. et al. (2016). Observation of gravitational waves from a binary black hole merger. *Phys. Rev. Lett.*, **116**(6), 061102.

Abreu, P., Adam, W., Adye, T., Adzic, P., Alekseev, G.D. et al. (2000). Cross-sections and leptonic forward-backward asymmetries from the Z^0 running of LEP. *Eur. Phys. J.C.*, **16**, 371.

Adams, D.Q., Alduino, C., Alfonso, K., Avignone III, F.T., Azzolini, O. et al. (2021). Measurement of the $2\nu\beta\beta$ decay half-life of ^{130}Te with CUORE. *Phys. Rev. Lett.*, **126**(17), 171801

Adli, E., Ahuja, A., Apsimon, O., Apsimon, R. Bachmann, A.-M. et al. (2018). Acceleration of electrons in the plasma wakefield of a proton bunch. *Nature*, **561**(7723), 363.

Adolphsen, C., Barone, M., Barish, B. Buesser, K, Burrows, P. et al. (2013). The international linear collider technical design report, Vol. 3.ii: Accelerator baseline design. ILC-REPORT-2013-040, arXiv:1306.6328.

Agafonova, N., Aleksandrov, A., Anokhina, A., Aoki, S., Ariga, A. et al. (2015). Discovery of τ neutrino appearance in the CNGS neutrino beam with the OPERA experiment. *Phys. Rev. Lett.*, **115**(12), 121802.

Aker, M., Altenmüller, K., Arenz, M., Babutzka, M., Barrett, J. et al. (2019). Improved upper limit on the neutrino mass from a direct kinematic method by KATRIN. *Phys. Rev. Lett.*, **123**(22), 221802.

Ali, S.T. and Englis, M. (2005). Quantization methods: A guide for physicists and analysts. *Rev. Math. Phys.*, **17**(4), 391–490.

Allison, J., Amakoc, A., Apostolakis, J., Arce, P., Asai, M. et al. (2016). Recent developments in GEANT4. *Nucl. Instrum. Meth. A*, **835**, 186.

Amaldi, U., de Boer, W., and Furstenau, H. (1991). Comparison of grand unified theories with electroweak and strong coupling constants measured at LEP. *Phys. Lett. B*, **260**(3), 447–55.

Amsler, C. (2018). *The Quark Structure of Hadrons*. Springer, Heidelberg.

Anderson, C.D. (1933). The positive electron. *Phys. Rev.*, **43**(6), 491.

Anderson, H. (1982). Early history of physics with accelerators. *Journal de Physique Colloques*, **43**, C8–101.

Andersson, N. (2020). *Gravitational-Wave Astronomy: Exploring the Dark Side of the Universe*. Oxford University Press, Oxford.

Annett, J.F. (2004). *Superconductivity, Superfluids, and Condensates*. Oxford University Press, Oxford.

Aprile, E. Aalbers, J., Agostini, F., Alfonsi, M., Amaro, F.D. et al. (2017*a*). The XENON1T dark matter experiment. *Eur. Phys. J. C.*, **77**(12), 881.

Aprile, E., Aalbers, J., Agostini, F., Alfonsi, M., Amaro, F.D. et al. (2017*b*). First dark matter search results from the XENON1T experiment. *Phys. Rev. Lett.*, **119**(18), 181301.

Arnison, G., Astbury, A., Aubert, B., Bacci, C., Bauer, G., Bezaguet A. et al. (1983). Experimental observation of isolated large transverse energy electrons with associated missing energy at $\sqrt{s}$=540 GeV. *Phys. Lett. B*, **122**(1), 103–16.

Atwood, W.B., Barczewski, T., Bauerdick, L.A.T., Bellantoni, L., Blucher, E. et al. (1991). Performance of the ALEPH time projection chamber. *Nucl. Instrum. Meth. A.*, **306**(3), 446.

Ballentine, L.E. (2014). *Quantum Mechanics: A Modern Development*. World Scientific, Singapore.

Banner, M., Battiston, R., Bloch, Ph., Bonaudi, F., Borer, K., et al. (1983). Observation of single isolated electrons of high transverse momentum in events with missing transverse energy at the CERN $\bar{p}p$ collider. *Phys. Lett. B*, **122**(5), 476–85.

Barate, R., Buskulic, D., Decamp, D., Ghez, P. Goy, C. et al. (1999). Dali, the ALEPH event display. Available at https://aleph.web.cern.ch/aleph/dali/.

Barr, G., Devenish, R., Walczak, R., and Weidberg, T. (2019). *Particle Physics in the LHC Era*. Oxford University Press, Oxford.

Barrow, J.D. and Tipler, F.J. (1988). *The Anthropic Cosmological Principle*. Oxford University Press, Oxford.

Bauer, M. and Plehn, T. (2019). *Yet Another Introduction to Dark Matter: The Particle Physics Approach*. Springer, Heidelberg.

Bell, M. and Gao S. (eds.) (2016). *Quantum Nonlocality and Reality (50 Years of Bell's Theorem)*. Cambridge University Press, Cambridge.

Benenson, W. and Kashy, E. (1979). Isobaric quartets in nuclei. *Rev. Mod. Phys.*, **51**(3), 527–40.

Bertone, G. (2013). *Particle Dark Matter (Observations, Models and Searches)*. Cambridge University Press, Cambridge.

Bertulani, C.A. (2007). *Nuclear Physics in a Nutshell*. Princeton University Press, Princeton, NJ.

Bettini, A. (2014). *Introduction to Elementary Particle Physics* (2nd edn). Cambridge University Press, Cambridge.

Betts, D. (1989). *An Introduction to Millikelvin Technology*. Cambridge University Press, Cambridge.

Beuthe, M. (2002). Towards a unique formula for neutrino oscillations in vacuum. *Phys. Rev. D.*, **66**(1), 013003.

Bilenky, S. (2018). *Introduction to the Physics of Massive and Mixed Neutrinos* (2nd edn). Springer, Heidelberg.

Binetruy, P. (2018). *Gravity! The Quest for Gravitational Waves*. Oxford University Press, Oxford.

Blackett, P.M.S. and Occhialini G.P.S., (1933). Some Photographs of the Tracks of Penetrating Radiation. *Proc. Roy. Soc. Lond.*, A, **139**(839), 699–720

Blasone, M., Capolupo, A., Terranova, F., and Vitiello, G. (2005). Lepton charge and neutrino mixing in pion decay processes. *Phys. Rev. D*, **72**(1), 013003.

Bogoliubov, N.N., Logunov, A.A., and Todorov, I.T. (1975). *Introduction to Axiomatic Quantum Field Theory*. W.A. Benjamin, Reading, MA.

Borsanyi, S.Z., Durr, S., Fodor, Z., Hoelbling, C., Katz, S.D. et al. (2015). Ab initio calculation of the neutron–proton mass difference. *Science*, **347**(6229), 1452.

Bromberg, J. (1976). The concept of particle creation before and after quantum mechanics. *Hist. Stud. Nat. Sci.*, **7**, 161.

Brooker, G. (2003). *Modern Classical Optics*. Oxford University Press, Oxford.

Brown, R.J. (ed.) (1937). World's biggest atom smasher uses 3,000,000 volts. *Popular Science, Monthly*, **131**, 53.

Brown, L.M., Dresden, M., Hoddeson, L. (1990). *From Pions to Quarks: Particle Physics in the 50s*. Cambridge University Press, Cambridge.

Bryan, J. C. (2021). *Introduction to Nuclear Science* (2nd edn). CRC, Boca Raton, FL.

Buchwald, J.Z., Fox, R. (2013). *The Oxford Handbook of the History of Physics*. Oxford University Press, Oxford.

Buskulic, D., Casper, D., De Bonis, I., Decamp, D., Ghez, P. et al. (1995). Measurement of α_s from scaling violations in fragmentation functions in e^+e^- annihilation. *Phys. Lett. B*, **357**(3), 487.

Byckling, E. and Kajantie, K. (1973). *Particle Kinematics*. Wiley, London.

Cahn, R.N. and Goldhaber, G. (2009). *The Experimental Foundations of Particle Physics* (2nd edn). Cambridge University Press, Cambridge.

Carlsmith, D. (2012). *Particle Physics*. Pearson, London.

Carroll, S.M. (2019). *Spacetime and Geometry (An Introduction to General Relativity)*. Cambridge University Press, Cambridge.

Carter, B. (1974). Large number coincidences and the anthropic principle in cosmology. *IAU Symp.*, **63**, 291.

Chandra, R. and Rahmim, A. (2017). *Nuclear Medicine Physics: The Basics* (8th edn). Wolters Kluwer, Philadelphia, PA.

Charitonidis, N., Longhin, A., Pari, M., Parozzi, E.G., and Terranova, F. (2021). Design and diagnostics of high-precision accelerator neutrino beams. *Appl. Sciences*, **11**(4), 1644.

Cheng, T.-P. (2010). *Relativity, Gravitation and Cosmology. A Basic Introduction* (2nd edn). Oxford University Press, Oxford.

Chinowsky, W. and Steinberger, J. (1954). Absorption of negative pions in deuterium: Parity of the pion. *Phys. Rev.*, **95**(6), 1561–1564.

Chou, C.W., Hume, D.B., Rosenband, T., and Winelands, D.J. (2010). Optical clocks and relativity. *Science*, **329**(5999), 1630–33.

Claessens, M. (2020). *ITER: The Giant Fusion Reactor: Bringing a Sun to Earth* Springer, Heidelberg.

Close, F. (2012). *Neutrinos*. Oxford University Press, Oxford.

CMS (2012). Candidate $h \to \gamma\gamma$ decay. CMS-PHO-EVENTS-2012-003-3. Available at https://cdsweb.cern.ch/record/1459459.

CMS (2014). SketchUpCMS. Available at https://twiki.cern.ch/twiki/bin/view/CMSPublic/SketchUpCMS.

CMS (2017). Collision events recorded by CMS in 2016. Available at: https://cds.cern.ch/record/2241144.

Cohen-Tannoudji, C., Diu, B., and Laloe, F. (1991). *Quantum Mechanics*. Wiley, New York, NY.

Costa, G. and Fogli, G. (2012). *Symmetry and Group Theory in Particle Physics*. Springer, Heidelberg.

Cribier, M. (2011). Reactor monitoring with neutrinos. *Nucl. Phys. B. Proc. Suppl.*, **221**, 57–61.

D'Auria, S. (2018). *Introduction to Nuclear and Particle Physics*. Springer, Heidelberg.

De Angelis, A. (2018). *Introduction to Particle and Astroparticle Physics: Multimessenger Astronomy and its Particle Physics Foundations* (2nd edn). Springer, Heidelberg.

de Gouvea, A. (2009). Neutrino masses and mixing — theory. arXiv:0902.4656.

Dodelson, S. and Schmidt, F. (2020). *Modern Cosmology* (2nd edn). Academic Press, London.

Donnelly, T.W., Formaggio, J.A., Holstein, B.R., Milner, R.G., and Surrow, B. (2017). *Foundations of Nuclear and Particle Physics*. Cambridge University Press, Cambridge.

Durr, S., Fodor, Z., Frison, J., Hoelbling, C., Hoffmann, R. et al. (2008). Ab-initio determination of light hadron masses. *Science*, **322**(5905), 1224.

Esarey, E., Sprangle, P., and Krall, J. (1995). Laser acceleration of electrons in vacuum. *Phys. Rev. E.*, **52**(5), 5443–5453.